W9-BMZ-189

APPLIED COMBINATORICS

APPLIED COMBINATORICS

ALAN TUCKER

SUNY Stony Brook

JOHN WILEY & SONS, INC.

Acquisitions Editor *Kimberly Murphy*
Marketing Manager *Julie Lindstrom*
Senior Production Editor *Petrina Kulek*
Director Design *Madelyn Lesure*
Illustration Coordinator *Gene Aiello*

This book was set in Times Roman by TechBooks and printed and bound by Von-Courier/Westford.
The cover was printed by Phoenix Color.

This book is printed on acid-free paper. ♾

ISBN 0-471-43809-X

Printed in the United States of America

10 9 8 7 6 5 4

PREFACE

Combinatorial reasoning underlies all analysis of computer systems. It plays a similar role in discrete operations research problems and in finite probability. Two of the most basic mathematical aspects of computer science concern the speed and logical structure of a computer program. Speed involves enumeration of the number of times each step in a program can be performed. Logical structure involves flow charts, a form of graphs. Analysis of the speed and logical structure of operations research algorithms to optimize efficient manufacturing or garbage collection entails similar combinatorial mathematics. Determining the probability that one of a certain subset of equally likely outcomes occurs requires counting the size of the subset. Such combinatorial probability is the basis of many nonparametric statistical tests. Thus, enumeration and graph theory are used pervasively throughout the mathematical sciences.

This book teaches students in the mathematical sciences how to reason and model combinatorially. It seeks to develop proficiency in basic discrete math problem solving in the way that a calculus textbook develops proficiency in basic analysis problem solving.

The three principal aspects of combinatorial reasoning emphasized in this book are: the systematic analysis of different possibilities, the exploration of the logical structure of a problem (e.g., finding manageable subpieces or first solving the problem with three objects instead of n), and ingenuity. Although important uses of combinatorics in computer science, operations research, and finite probability are mentioned, these applications are often used solely for motivation. Numerical examples involving the same concepts use more interesting settings such as poker probabilities or logical games.

Theory is always first motivated by examples, and proofs are given only when their reasoning is needed to solve applied problems. Elsewhere, results are stated without proof, such as the form of solutions to various recurrence relations, and then applied in problem solving. Occasionally, a few theorems are stated simply to give students a flavor of what the theory in certain areas is like.

Since 1980, collegiate curriculum recommendations from the Mathematical Association of America have included combinatorial problem solving as an important component of training in the mathematical sciences. Combinatorial problem solving underlies a wide spectrum of important subjects in the computer science curriculum. Indeed, it is expected that most students in a course using this book will be computer science majors. For both mathematics majors and computer science majors, this author believes that general reasoning skills stressed here are more important than mastering a variety of definitions and techniques.

This book is designed for use by students with a wide range of ability and maturity (sophomores through beginning graduate students). The stronger the students, the harder the exercises that can be assigned. The book can be used for a one-quarter, two-quarter, or one-semester course depending on how much material is used. It may also be used for a one-quarter course in applied graph theory or a one-semester or one-quarter course in enumerative combinatorics (starting from Chapter 5). A typical one-semester undergraduate discrete methods course should cover most of Chapters 1 to 3 and 5 to 8, with selected topics from other chapters if time permits.

Instructors are strongly encouraged to obtain a copy of the instructor's guide accompanying this book. The guide has an extensive discussion of common student misconceptions about particular topics, hints about successful teaching styles for this course, and sample course outlines (weekly assignments, tests, etc.).

The fourth edition of this book, like the second and third editions, puts the graph theory chapters first. The pictures associated with graphs have proven to be a valuable aid in introducing students to combinatorial reasoning. The reasoning required in enumeration problems quickly becomes very challenging—for too many students, overwhelming—without the preparation in combinatorial reasoning provided by graph theory. This edition has new examples, expanded discussions, and additional exercises throughout the text.

Many people gave useful comments about early drafts and the first edition of this text; Jim Frauenthal and Doug West were especially helpful. The idea for this book is traceable to a combinatorics course taught by George Dantzig and George Polya at Stanford in 1969, a course for which I was the grader. Many instructors who have used earlier editions of this book have supplied me with valuable feedback and suggestions that have, I hope, made this edition better. I gratefully acknowledge my debt to them. Ultimately, my interest in combinatorial mathematics and in its effective teaching rests squarely on the shoulders of my father, A. W. Tucker, who had long sought to give finite mathematics a greater role in mathematics as well as in the undergraduate mathematics curriculum. Finally, special thanks go to former students of my combinatorial mathematics courses at Stony Brook. It was *they* who taught *me* how to teach this subject.

Alan Tucker
Stony Brook, New York

CONTENTS

APPLIED COMBINATORICS

PART ONE
GRAPH THEORY

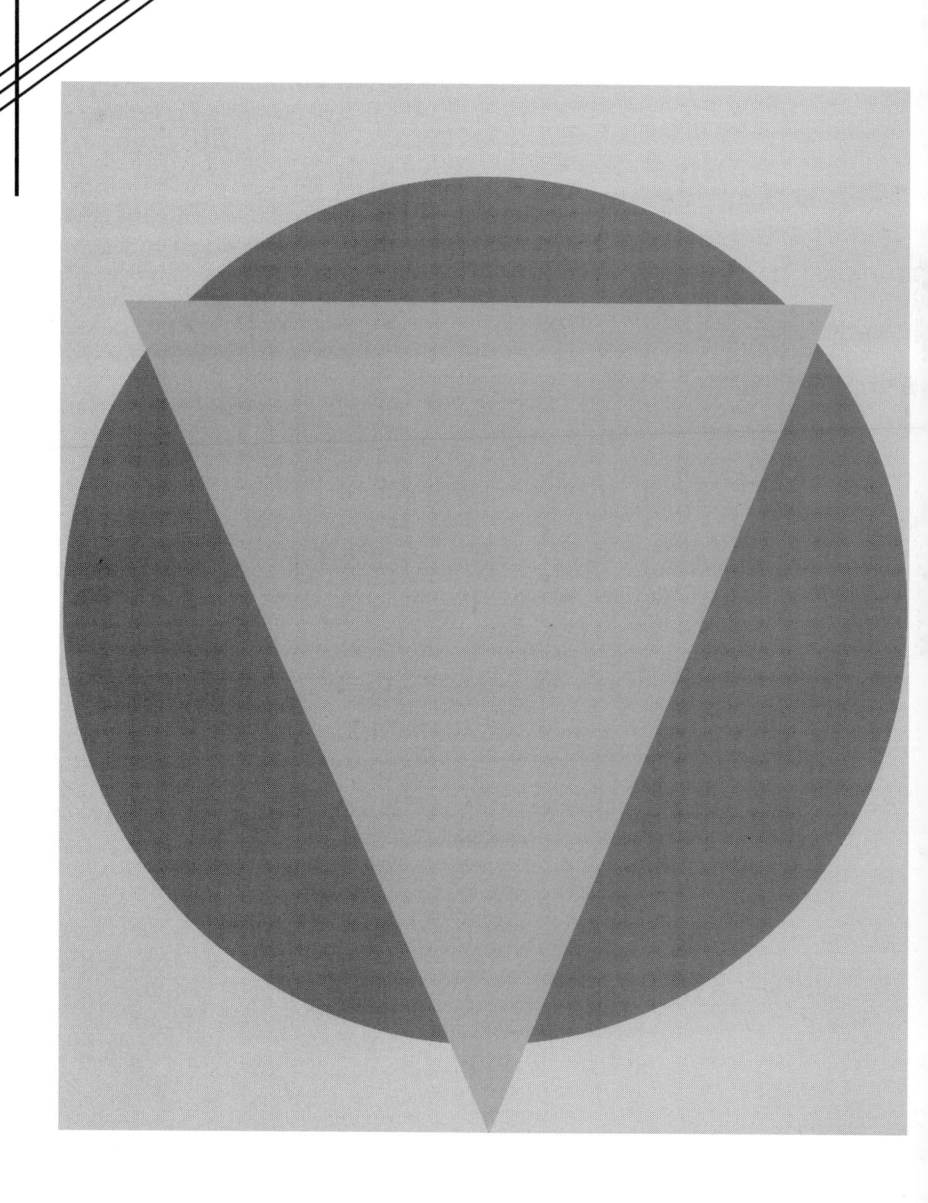

CHAPTER 1
ELEMENTS OF GRAPH THEORY

1.1 GRAPH MODELS

The first four chapters deal with graphs and their applications. A **graph** $G = (V, E)$ consists of a finite set V of **vertices** and a set E of **edges** joining different pairs of distinct vertices.* Figure 1.1a shows a depiction of a graph with $V = \{a, b, c, d\}$ and $E = \{(a, b), (a, c), (a, d), (b, d), (c, d)\}$. We represent vertices with points and edges with lines joining the prescribed pairs of vertices. This definition of a graph does not allow two edges to join the same two vertices. Also an edge cannot "loop" so that both ends terminate at the same vertex—an edge's end vertices must be distinct. The two ends of an undirected edge can be written in either order, (b, c) or (c, b). We say that vertex a is **adjacent** to vertex b when there is an edge from a to b.

Sometimes the edges are ordered pairs of vertices, called **directed edges**. In a **directed graph**, all edges are directed. See the directed graph in Figure 1.1b. We write $(\vec{b, c})$ to denote a directed edge from b to c. In a directed graph, we allow one edge in each direction between a pair of vertices. See edges $(\vec{a, c})$ and $(\vec{c, a})$ in Figure 1.1b.

The combinatorial reasoning required in graph theory, and later in the enumeration part of this book, involves different types of analysis than used in calculus and high school mathematics. There are few general rules or formulas for solving these problems. Instead, each question usually requires its own particular analysis. This analysis sometimes calls for clever model building or creative thinking but more often consists of breaking the problem into many cases (and subcases) that are easy enough to solve with simple logic or basic counting rules. A related line of reasoning is to solve a special case of the given problem and then to find ways to extend that reasoning to all the other cases that may arise. In graph theory, combinatorial arguments are made a little easier by the use of pictures of the graphs. For example, a case-by-case argument is much easier to construct when one can draw a graphical depiction of each case.

Graphs have proven to be an extremely useful tool for analyzing situations involving a set of elements in which various pairs of elements are related by some property. The most obvious examples of graphs are sets with physical links, such as electrical networks, where electrical components (transistors) are the vertices and

*What this book calls a graph is referred to in many graph theory books as a *simple graph*. In general, graph theory terminology varies a little from book to book.

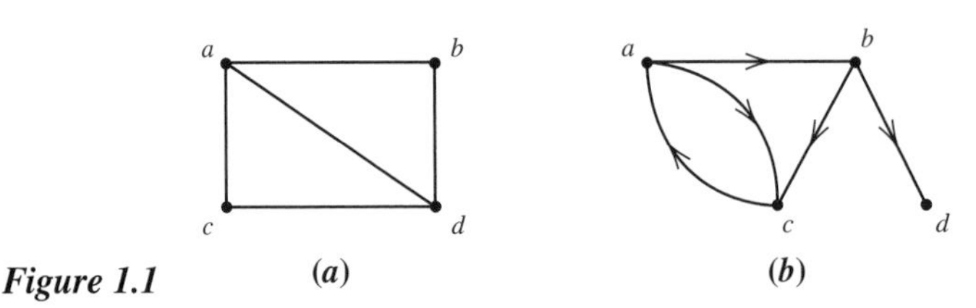

Figure 1.1 (*a*) (*b*)

connecting wires are the edges; or telephone communication systems, where telephones and switching centers are the vertices and telephone lines are the edges. Road maps, oil pipelines, and subway systems are other examples.

Another natural form of graphs is sets with logical or hierarchical sequencing, such as computer flow charts, where the instructions are the vertices and the logical flow from one instruction to possible successor instruction(s) defines the edges; or an organizational chart, where the people are the vertices and if person A is the immediate superior of person B then there is an edge $(\vec{A, B})$. Computer data structures, evolutionary trees in biology, and the scheduling of tasks in a complex project are other examples.

The emphasis in this book will be on problem solving, with problems about general graphs and applied graph models. Observe that we will not have any numbers to work with, only some vertices and edges. At first, this may seem to be highly nonmathematical. It is certainly very different from the mathematics that one learns in high school or in calculus courses. However, disciplines such as computer science and operations research contain as much graph theory as they do standard numerical mathematics.

This section consists of a collection of illustrative examples about graphs. We will solve each problem from scratch with a little logic and systematic analysis. Many, of these examples will be revisited in greater depth in subsequent chapters.

The following three graph theory terms are used in the coming examples. A **path** P is a sequence of distinct vertices, written $P = x_1–x_2–\cdots–x_n$, with each pair of consecutive vertices in P joined by an edge. If in addition, there is an edge (x_n, x_1), the sequence is called a **circuit**, written $x_1–x_2–\cdots–x_n–x_1$. For example, in Figure 1.1*a*, *b-d-a-c* forms a path, while *a-b-d-c-a* forms a circuit. A graph is **connected** if there is a path between every pair of vertices. The removal of certain edges or vertices from a connected graph G is said to *disconnect* the graph if the resulting graph is no longer connected; that is, if at least one pair of vertices is no longer joined by a path. The graph in Figure 1.1*a* is connected but the removal of edges (a, b) and (b, d) will disconnect it.

Example 1: Matching

Suppose that we have five people A, B, C, D, E and five jobs a, b, c, d, e, and various people are qualified for various jobs. The problem is to find a feasible one-to-one matching of people to jobs, or to show that no such matching can exist. We can represent this situation by a graph with vertices for each person and for each job, with edges joining people with jobs for which they are qualified. Does there exist a feasible matching of people to jobs for the graph in Figure 1.2?

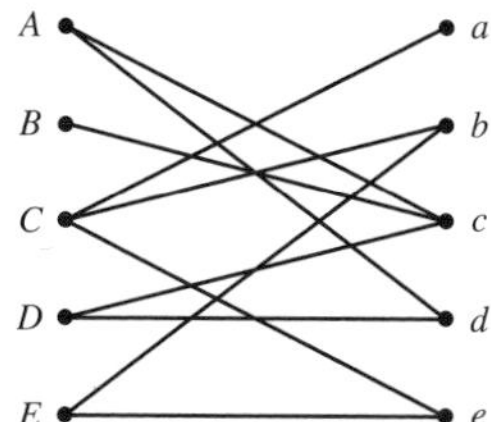

Figure 1.2

The answer is no. The reason can be found by considering people A, B, and D. These three people as a set are collectively qualified for only two jobs, c and d. Hence there is no feasible matching possible for these three people, much less all five people. An algorithm for finding a feasible matching, if any exists, will be presented in Chapter 4. Such matching graphs in which all the edges go horizontally between two sets of vertices are called **bipartite.** Bipartite graphs are discussed further in Section 1.3. ∎

Example 2: Spelling Checker

A spelling checker looks at each word X (represented in a computer as a binary number) in a document and tries to match X with some word in its dictionary. The dictionary typically contains close to 100,000 words. To understand how this checking works, we consider the simplified problem of matching an unknown letter X with one of the 26 letters in the English alphabet. In the spirit of the strategy humans use to home in on the page in a dictionary where a given word appears, the computer search procedure would first compare the unknown letter X with M, to determine whether X $\leq$ M or X $>$ M. The answer to this comparison locates X in the first 13 letters of the alphabet or the second 13 letters, thus cutting the number of possible letters for X in half. This strategy of cutting the possible matches in half can be continued with as many comparisons as needed to home in on X's letter. For example, if X $\leq$ M, then we could test whether or not X $\leq$ G; if X $>$ M, we could test whether X $\leq$ S.

This testing procedure is naturally represented by a directed graph called a **tree**. Figure 1.3 shows the first three rounds of comparisons for the letter-matching procedure. The vertices represent the different letters used in the comparisons. The left descending edge from a vertex Q points to the letter for the next comparison if X $\leq$ Q and the right descending edge from Q points to the next letter if X $>$ Q.

For our original spelling-checker problem, a word processor would use a similar, but larger, tree of comparisons. With just 12 rounds of comparisons, it could reduce

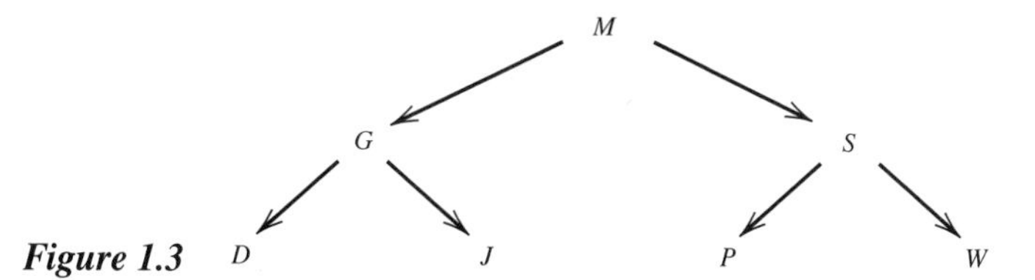

Figure 1.3

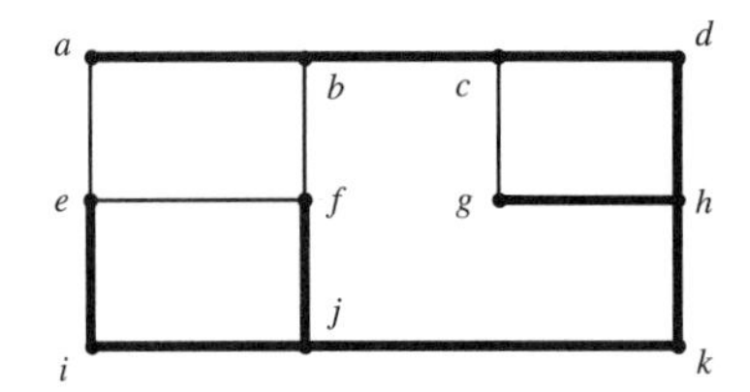

Figure 1.4

the number of possible matches for an unknown word X from 100,000 down to 25, about the number of words in a column of a page in a dictionary. (Once reduced to a list of about 25 possibilities, the search for X would usually run linearly down that list, just as a human would.) ■

Chapter 3 examines trees and their use in various search problems. Trees can be characterized as graphs that are connected and have a unique path between any pair of vertices (ignoring the directions of directed edges). The next example uses trees in a very different way.

Example 3: Network Reliability

Suppose the graph in Figure 1.4 represents a network of telephone lines (or electrical transmission lines). We are interested in the network's vulnerability to accidental disruption. We want to identify those lines and switching centers that must stay in service to avoid disconnecting the network.

There is no telephone line (edge) whose removal will disconnect the telephone network (graph). Similarly, there is no vertex whose removal disconnects the graph.

Is there any pair of edges whose removal disconnects the graph? There are several such pairs. For example, we see that if the two edges incident to a are removed, vertex a is isolated from the rest of the network. A more interesting disconnecting pair of edges is (b, c), (j, k). It is left as an exercise to find all disconnecting sets consisting of two edges for the graph in Figure 1.4.

Let us take a different tack. Suppose we want to find a minimal set of edges needed to link together the eleven vertices in Figure 1.4. There are several possible minimal connecting sets of edges. By inspection, we find the following one: (a, b), (b, c), (c, d), (d, h), (h, g), (h, k), (k, j), (j, f), (j, i), (i, e); the edges in this minimal connecting set are darkened in Figure 1.4. A minimal disconnecting set will always be a tree. One interesting general result about these sets is that if the graph G has n vertices, then a minimal connecting set for G (if any exists) always has $n - 1$ edges. ■

The number of edges incident to a vertex is called the **degree** of the vertex.

Example 4: Street Surveillance

Now suppose the graph in Figure 1.4 represents a section of a city's street map. We want to position police at corners (vertices) so that they can keep every block (edge) under surveillance, that is, every edge should have police at (at least) one of its end vertices. What is the smallest number of police that can do this job?

Let us try to get a lower bound on the number of police needed. The map has 14 blocks (edges). Corners *b*, *c*, *e*, *f*, *h*, and *j* each have degree 3, and corners *a*, *d*, *g*, *i*, and *k* each have degree 2. Since 4 vertices can be incident to at most $4 \times 3 = 12$ edges but there are 14 edges in all, we will need at least five police. If all five police were positioned at degree-3 vertices, then $5 \times 3 = 15$ edges are watched by the 5 police. Since there are only 14 edges, some edge would be covered by police at both end vertices. If four police are at degree-3 vertices and one at a degree-2 vertex, then exactly 14 edges are watched—and no edge need be covered at both ends. (If fewer than 4 of the 5 police are at degree-3 vertices, we could not watch all 14 edges). With these general observations, we are ready for a systematic analysis to try to find 5 vertices that are collectively adjacent to all 14 edges.

Consider edge (c, d). Suppose it is watched by an officer at vertex *d*. Then vertex *c* (the other end vertex of edge (c, d)) cannot also have an officer, since we noted above that if we use a degree-2 vertex, such as *d*, then no edge can be watched from both end vertices. However, if vertex *c* cannot be used, then edge (c, g) must be watched from its other end vertex *g*. But now we are using two degree-2 vertices, *d* and *g*. We noted above that at most one of the five police can be placed at a degree-2 vertex. We got into this trouble by assuming that edge (c, d) is watched from vertex *d*.

Now assume no officer is at vertex *d*. Then we must watch edge (c, d) with an officer at vertex *c*. Since vertex *d* cannot be used, edge (d, h) can be watched only by placing an officer at vertex *h*. Next look at edge (h, k). It is already watched by vertex *h*. Then we assert that (h, k) cannot also be watched by an officer at vertex *k*, since *k* has degree-2 and we noted above that if we use a degree-2 vertex, no edge can be watched from both ends. We conclude that there cannot be an officer at vertex *k*. Then edge (k, j) can be watched only by placing an officer at vertex *j*. We now have officers required to be at vertices *c*, *h*, and *j*.

Similar reasoning shows that with an officer at vertex *j*, there cannot be an officer at vertex *i;* then there must be an officer at vertex *e;* there cannot be an officer at vertex *a;* and there must be an officer at vertex *b*. In sum, we have shown that we should place police at vertices *c*, *h*, *j*, *e*, and *b*. A check shows that these five vertices do indeed watch all 14 edges. Since our reasoning forced us to use these five vertices, no other set of five vertices can work.

At the beginning of this example, we showed that at least five corners were needed to keep all the blocks (edges) under surveillance. Now we have produced a set of five corners that achieve such surveillance. It then follows that five is the minimum number of corners.

We conclude this example by noting that in this surveillance situation, one can also consider watching the vertices rather than the edges: How few officers are needed to watch all the vertices? We use the same type of argument as in the block surveillance problem to get a lower bound on the number of corners needed for corner surveillance. An officer at vertex *x* is considered to be watching vertex *x* and all vertices adjacent to *x*. There are 11 vertices, and 6 of these vertices watch 4 vertices (themselves and 3 adjacent vertices). Thus three is the theoretical minimum. This minimum can be achieved. Details are left as an exercise. ■

A set C of vertices in a graph G with the property that every edge of G is incident to at least one vertex in C is called an **edge cover**. The previous example was asking for an edge cover of minimal size in Figure 1.4. The reasoning in Example 4 illustrates the kind of systematic case-by-case analysis that is common in graph theory.

One goal of graph theory is to find useful relationships between seemingly unrelated graph concepts that arise from different settings. The following example introduces the concept of independent sets, which have an unexpected relation to edge covers.

Example 5: Scheduling Meetings

Consider the following scheduling problem. A state legislature has many committees that meet for one hour each week. One wants a schedule of committee meeting times that minimizes the total number of hours but such that two committees with overlapping membership do not meet at the same time.

This situation can be modeled with a graph in which we create a vertex for each committee and join two vertices by an edge if they represent committees with overlapping membership. Suppose that the graph in Figure 1.4 now represents the membership overlap of 11 legislative committees. For example, vertex c's edges to vertices b, d, and g in Figure 1.4 indicate that committee c has overlapping members with committees b, d, and g.

A set of committees can all meet at the same time if there are no edges between the corresponding set of vertices. A set of vertices without an edge between any two is called an **independent set** of vertices. Our scheduling problem can now be restated as seeking a minimum number of independent sets that collectively include all vertices. This problem is discussed in depth in Section 2.3.

If we want to know how many committees can meet at one time, we are asking the graph question: What is the largest independent set of the graph? Although it is very hard in general to find the largest independent set in a graph, for the graph in Figure 1.4, a little examination shows that there is one independent set of size 6, a, d, f, g, i, k. All other independent sets have five or fewer vertices.

Now for the unexpected link with edge covers. If V is the set of vertices in a graph G, then I will be an independent set of vertices if and only if $V - I$ is an edge cover! Why? Because if there are no edges between two vertices in I, then every edge involves (at least) one vertex not in I, that is, a vertex in $V - I$. Conversely, if C is an edge cover so that all edges have at least one end vertex in C, then there is no edge joining two vertices in $V - C$. So $V - C$ is an independent set. Check that in Figure 1.4, the vertices not in the independent set a, d, f, g, i, k form edge cover b, c, e, h, j.

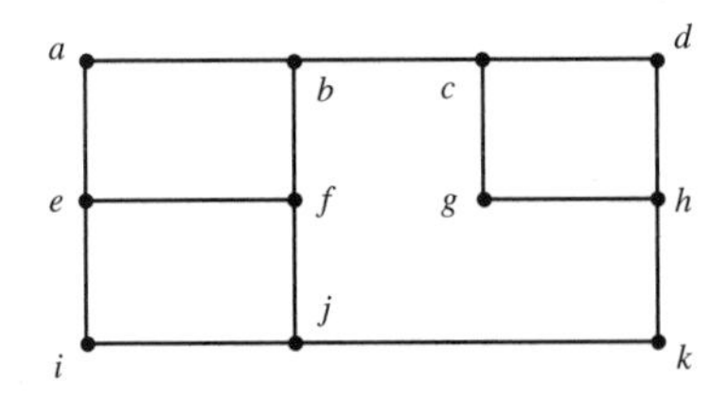

Figure 1.4

A consequence of this relationship is that if I is an independent set of largest possible size in a graph, then $V - I$ will be an edge cover of smallest possible size. So finding a maximal independent set is equivalent to finding a minimal edge cover. ■

Instead of modeling a geometric or other problem with a graph, it is possible to model a graph with a geometric configuration. A graph G is called an **interval graph** if there is a one-to-one correspondence between the vertices of G and a collection of intervals on the line so that two vertices of G are adjacent when the corresponding intervals overlap. We now consider two problems in which we model a real-world situation with a graph and then seek to model the resulting graph with an overlapping family of intervals. In Section 2.3 an interval graph model arises in VLSI (Very Large Scale Integrated) circuitry design.

Example 6: Interval Graph Modeling

A competition graph, used in ecology, has a vertex for each of a given set of species and an edge joining each pair of species that feed on a common prey (i.e., the two species compete for food). See the competition graph in Figure 1.5a. Ecologists have given considerable attention to viewing competition graphs as interval graphs for the following reason (see reference [3] for further details). Each species is thought of having a "niche" in its ecological system, its position in terms of food it eats and

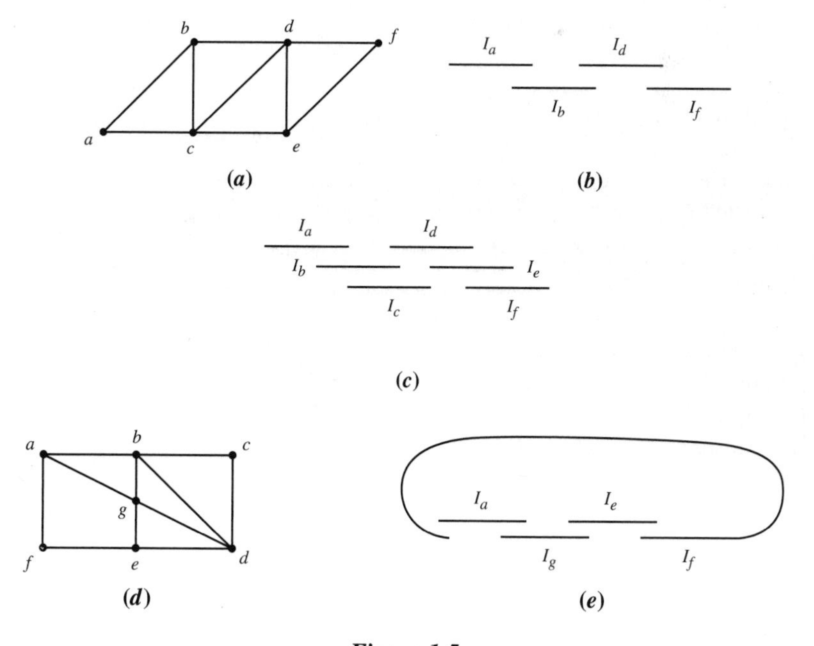

Figure 1.5

land where it lives and similar attributes. A niche may be regarded as a point in some suitable multidimensional parameter space.

We try to associate an interval with each species (vertex) such that competing species (adjacent vertices) correspond to overlapping intervals. An interval is a geometric realization of a species' niche and an interval graph model of a competition graph implies that the overall ecological system can be viewed in a simple, one-dimensional space.

Can the ecological niches of the species represented in the competition graph in Figure 1.5*a* be modeled by overlapping intervals on a line? That is, is the graph in Figure 1.5*a* an interval graph? The overlaps among vertices *a*, *b*, *d*, *f*, which form a 4-vertex path, can be modeled only by a set of four intervals I_a, I_b, I_d, I_f arranged as shown in Figure 1.5*b* (or its mirror image). With a little trial and error, we build the collection of intervals shown in Figure 1.5*c*, whose overlaps reflect exactly the adjacencies of the graph in Figure 1.5*a*.

Next consider the graph in Figure 1.5*d*. Does it have an interval model? A little thought shows that there is no way that four overlapping intervals can model the adjacencies among the four vertices *a, g, e, f*, which form a circuit in Figure 1.5*d*. Figure 1.5*e* shows that one of the intervals would have to wrap around to model the required adjacencies. ■

We next give an example involving directed graphs.

Example 7: Influence Model

Suppose psychological studies of a group of people determine which members of the group can influence the thinking of others in the group. We can make a graph with a vertex for each person and a directed edge $(\overrightarrow{p_1, p_2})$ whenever person p_1 influences p_2. Let the graph in Figure 1.6*a* represent a set of such influences. Now let us ask for a minimal subset of people who can spread an idea through to the whole group, either directly or by influencing someone who will influence someone else and so forth. In graph-theoretic terms, we want a minimal subset of vertices with directed paths to all other vertices (a *directed path* from p_1 to p_k is an edge sequence $(\overrightarrow{p_1, p_2}), (\overrightarrow{p_2, p_3}) \ldots (\overrightarrow{p_{k-1}, p_k})$). Such a subset of influential vertices is called a **vertex basis**.

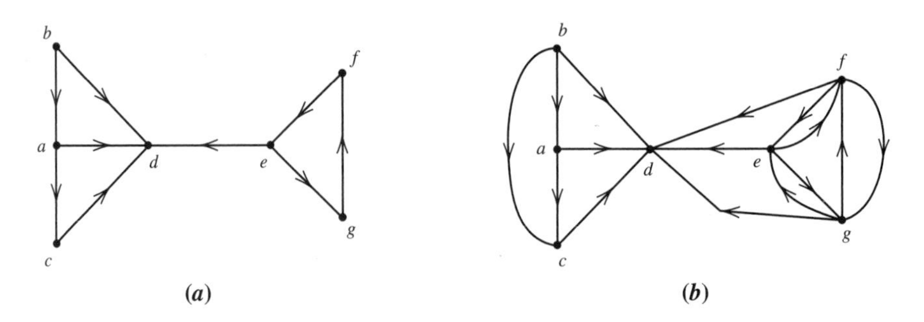

Figure 1.6

To aid us, we can build a *directed-path graph* for the original graph with the same vertex set and with a directed edge $(\vec{p_i, p_j})$ if there is a directed path from p_i to p_j in the original graph. Figure 1.6*b* shows the directed-path graph for the graph in Figure 1.6*a*. Now our original problem can be restated as: Find a minimal subset of vertices in the new graph with edges directed to all other vertices. This is just a directed-graph version of the vertex-covering problem mentioned at the end of Example 4. Observe that any vertex in Figure 1.6*b* with no incoming edges must be in this minimal subset (since no other vertices have edges to it); vertex b is such a vertex. Since b has edges to a, c, and d, then e, f, and g are all that remain to be "influenced." Either e, f, or g "influence" these three vertices. Then b, e, or b, f, or b, g are the desired minimal subsets of vertices. ■

1.1 EXERCISES

Summary of Exercises The first seven exercises involve simple graph models. Exercises 8–34 present examples and extensions of the models presented in the examples in this section. Exercises 35 and 36 involve other new graph models.

1. Suppose interstate highways join the six towns A, B, C, D, E, F as follows: I-77 goes from B through A to E; I-82 goes from C through D then through B to F; I-85 goes from D through A to F; I-90 goes from C through E to F; and I-91 goes from D to E.

(a) Draw a graph of the network with vertices for towns and edges for segments of interstates linking neighboring towns.

(b) What is the minimum number of edges whose removal prevents travel between some pair of towns?

(c) Is it possible to take a trip starting from town C that goes to every town without using any interstate highway for more than one edge (the trip need not return to C)?

2. (a) Suppose four teams, the Aces, the Birds, the Cats, and the Dogs, play each other once. The Aces beat all three opponents except the Birds. The Birds lost to all opponents except the Aces. The Dogs beat the Cats. Represent the results of these games with a directed graph.

(b) A ranking is a listing of teams such that the ith team on the list beat the $(i + 1)$ st team. Find all rankings for part (a).

3. (a) A schedule is to be made with five football teams. Each team is to play two other teams. Explain how to make a graph model of this problem.

(b) Show that except for interchanging names of teams, there is only one possible graph in part (a).

4. Suppose there are six people—John, Mary, Rose, Steve, Ted, and Wendy—who pass rumors among themselves. Each day John talks with Mary and Wendy; Mary talks with John, Rose, and Steve; Rose talks with Mary, Steve, and Ted; Steve talks with Mary, Rose, Ted, and Wendy; Ted talks with Rose, Steve, and Wendy; and Wendy talks with John, Steve, and Ted. Whatever people hear one day they pass on to others the next day.

(a) Model this rumor-passing situation with a graph.

(b) How many days does it take to pass a rumor from John to Steve? Who will tell it to Steve?

(c) Is there any way that if two people stopped talking to each other, then it must take three days to pass a rumor from one person to all the others?

5. **(a)** Give a direction to each edge in Figure 1.4 so that there are directed routes from any vertex to any other vertex.

(b) Do part (a) so as to minimize the length of the longest directed path between any pair of vertices. Explain why a smaller minimum is not possible.

6. **(a)** What is the length of the longest possible path (with the most vertices) in the graph in Figure 1.3, ignoring directions of edges?

(b) What is the length of the longest possible circuit (with the most vertices) in the graph in Figure 1.4?

7. The flow chart below displays the logical flow possibilities in a program. One possible execution sequence is *ABDFGBCEFGH*. How many execution sequences are there with 15 steps (vertices)?

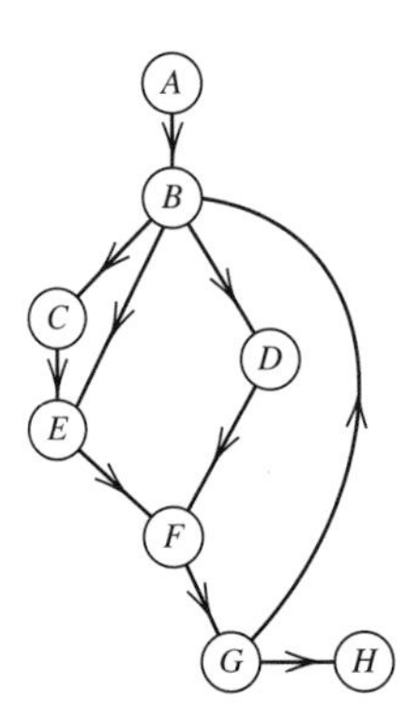

8. Give another reason why Figure 1.2 has no matching by considering the appropriate subset of jobs (showing they cannot all be filled).

9. Find a matching or explain why none exists for the following graphs:

(a) **(b)**

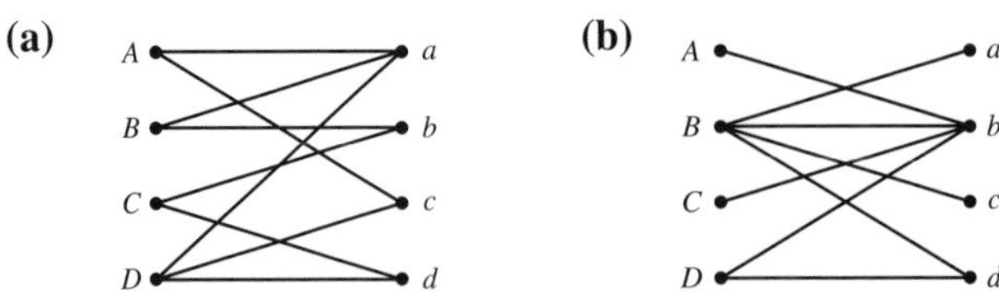

10. Find all matchings in Exercise 9(a).

11. We generalize the idea of matching in Example 1 to arbitrary graphs by defining a matching to be a pairing off of adjacent vertices in a graph. For example, one possible matching in Figure 1.1*a* is *a-b, c-d*. Which of the following graphs have a matching? If none exists, explain why.

(a) Figure 1.5*a* **(b)** Figure 1.4

(c)

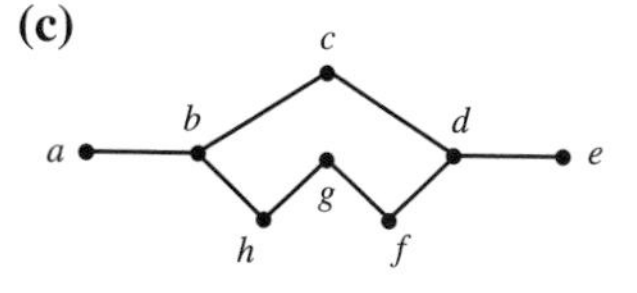

12. (a) Suppose a dictionary in a computer has a "start" from which one can branch to any of the 26 letters, and at any letter one can go to the preceding and succeeding letters. Model this data structure with a graph.

(b) Suppose additionally that one can return to "start" from letters c or k or t. Now what is the longest path between any two letters?

13. Build the complete testing tree in Example 2 to identify one of the 26 letters of the alphabet.

14. (a) Repeat Example 2 using three-way comparisons (less than; greater than or equal to) to identify one of the 26 letters.

(b) In general, how many more rounds of testing will a two-way scheme for n letters require over a three-way scheme (you do not need to justify your answer rigorously)?

15. Suppose eight current varieties of chipmunk evolved from a common ancestral strain through an evolutionary process in which at various stages one ancestral variety split into two varieties (none of the ancestral varieties survive when they split into two new varieties).

(a) Explain how one might model this evolutionary process with a graph.

(b) What is the total number of splits that must have occurred?

16. In Example 3, find a minimal connecting set of edges containing neither (a, b) nor (b, c).

17. (a) What are the other sets of 2 edges whose removal disconnects the graph in Figure 1.4 besides (a, b), (a, e) and (c, d), (d, h)? Either produce others or give an argument why no others exist.

(b) Find all sets of 2 vertices whose removal disconnects the remaining graph in Figure 1.4.

18. (a) For the following graph, find all sets of 2 vertices whose removal disconnects the graph of remaining vertices.

(b) Find all sets of 2 edges whose removal would disconnect the graph.

(c) What is the minimal number of vertex deletions required to disconnect the graph formed by the 12 edges of a cube?

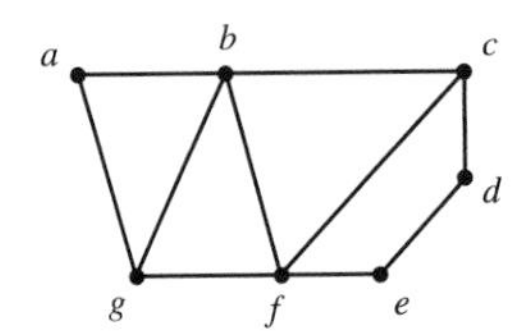

19. Find a minimal edge cover and a minimal set of vertices adjacent to all other vertices for the graph in Figure 1.2.

20. In Figure 1.4, find all sets of three corners that have all 11 corners under surveillance. Give a careful logical analysis.

21. Repeat Example 4 for minimal block and corner surveillance when the network in Figure 1.4 is altered by adding edges (f, g), (g, j) and deleting (b, f).

22. Repeat Example 4 for the edge cover and minimal corner surveillance when the network is formed by a regular array of north–south and east–west streets of size:

(a) 3 by 3 **(b)** 4 by 4 **(c)** 5 by 5

23. **(a)** A queen dominates any square on a chessboard in the same row, column, or diagonal as the queen. How few queens can dominate all squares on a 8 by 8 chessboard?

(b) Repeat this problem for bishops, which dominate only diagonals.

24. Solve the committee scheduling problem for the committee overlap graph in Figure 1.4. That is, what is the minimum number of independent sets needed to cover all vertices?

25. **(a)** Find a maximum independent set in the following graphs:

(i) Figure 1.1*a* **(ii)** Figure 1.2 **(iii)** Figure 1.5*a*

(b) Use your result in part (a) to produce a minimal edge cover in these graphs.

26. **(a)** Using a model similar to the schedule model in Example 5, describe how one might represent interlocking (overlapping) Boards of Directors of large corporations with a graph model, that is, a graph telling which Boards have overlapping members.

(b) Suppose Figure 1.5*d* is the graph of such interlocking boards. What is the maximal number of boards that could meet at the same time?

27. Which of the following graphs are interval graphs? If they are, give a model of overlapping intervals; if not, explain why not.

(a) Figure 1.1*a* **(b)** Figure 1.3 **(c)** Figure 1.4

(d)

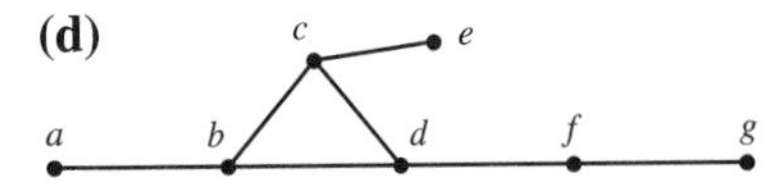

28. Find a graph G not containing a circuit of length four or more (without chords), such that G is not an interval graph but if any vertex is removed from G the resulting graph is always an interval graph.

29. A unit interval graph is an interval graph in which all intervals have the same length. Find unit interval models for the following graphs.

(a) Figure 1.1a **(b)** Figure 1.3
(c) Figure 1.5a with vertex e deleted

30. Show that the graph below on the left is an interval graph but not a unit-interval graph.

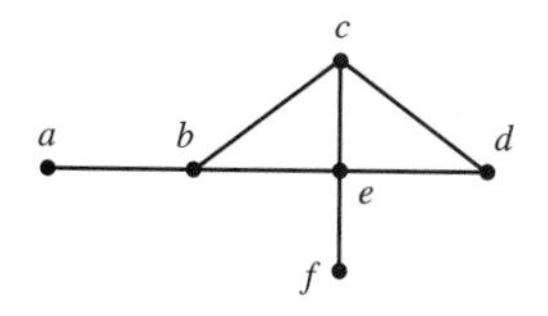

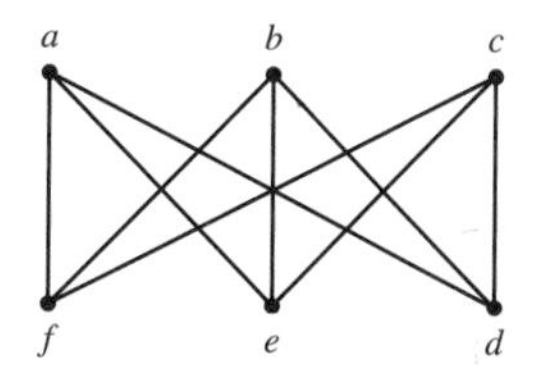

31. Show that adjacency in the graph above on the right can be modeled by overlapping rectangles in the plane.

32. Six students A, B, C, D, E, F went to the library once last Monday night. One of them stole a book. Each student is questioned about whom they saw at the library, and all tell the truth except the thief (it is possible that one student saw another without in turn being observed, but we assume that one must have seen the other if both were there at the same time). A saw B and E; B saw A and F; C saw D and F; D saw A and F; E saw B and C; and F saw C and E. Who is the thief? (*Hint:* Use an interval graph.)

33. Suppose a certain type of CB radio can be tuned to receive frequencies ranging from R to R' but can transmit only from T to T', where $R < T < T' < R'$. We make a "CB graph" for a set of CB radios with an edge between two vertices if they represent CB radios that can mutually receive each other.

(a) Show that any interval graph can be a CB graph.

(b) Find a CB graph that is not an interval graph.

34. Find a vertex basis in the following directed graphs:

(a) Figure 1.1b **(b)** Figure 1.3
(c) Figure 1.4 with edges directed by alphabetical order [e.g., edge (a, e) is directed from a to e]

35. Show that the vertex basis in a directed graph is unique if there is no sequence of directed edges that forms a circuit in the graph.

36. A game for two players starts with an empty pile. Players take turns putting 1 or 2 or 3 pennies in the pile. The winner is the player who brings the value of the pile up to 16¢.

(a) Make a directed graph modeling this game.

(b) Show that the second player has a winning strategy by finding a set of four "good" pile values, including 16¢, such that the second player can always move to one of the "good" piles (when the second player moves to one of the good piles, the next move of the first player must be to a non-good pile, and from this position the second player has a move to a good pile, etc.).

37. The parsing of a sentence can be represented by a directed graph, with a vertex S (for the whole sentence) having edges to vertices Su (subject) and P (predicate), then Su and P having edges to the parts into which they are decomposed into pieces, and so on.

Consider the abstract grammar with decomposition rules: $S \to AB$, $S \to BA$, $A \to ABA$, $B \to BAS$, and $B \to S$. For example, $BAABA$ can be "parsed" as shown below.

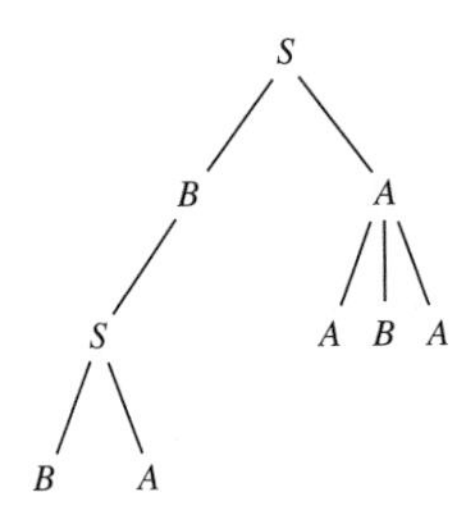

Find a parsing graph for each of the following (or explain why no parsing exists):

(a) $BABABABA$ **(b)** $BBABAABA$

38. Ask a friend to pose three subject areas in which you must find problems having graph models.

1.2 ISOMORPHISM

In this section we investigate some of the basic structure of graphs. We are interested in properties that distinguish one vertex in a graph from another vertex and, more generally, that distinguish one graph from another graph. We motivate this discussion with the question: How can we tell if two graphs are really the same graph, but drawn differently and with different names for the vertices? For example, are the two 5-vertex graphs in Figure 1.7 different versions of the same graph?

A graph can be drawn on a sheet of paper in many different ways. Thus, it is usually possible to draw a graph in two ways that would lead a casual viewer to consider the drawings to be "different" graphs. This motivates the following definition.

> Two graphs G and G' are called **isomorphic** if there exists a one-to-one correspondence between the vertices in G and the vertices in G' such that a pair of vertices are adjacent in G if and only if the corresponding pair of vertices are adjacent in G'.

Such a one-to-one correspondence of vertices that preserves adjacency is called an **isomorphism**. A useful way to think of isomorphic graphs is as follows: The first graph can be redrawn on a transparency that can be exactly superimposed over a drawing of the second graph.

To be isomorphic, two graphs must have the same number of vertices and the same number of edges. The two graphs in Figure 1.7 pass this initial test. Both graphs have one vertex, e and 5, respectively, at the end of just one edge. Then any isomorphism of these two graphs must match e with *5*. Also, the vertices at the other ends of the edge from e and *5* must be matched; that is, d matches with *4*. (Think of superimposing one graph over the other.) The remaining three vertices in each graph are mutually adjacent (forming a triangle) and also are all adjacent to d or *4*, respectively. Thus the matching a with *1*, b with *2*, and c with *3* (or any other matching of these two subsets of three vertices) will preserve the required adjacencies. The correspondence $a - 1$, $b - 2, c - 3, d - 4, e - 5$ is then an isomorphism, and the two graphs are isomorphic. To visualize how they can be made to look the same, think of moving vertices *4* and *5* in the right graph upward and to the right [past edge (*1,3*)], so that *1, 2, 3, 4* form a quadrilateral with crossing diagonals.

Recall that the *degree* $\deg(x)$ of a vertex is the number of edges incident to the vertex. Degrees are preserved under isomorphism, that is, two matched vertices must have the same degree. Then in Figure 1.7, e has to be matched with *5* and d matched with *4* because they are the unique vertices of degree 1 and 4 in their respective graphs. Further, two isomorphic graphs must have the same number of vertices of a given degree. For example, if they are to be isomorphic, the two graphs in Figure 1.7 must both have the same number of vertices of degree 3—they do; both have 3 vertices of degree 3.

A **subgraph** G' of a graph G is a graph formed by a subset of vertices and edges of G. If two graphs are isomorphic, then subgraphs formed by corresponding vertices and edges must be isomorphic. In Figure 1.7, removal of vertices e and *5* (and their incident edges) leaves two isomorphic subgraphs consisting of four mutually adjacent

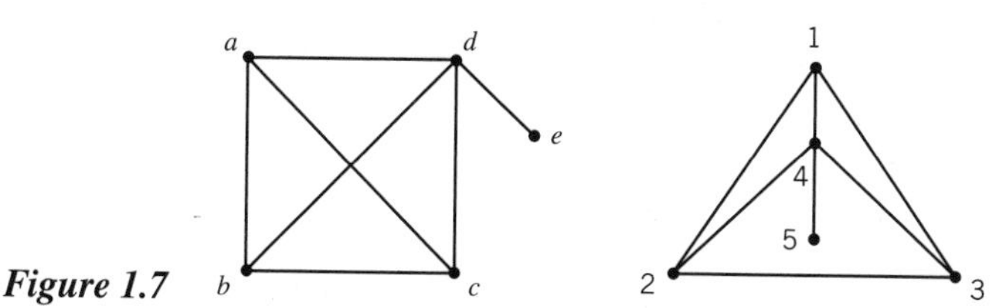

Figure 1.7

vertices. Once this subgraph isomorphism is noted, isomorphism of the whole graphs is easily demonstrated.

Subgraphs can be used to test for isomorphism in the following way. If a graph G has a set of 6 vertices forming a chordless circuit of length 6 (chordless means there are no other edges between these 6 vertices except the 6 edges forming the circuit), then any graph isomorphic to G must also have a set of 6 vertices generating such a chordless 6-circuit.

A graph with n vertices in which each vertex is adjacent to all the other vertices is called a **complete graph on n vertices**, denoted K_n. A complete graph on 2 vertices, K_2, is just an edge. Complete subgraphs are in a sense the building blocks of all larger graphs. For example, both graphs in Figure 1.7 consist of a K_4 and a K_2 joined at a common vertex. Conversely, every graph on n vertices is a subgraph of K_n.

Before examining other pairs of graphs for isomorphism, let us mention the practical importance of determining whether two graphs are isomorphic. Researchers working with organic compounds build up large dictionaries of compounds that they have previously analyzed. When a new compound is found, they want to know if it is already in the dictionary. Large dictionaries can have many compounds with the same molecular formula but differing in their structure as graphs (and possibly in other ways). Then one must test the new compound to see if its graph-theoretic structure is the same as the structure of one of the known compounds with the same formula (and the same in other ways), that is, whether the new compound is graph-theoretically isomorphic to one of a set of known compounds. A similar problem arises in designing efficient integrated circuitry for a computer. If the design problem has already been solved for an isomorphic circuit (or if a piece of the new network is isomorphic to a previously designed circuit), then valuable savings in time and money are possible.

Example 1: Simple Isomorphism

Are the two graphs in Figure 1.8 isomorphic?

Both graphs have 8 vertices and 10 edges. Let us examine the degrees of the different vertices. We see that *b, d, f, h* and *3, 4, 7, 8* have degree 2, while the other vertices have degree 3. Then the two graphs have the same number of vertices of degree 2 and the same number of degree 3. The respective subgraphs of the 4 vertices of degree 2 (and the edges between these degree-2 vertices) in each graph must be isomorphic if the whole graphs are isomorphic. However, there are no edges between

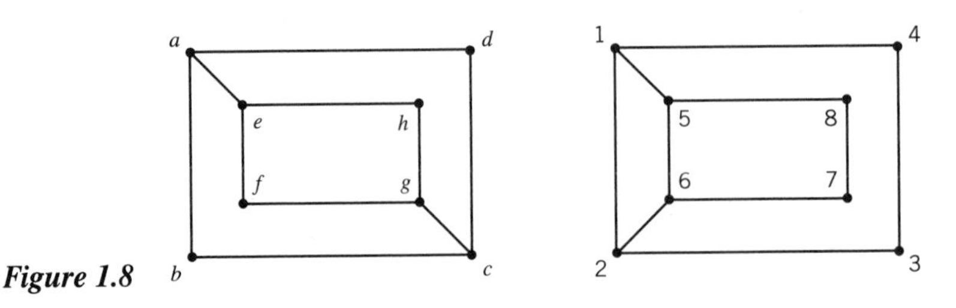

Figure 1.8

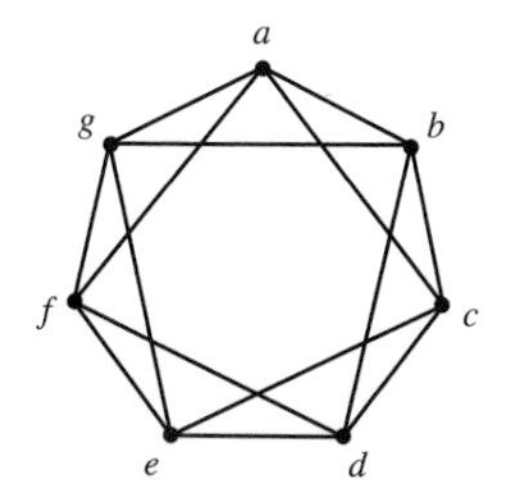

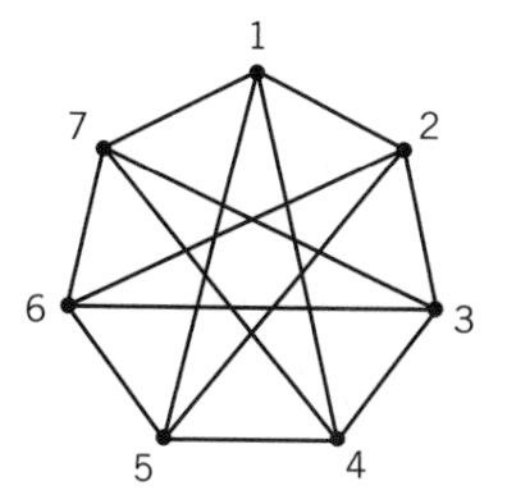

Figure 1.9

any pair of *b, d, f, h*, while the other subgraph of degree-2 vertices has two edges: (*4, 3*) and (*8, 7*). So the subgraphs of degree-2 vertices are not isomorphic, and hence the two full graphs are not isomorphic. The reader can also check that the two subgraphs of degree-3 vertices in each graph are not isomorphic. ■

The vertices of degree 2 in the left graph in Figure 1.8 form a subgraph of mutually nonadjacent vertices. Such a subgraph is called a set of **isolated vertices**.

Example 2: Isomorphism in Symmetric Graphs

Are the two graphs in Figure 1.9 isomorphic?

The two graphs both have 7 vertices and 14 edges. Every vertex in both graphs has degree 4. Further, both graphs exhibit all the symmetries of a regular 7-gon. With no distinctions possible among vertices within the same graph, our only option is to try to construct an isomorphism. Start with vertex *a* in the left graph. By symmetry, we can match *a* to any vertex in the right graph (that is, if the two graphs are isomorphic, there will exist an isomorphism with *a* matched to any vertex in the right graph). Let us use the match $a - 1$.

The set of neighbors of *a* (vertices adjacent to *a*) must be matched with the set of neighbors of *1*. Let us look at the subgraphs formed by these neighbors of *a* and *1*. See Figure 1.10. Both subgraphs are paths: One is *f* to *g* to *b* to *c*, and the other is *7* to *4* to *5* to *2*. The matching must make these path subgraphs isomorphic. Thus, *f* and *c* must be matched with *7* and *2* (matching ends of the two paths). By the horizontal symmetry of the graphs, it makes no difference which way *f* and *c* are matched—say $f - 7$ and $c - 2$. Then to complete the isomorphism of neighbors of *a* and *1*, we must match g with *4* and *b* with *5*. Now there remain only two unmatched vertices in each graph: *d, e* and *3, 6*. Vertex *g* is adjacent to *e* but not *d*, and its matched vertex *4* is adjacent to *3* but not to *6*. Thus we must match *e* with *3* and *d* with *6*.

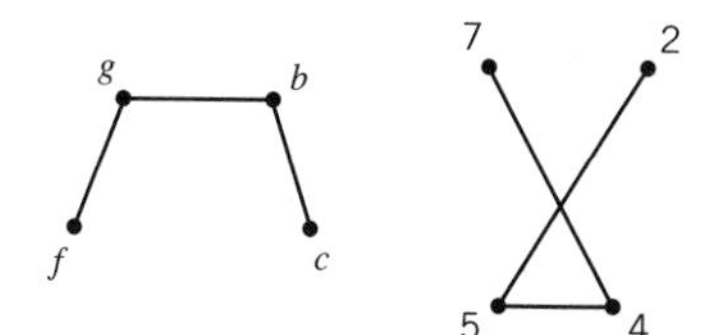

Figure 1.10

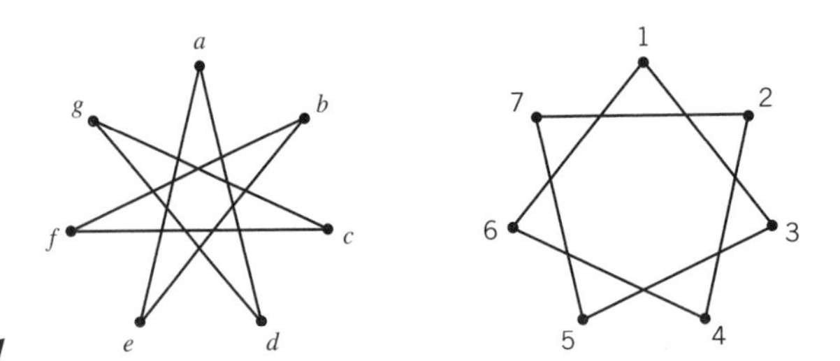

Figure 1.11

In sum, allowing for symmetries to match a with *1* and f with *7*, we conclude that if the graphs are isomorphic, one isomorphism must be $a-1, b-5, c-2, d-6, e-3, f-7, g-4$. Checking edges, we see that the graphs are indeed isomorphic with this matching (if this matching were found not to be an isomorphism, then the two graphs would not be isomorphic since the matches we made were all forced except for the symmetry involving the matches of a and f). ■

Given a graph $G = (V, E)$, its **complement** is a graph $\overline{G} = (V, \overline{E})$ with the same set of vertices but now with edges between exactly those pairs of vertices not linked in G. The union of the edges in G and $\overline{G}$ forms a complete graph. Two graphs G_1 and G_2 will be isomorphic if and only if $\overline{G}_1$ and $\overline{G}_2$ are isomorphic. The isomorphism problem in Example 2 is easy to answer using complements. Figure 1.9 shows the complements of the two graphs in Figure 1.11. Clearly, both these complementary graphs are just a (twisted) circuit of length 7 and hence are isomorphic.

In general, if a graph has more pairs of vertices joined by edges than pairs not joined by edges, then its complement will have fewer edges and thus will probably be simpler to analyze.

Example 3: Isomorphism of Directed Graphs

Are the two directed graphs in Figure 1.12 isomorphic?

Each graph has 8 vertices and 12 edges, and each vertex has degree 3. If we break the degree of a vertex into two parts, the **in-degree** (number of edges pointed in toward the vertex) and **out-degree** (number of edges pointed out), we see that

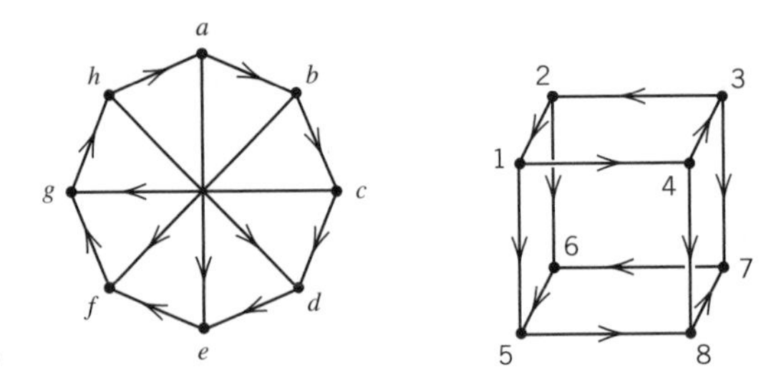

Figure 1.12

each graph has 4 vertices of in-degree 2 and out-degree 1, and each graph has 4 vertices of in-degree 1 and out-degree 2. We could try to build an isomorphism as in the previous example by starting with a match (by a symmetry argument) between a and 1 and then matching their neighbors (with edge directions also matched), and so forth.

However, there is a basic difference in the directed path structure of the two graphs. We will exploit this difference to prove nonisomorphism. In the left graph we can draw a directed path from any given vertex to any other vertex by going clockwise around the circle of vertices: The outer edges form a directed circuit through all the vertices in the left graph. But in the right graph, all edges between the vertex subsets $V_1 = \{1, 2, 3, 4\}$ and $V_2 = \{5, 6, 7, 8\}$ are directed from V_1 to V_2, and thus there can be no directed paths from any vertex in V_2 to any vertex in V_1 (nor is there a directed circuit through all the vertices). Thus, the two graphs are not isomorphic. ■

1.2 EXERCISES

1. List all nonisomorphic undirected graphs with four vertices.

2. List all nonisomorphic directed graphs with three vertices.

3. Draw two nonisomorphic graphs with:

(a) 6 vertices and 10 edges

(b) 9 vertices and 13 edges

4. If directions are ignored, are the two graphs in Figure 1.12 isomorphic?

5. Which of the following pairs of graphs are isomorphic? Explain carefully.

(a)

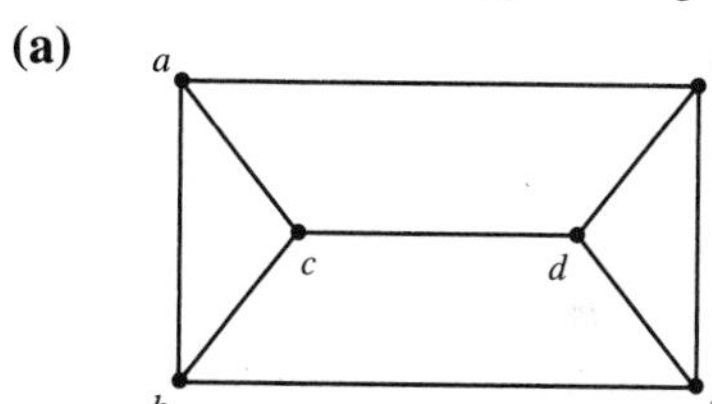

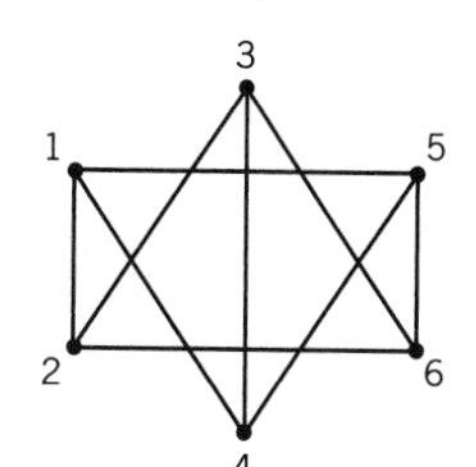

(b)

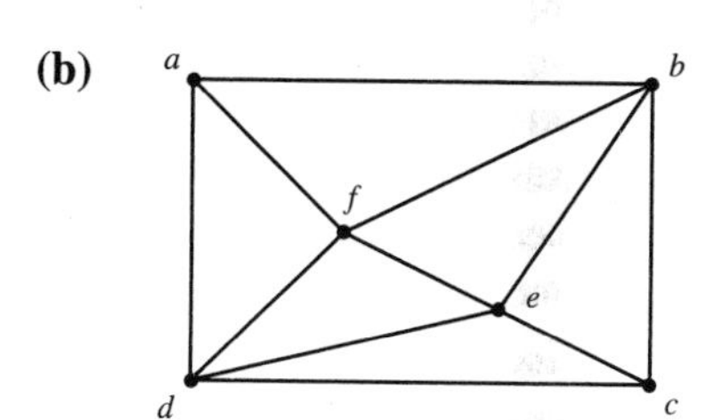

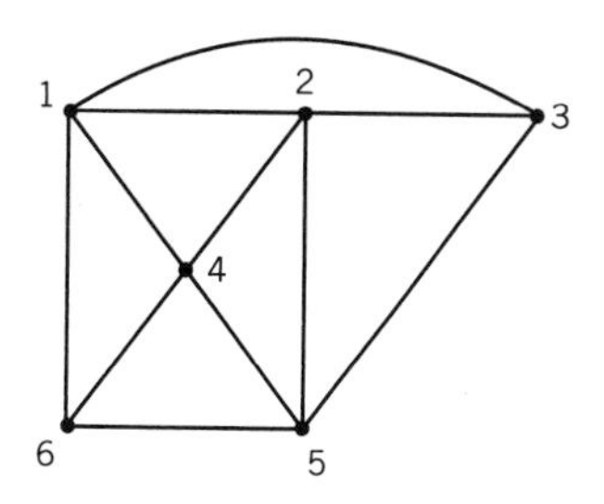

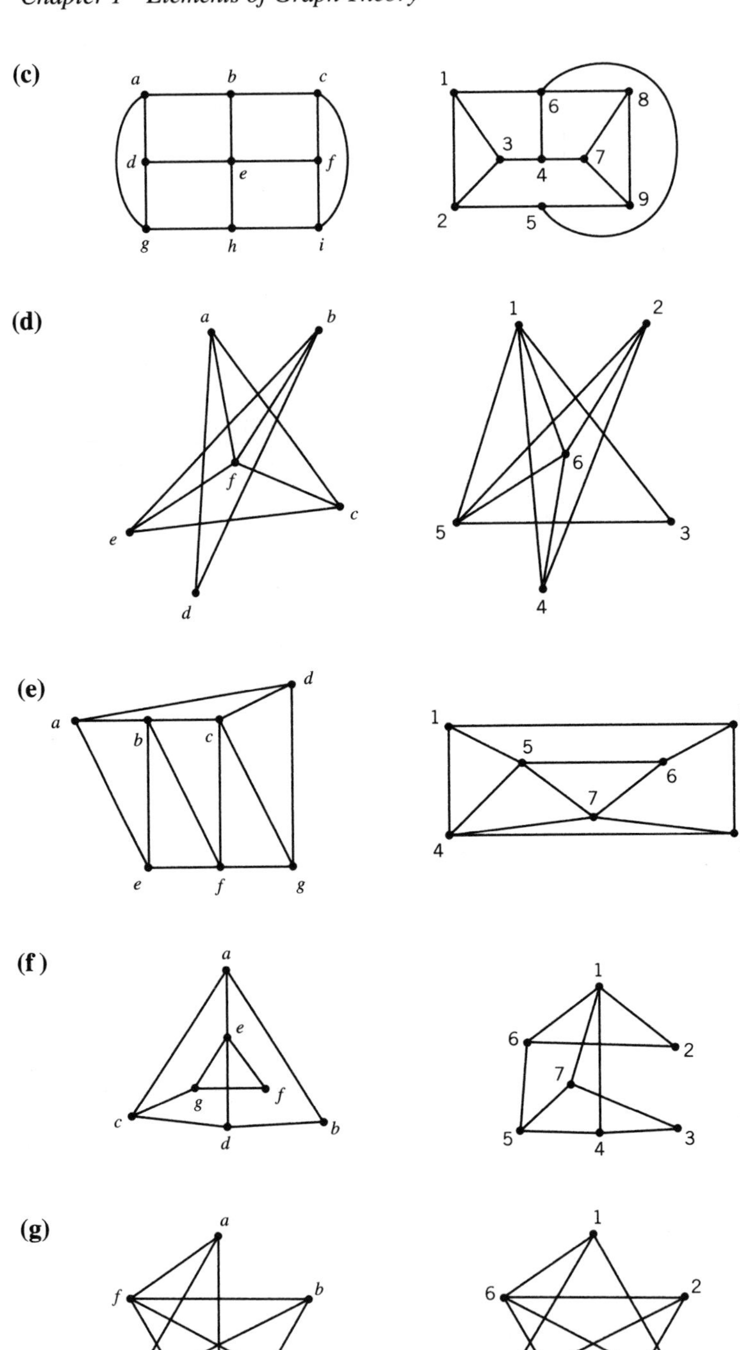
(c)
a
b
c
d
e
f
g
h
i
1
2
3
4
5
6
7
8
9
(d)
a
b
c
d
e
f
1
2
3
4
5
6
(e)
a
b
c
d
e
f
g
1
2
3
4
5
6
7
(f)
a
b
c
d
e
f
g
1
2
3
4
5
6
7
(g)
a
b
c
d
e
f
1
2
3
4
5
6

(h)

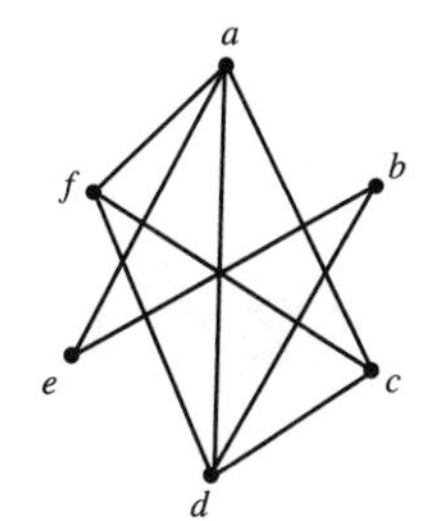

(i)

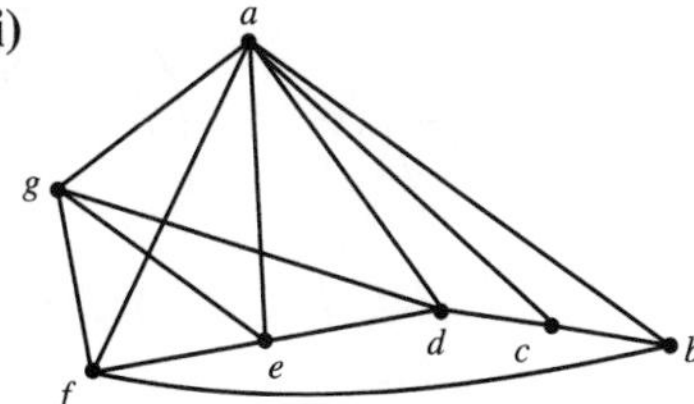

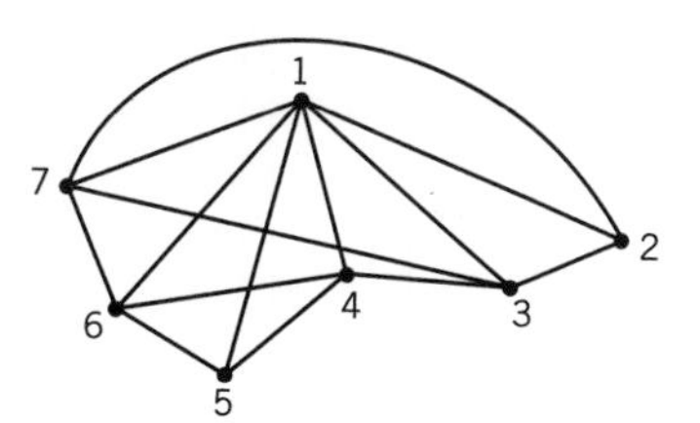

(j)

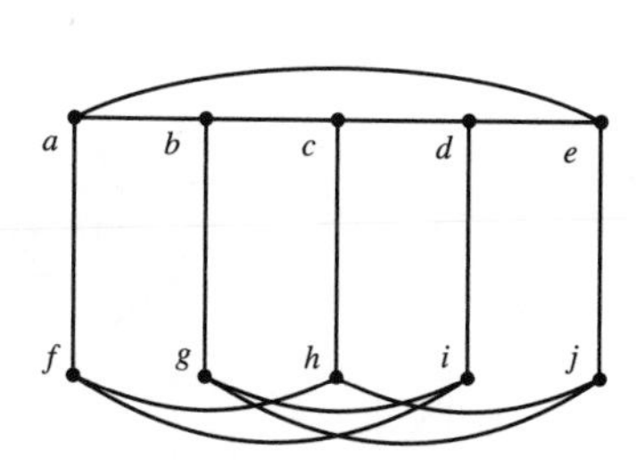

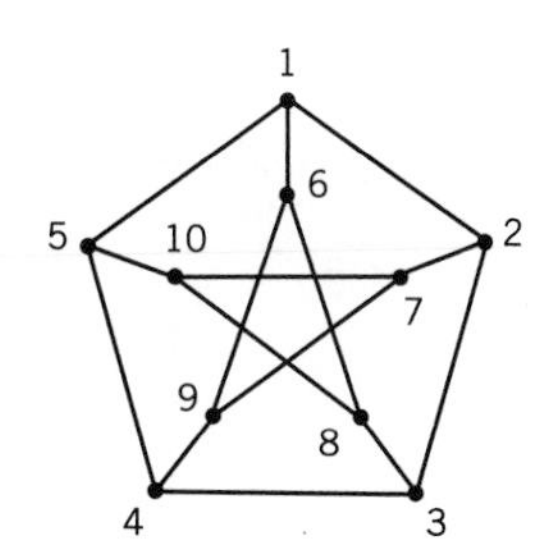

6. Which of the following pairs of graphs are isomorphic? Explain carefully.

(a)

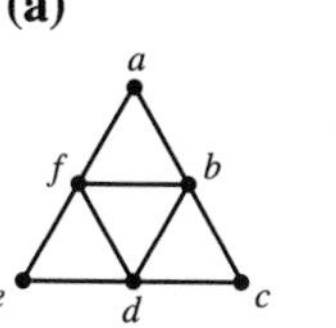

(b)

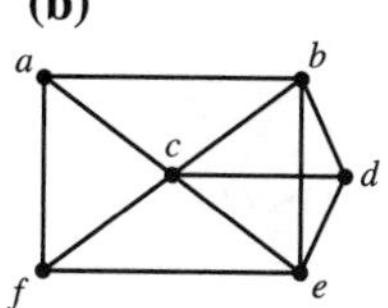

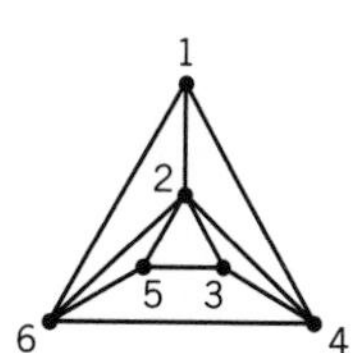

(c)

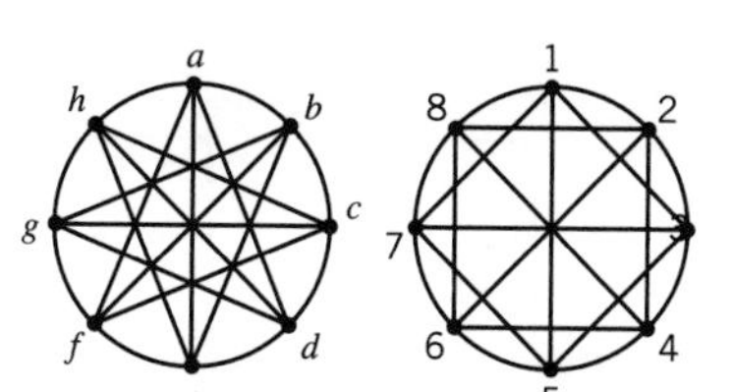

(d)

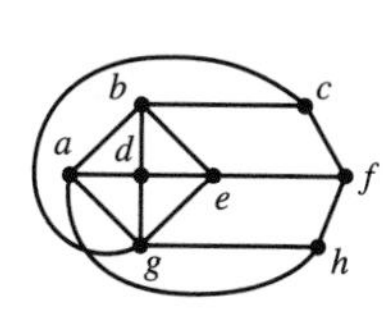

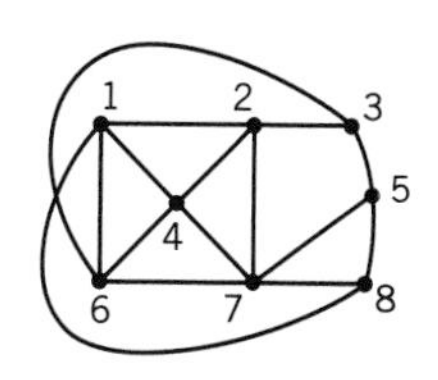

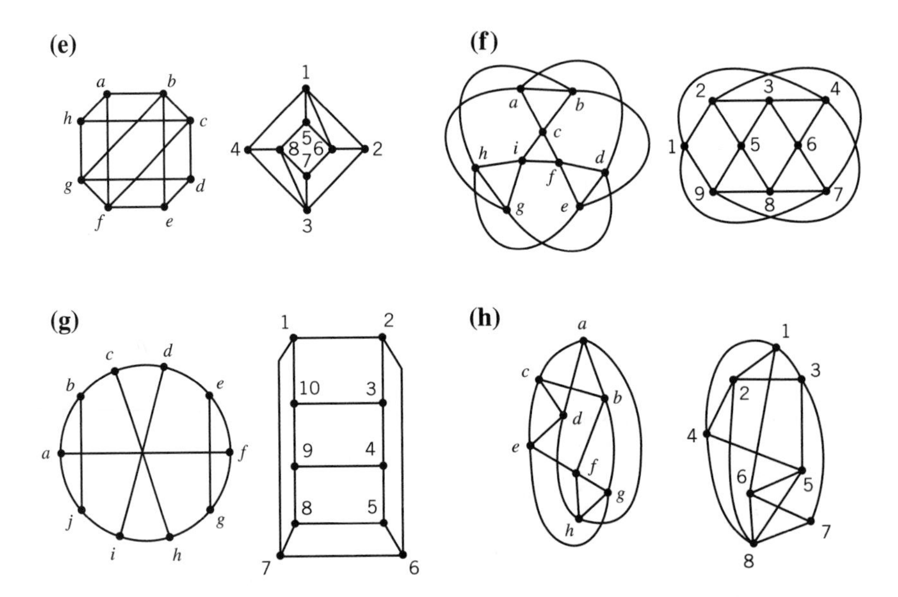

7. Which pairs of graphs in this set are isomorphic?

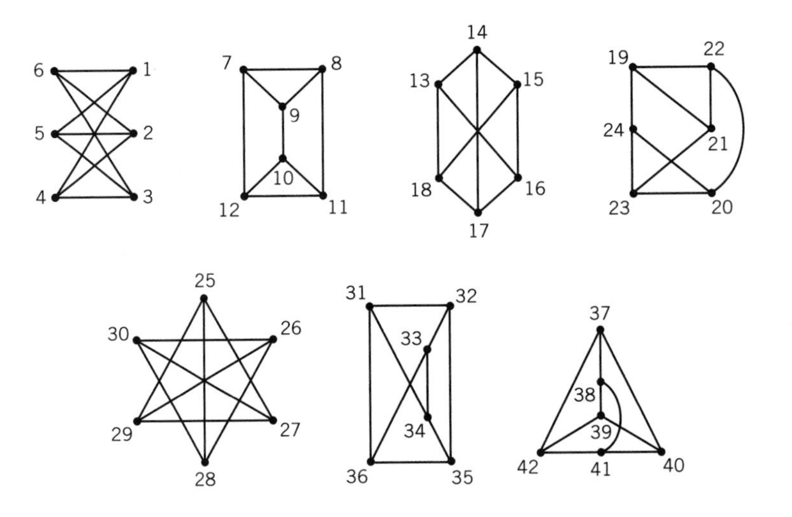

8. Which pairs of graphs in this set are isomorphic?

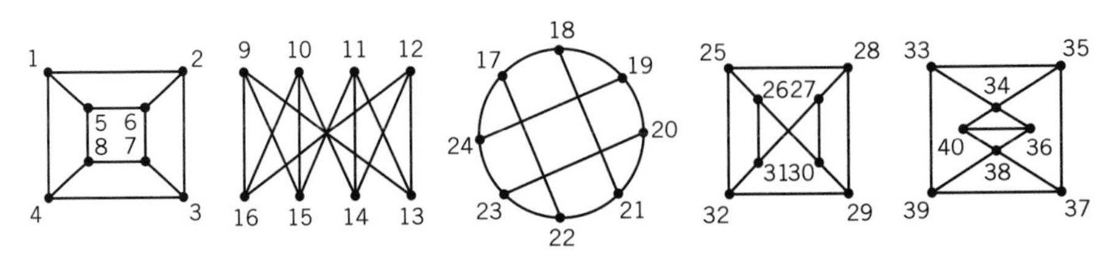

9. Suppose each edge in the graphs in Figure 1.9 is directed from the smaller (numerically or alphabetically) end vertex to the larger end vertex. Are the two resulting directed graphs isomorphic?

10. Are the following pairs of directed graphs isomorphic?

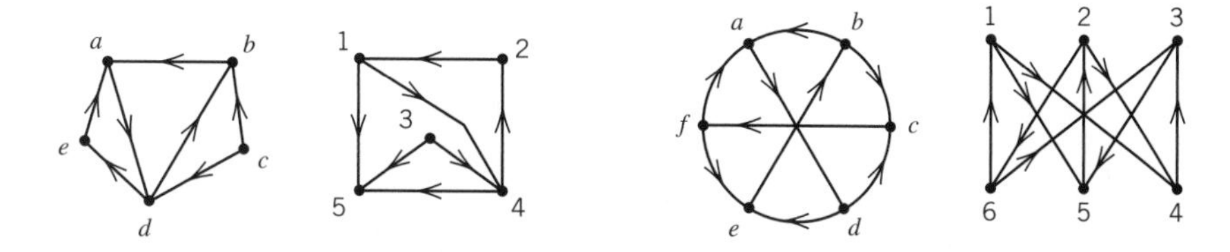

11. Show that all 5-vertex graphs with 2 edges incident to each vertex are isomorphic.

12. Are there any 6-vertex graphs with 3 edges incident to each vertex that are not isomorphic to one of the graphs in Exercise 7?

13. What is the minimum number of vertices that must be deleted from the two graphs in the following figures to get isomorphic subgraphs?

(a) Figure 1.8 **(b)** Figure 1.12 **(c)** Exercise 5(d)

14. What are the sizes of the largest complete subgraphs in the two graphs in Exercise 6(h)?

15. What are the sizes of the largest sets of isolated vertices in the two graphs in Exercise 6(g)?

16. Build 6-vertex graphs with the following degrees of vertices, if possible. If not possible, explain why not.

(a) Three vertices of degree 3 and 3 vertices of degree 1.

(b) Vertices of degrees 1, 2, 2, 3, 4, 5.

(c) Vertices of degrees 2, 2, 4, 4, 4, 4.

1.3 EDGE COUNTING

There is very little in the way of general assertions that can be made about all graphs. There is one useful general theorem, a formula for counting edges.

Theorem 1

In any graph, the sum of the degrees of all vertices is equal to twice the number of edges.

Proof

Summing the degrees of all vertices counts all instances of some edge being incident at some vertex. But each edge is incident with two vertices, and so the total number

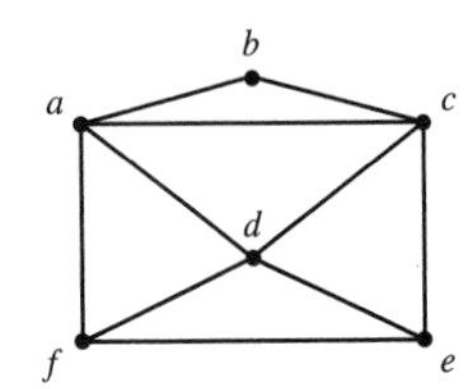

Figure 1.13

of such edge-vertex incidences is simply twice the number of edges. The theorem is now proved. ◆

As an illustration of Theorem 1, consider the graph in Figure 1.13 with 6 vertices, three of degree 4, two of degree 3, and one of degree 2. The sum of the degrees is $4+4+4+3+3+2=20$. This sum must equal twice the number of edges. The reader can check that the number of edges in this graph is 10.

For the sum of degrees to be an even integer, there must be an even number of odd integers in the sum. Thus we obtain:

Corollary

In any graph, the number of vertices of odd degree is even.

Let us now look at uses of this theorem and corollary.

Example 1: Use of Theorem 1

Suppose we want to construct a graph with 20 edges and have all vertices of degree 4. How many vertices must the graph have?

Let **v** denote the number of vertices. The sum of the degrees of the vertices will be 4**v**, and by the theorem this sum must be twice the number of edges: $4\mathbf{v} = 2 \times 20 = 40$. Hence $\mathbf{v} = 10$. ■

Example 2: Edges in a Complete Graph

How many edges are there in K_n, a complete graph on n vertices?

Recall that K_n has an edge between all possible pairs of vertices. At any given vertex, there will be edges going to each of the $n-1$ other vertices in K_n, and so each vertex has degree $n-1$. The sum of the degrees of all n vertices in K_n will be $n(n-1)$. Since this sum equals twice the number of edges, the number of edges is $n(n-1)/2$. ■

Example 3: Impossible Graph

Is it possible to have a group of seven people such that each person knows exactly three other people in the group?

If we try to model this problem using a graph with a vertex for each person and an edge between each pair of people who know each other, then we would have a graph with 7 vertices all of degree 3. But this is impossible by the Corollary—the number of vertices of odd degree must be even—and so no such set of seven people can exist. ■

We next present a puzzle that seems to have no relation to graphs. Recall that a graph G is connected if every pair of vertices in G is joined by a path in G. If G is not connected, its vertices can be partitioned into connected pieces, called **components**. Formally, a component H is a connected subgraph of G such that there is no path between any vertex in H and any vertex of G not in H. The component of G containing a particular vertex x consists of x and all vertices that may be reached from x by a path in G. Note that because each component of G is a graph in its own right, this section's Corollary applies to each component as well as to G.

Example 4: Mountain Climbers Puzzle

Two people start at locations A and Z at the same elevation on opposite sides of a mountain range whose summit is labeled M. See Figure 1.14*a*. We pose the following puzzle: Is it possible for the people to move along the range in Figure 1.14*a* to meet at M in a fashion so that they are always at the same altitude every moment? We shall show this is possible for *any* mountain range like Figure 1.14*a*. The one assumption we make is that there is no point lower than A (or Z) and no point higher than M.

We make a *range graph* whose vertices are pairs of points (P_L, P_R) at the same altitude with P_L on the left side of the summit and P_R on the right side, such that one of the two points is a local peak or valley (the other point might also be a peak or valley). The vertices for the range in Figure 1.14*a* are shown in the graph in Figure 1.14*b*. We make an edge joining vertices (P_L, P_R) and (P'_L, P'_R) if the two people can move constantly in the same direction (both going up or both going down) from point P_L to point P'_L and from P_R to P'_R, respectively. Our question is now, Is there a path in the range graph from the starting vertex (A, Z) to the summit vertex (M, M)? For the graph in Figure 1.14*b*, the answer is obviously yes.

We claim that vertices (A, Z) and (M, M) in any range graph have degree 1, whereas every other vertex in the range graph has degree 2 or 4. (A, Z) has degree 1

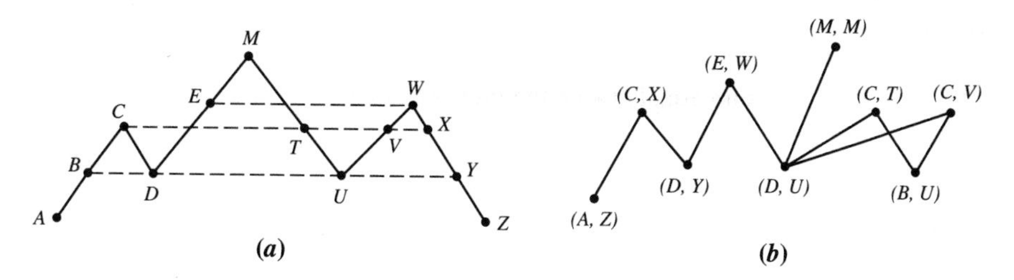

Figure 1.14

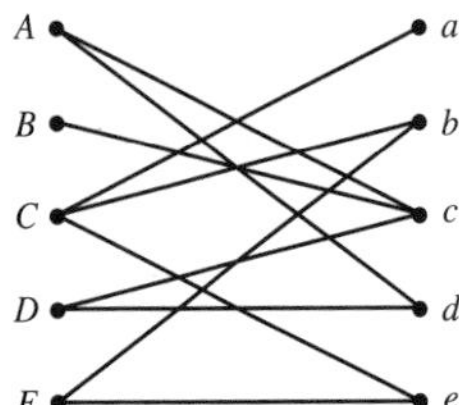

Figure 1.15

because when both people start climbing up the range from their respective sides, they have no choice initially but to climb upward until one arrives at a peak. In Figure 1.14*a*, the first peak encountered is C on the left, and so the one edge from (A, Z) goes to (C, X). A similar argument applies at (M, M). Next consider a vertex (P_L, P_R) where one point is a peak and the other point is neither peak nor valley, such as (E, W). From the peak we can go down in either direction: at W, we can go down toward Z or toward U. In either direction, the people go until one (or both) reaches a valley. At (E, W), the two edges go to (D, Y) and (D, U). So such a vertex has degree 2. A similar argument applies if one (but not both) point is a valley. It is left as an exercise for the reader to show that if a vertex (P_L, P_R) consists of two peaks or two valleys, such as (D, U), it will have degree 4. (A vertex consisting of a valley and a peak will have degree 0—why?)

Suppose there were no path from (A, Z) to (M, M) in the range graph. We use the fact that starting vertex (A, Z) and summit vertex (M, M) are the only vertices of odd degree. The component of the range graph consisting of (A, Z) and all the vertices that can be reached from (A, Z) would form a graph with just one vertex of odd degree, namely, (A, Z). This contradicts the Corollary and so any range graph must have a path from (A, Z) to (M, M). ■

Many interesting properties in graph theory are dependent on certain sets of edges having even size. Euler cycles, discussed in Section 2.1, arise when all vertices have even degree.

In Example 2 of Section 1.1, we considered a matching problem involving the graph shown in Figure 1.15. The vertices on the left represented people and the vertices on the right represented jobs. An edge links a left vertex to a right vertex to indicate that a certain person can perform a certain job. There can never be an edge between two vertices on the left or between two vertices on the right. Such a graph is called a bipartite graph. Formally, a graph G is **bipartite** if its vertices can be partitioned into two sets V_1 and V_2 and every edge joins a vertex in V_1 with a vertex in V_2.

Bipartite graphs can be characterized by the fact that all circuits in such graphs have even length (if there are no circuits, the graph is also bipartite), where the **length** of a circuit or path is the number of edges in it.

Theorem 2

A graph G is bipartite if and only if every circuit in G has even length.

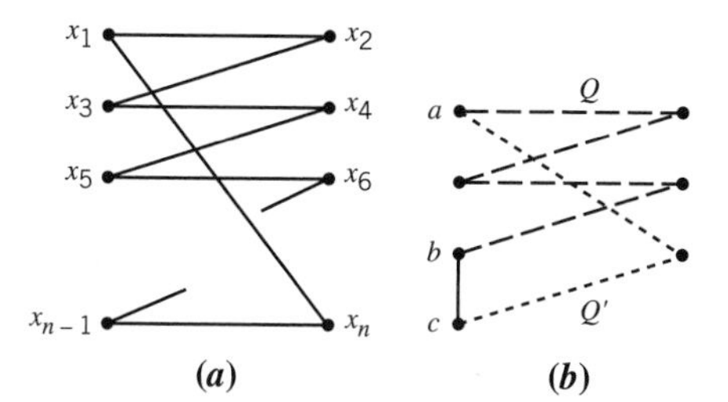

Figure 1.16 (*a*) (*b*)

Proof

Note that it is sufficient to prove this theorem for connected bipartite graphs. We claim that if the theorem is true for each connected component of a disconnected bipartite graph G, then it is true for G (components were formally defined just above Example 4). This claim follows from the facts that G is bipartite if and only if each of its components is bipartite and that any circuit in G has even length if and only if any circuit in each of its components has even length.

First we show that if G is bipartite, then any circuit has an even length. If G is bipartite so that it can be drawn with all edges connecting a left vertex with a right vertex, then any circuit $x_1–x_2–x_3 \cdots –x_n–x_1$ has alternately a left vertex, then a right vertex, then a left vertex, etc., assuming the first vertex x_1 is on the left. Odd-subscripted vertices are on the left and even-subscripted vertices on the right. See Figure 1.16*a*. Since x_n is adjacent to x_1, x_n must be on the right and so its subscript is even. That is, there are an even number of vertices in the circuit. Any circuit has the same number of edges as vertices, and thus this circuit has even length.

Suppose next that any circuit in G (there may be no circuits) has even length. We show how to construct a bipartite arrangement of G. Take any given vertex, call it a, and put it on the left. Put all vertices adjacent to a on the right. Next put all vertices that are two edges away from a, that is, at the end of some path of length 2 from a, on the left. In general, if there is a path of odd length between a and a vertex x, put x on the right. If there is a path of even length between a and x, put x on the left.

There cannot be distinct paths P and P' between a and x of odd and of even lengths, respectively, since taking P from a to x and then returning to a on P' yields an odd-length circuit. This is impossible, since all circuits have even length. (If P and P' have a vertex q in common besides a and x, then a further argument is needed to show that there is a circuit of odd length. See Exercise 13 for details.)

Similarly, we argue that there cannot be an edge between two vertices, say, b and c, both on the left. There must exist even-length paths Q, Q' joining a with b and c, respectively (since b and c are on the left). See Figure 1.16*b*, in which Q is dashed and Q' is dotted. Observe that Q' followed by the edge (c, b) yields an odd-length path from a to b. This is impossible, since we just proved that there cannot be both an even-length path (Q) and an odd-length plus (Q' plus (a, b)) from a to any other vertex in G. By similar reasoning, two vertices on the right cannot be adjacent. Thus, we have a bipartite arrangement of G. ◆

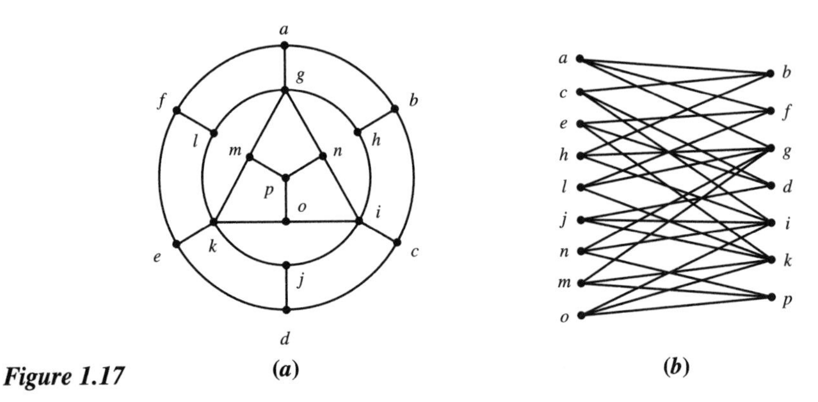

Figure 1.17

Example 5: Testing for a Bipartite Graph

Is the graph in Figure 1.17*a* bipartite?

Pick any vertex, say a, and put it on the left. We follow the approach in the second half of the proof of Theorem 2. Put vertices joined to a by an even-length path on the left and vertices joined to a by an odd-length path on the right. If all the circuits in this graph are even-length, then the reasoning in the above proof guarantees that our placement of vertices will yield a bipartite arrangement. If we end up with two vertices on the left (or on the right) being adjacent, then the graph cannot be bipartite. In this case, the construction succeeds, as shown in Figure 1.17*b*. ■

1.3 EXERCISES

1. How many vertices will the following graphs have if they contain:

(a) 12 edges and all vertices of degree 2.

(b) 15 edges, 3 vertices of degree 4, and the other vertices of degree 3.

(c) 20 edges and all vertices of the same degree.

2. For each of the following questions, describe a graph model and then answer the question.

(a) Must the number of people at a party who do not know an odd number of other people be even?

(b) Must the number of people ever born who had (have) an odd number of brothers and sisters be even?

(c) Must the number of families in Alaska with an odd number of children be even?

(d) For each vertex x in the following graph, let $s(x)$ denote the number of vertices adjacent to at least one of x's neighbors. Must the number of vertices with

$s(x)$ odd be even? Is this true in general?

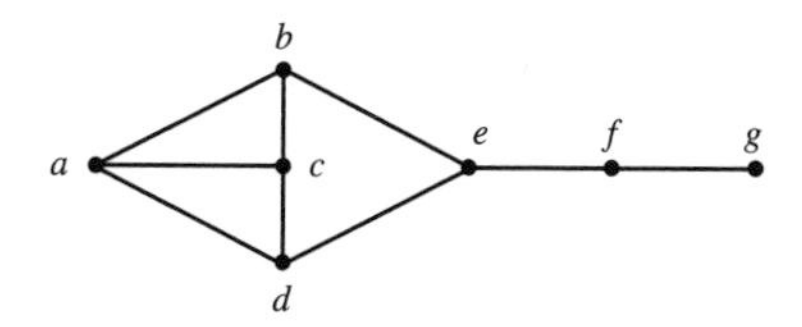

3. What is the largest possible number of vertices in a graph with 19 edges and all vertices of degree at least 3?

4. If a graph G has n vertices, all of which but one have odd degree, how many vertices of odd degree are there in $\overline{G}$, the complement of G?

5. Suppose all vertices of a graph G have degree p, where p is an odd number. Show that the number of edges in G is a multiple of p.

6. There used to be 26 football teams in the National Football League (NFL) with 13 teams in each of two conferences (each conference was divided into divisions, but that is irrelevant here). An NFL guideline said that each team's 14-game schedule should include exactly 11 games against teams in its own conference and 3 games against teams in the other conference. By considering the right part of a graph model of this scheduling problem, show that this guideline could not be satisfied!

7. Prove a directed version of Theorem 1: The sum of the in-degrees of vertices in a directed graph equals the sum of the out-degrees of vertices, and further, each sum equals the number of edges.

8. Build the range graph for each of the following mountain ranges and use the graph to find a solution to the problem in Example 4.

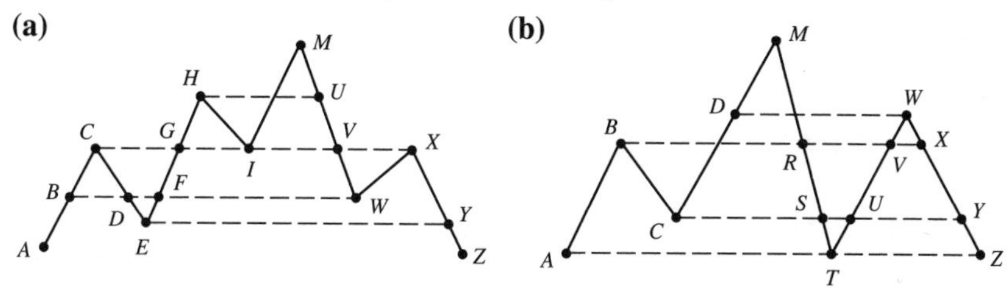

9. Prove in a range graph that if a vertex (P_L, P_R) consists of two peaks or two valleys, it will have degree 4.

10. Prove in a range graph that if a vertex (P_L, P_R) consists of a valley and a peak, it will have degree 0.

11. Determine whether the following graphs are bipartite. If so, give the partition into left and right vertices as in Figure 1.17*b*.

(a) Figure 1.4 **(b)** Figure 1.5*d* **(c)** Figure 1.8 (left graph)

12. Determine whether the following graphs are bipartite. If so, give the partition into left and right vertices as in Figure 1.17*b*.

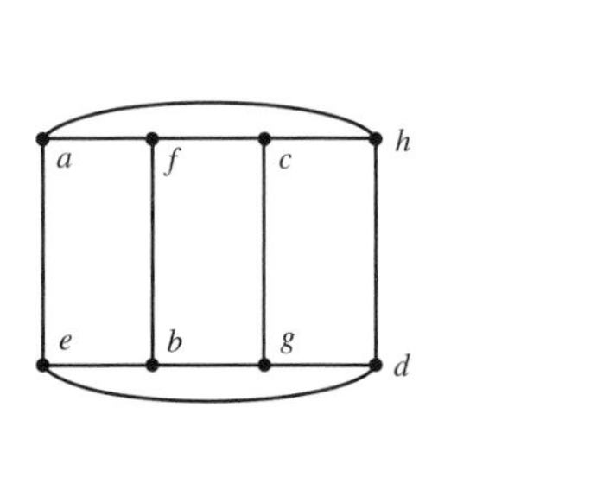

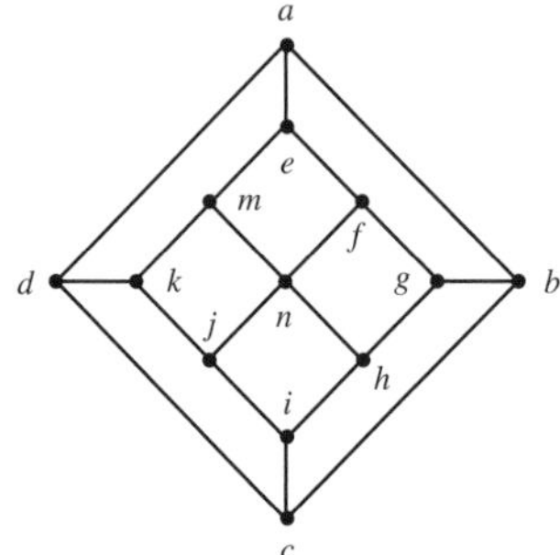

13. Suppose x and y are the only two vertices of odd degree in graph G, and x and y are not adjacent to each other. Show that G is connected if and only if the graph obtained from G by adding edge (x, y) is connected.

14. In the second part of the proof of Theorem 2, one can encounter the situation where there exist paths P and P' between a and x, P of odd length and P' of even length, and these two paths have one or more vertices in common. One must show that a subset of the edges on these two paths forms an odd-length circuit. Let q be the first vertex on P, starting from a, that also lies on P'. Show that either the circuit from a along P to q and then back on P' to a, or the edge sequence from q along P to x and then back on P' to q, has odd length. In the latter case, if the edge sequence is not a circuit, then it has a vertex q' on both P and P'. Repeat the same reasoning considering the circuit on P from q to q' and then back on P' to q or the edge sequence from q' along P to x and back along P' to q'.

1.4 PLANAR GRAPHS

The most natural examples of graphs are street maps and telephone networks. The graphs that arise from such physical networks usually have the property that they can be depicted on a piece of paper without edges crossing (different edges meet only at vertices). We say that a graph is **planar** if it can be drawn on a plane without edges crossing. We use the term **plane graph** to refer to a planar depiction of a planar graph. The two graphs in Figure 1.18 are both planar. The graph in Figure 1.18*b* is a plane graph. The graph in Figure 1.18*a* is planar since it can be redrawn in the form of the graph in Figure 1.18*b*.

Our principal focus in this section is determining whether a graph is planar. We take two approaches. The first approach involves a systematic method for trying to draw a graph edge-by-edge with no crossing edges, in the same spirit as when we tried to determine if two graphs are isomorphic. The second approach develops some

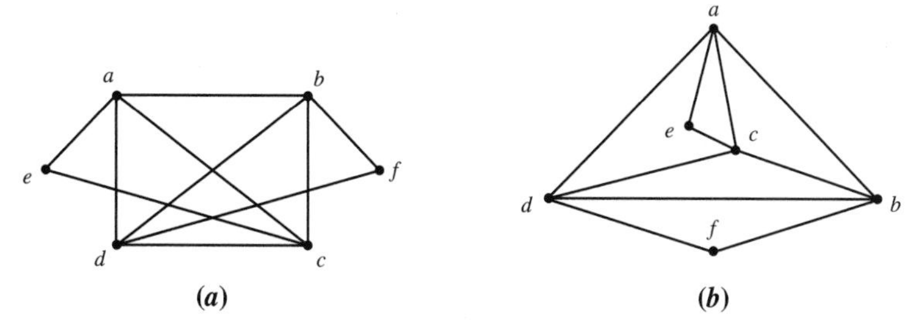

Figure 1.18

theory with a goal of finding useful properties of planar graphs. If a graph does not satisfy one or more of these properties, then we know that it cannot be planar.

Remember that if a graph G has been drawn with edges crossing, this does not mean the graph is nonplanar. There may be another way to draw the graph without edges crossing, as illustrated by the graph in Figure 1.18*a*, which can be redrawn to be the plane graph in Figure 1.18*b*.

Probably the most important need today for testing whether a graph is planar arises in designing electronic circuits. Complex integrated circuits are nonplanar and require several layers of (planar) circuit connections in their wiring. But the number of layers is limited and so a major problem in integrated circuit design is decomposing a large circuit into a minimal number of parts that are known to be planar.

A more mundane, but still important, use of planarity testing arises in checking data-entry errors in planar networks. When a large planar graph such as a city's street network is entered on data terminals for computerized analysis (say, for use by a computer program to determine routes for street sweepers), it is a common error-checking technique to test first whether the graph as typed in is indeed planar (most data-entry errors would make the graph nonplanar).

Planar graphs were first studied extensively by mathematicians over 100 years ago in connection with a map-coloring problem.

Example 1: Map Coloring

One of the most famous problems in mathematics concerns map coloring. The question is how many colors are needed to color countries on some map so that any pair of countries with a common border are given different colors. A map of countries is a planar graph with edges as borders and vertices where borders meet. See Figure 1.19*a*. However, a closely related planar graph called **a dual graph** of the map graph is more useful. The dual graph is obtained by making a vertex for each country and an edge between vertices corresponding to two countries with a common border. See Figure 1.19*b*. (Normally a vertex is also included for the unbounded region surrounding the map.)

The question in the dual graph now is how many colors are needed to "color" the vertices such that adjacent vertices have different colors. In Figure 1.19*b*, vertices

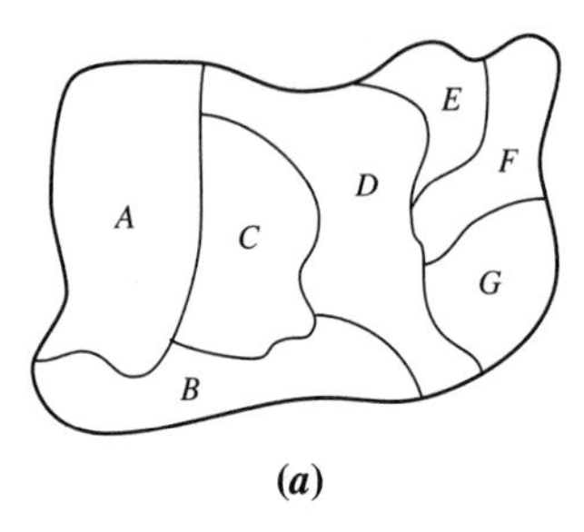

(a)

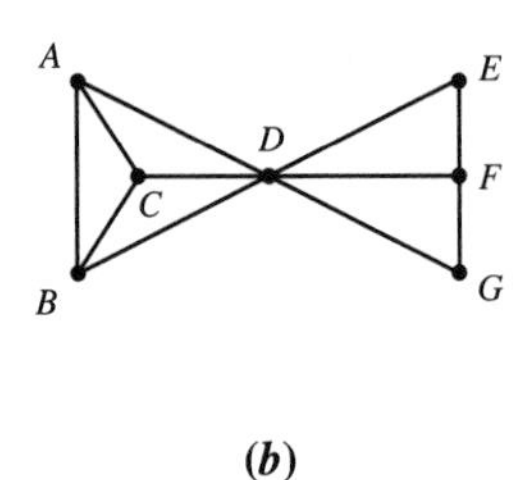

(b)

Figure 1.19

A, B, C, D form a complete subgraph on four vertices and so each requires a different color, four colors in all. With four colors, we can also properly color the remaining vertices.

One of the most famous unsolved problems in all of mathematics during the last century was the conjecture that *all* planar graphs can be properly colored with only four colors. In trying to resolve this conjecture, mathematicians developed a large theory about planar graphs. In 1976 the four-color conjecture was proven true by Appel and Haken using an immense computer-generated, case-by-case, exhaustive analysis (there were 1955 classes of graphical configurations to be considered, each involving numerous subcases). We will take a closer look at graph coloring in the next chapter. ■

We shall now to try to find a systematic way to draw a graph in the plane without edges crossing. As with isomorphism between two graphs, we want to be able to conclude that a graph is not planar if our construction fails. We shall call our approach the **circle–chord method.** It starts by finding a circuit that contains all the vertices of our graph (though such circuits do not exist for all graphs, they are common in the types of graphs we will be considering in this section). We draw this circuit as a large circle. The remaining noncircuit edges, which we will call *chords,* must be drawn either inside the circle or outside the circle in a planar drawing.

We choose a first chord and draw it, say, outside the circle. If properly chosen, this chord will force certain other chords to be drawn inside the circle (if also placed outside the circle, they would have to cross the first chord). These inside chords will force still other chords to be drawn outside, and so on. After the first chord is drawn, the choice of placing subsequent chords inside or outside is forced. Thus, if we reach a point where a new chord will have to cross some previous chord, whether the new chord is drawn inside or outside, we can claim that the graph must be nonplanar. If all the chords can be added without crossing other chords, then the graph is planar.

A critical decision is whether the first chord drawn should go inside or outside the circle. We claim that it makes no difference, because of the following inside–outside symmetry of a circle. Consider two maps of the earth, the first with the North Pole at the center of the map, the second with the South Pole at the center. Suppose each map has a path drawn in the Northern Hemisphere linking two cities on the equator.

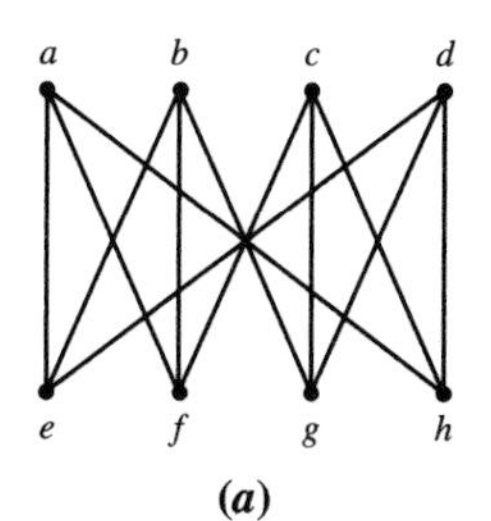

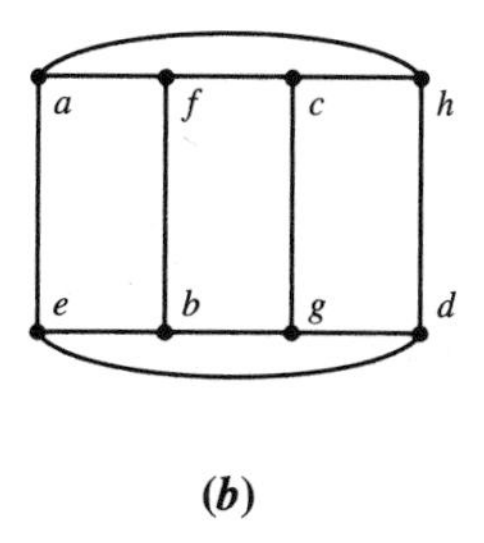

Figure 1.20 (*a*) (*b*)

In the first map (with the Northern Hemisphere inside the equator) the path is inside the equator's circle, whereas in the second map the path is outside the equator. Think of the circle formed by the circuit in the circle–chord method as the equator and the first chord as the path between two cities. Whether the chord is inside or outside the circle is just a matter of which "map" of the earth one uses.

We note that very efficient algorithms exist to test whether a graph is planar, whereas there is no efficient algorithm known to test whether two graphs are isomorphic. The planarity testing algorithms are fairly complicated and beyond the scope of this text.

Example 2: Circle–Chord Method

Use the circle–chord method to determine whether the graph in Figure 1.20*a* is planar. Let us look for a circuit with all 8 vertices. One possibility is *a-f-c-h-d-g-b-e-a*. Now try to add the other four edges (a, h), (b, f), (c, g), (d, e). By inside–outside symmetry, we can start by drawing (a, h) outside. See Figure 1.20*b*. Then (b, f) and (c, g) must go inside. Then (d, e) must go outside. So the graph is planar as shown in Figure 1.20*b*. ∎

Example 3: Showing $K_{3,3}$ is Nonplanar

Show that $K_{3,3}$, the graph in Figure 1.21*a*, is nonplanar. The notation $K_{3,3}$ indicates that this graph is a complete bipartite graph consisting of two sets of 3 vertices with each vertex in one set adjacent to all vertices in the other set. Applying the circle–chord method, we form a circuit containing all six vertices in $K_{3,3}$, and then try to add the remaining edges (not in the circuit) as inside and outside chords.

There are several choices for a 6-vertex circuit. Suppose we use the circuit *1–4–2–5–3–6–1* and draw it in a circle as shown in Figure 1.21*b*. Next the edges $(1, 5)$, $(2, 6)$, and $(3, 4)$ must be added. First draw chord $(1, 5)$. By the inside–outside symmetry of a circle discussed above, we can assume that $(1, 5)$ is drawn inside the circuit, as in Figure 1.21*b*. Then $(2, 6)$ must be drawn outside the circuit to avoid crossing chord $(1, 5)$. Finally, we must draw $(3, 4)$: If drawn outside the circuit, $(3, 4)$ would have to cross chord $(2, 6)$; if drawn inside the circuit, $(3, 4)$ would have to cross chord $(1, 5)$. Thus $K_{3,3}$ cannot be drawn in a planar depiction. Hence $K_{3,3}$ is nonplanar. ∎

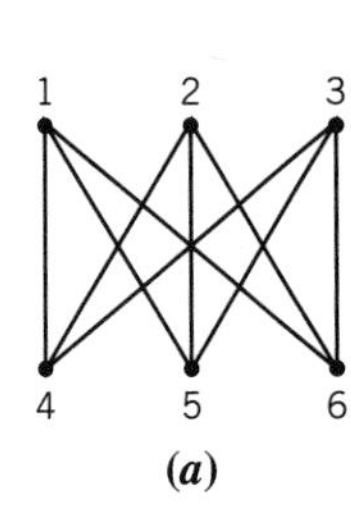

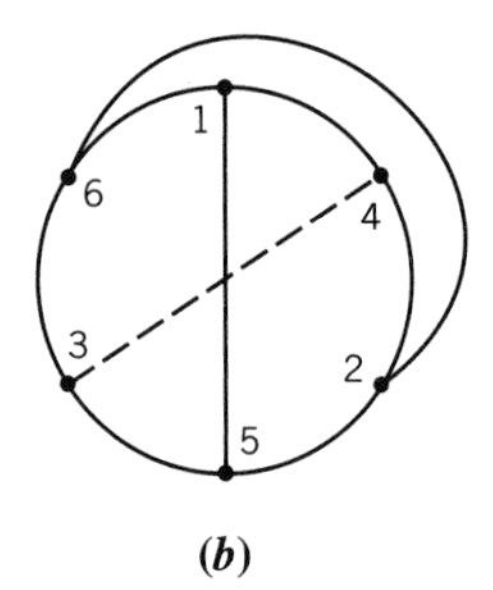

Figure 1.21 (*a*) (*b*)

Using a mixture of theory and careful, case-by-case analysis, it is possible to prove that any nonplanar graph always contains a $K_{3,3}$ or a K_5 (the complete graph on 5 vertices shown in Figure 1.22*a*) as a subgraph or a slight modification of these two graphs. It is left as an exercise to show that the circle–chord method (used in Example 3) can be used to show that K_5 is nonplanar. Thus, these two graphs are the "reason" that a graph cannot be drawn in a planar fashion.

We have to allow a slight variation in $K_{3,3}$ and K_5 in nonplanarity analysis. Figure 1.22*b* shows a $K_{3,3}$, graph that has been **subdivided** by adding vertices in the middle of some of its edges. The resulting graph is no longer a $K_{3,3}$ and does not contain $K_{3,3}$ as a subgraph, yet it is still nonplanar (repeatedly adding a vertex in the middle of an edge cannot make a nonplanar graph planar). We say that a subgraph is a $\boldsymbol{K_{3,3}}$ **configuration** if it can be obtained from a $K_{3,3}$ by adding vertices in the middle of some edges. A $\boldsymbol{K_5}$ **configuration** is defined similarly. The following planar graph characterization theorem was first proved by the Polish mathematician Kuratowski.

Theorem 1 **(Kuratowski, 1930)**

A graph is planar if and only if it does not contain a subgraph that is a K_5 or $K_{3,3}$ configuration.

If the circle–chord method shows that a graph is nonplanar, then by Theorem 1 this graph has a subgraph that is a K_5 or $K_{3,3}$ configuration. Finding such a configuration can sometimes be tricky. However, the following observation is helpful: *Most small nonplanar graphs contain a $K_{3,3}$ configuration.* All but one of the nonplanar graphs in the exercises have $K_{3,3}$ configurations. Note also that the depiction of a $K_{3,3}$ in Figure 1.21*b* as a 6-vertex circle with 3 chords joining pairs of opposite vertices is the way that a $K_{3,3}$ configuration normally arises in a nonplanar graph.

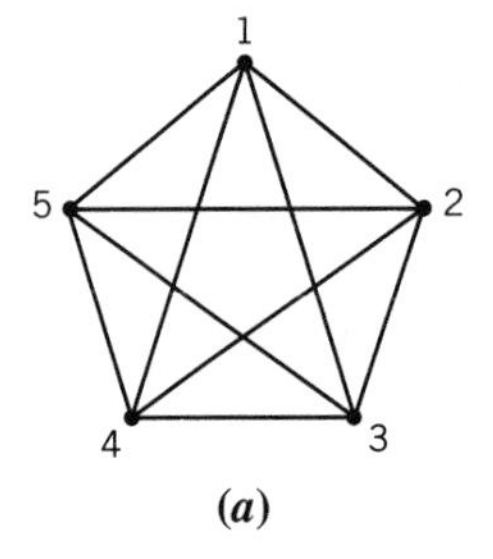

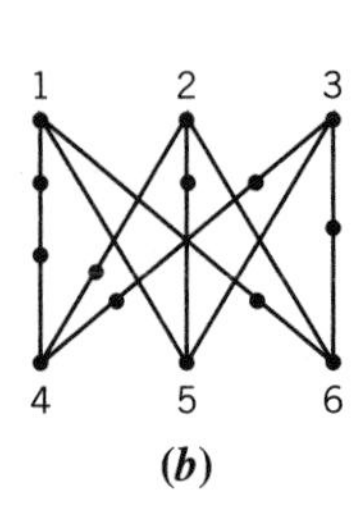

Figure 1.22 (*a*) (*b*)

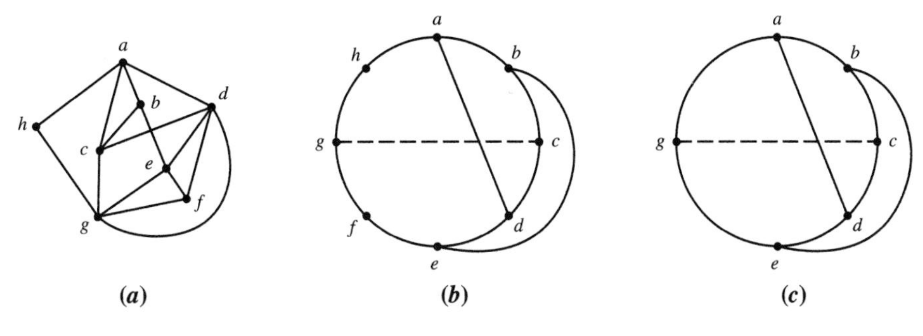

Figure 1.23

Example 4: Finding a $K_{3,3}$

Use the circle–chord method to determine whether the graph in Figure 1.23*a* is planar. If it is nonplanar, find a subgraph that is a $K_{3,3}$ configuration.

First we seek a circuit that visits all vertices. Many such circuits exist. Choose the circuit *a-b-c-d-e-f-g-h-a*, shown in Figure 1.23*b*. Say we pick (a, d) as our first chord to draw. By inside–outside symmetry, it makes no difference whether we draw (a, d) inside or outside the circuit. Put it inside. Then (b, e) must go outside to avoid intersecting (a, d). Next look for another chord reaching across the circle. Observe that chord (c, g) cannot be drawn inside without crossing (a, d) nor drawn outside without crossing (b, e). So the graph is nonplanar.

Next we look for a $K_{3,3}$ configuration. We can simplify the problem by restricting our attention to the nonplanar subgraph in Figure 1.23*b*, whose crossing edges proved that the full graph had to be nonplanar. A $K_{3,3}$ configuration has 6 vertices of degree 3 (corresponding to the vertices of a $K_{3,3}$) plus some number of vertices of degree 2 that subdivide the edges of a $K_{3,3}$. The way to find a $K_{3,3}$ configuration in a subgraph is to eliminate edge subdivisions in the graph (remove each vertex of degree 2 and combine the two edges at each such vertex into a single edge). Figure 1.23*c* shows the subgraph in Figure 1.23*b* with subdivisions removed. The graph in Figure 1.23*c* looks just like the depiction of a $K_{3,3}$ in Figure 1.21*b*. Thus, the subgraph in Figure 1.23*b* was a $K_{3,3}$ configuration. ■

Finding a $K_{3,3}$ configuration in the graph in Figure 1.23*a* (without using the subgraph in Figure 1.23*b*) would be difficult. The challenging problem in finding a $K_{3,3}$ configuration in a general nonplanar graph is the following. Let z be some vertex of degree 3 in the original graph and suppose that just two of the z's edges, say (z, r) and (z, q), are part of a $K_{3,3}$ configuration. Then z corresponds to a subdivision vertex in this $K_{3,3}$ configuration, and these two edges of z need to be fused into a single edge (r, q) to find the underlying $K_{3,3}$. That is, z disappears and a new edge (r, q) is created. Using the subgraph produced by the circle–chord method makes it much easier to identify vertices of degree 2 in a $K_{3,3}$ configuration whose two edges should be fused together.

There are many different plane graph depictions that can be drawn for a planar graph. For example, we can redraw the plane graph in Figure 1.24*a* by making the

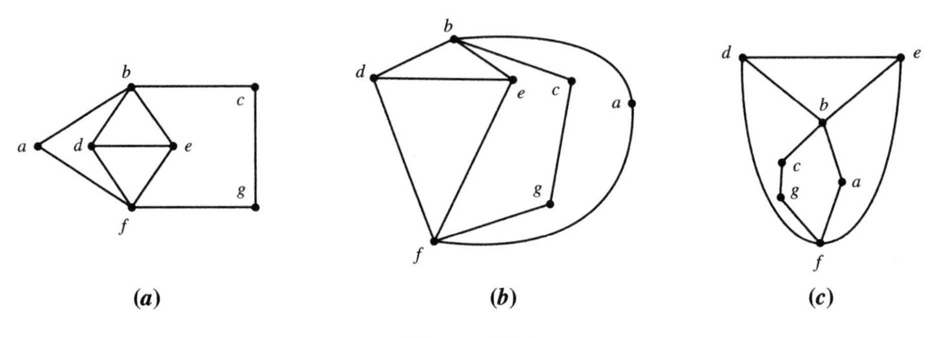

Figure 1.24

region bounded by the triangle (d, e, f) very large and bringing vertex a to the right side, as in Figure 1.24*b*. Now flip the part of the graph above and to the right of the triangle inside the triangle, obtaining the plane graph in Figure 1.24*c*. The triangle (d, e, f) has become the outside boundary of the whole graph. The boundary of any region can be converted to the outside boundary of the whole graph by a similar process.

Despite this variability in plane graph depictions of a planar graph, one important property of the plane depictions does not change. The number of regions is always the same. For simplicity, assume that G is a connected planar graph. (Recall that connected means having paths between every pair of vertices.) If $\mathbf{v}$ and $\mathbf{e}$ denote the number of vertices and edges, respectively, in G, then a plane graph depiction of G will always have a number of regions $\mathbf{r}$ given by the formula, $\mathbf{r} = \mathbf{e} - \mathbf{v} + 2$. Note that the unbounded region outside the graph is counted as a region. This remarkable formula for $\mathbf{r}$ was discovered by Euler 250 years ago.

Theorem 2 **Euler's Formula (1752)**

If G is a connected planar graph, then any plane graph depiction of G has $\mathbf{r} = \mathbf{e} - \mathbf{v} + 2$ regions.

Proof

Let us draw a plane graph depiction of G edge by edge. We choose successive edges so that at every stage we have a connected subgraph. Let G_n denote the connected plane graph obtained after n edges have been added, and let $\mathbf{v}_n$, $\mathbf{e}_n$, and $\mathbf{r}_n$ denote the number of vertices, edges, and regions in G_n, respectively. Initially we have G_1, which consists of one edge, its two end vertices, and the one (unbounded) region. Then $\mathbf{e}_1 = 1, \mathbf{v}_1 = 2, \mathbf{r}_1 = 1$, and so Euler's formula is valid for G_1, since $\mathbf{r}_1 = \mathbf{e}_1 - \mathbf{v}_1 + 2$: $1 = 1 - 2 + 2$. We obtain G_2 from G_1 by adding an edge at one of the vertices in G_1. In general, G_n is obtained from G_{n-1} by adding an nth edge at one of the vertices of G_{n-1}. The new edge might link two vertices already in G_{n-1}. If it does not, the other end vertex of the nth edge is a new vertex that must be added to G_n.

We will now use the method of induction (see Appendix A.2) to complete the proof. We have shown that the theorem is true for G_1. Next we assume that it is true

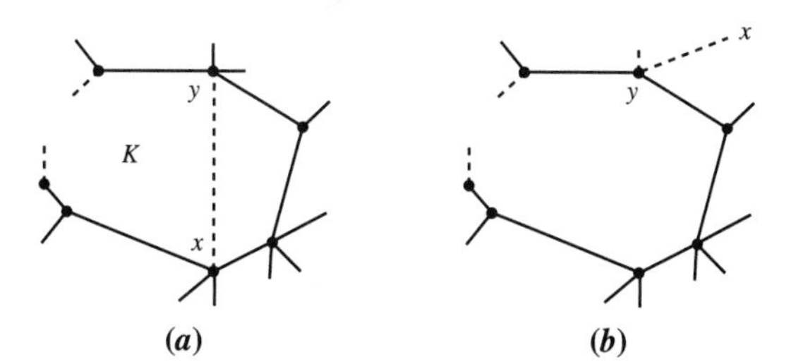

Figure 1.25

for G_{n-1} for any $n > 1$, and prove it is true for G_n. Let (x, y) be the nth edge that is added to G_{n-1} to get G_n. There are two cases to consider.

In the first case, x and y are both in G_{n-1}. Then they are on the boundary of a common region K of G_{n-1}, possibly the unbounded region [if x and y were not on a common region, edge (x, y) could not be drawn in a planar fashion, as required]. See Figure 1.25*a*. Edge (x, y) splits K into two regions. Then $\mathbf{r}_n = \mathbf{r}_{n-1} + 1$, $\mathbf{e}_n = \mathbf{e}_{n-1} + 1$, $\mathbf{v}_n = \mathbf{v}_{n-1}$. So each side of Euler's formula grows by 1. Hence if the formula was true for G_{n-1}, it will also be true for G_n.

In the second case, one of the vertices x, y is not in G_{n-1}, say it is x. See Figure 1.25*b*. Then adding (x, y) implies that x is also added, but no new regions are formed (i.e., no existing regions are split). Thus $\mathbf{r}_n = \mathbf{r}_{n-1}$, $\mathbf{e}_n = \mathbf{e}_{n-1} + 1$, $\mathbf{v}_n = \mathbf{v}_{n-1} + 1$, and the value on each side of Euler's formula is unchanged. The validity of Euler's formula for G_{n-1}, implies its validity for G_n.

So each increase in $\mathbf{r}$ is balanced in Euler's formula by an increase in e or v. By induction, the formula is true for all G_n's and hence true for the full graph G. ◆

Example 5: Using Euler's Formula

How many regions would there be in a plane graph with 10 vertices each of degree 3?

By the Theorem in Section 1.3, the sum of the degrees, 10×3, equals $2\mathbf{e}$, and so $\mathbf{e} = 15$. By Euler's formula, the number of regions $\mathbf{r}$ is

$$\mathbf{r} = \mathbf{e} - \mathbf{v} + 2 = 15 - 10 + 2 = 7 \quad \blacksquare$$

Theorem 2 has the following corollary that can often be used to show quickly that a graph is nonplanar.

Corollary

If G is a connected planar graph with $\mathbf{e} > 1$, then $\mathbf{e} \leq 3\mathbf{v} - 6$.

Proof

Let us define the *degree of a region* analogously to the degree of a vertex to be the number of edges incident to a region, that is, the number of edges on its boundary. If an edge occurs twice along a boundary, as does (x, y) in region K in Figure 1.26*a*, the edge is counted twice in region K's degree; for example, region K has degree 10

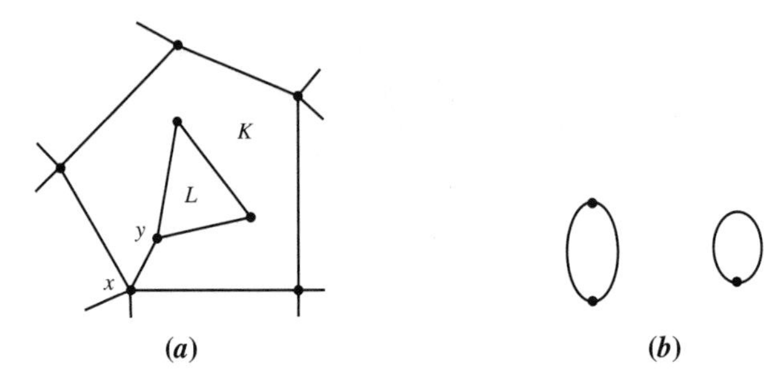

Figure 1.26

and region L has degree 3 in Figure 1.26*a*. Observe that each region in a plane graph must have degree ≥ 3, for a region of degree 2 would be bounded by 2 edges joining the same pair of vertices and a region of degree 1 would be bounded by a loop edge (see Figure 1.26*b*), but parallel edges and loops are not allowed in graphs.

Since each region has degree ≥ 3, the sum of the degrees of all regions will be at least $3\mathbf{r}$. But this sum of degrees of all regions must equal $2\mathbf{e}$, since this sum counts each edge twice, that is, each of an edge's two sides is part of some boundary (this is the same type of argument as used to show that the sum of the vertices' degrees equals $2\mathbf{e}$). Thus, $2\mathbf{e}$ = (sum of regions' degrees) $\geq 3\mathbf{r}$, or $\frac{2}{3}\mathbf{e} \geq \mathbf{r}$. Combining this inequality with Euler's formula (Theorem 2), we have

$$\tfrac{2}{3}\mathbf{e} \geq \mathbf{r} = \mathbf{e} - \mathbf{v} + 2 \qquad \text{or} \qquad 0 \geq \tfrac{1}{3}\mathbf{e} - \mathbf{v} + 2$$

Solving for $\mathbf{e}$, we obtain $\mathbf{e} \leq 3\mathbf{v} - 6$. ◆

Example 6: Showing K_5 is Nonplanar

Use the Corollary to prove that K_5, the complete graph on 5 vertices, is nonplanar.

The graph K_5 has $\mathbf{v} = 5$ and $\mathbf{e} = 10$ (see Example 2 in Section 1.3 for how to find the number of edges in a complete graph). Then $3\mathbf{v} - 6 = 3 \times 5 - 6 = 9$. But the Corollary says that $\mathbf{e} \leq 3\mathbf{v} - 6$ must be true in a connected planar graph, and so K_5 cannot be planar. ■

The corollary *should not be misinterpreted to mean* that if $\mathbf{e} \leq 3\mathbf{v} - 6$, then a connected graph is planar. Many nonplanar graphs also satisfy this inequality. For example, $K_{3,3}$ with $\mathbf{v} = 6$ and $\mathbf{e} = 9$ satisfies it.

Our two theorems and corollary have laid the foundation for a mathematical theory of planar graphs. In the process, we have acquired a practical aid for showing that a graph is nonplanar. One way to extend this theory is to make the inequality in the corollary "stronger," that is, get a smaller upper bound on $\mathbf{e}$. Recall that the key step in proving the corollary was the observation that every region has degree at least 3. This led to the inequality $2\mathbf{e} \geq 3\mathbf{r}$. Suppose that a certain connected graph G (with at least 2 edges) is known to be bipartite. By Theorem 2 in Section 1.3, all the circuits in a bipartite graph have even length. Then no region in this graph can have degree 3 (since this would imply a boundary circuit of length 3). Then every region in a bipartite planar graph must have degree ≥ 4. Summing the degrees of all regions, we now obtain the inequality $2\mathbf{e} =$ (sum of degrees of regions) $\geq 4\mathbf{r}$. Reworking the Corollary with this inequality $2\mathbf{e} \geq 4\mathbf{r}$, we have $\frac{2}{4}\mathbf{e} \geq \mathbf{r} = \mathbf{e} - \mathbf{v} + 2$ and hence

$$\mathbf{e} \leq 2\mathbf{v} - 4$$

Every connected planar graph that is bipartite must satisfy this inequality. Consider our "favorite" bipartite graph $K_{3,3}$. $K_{3,3}$ has $\mathbf{v} = 6$ and $\mathbf{e} = 9$ and, as noted above, satisfies the corollary inequality $\mathbf{e} \leq 3\mathbf{v} - 6$. But since it is bipartite, $K_{3,3}$ would also have to satisfy the new inequality $\mathbf{e} \leq 2\mathbf{v} - 4$ if it were planar. It does not: $9 \nleq 2 \times 6 - 4$.

1.4 EXERCISES

Summary of Exercises The first six exercises involve determining whether various graphs are planar and drawing planar graphs in different ways. Exercise 10 involves duality. Exercises 15–26 build on Euler's formula and the corollary $\mathbf{e} \leq 3\mathbf{v} - 6$. The other exercises introduce new concepts.

1. Draw a dual graph of the planar graph in (include a vertex for the unbounded region outside the graph):

   (a) Figure 1.18*b*

   (b) Figure 1.19*b*

2. Show that K_5 is nonplanar by the method in Example 2.
3. Which of the following graphs are planar? Find $K_{3,3}$ configurations in the nonplanar graphs.

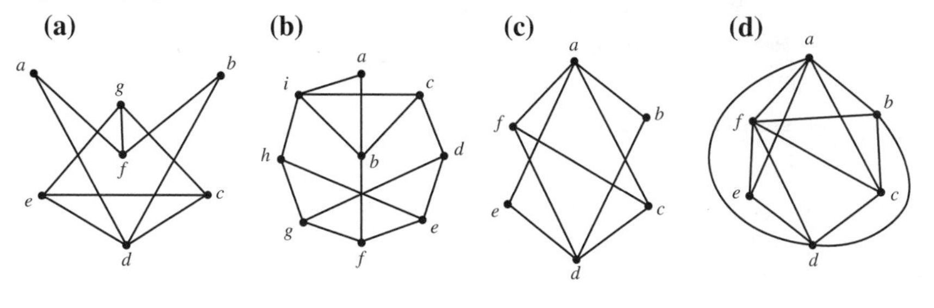

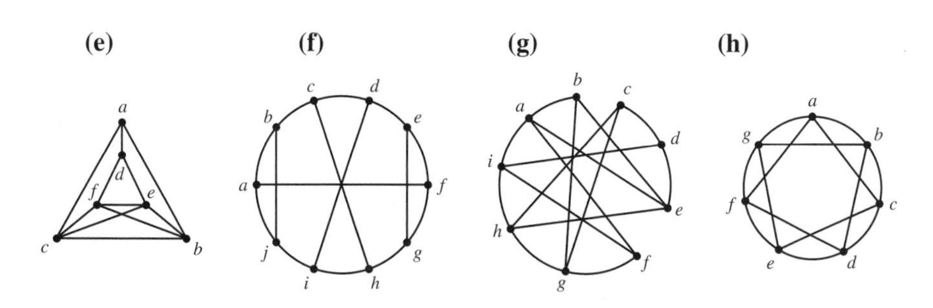

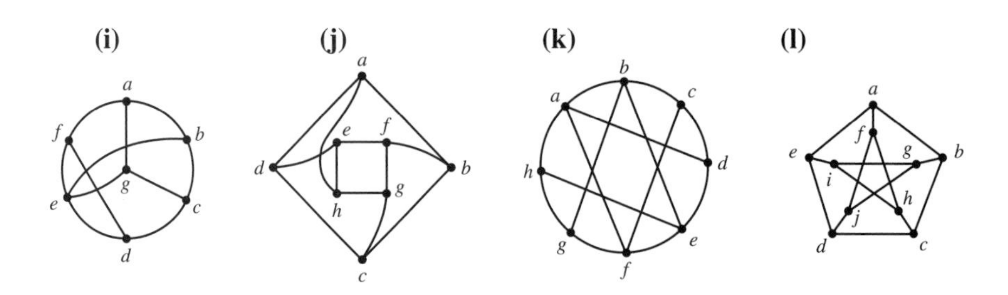

4. Re-draw the graph in Figure 1.20*b* so that the infinite region is bounded by the circuit *a*-*f*-*c*-*h*-*a*.

5. **(a)** For what values of n is K_n planar?

   (b) For what values of r and s is the complete bipartite graph $K_{r,s}$ planar? ($K_{r,s}$ is a bipartite graph with r vertices on the left side and s vertices on the right side and edges between all pairs of left and right vertices.)

6. A complete tripartite $K_{r,s,t}$ is a generalization of a complete bipartite graph (see part **(b)** of the previous exercise). There are three subsets of vertices, r in the first subset, s in the second subset, and t in the third subset. Every vertex in one particular subset is adjacent to every vertex in the other two subsets; that is, a vertex is adjacent to all vertices except those in its own subset. Determine all the triples r, s, t for which $K_{r,s,t}$ is planar.

7. In each case, give the values of **r, e,** or **v** (whichever is not given) assuming that the graph is planar. Then either draw a connected, planar graph with the property, if possible, or explain why no such planar graph can exist.

   (a) 7 vertices and 13 edges
   (b) 6 vertices and 8 regions
   (c) 13 edges and 9 regions
   (d) 6 vertices and 14 edges
   (e) 5 regions and 10 edges
   (f) 6 vertices all of degree 4
   (g) 7 vertices all of degree 3
   (h) 5 regions all with 4 boundary edges
   (i) 12 vertices and every region has 4 boundary edges
   (j) 17 regions and every vertex has degree 5

8. If a connected planar graph with n vertices all of degree 4 has 10 regions, determine n.

9. A *line graph* $L(G)$ of a graph G has a vertex of $L(G)$ for each edge of G and an edge of $L(G)$ joining each pair of vertices corresponding to two edges in G with a common end vertex.

   (a) Show that $L(K_5)$ is nonplanar.

   (b) Find a planar graph whose line graph is nonplanar.

10. The construction of a dual $D(G)$ can be applied to any plane graph G: Draw a vertex of $D(G)$ in the middle of each region of G and draw an edge e^* of $D(G)$ perpendicular to each edge e of G; e^* connects the vertices of $D(G)$ representing the regions on either side of e.

(a) A dual need not be a graph. It might have two edges between the same pair of vertices or a self-loop edge (from a vertex to itself). Find two planar graphs with duals that are not graphs because they contain these two forbidden situations.

(b) Show that the duals of the two different plane depictions of the graph in Figures 1.24*a* and 1.24*c* are isomorphic.

(c) Show that the degree of a vertex in the dual graph $D(G)$ equals the number of boundary edges of the corresponding region in the planar graph G.

(d) Find a planar graph that is isomorphic to its own dual.

(e) Show for any plane depiction of a graph G that the vertices of G correspond to regions in $D(G)$.

11. **(a)** Show that if a circuit in a planar graph encloses exactly two regions, each of which has an even number of boundary edges, then the circuit has even length.

(b) Show that if a circuit in a planar graph encloses a collection of regions, each of which has an even number of boundary edges, then the circuit has even length.

12. The *crossing number* $c(G)$ of a graph G is the minimum number of pairs of crossing edges in a depiction of G. For example, if G is planar then $c(G) = 0$. Determine $c(G)$ for the following graphs:

(a) $K_{3,3}$ **(b)** K_5 **(c)** Figure 1.23*a*

(d) An 8-gon with diagonals joining opposite pairs of corners

13. A graph G is *critical nonplanar* if G is nonplanar but any subgraph obtained by removing a vertex is planar.

(a) Which of the following graphs are critical planar?

(i) $K_{3,3}$ (ii) K_5 (iii) Figure 1.23*a*

(b) Show that critical nonplanar graphs must be connected and cannot have a vertex whose removal disconnects the graph.

14. A graph G is called *maximal planar* if G is planar but the addition of another edge will make the graph nonplanar.

(a) Show that every region of a connected, maximal planar graph will be triangular.

(b) If a connected, maximal planar graph has n vertices, how many regions and edges does it have?

15. Suppose G is a planar graph that is not necessarily connected, as required in Euler's formula (Theorem 2). Recall that a *component* H is a connected subgraph of G with the property that there is no path between any vertex in H and any vertex of G not in H.

(a) Find the appropriate modification of Euler's formula for a planar graph with c components.

(b) Show that the corollary is valid for unconnected planar graphs.

16. Prove that if a graph G has 11 vertices, then either G or its complement $\overline{G}$ must be nonplanar. *Hint*: Determine the total number N_{11} of edges in a complete graph on 11 vertices; if the result were false and G and its complement were each planar, how many of the N_{11} edges could be in each of these two graphs?

17. Mimic the argument in the corollary to prove that $\mathbf{e} \leq 3\mathbf{r} - 6$ in a planar graph with each vertex of degree ≥ 3.

18. **(a)** Prove that every connected planar graph has a vertex of degree at most 5. *Hint:* Assume that every vertex has degree at least 6 and obtain a contradiction as follows. Get a lower bound, involving $\mathbf{v}$, on the sum of the degrees of vertices (similar to the lower bound on the sum of the degrees of regions obtained in the proof of the corollary). Since this sum of degrees equals $2\mathbf{e}$, the corollary's bound on $\mathbf{e}$ can be used to get an upper bound, also involving $\mathbf{v}$, on this sum of degrees. Combining these two bounds yields the desired contradiction.

(b) Show that part (a) immediately generalizes to any (unconnected) planar graph.

19. Prove that every connected planar graph with less than 12 vertices has a vertex of degree at most 4. [*Hint:* Assume that every vertex has degree at least 5 to obtain a lower bound one $\mathbf{e}$ (together with the upper bound on $\mathbf{e}$ in the corollary) that implies $\mathbf{v} \geq 12$.] If instead of less than 12 vertices there are fewer than 30 edges, then show again that there is a vertex of degree at most 4.

20. If G is a connected planar graph with all circuits of length at least k, show that the inequality $\mathbf{e} \leq 3\mathbf{v} - 6$ can be strengthened to $\mathbf{e} \leq \frac{k}{k-2}(\mathbf{v} - 2)$.

21. **(a)** Show that every circuit in the graph in Exercise 3(a) has at least 5 edges.

(b) Use part (a) and the result of Exercise 20 to show that this graph is nonplanar.

22. **(a)** Give an example of a graph with regions consisting solely of squares (regions bounded by 4 edges) and hexagons, and with vertices of degree at least 3.

(b) Use Exercise 17 to show that any graph of the sort defined in part (a) has at least 6 squares.

(c) If each vertex has degree 3, show that any graph of the sort defined in part (a) has exactly 6 squares.

23. If G is a connected planar graph where $\mathbf{e} = 3\mathbf{v} - 6$, show that every region is triangular (has three boundary edges).

24. A *Platonic graph* is a planar graph in which all vertices have the same degree d_1 and all regions have the same number of bounding edges d_2, where $d_1 \geq 3$ and $d_2 \geq 3$. A Platonic graph is the "skeleton" of a Platonic solid, for example, an octahedron.

(a) If G is a Platonic graph with vertex and face degrees d_1 and d_2, respectively, then show that $\mathbf{e} = \frac{1}{2}d_1\mathbf{v}$ and $\mathbf{r} = (d_1/d_2)\mathbf{v}$.

(b) Using part (a) and Euler's formula, show that $\mathbf{v}(2d_1 + 2d_2 - d_1d_2) = 4d_2$.

(c) Since **v** and $4d_2$ are positive integers, we conclude from part (b) that $2d_1 + 2d_2 - d_1d_2 > 0$. Use this inequality to prove that $(d_1 - 2)(d_2 - 2) < 4$.

(d) From part (c), find the five possible pairs of positive (integral) values of d_1, d_2.

25. Suppose l lines are drawn through a circle and these lines form p points of intersection (involving exactly two lines at each intersection). How many regions r are formed inside the circle by these lines? Assume that the lines end at the edge of the circle at $2l$ distinct points.

26. Consider an overlapping set of 4 circles A, B, C, D. One would like to position the circles so that every possible subset of the circles forms a region, e.g., 4 regions each contained in just one (different) circle, 6 regions formed by the intersection of 2 circles (AB, AC, AD, BC, BD, CD), 4 regions formed by the intersection of 3 of the 4 circles, and 1 region formed by the intersection of all 4 circles. Prove that it is not possible to have such a set of 15 bounded regions.

27. Show that the following graphs can be drawn on the surface of a doughnut (torus) without crossing edges:

(a) $K_{3,3}$ (b) K_5 (c) K_6

1.5 SUMMARY AND REFERENCES

This chapter introduced graphs, their applications, and some of their basic structures. This text takes a user-oriented approach to graph theory. Readers interested in a more formal graph theory text presenting the subject as an interesting area of pure mathematics should see the books by Bondy and Murty [1], Harary [4], or Wilson [5]. Section 1.1 introduced a set of illustrative graph models. The basic structure of graphs was explored in Section 1.2 under the guise of determining what makes two graphs different. Section 1.3 presented some useful edge-counting results. The final section introduced the important class of planar graphs. It surveyed ad hoc and theoretical approaches for determining whether a graph is planar.

The history of graph theory begins with the work of L. Euler in 1736 on Euler circuits (discussed in Section 2.1). Euler's formula for planar graphs, originally stated in terms of polyhedra, was proved in 1752. Bits of graph theory appeared in papers about topology and geometric games, but it was not until around 1850 that formal studies of graphs began to appear. One was A. Cayley's 1857 paper counting the number of trees (discussed in Chapter 3). Another was G. Kirchhoff's 1847 paper presenting an algebra of circuits and introducing graphs in the study of electrical circuits. This same paper contains Kirchhoff's famous current and voltage laws (Kirchhoff was 21 when he wrote this historic paper). The term graph was first used by J. Sylvester in 1877. The first book on graph theory, by D. Konig, did not appear until 1936. An excellent sourcebook on the history of graph theory is *Graph Theory 1736–1936* by Biggs, Lloyd, and Wilson [2].

See the General References (at the end of the book) for a list of other introductory texts on graph theory.

1. J. A. Bondy and U. S. R. Murty, *Graph Theory with Applications,* American Elsevier, New York, 1976.
2. N. Biggs, E. Lloyd, and R. Wilson, *Graph Theory 1736–1936,* Cambridge University Press, Cambridge, 1999.
3. J. Cohen, *Food Webs and Niche Space,* Princeton University Press, Princeton, N. J., 1978.
4. F. Harary, *Graph Theory,* Perseus Press, New York, 1995.
5. R. Wilson, *Introduction to Graph Theory,* 4th., John Wiley and Sons, New York, 1997.

SUPPLEMENT I: REPRESENTING GRAPHS INSIDE A COMPUTER

This supplement is meant as a brief overview. For a fuller discussion of data structures, the reader should consult a data structures text, such as R. Kruse, Data Structures and Program Design in C++, *Prentice-Hall, Upper Saddle River, NJ, 1998 or R. Sedgewick,* Algorithms in C, *3rd ed., Addison-Wesley, Reading, MA, 1997.*

Graph problems that involve large graphs or that must be solved repeatedly need to be programmed for computer solution. Unlike numerical computations, such as inverting a matrix or numerically approximating the solution of a differential equation, most graph problems require no arithmetic operations. Instead, the essence of these problems is systematic search and the use of the right data structures for organizing the information about vertices and edges of a graph inside a computer. In this supplement, we present the most common ways of representing graphs in a computer.

The simplest but least efficient way to represent a graph is with an **adjacency matrix**. This matrix has a row and a column for each vertex. Entry $(i, j) = 1$ if vertex x_i is adjacent to vertex x_j and entry $(i, j) = 0$ otherwise. Figure 1.27*b* shows the adjacency matrix for the graph in Figure 1.27*a*. In a directed graph, entry $(i, j) = 1$ if and only if there is edge $(\vec{x_i, x_j})$. Because the adjacency matrix often has 0s in most entries, it is an inefficient way to represent a graph.

A better approach is to store only those adjacencies that do occur. A list of all the edges is compact but it requires extensive searching to determine the set of vertices adjacent to a given vertex. A compromise between an adjacency matrix and an edge list is a set of adjacency lists—for each vertex, we have a list of the vertices adjacent to it. Figure 1.28 contains the adjacency lists for the graph in Figure 1.27a. The first element in each list is the length of the list.

It should be noted that these adjacency lists can often be quite wasteful of space if one does not know in advance how long an adjacency list might be. In some graph problems the number of adjacencies will vary over time. Then each list may initially

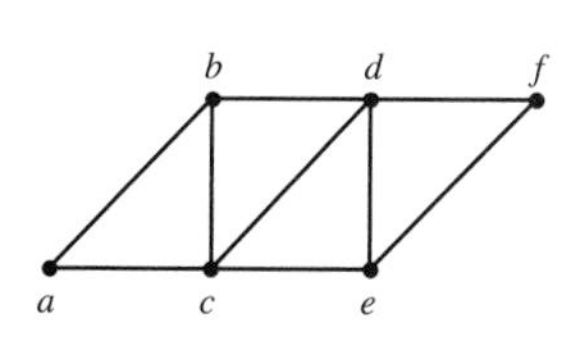

(*a*)

	a	*b*	*c*	*d*	*e*	*f*
a	0	1	1	0	0	0
b	1	0	1	1	0	0
c	1	1	0	1	1	0
d	0	1	1	0	1	1
e	0	0	1	1	0	1
f	0	0	0	1	1	0

(*b*)

Figure 1.27

Vertex	*Adjacency List*
a	2: *b*, *c*
b	3: *a*, *c*, *d*
c	4: *a*, *b*, *d*, *e*
d	4: *b*, *c*, *e*, *f*
e	3: *c*, *d*, *f*
f	2: *d*, *e*

Figure 1.28

have to be given a length equal to the maximum possible number of adjacencies at a vertex (for an n-vertex graph, the maximum possible number of adjacencies is $n - 1$).

The following more sophisticated data structure, called a **linked list**, efficiently handles this difficulty. A linked list is a set of arrays. Figure 1.29 has a linked list for the graph in Figure 1.27*a*. The primary array contains the adjacency lists of each vertex, but these lists are not stored in order. The secondary, or *pointer*, array contains "pointers" telling where the next entry in an adjacency list appears in the primary array. For example, if entry 17 in the main array contains the name of one of the vertices adjacent to c, then entry 17 in the pointer array has the entry number in the primary array of the next vertex in c's adjacency list. In Figure 1.29 entry 17 in the pointer array is 21, telling us that the next vertex in c's adjacency list is in entry 21 of the primary array. Entry 21 in the primary array is the last vertex in c's adjacency list, and this is indicated by having entry 21 in the pointer array blank.

There are two other arrays in this linked list. We need an array, called the Index Array, telling us where the first entry in each adjacency list appears (in the primary array). We also maintain an array, called the Unused Entry Array, with the numbers of entries not currently being used. The reason for this array will be explained shortly.

Vertex c's adjacency list is stored in entries $3 \to 4 \to 17 \to 21$. Suppose we want to delete the adjacency in entry 17 of the primary array. Then we change entry 4 in the pointer array from 17 to 21, and now c's adjacency list is stored in entries $3 \to 4 \to 21$. We also place the number 17 in the Unused Entries array.

On the other hand, if we wanted to add a new entry in c's original adjacency list following entry 4, then we get the location of an unused entry from the Unused Entry Array, say entry 23, and change entry 4 in the pointer array from 17 to 23. We store the name of the new vertex adjacent to c in entry 23 of the primary array

	Primary Array	*Pointer Array*
1	*b*	*6*
2	*a*	*11*
3	*a*	*4*
4	*b*	*17*
5	—	—
6	*c*	—
7	*b*	*19*
8	*e*	—
9	—	—
10	*f*	—
11	*c*	*16*
12	*d*	*8*
13	*c*	*22*
14	*f*	—
15	—	—
16	*d*	—
17	*d*	*21*
18	*e*	*14*
19	*c*	*18*
20	—	—
21	*e*	—
22	*d*	*10*
23	—	—

Index Array	
a	1
b	2
c	3
d	7
e	13
f	12

Unused Entry Array
5
15
20
9
23

Figure 1.29

and set entry 23 in the pointer array to 17. Now c's adjacency list is stored in entries $3 \to 4 \to 23 \to 17 \to 21$.

SUPPLEMENT II: SUPPLEMENTARY EXERCISES

Summary of Exercises Graph theory is a field famous for its interesting problems. Several exercises introduce new graph concepts, such as strong connectedness (Exercise 14) and cut-set (Exercise 25). Exercise 30 is a very famous problem in Ramsey Theory (see Appendix A.4). For more problems in graph theory, see any of the graph theory texts listed in the References at the end of this text.

1. Suppose that there are seven committees with each pair of committees having a common member and each person being on two committees. How many people are there?
2. Show that the complement of K_n, a complete graph on n vertices, is a set of n isolated vertices.

3. A graph is *regular* if all vertices have the same degree. If a graph with n vertices is regular of degree 3 and has 18 edges, determine n.

4. If a graph has 50 edges, what is the least number of vertices it can have?

5. Show that at least two vertices have the same degree in any graph with at least two vertices. (*Hint:* Be careful about vertices of degree 0.)

6. Show that an undirected graph with all vertices of degree ≥ 2 must contain a circuit (edges cannot be repeated in a circuit).

7. If every vertex in the graph G has degree 2, does every vertex lie on a circuit? Prove or give a counterexample.

8. If every vertex in a graph G has degree $\geq d$, then show that G must contain a circuit of length at least $d + 1$.

9. If every vertex in a directed graph G has positive out-degree (at least one outwardly directed edge), then:

(a) Must G contain a directed circuit?

(b) Must every vertex of G be on a directed circuit?

10. If G is a connected graph that is not a complete graph, show that some vertex, call it x, has two neighbors, call them y, z, that are not adjacent to each other [that is, there are edges (x, y) and (x, z) but not edge (y, z)].

11. Show that if a graph is not connected, then its complement must be connected.

12. Show that removal of some vertex x disconnects the connected undirected graph G if and only if there are two vertices a and b in G such that all paths in G from a to b pass through x.

13. Let G be a connected graph such that G-x is not connected for all but two vertices x of G. Show that G is a path.

14. A directed graph G is called *strongly connected* if there is a directed path from x to y for any two vertices x, y in G. Direct the edges in the following graphs to make the graphs strongly connected. If not possible, explain why. (An application of this problem is making streets in a city one-way.)

(a) Figure 1.4 **(b)** Figure 1.2 **(c)** Figure 1.8 (left graph)

15. Prove that G is strongly connected (see Exercise 14) if and only if G's vertices cannot be partitioned into two sets V_1, V_2 such that there are no edges from a vertex in V_1 to a vertex in V_2.

16. A *bridge* is an edge in a connected graph whose removal disconnects the graph. Show that if a graph G contains a bridge, then it cannot have a circuit that contains all vertices of G.

17. Show that if every edge in a connected graph G lies on a circuit, then G cannot have a bridge (a bridge is defined in Exercise 16).

18. Prove that the edges of a connected undirected graph G can be directed to create a strongly connected graph (see Exercise 14) if and only if there is no bridge in G (see Exercise 16).

19. Draw planar graphs with the following types of vertices, if possible:

(a) Six vertices of degree 3.

(b) i vertices of degree i, $i = 1, 2, 3, 4, 5$.

(c) Two vertices of degree 3 and four vertices of degree 5.

20. Mr. Megabucks invites three married couples to his penthouse for dinner. Upon arrival, Mr. Megabucks and the six guests shake the hands of some of the other people (none of the guests shakes hands with his or her spouse). Suppose each of the six guests shakes a different number of hands (possibly one person shakes no hands). By building a graph model of this situation (with 7 vertices for the 6 guests and Mr. Megabucks), determine exactly how many hands Mr. Megabucks must have shaken.

21. As four brothers leave for a dance, each of them accidentally takes a hat belonging to another brother and a coat belong to yet another brother. Use a graph model with the following information to determine whose coat and hat brothers A and B each took:

(a) Brother A took the coat belonging to the brother whose hat was taken by brother B.

(b) Brother B's coat was taken by the brother who took brother A's hat.

(c) Brother C took brother D's hat.

22. Suppose that there are three farms each with a child, a goat, and a rabbit. Arbitrarily name the three sets of inhabitants C_a, C_b, C_c, G_a, G_b, G_c, and R_a, R_b, R_c. Use the following information and a graph model to determine the three groups of a child, goat, and rabbit on each farm:

(a) The male child on the farm with goat G_a, and the male child on the farm with rabbit R_c are competing for the attention of the female child C_b on the third farm.

(b) Goat G_c and rabbit R_b are not on the same farm.

(c) The boy on the farm with rabbit R_a is not C_a.

23. A *trail* is a sequence of consecutively linked edges in which no edge can appear more than once. Unlike a path, a vertex can be visited any number of times in a trail. A *cycle* is a trail that starts and ends at the same vertex. A cycle that repeats no vertices is a circuit.

(a) Show that a subset of the vertices on a trail from x to y can be used to make a path from x to y.

(b) Prove or give a counterexample: If x and y lie on a cycle, then they must lie on a circuit.

(c) Show that the edges in a cycle can be partitioned into a collection of circuits.

(d) Show that if C is an odd-length cycle, then a subset of C's edges forms an odd-length circuit.

24. Consider a collection of circles (of varying sizes) in the plane. Make a *circle graph* with a vertex for each circle and an edge between two vertices when they correspond to two circles that cross (if one circle properly contains another, there would be no edge).

(a) Draw a family of circles whose circle graph is isomorphic to K_4.

(b) Draw a family of circles whose circle graph is the graph in Figure 1.3 (ignoring edge directions).

(c) Draw a family of circles whose circle graph is isomorphic to $K_{3,3}$.

25. A *cut-set* S is a set of edges in a connected undirected graph G whose removal disconnects G, but no proper subset of S can disconnect G.

(a) Find a cut-set of minimal size in Figure 1.23*a*.

(b) Show that every cut-set has an even number of edges in common with any circuit (remember that 0 is an even number).

26. A graph with n vertices and $n + 2$ edges must contain two edge-disjoint circuits. Prove or give a counterexample.

27. Show that if an n-vertex graph has more than $\frac{1}{2}(n - 1)(n - 2)$ edges, then it must be connected. (*Hint:* The most edges possible in a disconnected graph will occur when there are two components, each complete subgraphs.)

28. Show that an n-vertex graph cannot be a bipartite graph if it has more than $\frac{1}{4}n^2$ edges.

29. Suppose G is a connected graph containing no triangles and every pair of two non-adjacent vertices in G have exactly two neighbors in common. Show that every vertex of G must have the same degree. (*Hint:* Show that any pair of adjacent vertices must have the same degree.)

30. (*Famous Ramsey Theory problem*) Let each edge of a complete graph on 6 vertices be painted red or white. Show that there must always be either a red triangle of 3 edges or a white triangle of 3 edges.

31. **(a)** Find a graph that is isomorphic to its own complement.

(b) Show that any self-complementary graph [as in part (a)] must have either $4k$ or $4k + 1$ vertices, for some integer k. (*Hint:* Use the fact that G and $\overline{G}$ both must have the same number of edges.)

32. Suppose that each path in a certain 7-vertex planar graph contains an even number of edges (0 edges or 2 edges or 4 edges, etc.). Draw the graph. (*Hint:* This is a "trick" problem.)

33. Show that a directed graph has no directed circuits if and only if its vertices can be indexed $x_1, x_2, \ldots, x_n$, so that all edges are of the form $(\vec{x_i, x_j})$, $i < j$.

34. **(a)** A *clique* is a complete subgraph not contained as a subset of a larger complete subgraph. Find all cliques in the following interval graphs:

(i) Figure 1.1 **(ii)** Figure 1.5*a*

(b) Show that the cliques in an interval graph correspond to points of maximal overlap in an interval model and that when cliques are indexed according to the order of the points in an interval model, each vertex will be in a consecutive set of cliques.

35. A *line graph* $L(G)$ of a graph G has a vertex of $L(G)$ for each edge in G and an edge between 2 vertices in $L(G)$ corresponding to 2 edges of G with a common end vertex.

(a) Draw a line graph of the left graph in Figure 1.7.

(b) Show that each vertex in $L(K_n)$ has degree $2(n-2)$.

(c) Find all graphs that are isomorphic to their own line graph.

36. Show that if a graph H is the line graph (see Exercise 35) of some graph, then the edges of H can be partitioned into a collection of complete subgraphs such that each vertex of H is in exactly two such complete subgraphs.

37. An *automorphism* of a graph is an isomorphism ($1-1$ mapping preserving adjacency) of the vertices of a graph with themselves. Find an automorphism of the graph in

(a) Figure 1.1*a* **(b)** Figure 1.4 **(c)** Figure 1.13

38. **(a)** Show that there is no way to pair off the 14 vertices in the graph below with 7 edges.

(b) Generalize part (a) to the problem of trying to use 31 dominoes to cover the 62 squares of an 8×8 chessboard with two opposite corner squares removed.

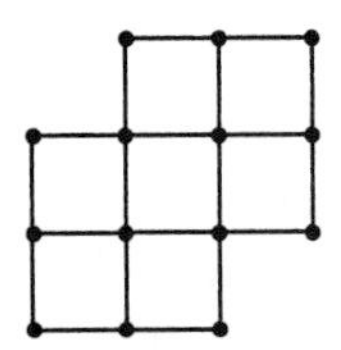

39. A round-robin tournament can be represented by a complete directed graph with vertices for competitors and an edge $(\vec{a, b})$ if a beats b. Each competitor plays every other competitor once. A vertex's (competitor's) score will be its out-degree (number of victories). Show that if vertex x has a maximum score among vertices in a round-robin tournament, then for any other vertex y, either there is an edge $(\vec{x, y})$ or for some w there are edges $(\vec{x, w})$ and $(\vec{w, y})$.

40. Suppose the round-robin tournament graph (see Exercise 37) has no directed circuits. We define a ranking of vertices (competitors) as follows. A vertex with no outward edge has a rank of 0. In general, a vertex has rank k if it has an edge directed to a rank $k-1$ vertex and all other edges directed to lower ranks ($\leq k-1$). Show that a directed complete graph with no directed circuits always has such a ranking and that each vertex will have a different rank.

41. Suppose circuits C_1 and C_2 have common edges (but $C_1 \neq C_2$). Show that the edges in $(C_1, \cup C_2) - (C_1, \cap C_2)$ form a circuit (or collection of circuits).

42. If the graph G has $2n$ vertices and no triangles, then show that G cannot have more than n^2 edges.

CHAPTER 2
COVERING CIRCUITS AND GRAPH COLORING

2.1 EULER CYCLES

In this chapter we examine two graph-theoretic concepts that have many important applications. One is edge sequences, which visit either all the edges in a graph once or all the vertices once. The other is graph coloring, a concept introduced in Example 1 of Section 1.4.

In some applications, such as a highway network linking a group of cities, it is natural to permit several edges to join the same pair of vertices. Such generalized graphs are called **multigraphs**; loop edges, of the form (a, a), are also allowed in multigraphs. Figure 2.1*b* displays a multigraph.

At the beginning of Chapter 1, we defined a *path* $P = x_1–x_2–\cdots–x_n$ to be a sequence of distinct vertices with each pair of consecutive vertices in P joined by an edge. When there is also an edge (x_n, x_1), the sequence is called a *circuit,* written $x_1–x_2–\cdots–x_n–x_1$. Just as we generalized our notion of a graph, now we generalize our definitions of path and circuit in order to allow repeated vertices. A **trail** $T = x_1–x_2–\cdots–x_n$ is a sequence of vertices (not necessarily distinct) in which, like a path, consecutive vertices are joined by an edge. However, *no edge can be repeated in a trail.* In Figure 2.1*b*, *A-B-D-B-C* is a trail (we assume segments *B-D* and *D-B* on this trail are using the two different edges joining *B* and *D*). When there is also an edge (x_n, x_1), the sequence of vertices is called a **cycle,** written $x_1–x_2–\cdots–x_n–x_1$.

Example 1: Konigsberg Bridges

The old Prussian city of Konigsberg was located on the banks of the Pregel River. Part of the city was on two islands that were joined to the banks and each other by seven bridges, as shown in Figure 2.1*a*. The townspeople liked to take walks, or *Spaziergangen,* about the town across the bridges. Several people were apparently bothered by the fact that no one could figure out a walk that crossed each bridge just once, for they brought this problem to the attention of the famous mathematician Leonhard Euler. He solved the *Spaziergangen* problem, thereby giving birth to graph theory and immortalizing the Seven Bridges of Konigsberg in mathematics texts.

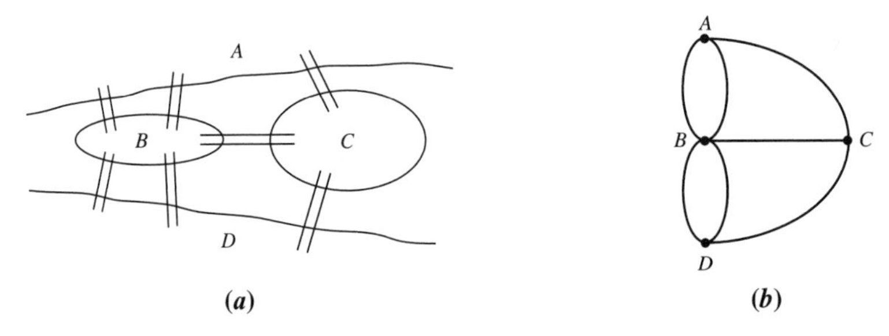

Figure 2.1

We can model this walk problem with a multigraph having a vertex for each body of land and an edge for each bridge. See Figure 2.1*b*. The desired type of walk corresponds to what we now call an Euler cycle. An **Euler cycle** is a cycle that contains all the edges in a graph (and visits each vertex at least once). Readers should convince themselves that no Euler cycle exists for this multigraph. ∎

A multigraph possessing an Euler cycle will have to have an even degree at each vertex, since each time the cycle passes through a vertex it uses two edges. A second obvious property is that the multigraph must be connected. Euler showed that these two properties are also sufficient to guarantee the existence of an Euler cycle. We will prove this theorem for undirected multigraphs. Let us first build an Euler cycle by ad hoc methods in a graph that is connected and has even-degree vertices. Then we extend our construction to a general proof of the Euler cycle theorem.

Example 2: Building an Euler Cycle

Build an Euler cycle for the graph in Figure 2.2*a* that is connected and has even-degree vertices.

Let us start by blindly tracing out a trail from vertex *a*. Suppose we go *a-d-j-n-o-k-l-h-f-e-b-a*. Now we are back to *a*. Note that the even degree property means that we can always leave any vertex we enter except *a*. There are no vertices of degree 1 or that are reduced to degree 1 after being visited several times. Thus, any trail we trace from *a* will never be forced to end at another vertex and so must eventually come back to *a*, forming a cycle; in this case the cycle is a circuit (no vertex is visited twice).

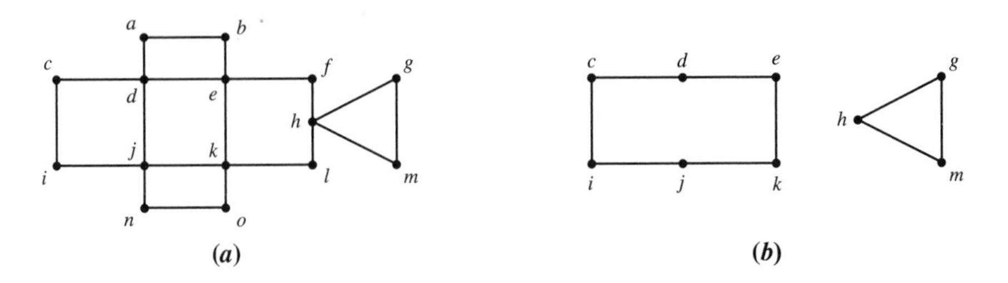

Figure 2.2

Next consider the graph of remaining edges, shown in Figure 2.2*b*. It is no longer connected, but all vertices still have even degree (removing the cycle reduced each degree by an even amount). Each connected piece of the remaining graph has an Euler cycle: *d-c-i-j-k-e-d* and *h-g-m-h*. These two cycles can be inserted into the original cycle at vertices *d* and *h*, respectively, yielding the Euler cycle *a-d-c-i-j-k-e-d-j-n-o-k-l-h-g-m-h-f-e-b-a*. ■

Theorem 1 **(Euler, 1736)**

An undirected multigraph has an Euler cycle if and only if it is connected and has all vertices of even degree.

Proof

The proof generalizes the construction in Example 2. As noted above, when an Euler cycle exists, the multigraph must be connected and have all vertices of even degree.

Suppose a multigraph G satisfies these two conditions. Then pick any vertex a and trace out a trail. As in Example 2, the even degree condition means that we are never forced to stop at some other vertex, and so eventually the trail must terminate at a (possibly it passes through a several times before finally being forced to stop at a). Let C be the cycle thus generated and let G' be the multigraph consisting of the remaining edges of G-C. As in Example 2, G' may not be connected but it must have all vertices of even degree.

Since the original graph was connected, C and G' must have a common vertex or else there is no path from vertices in C to vertices in G'. Let a' be such a common vertex. Now build a cycle C' tracing through G' from a' just as C was traced in G from a. Incorporate C' into the cycle C at a', as done in Example 2, to obtain a new larger cycle C'. Repeat this process by tracing a cycle in the graph G'' of still remaining edges and incorporate the cycle into C' to obtain C''. Continue until there are no remaining edges. ◆

Let us now consider an application of Euler cycles to an urban systems analysis problem.

Example 3: Routing Street Sweepers

Suppose the graph of solid edges in Figure 2.3 represents a collection of blocks to be swept by a street sweeper in a certain district of some city from 10 A.M. to 11 A.M.

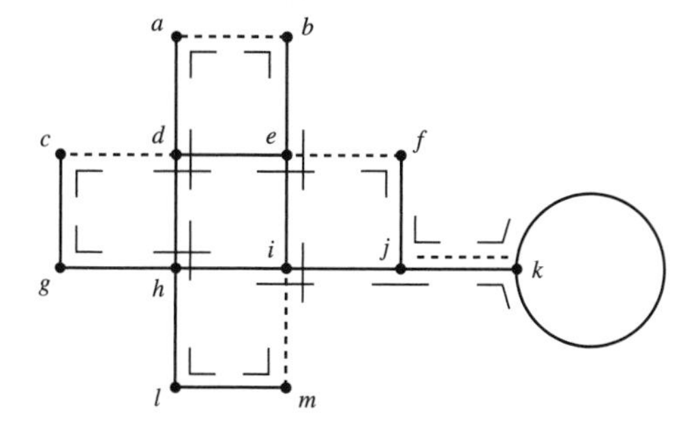

Figure 2.3

(when parking on these blocks is forbidden). The looping edge at k represents a circle. We want a tour that sweeps each solid edge once. That is, we want an Euler cycle of the solid edges. Unfortunately, the graph of edges to be swept in such applications rarely has all vertices of even degree. Frequently the graph is also not connected. In such cases we must traverse extra edges, called *deadheading edges,* to obtain the desired tour. This creates a new problem: minimizing the number of deadheading edges.

The desired tour will be an Euler cycle of the multigraph of sweeping and deadheading edges—a multigraph because some edges may be repeated in deadheading. By the theorem, this multigraph must be connected and have all vertices of even degree. Thus to minimize deadheading, the deadheading edges should be a minimal set of extra edges with the property that when added to the original graph all vertices have even degree and the new multigraph is connected. Suppose the dashed edges in Figure 2.3 are such a minimal set of extra edges.

There is one more problem to be faced once we have a multigraph possessing an Euler cycle. We want to build an Euler cycle that minimizes turns at corners, especially U-turns. Even right-hand turns can tie up traffic in busy cities. We can successively look at each corner (vertex) and pair off the edges at the corner so as to minimize disruption on the tour's passes through that corner. For the multigraph in Figure 2.3, we would go straight through corners d, e, h, i on visits to these corners; we would go straight through j once and turn once [going between (f, j) and (k, j)]; we must turn both times at k; and at all other corners (of degree 2) we would have forced turns. The short lines near each vertex in Figure 2.3 indicate these edge pairings. Of course, such edge pairings are very unlikely to produce a single Euler cycle. In Figure 2.3, we get two cycles from these pairings: *a-b-e-i-m-l-h-d-a* and *c-d-e-f-j-k-k-j-i-h-g-c*. Now we break the two cycles at a common vertex and fuse them together. For example, at d, change the edge pairings to

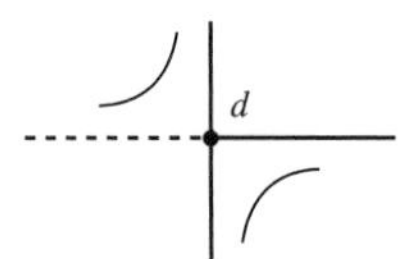

The result is an Euler cycle with optimal pairings at all but one corner.

A large street network with 200 edges typically forms only three or four cycles when minimal-disruption edge pairing is performed at each corner, and so only a few pairings need to be changed. In practice, it is also necessary to use directed graphs since street sweepers cannot move against the traffic on their side (curb) of the street. ∎

It is interesting to note that the street sweeping problem gives rise to an alternative proof of the Euler cycle theorem:

Alternative Proof of Theorem

The even-degree condition means that we can pair off (and link together) the edges at each vertex. These linked edges form a set of cycles. The connectedness condition means that these cycles have common vertices where they can be joined to form a single cycle—an Euler cycle. ◆

This proof is not as intuitive or pictorial as our original path-tracing method, but it is both simpler and more applicable.

We conclude this section by extending the concept of an Euler cycle to an Euler trail. An **Euler trail** is a trail that contains all the edges in a graph (and visits each vertex at least once).

Corollary

A multigraph has an Euler trail, but not an Euler cycle, if and only if it is connected and has exactly two vertices of odd degree.

Proof

Suppose a multigraph G has an Euler trail but not an Euler cycle. Call it T. Then the starting and ending vertices of T must have odd degree while all other vertices have even degree (by the same reasoning that showed all vertices have even degree in an Euler cycle). Also the graph must be connected.

On the other hand, suppose that a multigraph G is connected and has exactly two vertices, p and q, of odd degree. Add a supplementary edge (p, q) to G to obtain the graph G'. G' is connected and has all vertices of even degree. Hence by the Euler cycle theorem, G' has an Euler cycle, call it C. Now remove the edge (p, q) from C. This removal reduces the Euler cycle to an Euler trail that includes all edges of G. ◆

2.1 EXERCISES

Summary of Exercises The first four exercises involve trying to build Euler cycles. Exercises 5–8 and 11–14 present extensions and other questions related to the Euler cycle theorem. The remaining questions involve some modeling and further theory, and the last two exercises ask for computer programs.

1. (a) Build an Euler cycle for the right graph in Figure 1.9.

(b) Build an Euler trail for the graph in Figure 1.23*a* with edge (d, g) removed.

2. (a) For which values of n does K_n, the complete graph on n vertices, have an Euler cycle?

(b) Are there any K_n that have Euler trails but not Euler cycles?

(c) For which values of r and s does the complete bipartite graph $K_{r,s}$ have an Euler cycle?

3. Find a graph G with 7 vertices such that G and its complement both have an Euler cycle.

4. What is the minimum number of times one must raise one's pencil in order to draw the graph in Figure 1.4?

5. (a) Can a graph with an Euler cycle have a bridge (an edge whose removal disconnects the graph)? Prove or give a counterexample.

(b) Give an example of a 10-edge graph with an Euler trail that has a bridge.

6. Give an argument for and an argument against the statement that a 1-vertex graph (with no edges) has an Euler cycle.

7. Suppose that in the definition of an Euler cycle, we drop the seemingly superfluous requirement that the Euler cycle visit every vertex and require only that the cycle include every edge. Show that now the theorem is false. Draw a graph that illustrates why the theorem is now false.

8. Prove that if a connected graph has a $2k$ vertices of odd degree, then there are k disjoint trails that contain all the edges.

9. The matrix below marks with a 1 each pair of the set of racers A, B, C, D, E, F who are to have a drag race together. It is most efficient if a racer can run in two races in a row (but not three in a row). Is it possible to design a sequence of races such that one of the racers in each race (except the last race) also runs in the following race (but not three in a row)? If possible, give the sequences of races (pairs of racers); if not, explain why not.

	A	B	C	D	E	F
A	—	1	0	1	1	0
B	1	—	1	1	0	1
C	0	1	—	0	1	0
D	1	1	0	—	1	1
E	1	0	1	1	—	1
F	0	1	0	1	1	—

10. Is it possible for a knight to move around an 8 by 8 chessboard so that it makes every possible move exactly once (consider a move between two squares connected by a knight to be completed when the move is made in either direction)?

11. Show that in Example 3 the minimum set of deadheading edges in any sweeping problem will be a collection of edges forming paths between pairs of different odd-degree vertices.

12. Prove the directed version of the Euler Cycle Theorem: A directed multigraph has a directed Euler cycle if and only if the multigraph is connected (when directions are ignored) and the in-degree equals the out-degree at each vertex.

(a) Model your proof after the argument in the proof of the theorem.

(b) Model your proof after the argument in the alternative proof of the theorem.

(c) Build a directed Euler cycle for the graph below.

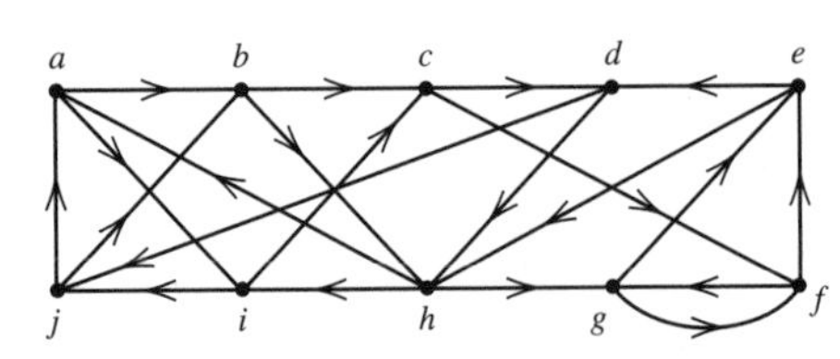

13. State and prove a directed multigraph version of the corollary.

14. A directed graph is called *strongly connected* if there is a directed path from any given vertex to any other vertex. Show that if a directed graph possesses a directed Euler cycle, then it must be strongly connected.

15. Try to find a minimal set of edges in the graph below whose removal produces an Euler cycle. (*Hint:* Tricky.)

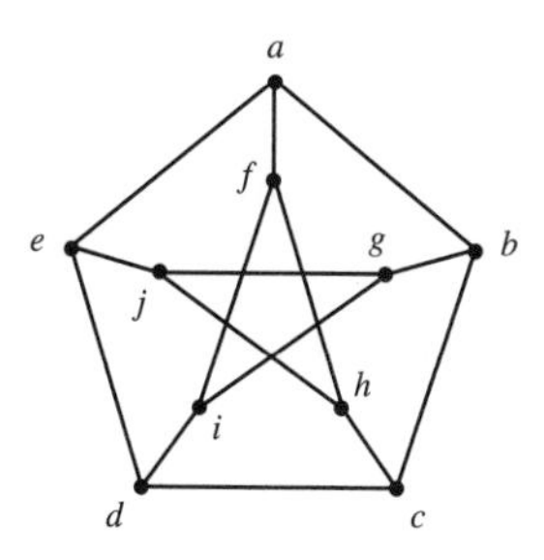

16. The *line graph* $L(G)$ of a graph G has a vertex for each edge of G, and two of these vertices are adjacent if and only if the corresponding edges in G have a common end vertex.

(a) Show that $L(G)$ has an Euler cycle if G has an Euler cycle.

(b) Find a graph G that has no Euler cycle but for which $L(G)$ has an Euler cycle.

17. Consider the following algorithm due to Fleury for building an Euler cycle, when one exists, in a single pass through a graph (without later adding side cycles as in the proof of the theorem). Starting at a chosen vertex a, build a cycle and erase edges after they are used (also erase vertices when they become isolated points). The one rule to follow in the cycle building is that one never chooses an edge whose erasure will disconnect the resulting graph of remaining edges.

(a) Apply this algorithm to build Euler cycles for the graph in:

(i) Figure 2.2a **(ii)** Figure 2.3 (including deadheading edges)

(b) Prove that this algorithm works.

(c) Does this algorithm work for Euler trails? Explain.

18. Suppose we are given an undirected connected graph representing a network of two-way streets.

(a) Show that there always exists a tour of the network in which a person drives along each side of every street once.

(b) Show that the tour in part (a) can be generated by the following rule: At any intersection do not leave by the street first used to reach this intersection unless all other streets from the intersection have been used.

19. A set of 8 binary digits (0 or 1) are equally spaced about the edge of a disk. We want to choose the digits so that they form a circular sequence in which every subsequence of length three is different. Model this problem with a graph with 4 vertices, one for each different subsequence of two binary digits. Make a directed edge for each subsequence of three digits whose origin is the vertex with the first two digits of the edge's subsequence and whose terminus is the vertex with the last two digits of the edge's subsequence.

(a) Build this graph.

(b) Show how an Euler cycle (which exists for this graph) will correspond to the desired 8-digit circular sequence.

(c) Find such an 8-digit circular sequence with this graph model.

(d) Repeat the problem for 4-digit binary sequences.

20. Write computer programs for finding an Euler cycle, when one exists, in a multigraph:

(a) Use the method in the proof of the theorem.

(b) Repeat part (a) for a directed graph.

(c) Use the method in the alternative proof of the theorem.

(d) Modify the program in parts (a) or (c) to build the k trails described in Exercise 8.

21. Write a program to implement the algorithm in Exercise 17.

2.2 HAMILTON CIRCUITS

In this section we explore **Hamilton circuits** and **paths,** circuits and paths that visit each vertex in a graph exactly once. Hamilton circuits arise in operations research problems involving the route of a delivery truck that must supply a set of stores or motion planning for a drill press making holes at specified locations on printed circuit boards. In these applications, the most efficient solution will be obtained by finding a minimal-cost Hamilton circuit (this problem is discussed in Section 3.4).

The problem of determining whether a graph has an Euler cycle was answered by the Euler cycle theorem, which tells us that a graph has an Euler cycle if and only if all vertices have even degree and the graph is connected. Such a nice, simple answer is very unusual in graph theory (this is probably why the Euler cycle theorem was the first result proved in the field of graph theory). In this section, we return to graph theory normalcy: There is no simple way to determine whether or not an arbitrary graph has a Hamilton circuit or a Hamilton path.

Finding Hamilton circuits by inspection, when they exist, is usually not too hard in moderate-sized graphs, but proving that no Hamilton circuit exists in a given graph can be very difficult. Such a proof typically involves the same type of reasoning needed to show that two graphs are not isomorphic or that a graph is nonplanar. At the end of this section we present a sampling of the theory that has been developed about the existence of Hamilton circuits. These theorems give various special conditions on a graph that guarantee the existence of a Hamilton circuit. Most graphs satisfy none of these conditions.

Our focus will be proving that a Hamilton circuit does not exist in particular graphs. This nonexistence problem requires the type of systematic logical analysis that is the essence of most applied graph theory. To prove nonexistence, we must begin building parts of a Hamilton circuit and then show that the construction must always fail. This is similar to the way in Section 1.4 that we showed a graph to be nonplanar by building a circuit and then adding chords in a structured fashion that forced two edges to cross.

Our analysis is built on three simple rules that must be satisfied by the set of edges forming a Hamilton circuit. The idea underlying these rules is that *any Hamilton circuit must contain exactly two edges incident to each vertex.*

Rule 1. If a vertex x has degree 2, both of the edges incident to x must be part of any Hamilton circuit.

Rule 2. No proper subcircuit, that is, a circuit not containing all vertices, can be formed when building a Hamilton circuit.

Rule 3. Once the Hamilton circuit is required to use two edges at a vertex x, all other (unused) edges incident at x must be removed from consideration.

Example 1: Nonexistence of Hamilton Circuit I

Show that the graph in Figure 2.4 has no Hamilton circuit.

We can apply Rule 1 at vertices a, b, d, and e. To indicate that the two edges at each of these vertices must be used, we draw a little line segment from one edge to the other, as shown in Figure 2.4. There are two types of contradictions that have now arisen. First, using Rule 1 for vertices a and d forces the Hamilton circuit to use edges (a, c), (a, d), (d, c) in forming a triangle; this violates Rule 2. Second, using Rule 1 for all four vertices of degree 2 forces the Hamilton circuit to contain more than two edges incident at c. Either of these difficulties proves that this graph has no Hamilton circuit. ■

In the following two examples, our three rules will be used in the following sort of sequential reasoning. When Rule 3 requires us to delete edges at some vertex q, one of q's deleted edges may go to a vertex r of degree 3. After the deletion of (q, r), r has degree 2, requiring by Rule 1 that its remaining two edges be used. Possibly, one of these two edges, say (r, s), causes two edges to be used at neighboring vertex s, and so now Rule 3 can be invoked at s, continuing the sequence of forced moves.

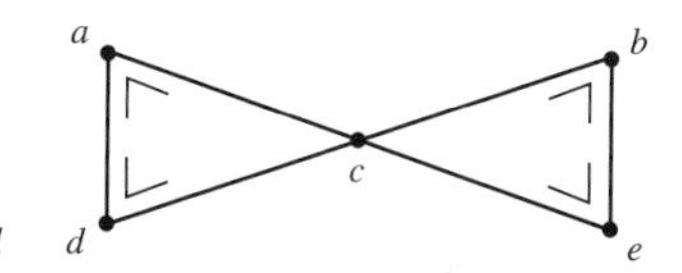

Figure 2.4

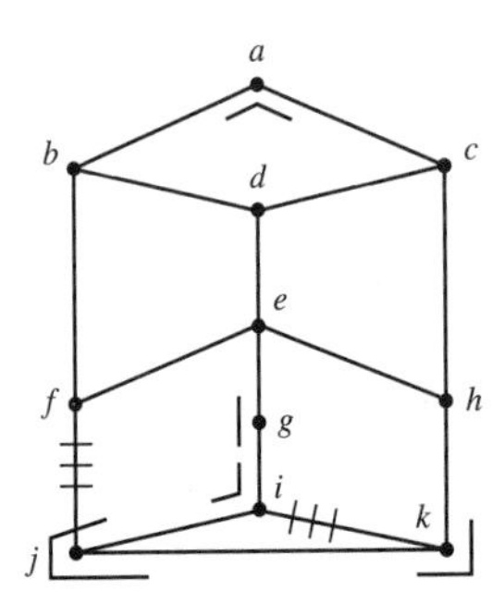

Figure 2.5

Example 2: Nonexistence of Hamilton Circuit II

Show that the graph in Figure 2.5 has no Hamilton circuit.

We can apply Rule 1 at vertices a and g, so that the subpaths b-a-c and e-g-i must be part of any Hamilton circuit. Next consider vertex i. We already know that (g, i) must be part of the circuit. Since the graph is symmetric with respect to edges (i, j) and (i, k), it does not matter which of these two edges we choose as the other edge incident to i on the Hamilton circuit. Suppose we pick (i, j). If we obtain a contradiction using (i, j), then by symmetry we would also obtain a contradiction with (i, k). The situation can be visualized as follows: If we marked (i, j) and then held the paper with the graph in front of us and viewed it in a mirror, it would appear that we had chosen (i, k); every subsequent move starting with (i, j) would have a mirror image move starting from (i, k).

Having chosen to use (i, j) as the second edge at i, we delete the other edge at i, (i, k), by Rule 3. See Figure 2.5. Each time we apply a rule, we need to check how other edges and vertices are affected. Does any vertex now have only two remaining edges at it, allowing Rule 1 to be applied? Or are two edges now required to be used at some vertex, allowing Rule 3 to be applied? Also, is there some edge that, if used, would complete a subcircuit, allowing Rule 2 to be used to delete it?

Deleting (i, k) reduces the degree of k to 2, and so Rule 1 requires that we use both remaining edges incident at k, (j, k) and (h, k). Edge (j, k) is the second edge used that is incident to j. Then by Rule 3, j's other edge (f, j) must be deleted. This deletion reduces the degree of f to 2. See Figure 2.5. So we must use the two remaining edges at f, (b, f) and (f, e). Edge (b, f) is the second edge used at b, and so edge (b, d) at b must be deleted. Similarly, edge (f, e) is the second edge used at e, and so e's other edges, (e, d) and (e, h), must be deleted.

Deleting (e, h) forces the use at h of (c, h). Using (c, h) forces the deletion at c of (c, d). However, now we have deleted all the edges incident to d from consideration on a Hamilton circuit. This contradiction implies that G cannot have a Hamilton circuit.

Note that this graph does have Hamilton paths, for example, *a-b-f-e-g-i-j-k-h-c-d*. ■

It should be emphasized that the following is *not* a useful line of reasoning to show that a graph has no Hamilton circuit: Start from some vertex and construct a route visiting successive vertices and show that several attempts to find a circuit through all vertices fail. Even in a simple graph like the one in Figure 2.5, there are hundreds of possible beginning subpaths for a Hamilton circuit that would all have to be checked. The approach presented here requires much less work than a rigorous trial-and-error effort.

Example 3: Nonexistence of Hamilton Circuit III

Show that the graph in Figure 2.6 has no Hamilton circuit.

Note that this graph has vertical and horizontal symmetry (although vertex n is off to one side, its adjacencies have a square-like symmetry). It sometimes takes a bit of trial-and-error experimenting to find a good vertex at which to start trying to build a Hamilton circuit when there are no vertices of degree 2. We seek a vertex with the property that once two edges are chosen at the vertex, then the use of Rules 1 and 3 will force the successive deletion and inclusion of many edges. Vertex e is such a vertex. We can either use two edges incident at e from opposite sides (180° apart) or use two edges incident at e that form a 90° angle. We must examine both cases to show that no Hamilton circuit can exist. (Hamilton circuits frequently have many such subcases that must all be checked out.)

Case I Suppose we use two edges incident at e from opposite sides. By symmetry, they can either be edges from d and f or from b and h. Suppose we choose (d, e) and (e, f). Then by Rule 3, we can delete edges (e, b) and (e, h). Then at b and at h we must use both remaining edges, getting subpaths a-b-c and g-h-i. Now at d we can use either edge (d, a) or edge (d, g). The two cases are symmetrical with respect to the edges chosen for the circuit thus far. So without loss of generality, we can choose edge (d, a).

At f, we cannot use (f, c) or else subcircuit a-b-c-f-e-d-a results. So we must use (f, i). See Figure 2.6. Since we have used two edges at vertices a and i, the other edges at these vertices can be deleted by Rule 3. We now obtain several inconsistencies. Vertices j, k, l, and m each now have degree 2. But each is incident to n, and so using the two remaining edges at each causes 4 edges to be used at n. We conclude that there is no Hamilton circuit in Case I.

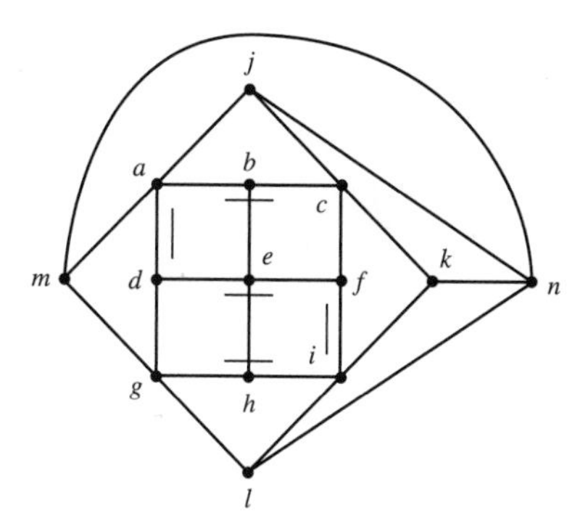

Figure 2.6

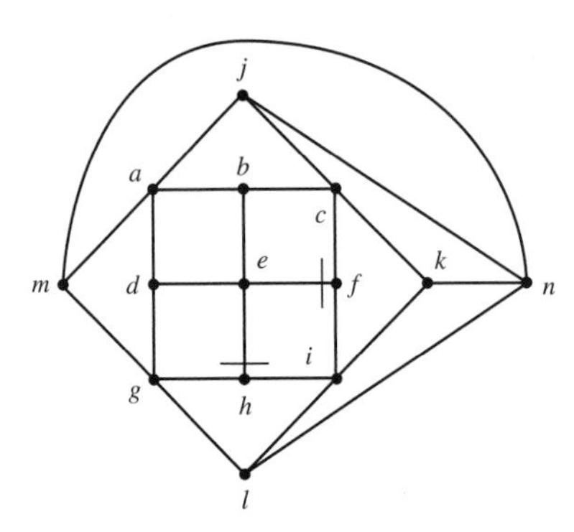

Figure 2.7

Case II Suppose we use two edges incident at e that form a 90° angle. By symmetry it does not matter which of the four 90° angle pairs of edges we choose. Suppose we choose (b, e) and (d, e). See Figure 2.7. Then by Rule 3 we can delete edges (e, f) and (e, h). Then at f and h we must use both remaining edges getting subpaths c-f-i and g-h-i. See Figure 2.7. By Rule 3 at i, we can delete edges (i, k) and (i, l). By Rule 1, we must use the remaining edges at k and l, forming the subcircuit i-f-c-k-n-l-g-h-i.

Having obtained contradictions in both cases, we have proved that the graph has no Hamilton circuit. ∎

We now present a few of the theoretical results about the existence of Hamilton circuits and paths. (See [4] for proofs of Theorems 1, 2, and 3.)

Theorem 1 **(Dirac, 1952)**

A graph with n vertices, $n > 2$, has a Hamilton circuit if the degree of each vertex is at least $n/2$.

Theorem 2 **(Chvatal, 1972)**

Let G be a connected graph with n vertices, and let the vertices be indexed $x_1, x_2, \ldots, x_n$, so that $\deg(x_i) \leq \deg(x_{i+1})$. If for each $k \leq n/2$, either $\deg(x_k) > k$ or $\deg(x_{n-k}) \geq n - k$, then G has a Hamilton circuit.

Theorem 3 **(Grinberg, 1968)**

Suppose a planar graph G has a Hamilton circuit H. Let G be drawn with any planar depiction, and let r_i denote the number of regions inside the Hamilton circuit bounded by i edges in this depiction. Let r_i' be the number of regions outside the circuit bounded by i edges. Then the numbers r_i and r_i' satisfy the equation

$$\sum_i (i-2)(r_i - r_i') = 0 \qquad (*)$$

Just as the inequality $\mathbf{e} \leq 3\mathbf{v} - 6$ for planar graphs in the corollary in Section 1.4 could be used to prove that some graphs are not planar, Theorem 3 can be used to show that some planar graphs cannot have Hamilton circuits.

Example 4: Application of Theorem 3

Show that the planar graph in Figure 2.8 has no Hamilton circuit.

We have indicated the number of bounding edges inside each of the regions in the planar depiction of the graph in Figure 2.8. There are three regions with four edges and six regions with six edges. Thus, no matter where a Hamilton circuit is drawn (if it exists), we know that $r_4 + r_4' = 3$ and $r_6 + r_6' = 6$. Observe that for this graph, Eq. (*) reduces to

$$2(r_4 - r_4') + 4(r_6 - r_6') = 0$$

We cannot have $r_6 - r_6' = 0$, that is, $r_6 = r_6' = 3$, for then Eq. (*) would require $r_4 - r_4' = 0$ or $r_4 = r_4'$—which is impossible since $r_4 + r_4' = 3$. If $r_6 - r_6' \neq 0$, then $|r_6 - r_6'| \geq 2$ and so $|4(r_6 - r_6')| \geq 8$. Now it is impossible to satisfy Eq. (*) since even if $r_4 = 3$, $r_4' = 0$ (or $r_4 = 0$, $r_4' = 3$), $|2(r_4 - r_4')| \leq 6$. Thus, it is impossible for Eq. (*) to be valid for this graph, and so no Hamilton circuit can exist. ∎

We next present a theorem involving directed graphs and Hamilton paths. A **tournament** is a directed graph obtained from a complete (undirected) graph by giving a direction to each edge.

Theorem 4

Every tournament has a Hamilton path.

Proof

The proof is by induction. For a 2-vertex tournament, a directed Hamilton path trivially exists. Next assume by induction that any tournament with $n - 1$ vertices, for $n \geq 3$, has a directed Hamilton path and let us prove that a n-vertex tournament G has a directed Hamilton path.

Remove a vertex z from G, leaving a tournament G' with $n - 1$ vertices. By the induction assumption, G' has a Hamilton path $H = x_1 – x_2 – x_3 – \cdots – x_{n-1}$. If the edge between z and x_1 is $(\vec{z, x_1})$, then z can be added to the front of H to obtain a Hamilton path for G. Similarly, if the edge between z and x_{n-1} is $(\vec{x_{n-1}, z})$, then z can be added to the end of H to obtain a Hamilton path for G. So assume

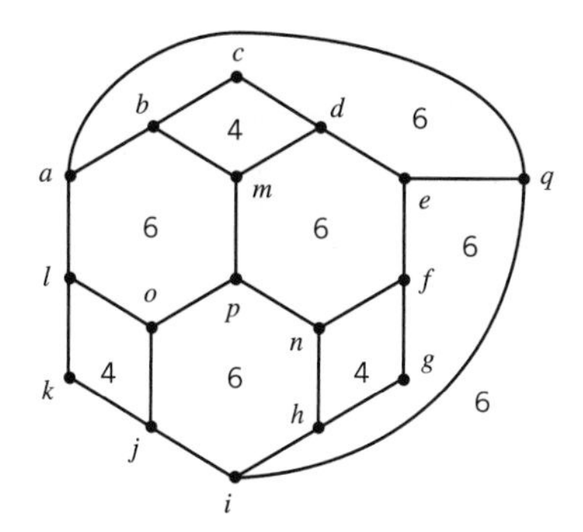

Figure 2.8

the edge from the first vertex x_1 of H points toward z and the edge from the last vertex x_{n-1} of H points from z. Then for some consecutive pair on H, x_{i-1}, x_i, the edge direction must change, that is, we have edges $(\overrightarrow{x_{i-1}, z})$ and $(\overrightarrow{z, x_i})$. We can insert z between x_{i-1} and x_i in H to obtain a Hamilton path $x_1–x_2–\cdots–x_{i-1}–z–x_i–\cdots–x_{n-1}$. ◆

We conclude this section with an application of Hamilton paths to a problem in coding theory.

Example 5: Gray Code

When a spacecraft is sent to distant planets and transmits pictures back to earth, these pictures are transmitted as a long sequence of numbers, each number being a darkness value for one of the dots in the picture. For simplicity, assume the darkness numbers range between 1 and 8. These numbers are actually sent as a sequence of 0s and 1s. A straightforward encoding scheme would be to express each number in its binary representation, that is, 1 as 001, 2 as 010, 3 as 011, and so on, ending with 8 as 000.

However, a better scheme, called a *Gray code,* uses an encoding with the property that *two consecutive numbers are encoded by binary sequences that are almost the same, differing in just one position.* For example, a fragment of a Gray code might be 4 as 010, 5 as 011, 6 as 001. The advantage of such an encoding is that if an error from "cosmic static" causes one binary digit in a sequence to be misread at a receiving station at Earth, then the mistaken sequence will often be interpreted as a darkness number that is almost the same as the true darkness number. For example, in the preceding fragment of a Gray code, if 011 (5) were transmitted and an error in the last position caused 010 (4) to be received, the resulting small change in darkness would not seriously affect the picture. (Of course, some errors will cause substantial inaccuracies.)

With this background, we now translate the problem of finding a Gray code for the 8 darkness numbers into the problem of finding a Hamilton circuit in a graph. We define the graph as follows: Each vertex corresponds to a 3-digit binary sequence, and two vertices are adjacent if their binary sequences differ in just one place. The graph is shown in Figure 2.9*a*. Observe that it is a cube (the binary sequences can be

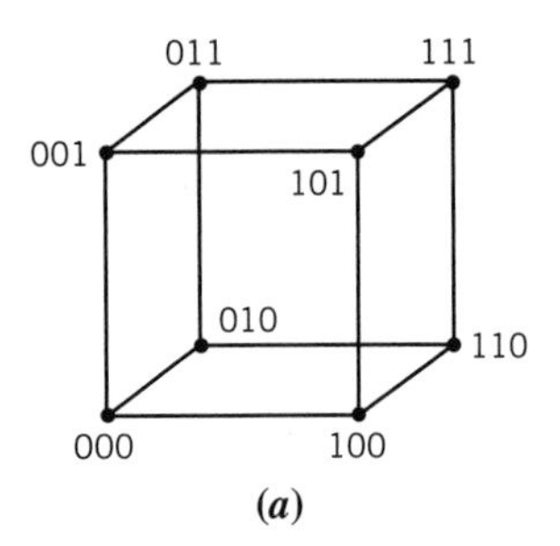

(*a*)

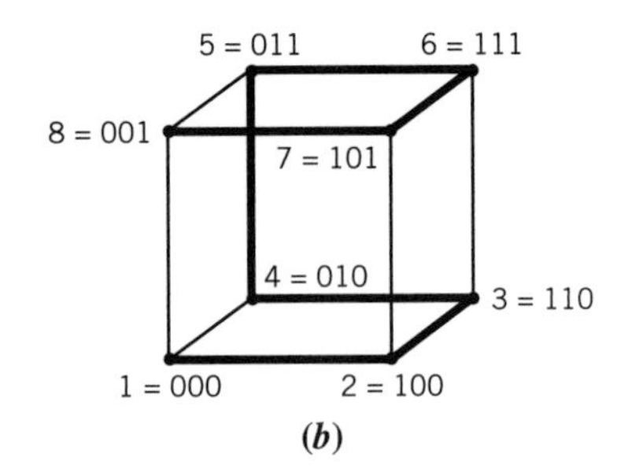

(*b*)

Figure 2.9

thought of as the coordinates of a cube drawn in three dimensions). A similar graph can be drawn for the 16 4-digit binary sequences, or for any given n, the 2^n n-digit binary sequences.

We claim that the order in which vertices (binary sequences) occur along a Hamilton path in this graph produces a Gray code. That is, 1 is encoded as the first binary sequence (vertex) in a given Hamilton path, 2 is encoded as the second binary sequence, and so on. This process yields a Gray code because consecutive vertices in the Hamilton path, which encode consecutive darkness numbers, will correspond to binary sequences that differ in just one position. Figure 2.9*b* illustrates how a Hamilton path in the graph produces a Gray code. ■

The graph with one vertex for each n-digit binary sequence and an edge joining vertices that correspond to sequences that differ in just one position is called an n-dimensional cube, or *hypercube*. The graph in Figure 2.9*a* is a 3-dimensional cube (or standard cube). An n-dimensional cube has 2^n vertices, each of degree n. It has the property that the longest path between any two vertices has length n. Hypercubes arise in the design of supercomputers. In a parallel supercomputer with 2^n processors, there are too many processors to permit a direct connection between every pair of processors. Instead, one design for supercomputers interconnects the processors with the structure of a hypercube graph.

A generalization of a Hamilton circuit is used to solve the Instant Insanity puzzle in the supplement to this chapter.

2.2 EXERCISES

Summary of Exercises The first ten exercises involve the existence or nonexistence of Hamilton paths and circuits. Exercises 9 and 10 introduce other potential aids for proving nonexistence. Exercises 15–19 involve applications of Hamilton circuits. The last five exercises involve theory.

1. **(a)** Draw a graph with a Hamilton circuit but no Euler cycle.

   (b) Draw a graph with an Euler circuit but no Hamilton cycle.

2. Find a Hamilton circuit in each of the following graphs.

(a)

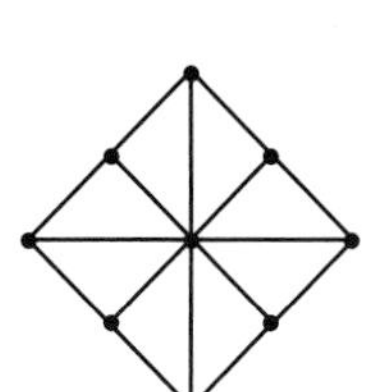

(b)

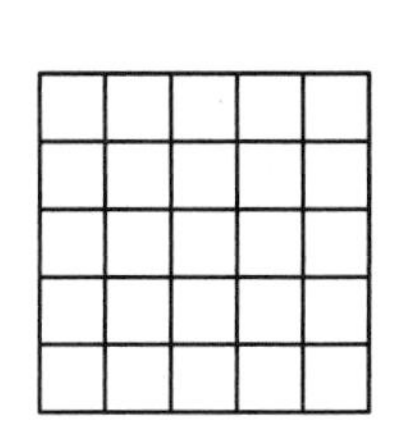

(c)

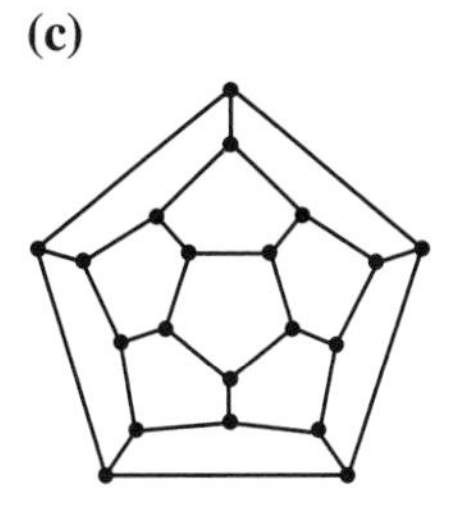

3. Find a Hamilton circuit in the following graph.

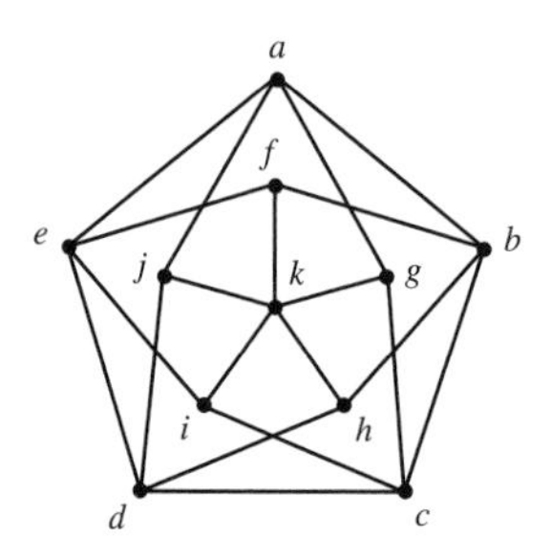

4. Find a Hamilton path in each of the following graphs and prove that no Hamilton circuit exists:

(a)

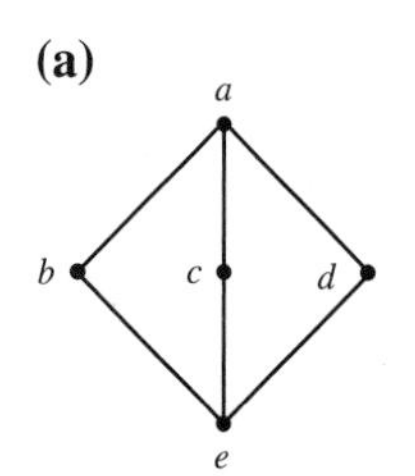

(b)

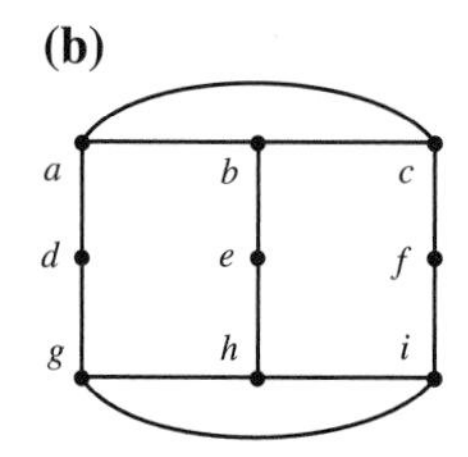

(c)

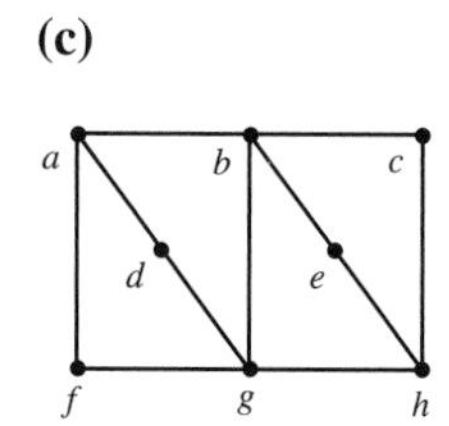

(d)

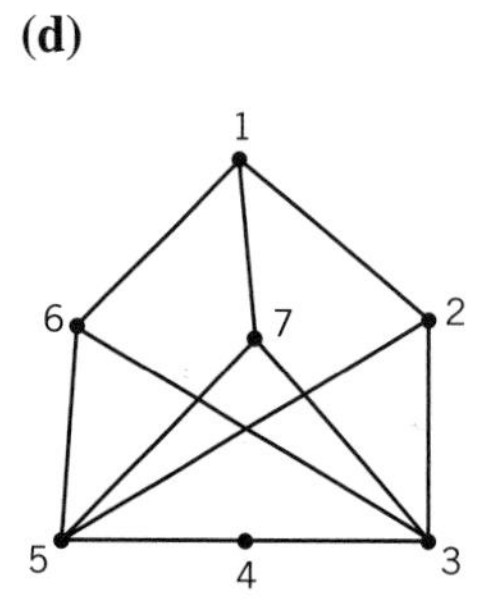

(e)

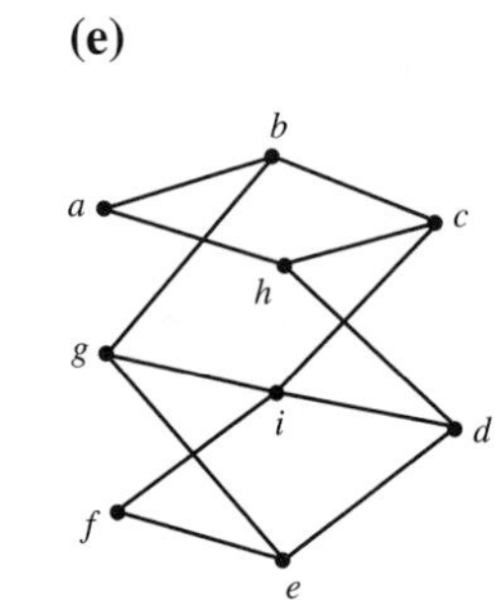

(f)

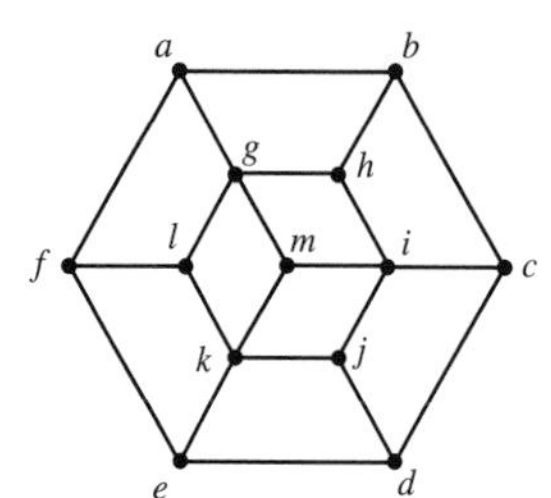

(g)

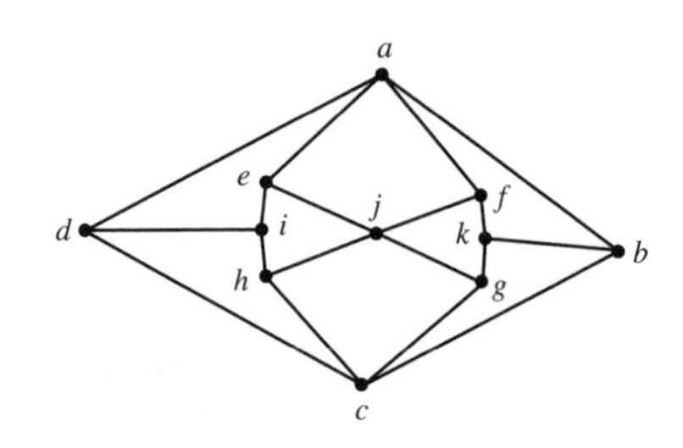

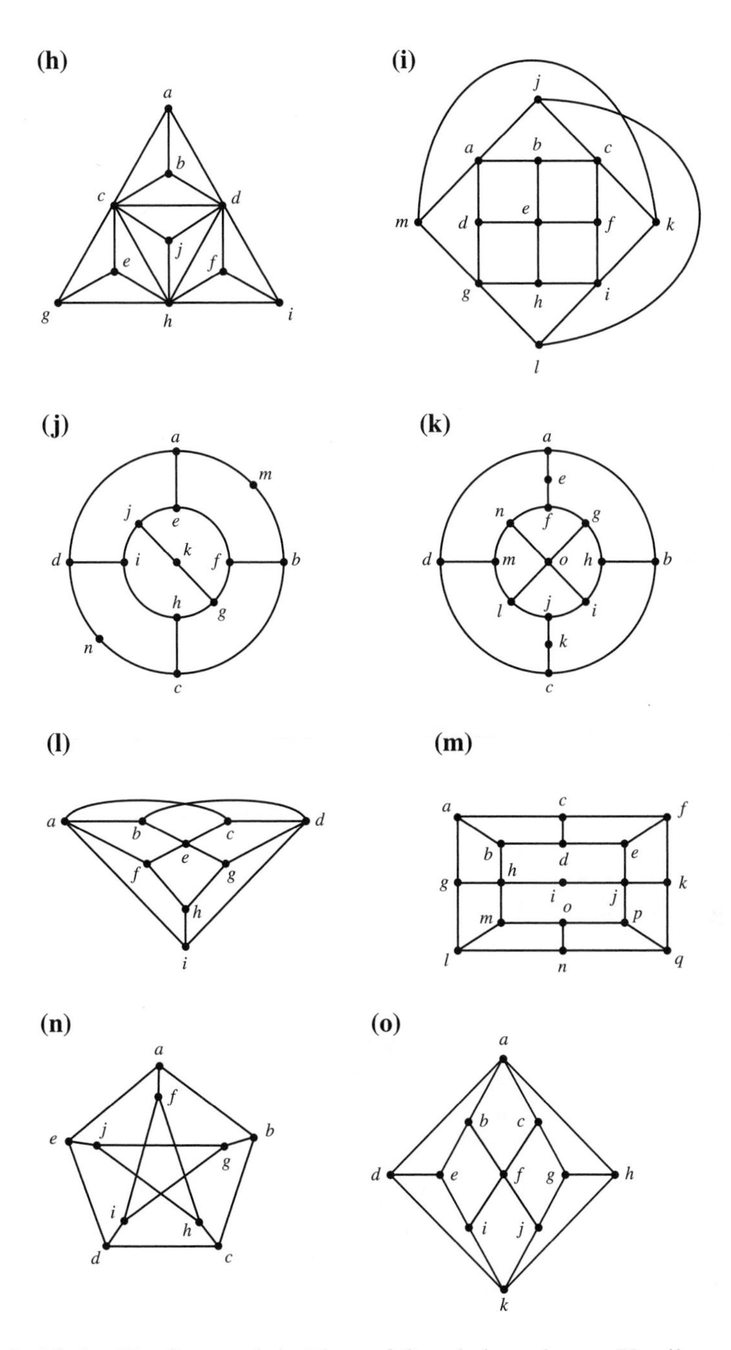

5. Find a Hamilton path in Figure 2.8 and show that no Hamilton circuit exists (using the reasoning in Examples 1, 2, 3).

6. Show that there can be no Hamilton circuit in the following graph using both edges (a, f) and (c, h).

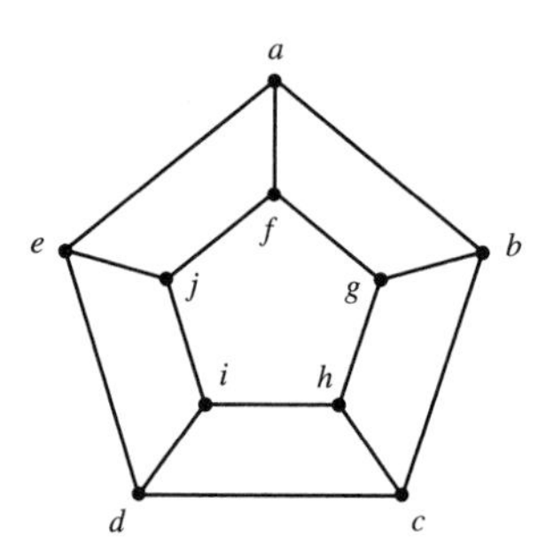

7. Prove that the following graphs have no Hamilton circuits:

(a)

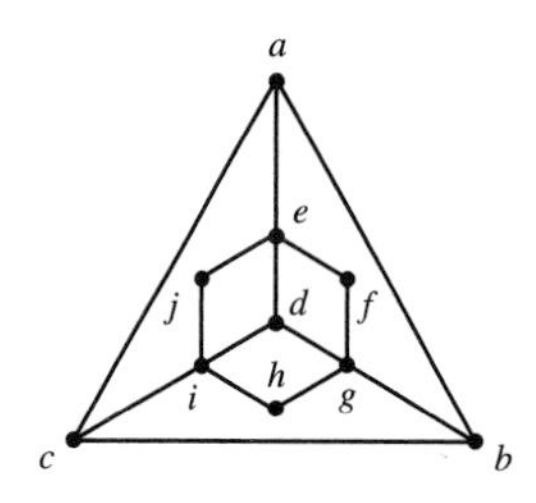

(b)

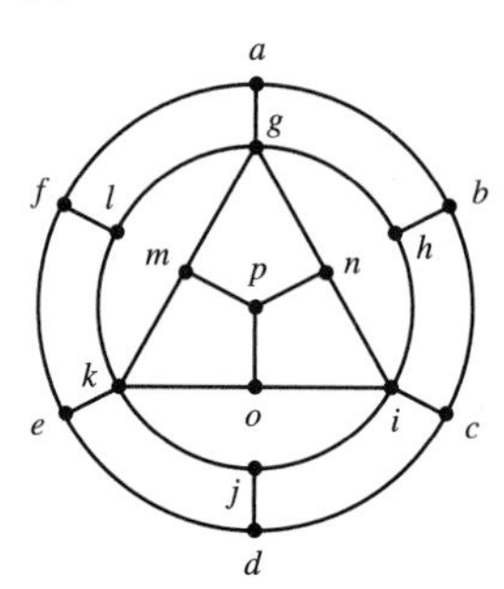

8. Use Theorem 3 to show that the following planar graphs have no Hamilton circuit:

(a) Exercise 4(a) **(b)** Exercise 4(b) **(c)** Exercise 4(o)

9. Recall from Example 1 in Section 1.1 that a graph is *bipartite* if the vertices can be divided into two sets; for convenience call them blue vertices and red vertices, such that every edge connects a blue and a red vertex.

(a) Show that if a connected bipartite graph has a Hamilton circuit, then the numbers of red and blue vertices must be equal. Further, if a bipartite graph has an odd number of vertices, then it has no Hamilton circuit.

(b) Show that if a connected bipartite graph has a Hamilton path, then the numbers of reds and blues can differ by at most one.

(c) Use part (a) to show that the following graphs have no Hamilton circuit:

(i) Figure 2.8 **(ii)** Exercise 4(l) **(iii)** Exercise 7(b)

10. Suppose a set I of k vertices in a graph G is chosen so that no pair of vertices in I are adjacent. Then for each x in I, $\deg(x) - 2$ of the edges incident to x will not be used in a Hamilton circuit. Summing over all vertices in I, we have $\mathbf{e}' = \sum_{x \in I}(\deg(x) - 2) = \{\sum_{x \in I}(\deg(x)\} - 2k$ edges that cannot be used in a Hamilton circuit.

(a) Let $\mathbf{v}$ and $\mathbf{e}$ be the numbers of vertices and edges in G, respectively. Show that if $\mathbf{e} - \mathbf{e}' < \mathbf{v}$, then G can have no Hamilton circuit.

(b) Why is the claim in part (a) valid only when I is a set of nonadjacent vertices?

(c) With a suitably chosen set I, use part (a) to show that the following graphs have no Hamilton circuits:

(i) Figure 2.6 (ii) Exercise 4(o) (iii) Exercise 7(b)

11. (a) Draw a 4-dimensional hypercube graph.

(b) Use the graph in part (a) to find a Gray code for encoding the numbers 1 through 16 as 4-bit binary sequences.

12. Let the distance between two vertices in a connected graph be defined as the number of edges in the shortest path connecting those two vertices. Then the *diameter* of a graph is defined to be greatest distance between any two vertices in the graph. Show that a 4-dimensional hypergraph has diameter 4. In general, show that a k-dimensional hypergraph has diameter k. Note: No graph with 2^k vertices, all of degree k, has a smaller diameter than k. This minimum is achieved by a k-dimensional hypergraph.

13. Find a connected, cubic graph (all vertices have degree 3) with no Hamilton circuit.

14. Show without citing any theorems stated in this section that any 6-vertex, undirected graph with all vertices of degree 3 has a Hamilton circuit.

15. Find a path of knight's moves visiting all squares exactly once on an 8×8 chessboard.

16. Suppose a classroom has 25 students seated in desks in a square 5×5 array. The teacher wants to alter the seating by having every student move to an adjacent seat (just ahead, just behind, on the left, or on the right). Show that such a move is impossible.

17. (a) Describe how to construct a circuit including all squares of an $n \times n$ chessboard, n even, using a rook. Using a king.

(b) Repeat part (a) for n odd.

18. Consider 27 little cubes arranged in a 3 by 3 by 3 array (as in Rubik's Cube). Form an associated graph with 27 vertices, one for each little cube, and with two vertices adjacent if they have touching faces (not just edges). Does this graph have a Hamilton path starting at the vertex corresponding to the middle inside cube and ending at one of the vertices corresponding to a corner cube?

19. (a) How many different Hamilton circuits are there in K_n, a complete graph on n vertices?

(b) Show that K_n, n prime ≥ 3, can have its edges partitioned into $\frac{1}{2}(n-1)$ disjoint Hamilton circuits.

(c) If 17 professors dine together at a circular table during a conference, and if each night each professor sits next to a pair of different professors, how many days can the conference last?

20. (a) If a graph G has an Euler cycle, show that $L(G)$, the line graph of G (see Exercise 16 of Section 2.1 for the definition of a line graph), has a Hamilton circuit.

(b) If G has a Hamilton circuit, show that $L(G)$ has a Hamilton circuit.

(c) Show that the converses of parts (a) and (b) are false by finding counter-examples.

21. Show that if G is not a complete graph, then it is possible to direct the edges of G so that there is no directed Hamilton path.

22. Show that in a tournament (defined preceding Theorem 4) it is always possible to rank the contestants so that the person ranked ith beats the person ranked $(i+1)$st. (*Hint:* Use Theorem 4.)

23. Show that Theorem 1 is false if the requirement of degree $\geq \frac{1}{2}n$ is relaxed to just $\geq \frac{1}{2}(n-1)$.

24. **(a)** Prove for $n \geq 3$ that an undirected graph with n vertices and at least $\binom{n-1}{2}+2$ edges must have a Hamilton circuit.

(b) Show that part (a) is false if there are only $\binom{n-1}{2}+1$ edges.

2.3 GRAPH COLORING

In Example 1 of Section 1.4 we introduced the problem of map coloring—coloring the countries of a map so that two countries with a common border are assigned different colors. The problem of showing that any map can be 4-colored tantalized mathematicians for 100 years until a computer-assisted proof was obtained by Appel and Haken in 1976. More recently, graph coloring has been applied to a variety of problems in computer science, operations research, and design of experiments. Recall that coloring countries in a map is equivalent to coloring vertices, with adjacent vertices getting different colors, in the dual graph obtained by making a vertex for each country and an edge between vertices representing countries with a common border; see Example 1 in Section 1.4. In general, a **coloring** of a graph G assigns colors to the vertices of G so that adjacent vertices are given different colors.

In this section we show how to determine the minimal number of colors required to color a given graph. This minimal number of colors is called the **chromatic number** of a graph. We also give some applications of graph coloring. In the next section we will present some theorems about graph coloring. In a coloring of a graph, the vertices that have a common color will be mutually nonadjacent (no pair is joined by an edge). In Example 5 of Section 1.1 we introduced the term *independent set* to refer to such a set of mutually nonadjacent vertices. We shall revisit that example later in this section.

For graphs with 15 or fewer vertices, it is usually not difficult to guess a graph's chromatic number. To verify rigorously that the chromatic number of a graph is a number k, we must also show that the graph cannot be properly colored with $k-1$ colors. Proving that a graph cannot be $(k-1)$-colored is similar to proving that a graph has no Hamilton circuit or cannot be isomorphic to another particular graph.

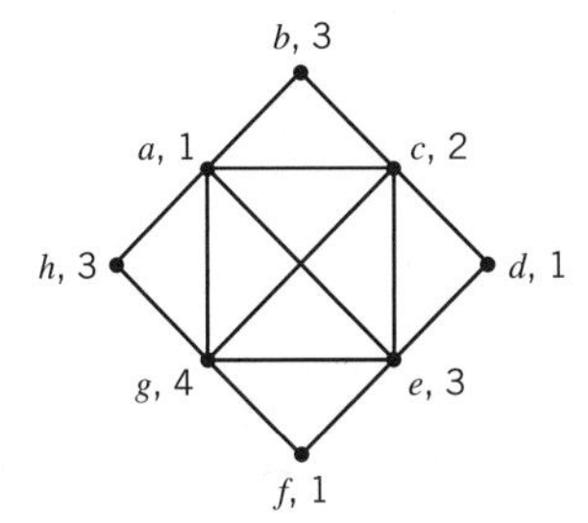

Figure 2.10

In this case, the goal is to show that any $(k - 1)$-coloring we might construct for the graph must force two adjacent vertices to have the same color.

Example 1: Simple Graph Coloring

Find the chromatic number of the graph in Figure 2.10.

Looking at the inner square with crossing diagonals, we see that vertices a, c, e, g are mutually adjacent, that is, they form a complete subgraph. They each require a different color in a proper coloring, four colors in all. Once four colors are available, it is easy to properly color the remaining vertices b, d, f, h. Each of them is adjacent to only two other vertices, and so at most two out of the four colors need ever be avoided with these vertices. Let us use the numbers 1, 2, 3, 4 as the "names" of our colors. Then one possible 4-coloring of the graph is shown in Figure 2.10.

In this problem it is immediately clear that the graph cannot be 3-colored, since some adjacent pair of vertices in the complete subgraph formed by a, c, e, g would have the same color in a 3-coloring. So the chromatic number of this graph is 4. ■

Example 1 points up two important rules. First, a complete subgraph on k vertices requires k colors [cannot be $(k - 1)$-colored]. Second, when building a k-coloring of some graph, we can ignore all vertices of degree $<k$ (and their incident edges), since once the other vertices are colored, there will always be at least one color available (not used by any adjacent vertex) to properly color each such vertex.

Example 2: Coloring a Wheel

Find the chromatic number of the graph in Figure 2.11.

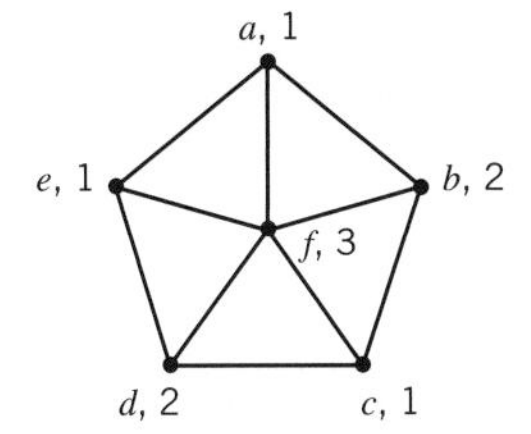

Figure 2.11

We note that a graph of this form is called a **wheel.** The largest complete subgraph in this graph is a triangle. Let us try to build a 3-coloring of this graph with "colors" 1, 2, and 3. We will start by coloring the vertices of a triangle. Suppose we choose the triangle a, b, f: Let a be 1, b be 2, and f be 3 (the order of colors is arbitrary). See Figure 2.11. Since c is adjacent to vertices b and f of colors 2 and 3, respectively, c is forced to be color 1. Similarly, d is forced to be 2, and then e is forced to be 1. However, now the adjacent vertices a and e both have color 1. Thus the graph cannot be 3-colored. On the other hand, using a fourth color for e yields a proper coloring. So the chromatic number of this graph is 4. ∎

Observe that in Example 2 if vertex e were missing and d were adjacent to a instead, then three colors would work. In general, wheel graphs with an even number of "spokes" can be 3-colored, whereas wheels with an odd number of spokes require four colors.

The key to the impossibility of finding a 3-coloring in Example 2 is the sequence c, d, e of forced vertex colors. In general, when attempting to build a k-coloring graph, it is desirable to start by k-coloring a complete subgraph of k vertices and then successively finding an uncolored vertex adjacent to vertices of $k - 1$ different colors, thereby forcing the color choice for this vertex.

The following example involves a graph where vertex colors are not forced.

Example 3: Unforced Coloring

Find the chromatic number of the graph in Figure 2.12.

The largest complete subgraph is again a triangle, and so we want to try building a 3-coloring. The only triangles whose coloring will force the color of another vertex are (d, e, f) and (e, f, g). Suppose we color d 1, e 2, and f 3. Then g is forced to be 1. Now we are in trouble, since no more uncolored vertices are adjacent to two colors.

Observe that b and c are both adjacent to a vertex of color 1 and are adjacent to each other. Thus, one of b and c must be color 2 and the other color 3. By the symmetry of the graph, we can assume b is 2 and c 3, although it actually does not matter. What is important is that one of b, c is color 2 and the other color 3, for this forces the color of a to be 1. Similarly h and i will be 2 and 3, collectively, forcing j to be 1. But the adjacent pair a, j are both 1. Thus the graph cannot be 3-colored. Making a or j a fourth color yields a proper 4-coloring, and so the graph's chromatic number is 4. ∎

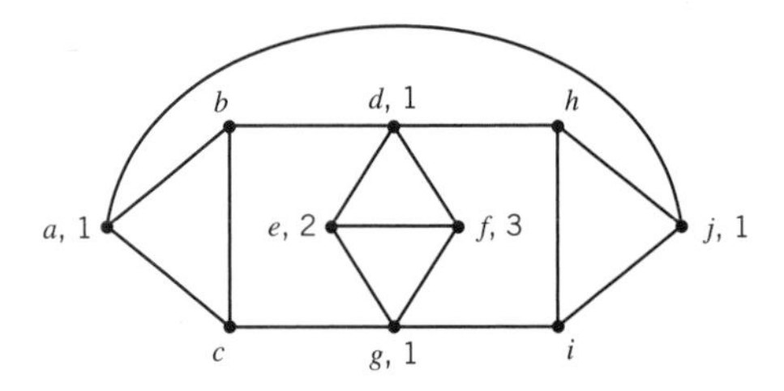

Figure 2.12

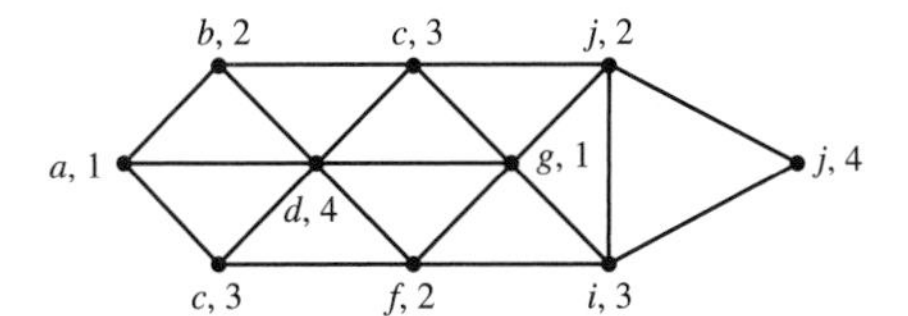

Figure 2.13

Let us now look at some applications of graph coloring. We start by repeating Example 5 of Section 1.1, which involved the problem of scheduling committee meeting times.

Example 4: Committee Scheduling

A state legislature has many committees that meet for one hour each week. One wants a schedule of committee meeting times that minimizes the total number of hours but such that two committees with overlapping membership do not meet at the same time. We show that this is a graph-coloring problem.

The key information about the committees is which committees have overlapping membership. Let us create a graph with a vertex corresponding to each committee and with an edge joining 2 vertices if they represent committees with overlapping membership. For concreteness, suppose that the graph in Figure 2.13 represents the membership overlap of 10 legislative committees. We must schedule the vertices so that adjacent vertices (overlapping committees) get different meeting hours. A coloring of this graph performs exactly this type of "scheduling." The colors will represent different meeting times. The graph in Figure 2.13 has a chromatic number of 4 (a minimal coloring is shown in Figure 2.13). Thus, four hours suffice to schedule committee meetings without conflict. ■

Example 5: Integrated Circuit Design

In a simplified (one-dimensional) form of Very Large Scale Integrated circuit (VLSI) design, one has a row of logical subcircuits called *gates* $G_1, G_2, \ldots, G_n$ with specified connections between certain pairs of these gates. These connections are laid in a set of parallel tracks, as illustrated in Figure 2.14*b*. Two connections must go in different tracks if they would overlap, even if just at a common endpoint.

What is the minimum number of tracks needed for the following six connections among a sequence of six gates: (*1*, *3*), (*1*, *3*)—two copies of this connection—(*2*, *3*), (*2*, *4*), (*2*, *5*), (*4*, *5*)? See Figure 2.1*b*. The set of connections can be treated as a family of intervals on a line. Two intervals (connections) can have the same track if they do not overlap. We model this problem as a graph-coloring problem.

We form a graph indicating which connections overlap. Since the connections are intervals along the line from 1 to 5, this connection graph is just a graph of overlapping intervals. Interval graphs were discussed in Example 6 in Section 1.1. We make a vertex for each interval and join two vertices with an edge if they represent intervals that overlap. See Figure 2.14*a*. Now let colors stand for tracks. A minimal

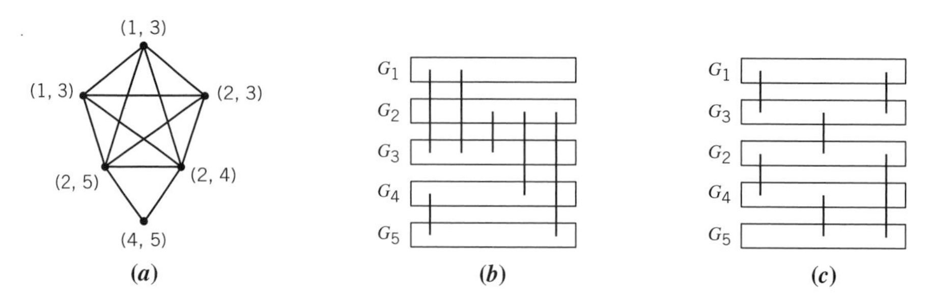

Figure 2.14

coloring of the interval graph will correspond to a minimal track assignment. Since the interval graph contains a complete graph on five vertices—all vertices except (*4, 5*)—five colors (tracks) are needed, as shown in Figure 2.14*b*.

Frequently the order of the gates is under the designer's control. Is it possible to order the gates differently so as to reduce the number of tracks in the previous design problem? Changing the gate order produces a different set of connection intervals and so may change the chromatic number of the associated interval graph. By inspection, we observe that with the gate order *1, 3, 2, 4, 5*, only three tracks are needed. See Figure 2.14*c*. (No efficient procedure is known for finding the best gate order; a huge amount of computer time in VLSI design is taken up searching for the best order.) ■

Example 6: Garbage Truck Scheduling

We now consider a complex optimization problem in which graph coloring plays only a secondary role. There is a set of sites S_i that must be serviced (visited) k_i times each week ($1 \leq k_i \leq 6$). We seek a minimal set of day-long truck tours for a week such that site S_i is visited on k_i of the tours. In addition, we require that these tours can be partitioned among the six days of the week (Sunday is excluded) in a manner so that no site is visited twice on one day. This is an extremely difficult problem, which cannot be solved exactly.

When a problem of this type (involving garbage collection) was analyzed for the New York City Department of Environmental Protection, an algorithm was used that started with an inefficient set of tours and successively tried to improve the set of tours. Suppose we had a simplified situation where the week contained just three workdays, and our algorithm had generated the set of tours shown in Figure 2.15*a*. Can these six tours be so partitioned among the three workdays so that no site is visited twice on the same day? Another way to state the constraint is: If two tours have a site in common, then the tours are assigned to different workdays. This is the type of constraint handled by a coloring model.

Given a set of tours, we form an associated tour graph with one vertex for each tour and two vertices adjacent if they correspond to two tours that visit a common site. We assign colors, representing the different workdays, to the vertices (tours) of the tour graph. In the general six-day problem, partitioning the tours among the six

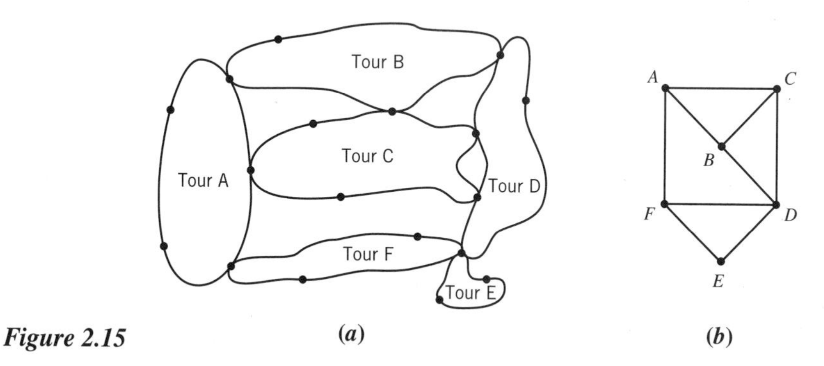

Figure 2.15 (*a*) (*b*)

days is equivalent to 6-coloring the tour graph. In the tour graph in Figure 2.15*b*, we desire just a 3-coloring, which indeed exists. If the optimizing algorithm were next to combine tours *A* and *F* in Figure 2.15*a* to get a smaller set of tours, such a move would have to be blocked (and other optimizing moves tried instead) because the resulting tour graph would have a complete graph on the four vertices *A*, *B*, *C*, *D*; this would require four colors (days). ∎

We close this section with a brief introduction to chromatic polynomials. The chromatic polynomial $P_k(G)$ of a graph G gives a formula for the number of ways to properly color G with k colors. The formula is a polynomial in k. If k is so small that G cannot be colored with only k colors, then $P_k(G)$ will have to equal 0 for that value of k.

Example 7: Chromatic Polynomial

What is the chromatic polynomial of:

(a) A complete graph k_5 on five vertices (all vertices adjacent to each other)?

(b) The graph C_4 of a circuit of length 4?

(a) In a complete graph, each vertex must be a different color. Thus, $P_k(K_5) = k(k-1)(k-2)(k-3)(k-4)$, since there are k possible choices for the first vertex to be colored; then that color cannot be used again, and so the second vertex to be colored has $k-1$ choices, and so on.

(b) Let the vertices on the circuit C_4 be named x_1, x_2, x_3, x_4 with edges $(x_1, x_2), (x_2, x_3), (x_3, x_4), (x_4, x_1)$. We break the computation of $P_k(C_4)$ into two cases, depending on whether or not x_1 and x_3 are given the same color.

If x_1 and x_3 have the same color, there are k choices for the color of these two vertices. Then x_2 and x_4 each must only avoid the common color of x_1 and x_3—$k-1$ color choices each. So the number of k-colorings of C_4 in this case is $k(k-1)^2$.

If x_1 and x_3 have different colors, there are $k(k-1)$ choices for the two different colors for x_1 and then x_3. Now x_2 and x_4 each have $k-2$ color choices. So in this case the total number of k-colorings of C_4 is $k(k-1)(k-2)^2$. Combining the two cases, we obtain $P_k(C_4) = k(k-1)^2 + k(k-1)(k-2)^2$. ■

2.3 EXERCISES

Summary of Exercises Exercises 1–8 involve finding minimal vertex colorings and associated problems. Exercises 9–11 require minimal colorings of maps and a geometric array. Exercises 12–18 are color-modeling problems.

1. Find the chromatic number of each of the following graphs. Give a careful argument to show that fewer colors will not suffice.

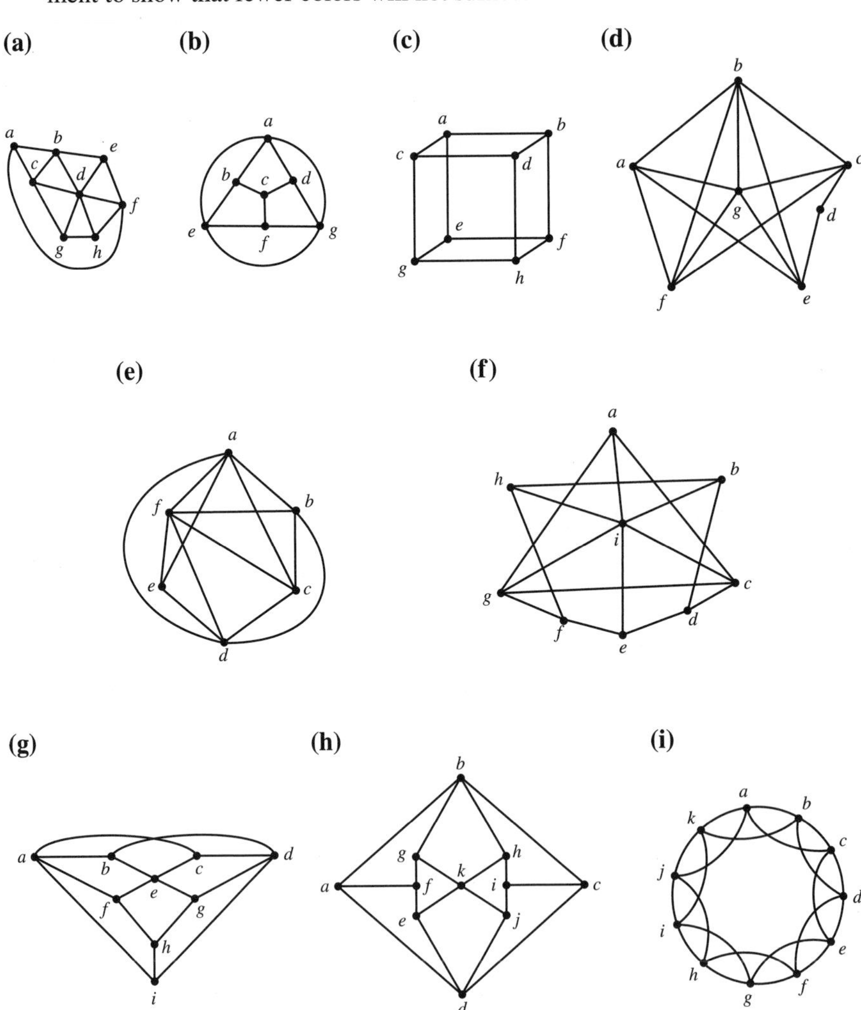

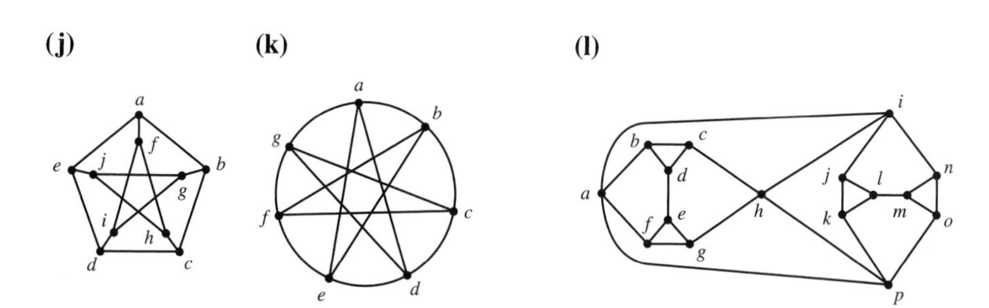

2. Find a minimal edge coloring of the following graphs (color edges so that edges with a common end vertex receive different colors).

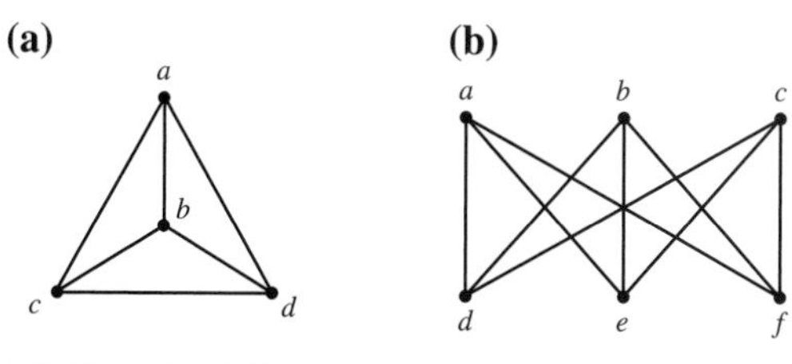

(c) Exercise 1(j)

3. A graph G is *color critical* if the removal of any vertex of G decreases the chromatic number. Which graphs in Exercise 1 are color critical?

4. Find all sets of three or more vertices that could have the same color in a proper coloring of the graph in:

   (a) Exercise 1(a) **(b)** Exercise 1(b)

5. A coloring partitions a graph G into sets of mutually nonadjacent vertices. In the complement $\overline{G}$ of G, this partition becomes a partition of G into sets of mutually adjacent vertices, that is, complete subgraphs. Find such a minimal set of complete subgraphs partitioning the vertices in the graph in:

   (a) Exercise 1(a) **(b)** Exercise 1(b) **(c)** Exercise 1(d)

6. An *equitable coloring* is a minimal coloring in which the numbers of vertices of each color differ by at most one. Which of the following graphs have minimal colorings that are equitable?

   (a) Exercise 1(a) **(b)** Exercise 1(c)

7. **(a)** Determine the chromatic polynomial $P_k(G)$ for the following graphs:

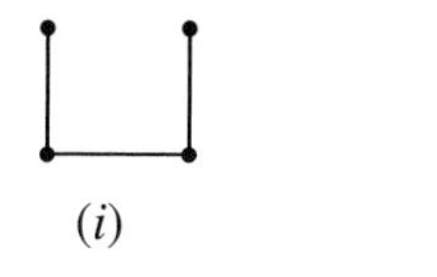

(*i*)

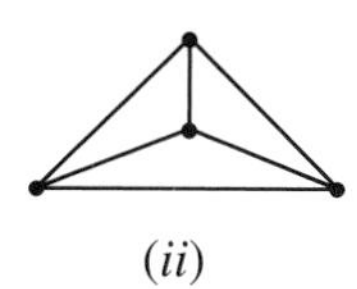

(*ii*)

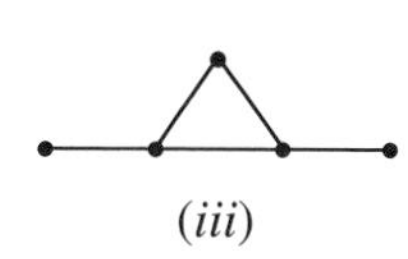

(*iii*)

(b) If we know $P_k(G)$, how can we use it to determine the chromatic number of G?

8. An edge coloring assigns colors to edges so that edges with a common end vertex receive different colors. Describe graphs that can be edge colored using just two colors.

9. Can the 50 states in a map of the United States be properly 3-colored? (Note that states meeting only at a corner, such as Colorado and Arizona, are not considered adjacent.)

10. Suppose a map is made by drawing n intersecting circles. Show that the regions in this map can be properly 2-colored.

(a) Solve by an inductive argument.

(b) Solve by assigning colors based on the number of circles that contain a region.

11. How many colors are needed to color the 15 billiard balls in this triangular array with touching balls different colors?

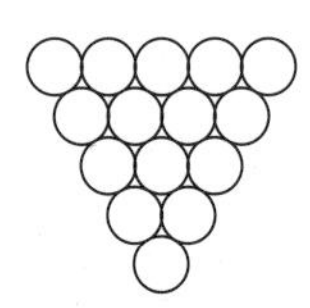

12. A zoo is going to place its animals in a set of large open areas, instead of having them in individual cages. If two different animals cannot live together peacefully (e.g., a tiger and deer cannot live together because the tiger will eat the deer), then they must be put in different open areas. The zoo wants to determine the minimum number of open areas needed to safely house all its animals. Model this problem of assigning animals to a minimal number of open areas as a graph-coloring problem. What are the vertices, what are the edges, what are the colors?

13. The Applied Math Department is scheduling the times for classes for next semester. Each student has already decided which subset of ApMath classes he/she wants to take. Classes must be scheduled so that every student can have the courses requested without conflict. Model this scheduling problem as a graph-coloring problem. What are the vertices, what are the edges, what are the colors?

14. **(a)** A set of solar experiments is to be made at observatories. Each experiment begins on a given day of the year and ends on a given day (each experiment is repeated for several years). An observatory can perform only one experiment at a time. The problem is, What is the minimum number of observatories required to perform a given set of experiments annually? Model this scheduling problem as a graph-coloring problem.

(b) Suppose experiment A runs from Sept. 2 to Jan. 3, experiment B from Oct. 15 to March 10, experiment C from Nov. 20 to Feb. 17, experiment D from Jan. 23 to May 30, experiment E from April 4 to July 28, experiment F from April 30 to July 28, and experiment G from June 24 to Sept. 30. Draw the associated graph and find a minimal coloring (show that fewer colors will not suffice).

15. A banquet center has 8 different special rooms. Each banquet requires some subset of these 8 rooms. Suppose that there are 12 evening banquets that we wish to schedule in a given week (7 days). Two banquets that are scheduled on the same evening must use different special rooms. Model and restate this scheduling problem as a graph-coloring problem.

16. Which of the following pairs of tours in Figure 2.15*a* can be combined without violating the 3-colorability requirement of the tour graph in Example 6?

(a) Tours D and E **(b)** Tours C and D

17. In a round-robin tournament where each pair of n contestants plays each other, a major problem is scheduling the play over a minimal number of days (each contestant plays at most one match a day).

(a) Restate this problem as an edge-coloring problem (see Exercise 8).

(b) Solve this problem for $n = 4$ and $n = 5$.

18. Consider a graph representing games played between a set of football teams with a directed edge from vertex A to vertex B if team A beats team B. Suppose that it is known that this graph has *no directed circuits* (i.e., no situation such as A beats B, B beats C, and C beats A). We can define a set of levels in this football graph as follows: A vertex with no outward edges (the team beat no other team) is at level 0; if all a vertex's outward edges go to level-0 vertices (no-win teams), then the vertex is at level 1; and in general, a vertex is at level k if the greatest level of a team it beat is level $k - 1$. Show that the level number of each vertex is a proper "coloring" of the vertices.

2.4 COLORING THEOREMS

In the previous section we studied strategies for finding a minimal coloring of a graph and gave some applications that could be modeled as graph-coloring problems. In this section we present some theorems about graph coloring.

We begin with a theorem about coloring the corners of a polygon and use it to obtain a simple solution to an interesting problem in computational geometry. We treat a polygon as a plane graph consisting of a single circuit with edges drawn as straight lines. The polygon need not be a convex figure. See the sample polygon in Figure 2.16*a*. By a *triangulation of a polygon,* we mean the process of adding a set of straight-line chords between pairs of vertices of a polygon so that all interior regions of the graph are bounded by a triangle (these chords cannot cross each other nor can they cross the sides of the polygon). Figure 2.16*b* shows one possible triangulation of the polygon in Figure 2.16*a*.

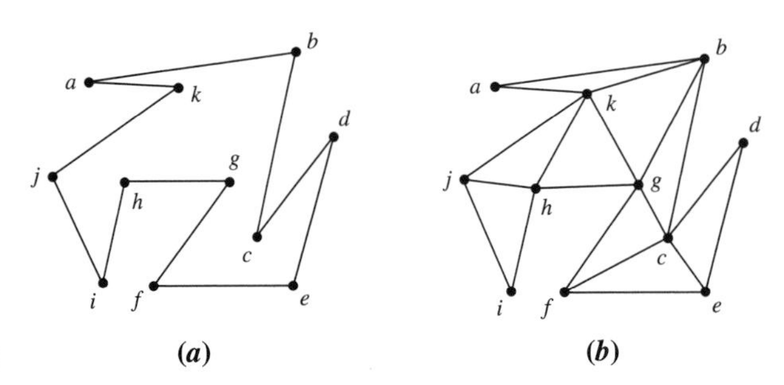

Figure 2.16

Theorem 1

The vertices in a triangulation of a polygon can be 3-colored.

Proof

Our proof is by induction on n, the number of edges of the polygon. For $n = 3$, give each corner a different color. Assume that any triangulated polygon with less than n boundary edges, $n \geq 4$, can be 3-colored and consider a triangulated polygon T with n boundary edges.

Pick a chord edge e, as illustrated by chord (g, k) in Figure 2.17a. We note that T must have at least one chord edge, or else since $n \geq 4$, T would not be triangulated. This chord e splits T into two smaller triangulated polygons, as shown in Figure 2.17b, each of which can be 3-colored by the induction assumption. The 3-colorings of the two subgraphs can be combined to yield a 3-coloring of the original triangulated polygon by picking the names for the colors in the two subgraphs so that the end vertices of chord e have the same colors in each subgraph. In Figure 2.17b, this would mean making the color of k be the same in the two subgraphs and making the color of g be the same. ◆

In practice, it is easy to produce a 3-coloring of a triangulation of a polygon. Moreover, this 3-coloring is unique. For details, see Exercise 5. We now present an interesting application of Theorem 1 to a problem that seems to have nothing to do with coloring.

The Art Gallery Problem asks what is the smallest number of guards needed to watch paintings along the n walls of an art gallery. The walls are assumed to

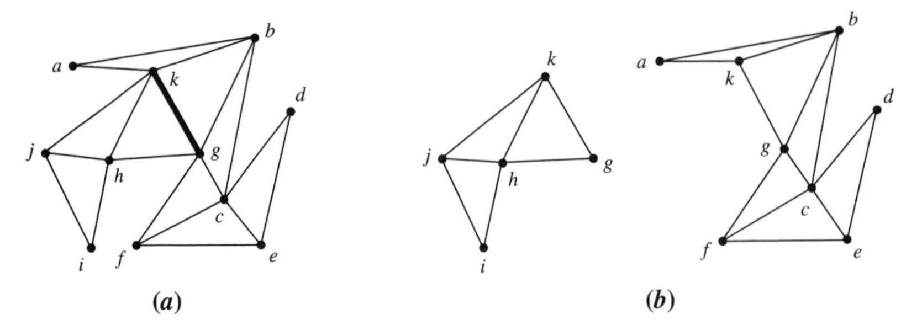

Figure 2.17

form a polygon. The guards need to have a direct line of sight to every point on the walls. A guard at a corner is assumed to be able to see the two walls that end at that corner. The expression $[r]$ denotes the largest integer $\leq r$. For further details, see O'Rourke [5].

Corollary **(Fisk, 1978)**

The Art Gallery Problem with n walls requires at most $\lfloor n/3 \rfloor$ guards.

Proof

Make a triangulation of the polygon formed by the walls of the art gallery. Observe that a guard at any corner of a triangle has all sides of the triangle under surveillance. Now obtain a 3-coloring of this triangulation. Note that each triangle will have one corner of each color. Take one of the colors, say "red," and place a guard at every corner colored "red." This places a guard at a corner on every triangle. Hence the sides of all triangles, and, in particular all the gallery walls, will be watched. A polygon with n walls has n corners. If there are n corners and 3 colors, some color is used at $\lfloor n/3 \rfloor$ or fewer corners. ◆

This bound is the best possible. For example, Figure 2.18 gives an example of a polygon with 12 corners that requires 4 guards.

We now present three representative coloring theorems. We will prove only the last one, the 5-color theorem for planar graphs. For proofs of Theorems 2 and 3 see [4]. Note that Theorem 2 in Section 1.3 was a coloring theorem. As stated there, it said that a connected graph is bipartite if and only if all circuits have even length. Observe that being bipartite is the same as being 2-colorable. Thus, a graph is 2-colorable if and only if all circuits have even length. (We can drop the condition that the graph be connected, because if each component is 2-colorable, then the whole graph is 2-colorable.) Let $\chi(G)$ denote the chromatic number of the graph G.

Theorem 2 **(Brooks, 1941)**

If the graph G is not an odd circuit or a complete graph, then $\chi(G) \leq d$, where d is the maximum degree of a vertex of G.

The maximum degree is for most graphs a poor upper bound on $\chi(G)$. The examples in Section 2.3 had $\chi(G)$ closely related to the size of the largest complete subgraph. Thus, it seems natural that there should be a good bound on $\chi(G)$ in terms of the size of the largest complete subgraph. However, for any positive integer k, there exists a triangle-free graph G with $\chi(G) = k$; see Exercise 16 for details.

Instead of coloring vertices, we can color edges so that edges with a common end vertex get different colors. A very good bound on the edge chromatic number of

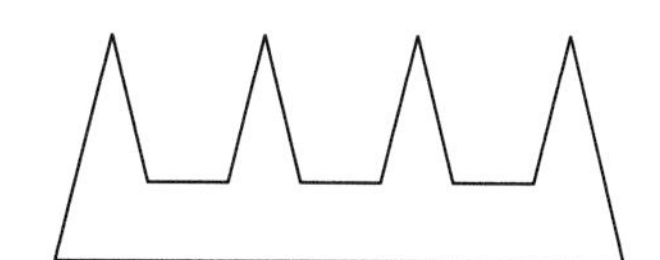

Figure 2.18

a graph in terms of degree is possible. All edges incident at a given vertex must have different colors, and so the maximum degree of a vertex in a graph is a lower bound on the edge chromatic number. Even better, one can prove:

Theorem 3 **(Vizing, 1964)**

If the maximum degree of a vertex in a graph G is d, then the edge chromatic number of G is either d or $d + 1$.

Finally we would like to present and prove a theorem about coloring planar graphs. As noted earlier, in 1976 Appel and Haken proved that all planar graphs can be 4-colored. But their proof is incredibly long and requires thousands of hours of computer analysis involving 1955 cases, each of which in turn involves many pages of analysis. We will state and prove an easier "second best" theorem.

Theorem 4

Every planar graph can be 5-colored.

Proof

We need to consider only connected planar graphs, since we can 5-color unconnected planar graphs by 5-coloring each connected component. A key step in this proof uses a fact about planar graphs proved in Exercise 18 in Section 1.4: Any connected planar graph has a vertex of degree at most 5. We prove this theorem by induction on the number of vertices. Trivially a 1-vertex graph can be 5-colored.

Next we assume that all connected planar graphs with $n - 1$ vertices ($n \geq 2$) can be 5-colored. We will prove that a connected planar graph G with n vertices can be 5-colored. As noted above, G has a vertex x of degree at most 5. Delete x from G to obtain a graph with $n - 1$ vertices, which by assumption can be 5-colored. Now we reconnect x to the rest of the graph and try to properly color x. If x has degree ≤ 4, then we can simply assign x a color different from the colors of its neighbors. The same approach works if the degree of x is 5 but two neighbors have the same color. Thus, it remains to consider the case where x has five adjacent vertices each with a different color. See Figure 2.19, where we label the neighbors of x as a, b, c, d, e according to their clockwise order about x in some planar depiction of G. Let the colors be the numbers shown in Figure 2.19.

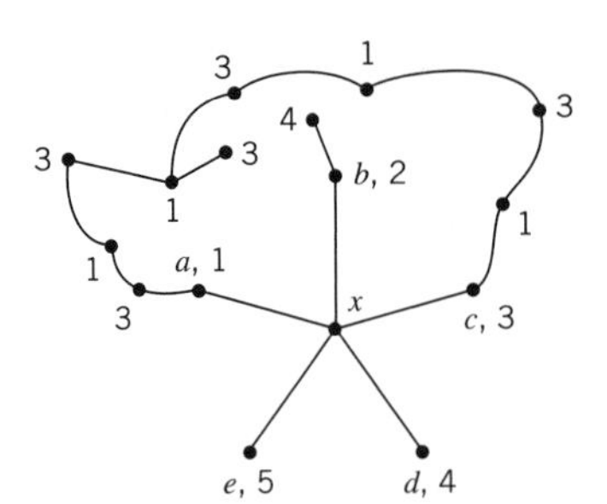

Figure 2.19

First consider all paths starting from a whose vertices are colored 1 and 3. See Figure 2.19. Suppose there is no path consisting of 1 and 3 vertices from a to c. Then we can change the color of a from 1 to 3, change the neighbors of a colored 3 to 1, and so on along all paths of 1 and 3 vertices emanating from a. This 1–3 interchange will not affect c since there is no path of 1s and 3s from a to c. After this 1–3 interchange from a, a and c are both color 3, and x can be properly colored with 1.

On the other hand, if there is a 1–3 path from a to c, then we consider all paths starting from b, whose vertices are colored 2 and 4. The 1–3 path from a to c, together with edges (x, a), (x, c), form a circuit that blocks the possibility of any 2–4 path going from b to d. Thus, we can perform a 2–4 interchange along all paths of 2s and 4s emanating from b without changing the color of d. After this 2–4 interchange, b and d are both color 4, and x can be properly colored with 2.

This completes the induction step of the proof that any n-vertex connected planar graph can be 5-colored. ◆

2.4 EXERCISES

Summary of Exercises These exercises are all proofs of results in coloring theory.

1. Use the fact that every planar graph has a vertex of degree ≤ 5 to give a simple induction proof that every planar graph can be 6-colored. Follow the argument in the beginning of the proof of Theorem 5.

2. Show that a planar graph G with 8 vertices and 13 edges cannot be 2-colored. (*Hint*: Use results in Section 1.4 to show that G *must* contain a triangle.)

3. For any two nonadjacent vertices x, y in a graph G, define graphs G^+_{xy} and G^c_{xy} as follows. G^+_{xy} is obtained by adding the edge (x, y) to G, and G^c_{xy} is obtained from G by coalescing vertices x and y into a single vertex. Show that $\chi(G) = \min\{\chi(G^+_{xy}), \chi(G^c_{xy})\}$.

4. Prove by induction that the graph of any triangulation of a polygon will have at least two vertices of degree 2. (*Hint:* Split the triangulation graph into two triangulation graphs at some chord e.)

5. **(a)** Using the type of reasoning in Section 2.3, explain how to 3-color any triangulation of a polygon.

   (b) Use the argument in part (a) to show that the 3-coloring of any triangulation of a polygon is unique, except for changing the names of the colors.

6. **(a)** If q is the size of the largest independent set in a graph G, show that $\chi(G)q \geq n$, where n is the number of vertices in G.

   (b) If the minimum degree of a vertex is d in an n-vertex graph G, then use the result in part (a) to show that $\chi(G)(n - d) \geq n$, and hence $\chi(G) \geq n/(n - d)$.

7. A graph is *color critical* if the removal of any vertex decreases the graph's chromatic number. Show that every k-chromatic color critical graph G has the

following properties:

(a) G is connected.

(b) Every vertex of G has degree $\geq k-1$.

(c) G has no vertex whose removal disconnects G.

8. Show that G can be edge k-colored if and only if $L(G)$, the line graph of G (see Exercise 16 in Section 2.1), can be vertex k-colored.

9. If $\overline{G}$ is the complement of G, then show that:

(a) $\chi(G)+\chi(\overline{G}) \leq n+1$ (*Hint:* Use induction.)

(b) $\chi(G)\chi(\overline{G}) \geq n$

(c) $\chi(G)+\chi(\overline{G}) \geq 2\sqrt{n}$

10. Show that if every region in a planar graph has an even number of bounding edges, then the vertices can be 2-colored.

11. Show that no planar graph G has a chromatic polynomial of the form $P_k(G) = (k^2-6k+8)Q(k)$, where $Q(k)$ is positive for $k>0$.

12. Show that a graph with at most two odd-length circuits can be 3-colored.

13. Show that a graph is k-colorable if and only if its edges can be directed so that it has no directed circuits and its longest path has length $k-1$.

14. Use the fact that every planar graph with fewer than 12 vertices has a vertex of degree ≤ 4 (Exercise 19 in Section 1.4) to prove that every planar graph with less than 12 vertices can be 4-colored.

15. Show that if G is an interval graph (see Example 6 in Section 1.1), then $\chi(G)$ equals the size of the largest complete subgraph in G.

16. Prove that for any positive integer k, there exists a triangle-free graph G with $\chi(G)=k$.

(a) The proof should be by induction on k, the chromatic number. Initially for $k=3$, we use the graph G_3 consisting of a 5-circuit.

(b) Assuming one can construct G_k, a triangle-free graph with $\chi(G_k)=k$, one constructs G_{k+1} by making k copies of G_k and then adding $(n_k)^k$ vertices, where n_k is the number of vertices in G_k. Each new vertex has as its set of neighbors a different k-tuple consisting of one vertex from each copy of G_k. Confirm that this new graph is the desired G_{k+1}.

2.5 SUMMARY AND REFERENCES

This chapter presented two important graph-theoretic concepts: covering cycles or circuits and coloring. Section 2.1 discussed Euler cycles—cycles that traverse every edge exactly once. Section 2.2 discussed Hamilton circuits—circuits that visit every vertex exactly once. Both types of covering edge sets arise naturally in operations research routing problems. Despite the similarity in the definitions of Euler cycles and Hamilton circuits, determining the existence of Euler cycles and Hamilton circuits in a graph are as different as graph-theoretic problems can be. Euler's theorem allows one quickly to decide whether an Euler cycle exists. On the other hand, except in special cases, the existence or nonexistence of a Hamilton circuit can be determined only by a laborious systematic search to try all possible ways of constructing a Hamilton circuit.

Section 2.3 introduced graph coloring with ad hoc coloring schemes and some applications of coloring. Section 2.4 gave a sampling of coloring theory, highlighted by a proof of the fact that any planar graph can be 5-colored. The stronger theorem proved in 1976 by Appel and Haken [1], that planar graphs are 4-colorable, was the motivation of much of the research in graph theory over the last 100 years. The search for a proof of the Four Color Theorem led to reformulations of this theorem in terms of Hamilton circuits and other graph concepts whose properties were then examined.

Euler's 1736 analysis of Euler circuits was the first paper on graph theory. Euler's paper (translated in Biggs, Lloyd, and Wilson [3]) makes very interesting reading. It is instructive to see how awkward Euler's writing was when he lacked the modern terminology of graph theory. The first use of the concept of a Hamilton circuit occurred in a 1771 paper by A. Vandermonde that presented a sequence of moves by which a knight could tour all positions of a chessboard (without repeating a position). The name "Hamilton" refers to W. Hamilton, whose algebraic research led him to consider special types of circuits and paths on the edges of a dodecahedron [the graph in Exercise 2(c) in Section 2.2]. Hamilton even had a game marketed that involved finding a Hamilton circuit on a dodecahedron (Hamilton's instructions for this game are reprinted in [3]). See Barnette [2] for a good history of the Four Color Problem, its restatements, and final solution by Appel and Haken.

1. K. Appel and W. Haken, "Every planar map is 4-colorable," *Bull. Am. Math. Soc.* 82 (1976), 711–712.
2. D. Barnette, *Map Coloring and The Four Color Problem,* Mathematical Association of America, Washington, DC, 1984.
3. N. Biggs, E. Lloyd, and R. Wilson, *Graph Theory 1736–1936,* Cambridge University, Cambridge, 1999.
4. J. Bondy and U. Murty, *Graph Theory with Applications,* American Elsevier, New York, 1976.
5. J. O'Rourke, *Art Gallery Theorems and Algorithms,* Oxford University Press, New York, 1987.

SUPPLEMENT: GRAPH MODEL FOR INSTANT INSANITY

This supplement presents a clever solution to the Instant Insanity puzzle devised by Blanche Decartes (an alleged pseudonym for the famous graph theorist W. Tutte). The solution uses a

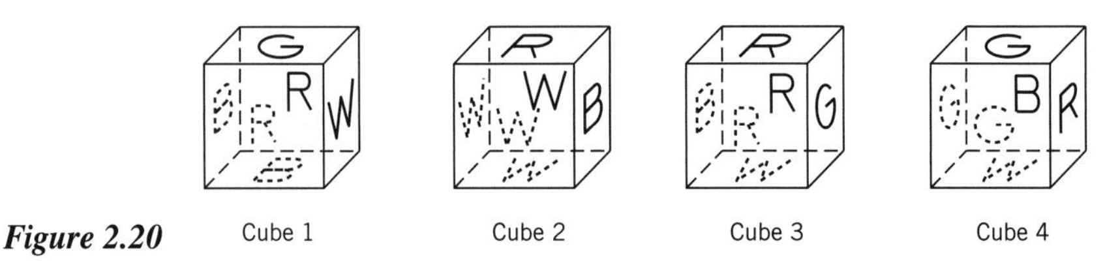

Figure 2.20

generalized form of Hamilton circuit in which all vertices are covered by a collection of vertex-disjoint circuits rather than a single circuit.

The Instant Insanity puzzle consists of four cubes whose faces are colored with one of the four colors: red (*R*), white (*W*), blue (*B*), and green (*G*). The six faces on the *i*th cube are denoted: f_i—front face, l_i—left face, b_i—back face, r_i—right face, t_i—top face, and u_i—under face. In Figure 2.20, the first cube is colored: $l_1 = B, r_1 = W, f_1 = R, b_1 = R, t_1 = G, u_1 = B$. The objective of this puzzle is to place the four cubes in a pile (cube 1 on top of cube 2 on top of... etc.) so that each side of the pile has one face of each color. For example, Figure 2.21 shows one solution for the pile of cubes given in Figure 2.20.

We shall work with the four cubes shown in Figure 2.20. Determining how to arrange these four cubes in a pile that is an Instant Insanity solution is a very difficult task. Observe that there are 24 symmetries of a cube, and thus $24^4 = 331{,}776$ different piles that can be built. An enumeration tree search will be immense, although symmetries of the face colors and the constraint of no repeated colors on a side will eliminate many possibilities. A computer program to do this search is easily written but may require a lot of computer time. Fortunately, we can model this puzzle with a 4-vertex graph in such a fashion that the graph-theoretic restatement of the puzzle can be solved by inspection in a few minutes.

Before presenting the graph model, we need to discuss a simple decomposition principle for this puzzle. Arranging the cubes in a pile so that left and right sides of the pile have one face of each color is "independent" of arranging the front and back sides with one face of each color. By "independent" we mean that once cube

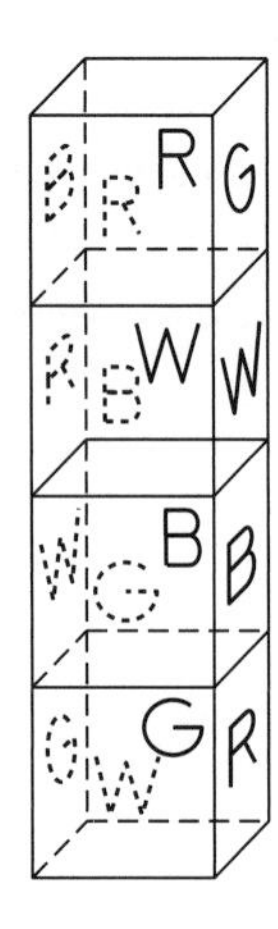

Figure 2.21

i is arranged so that a given pair of opposite faces are on the left and right side of the pile, then any remaining pair of opposite faces on cube i can be on the front and back sides of the pile.

Let $l_i^*, r_i^*, f_i^*, b_i^*$ denote the colors of the four respective faces of cube i visible on the four sides of the pile when cube i is reoriented to obtain an Instant Insanity solution. Suppose that $l_i^* = t_1 = G$ and $r_1^* = u_1 = B$, that is, the top face t_1 of cube 1 in Figure 2.20, which is green, is reoriented to be the left side l_1^*, and the under face u_1, which is blue, is reoriented to be the right side r_1^*. Then by rotating cube 1 about the centers of these two faces, we can get $f_1^* = B, b_1^* = W$, or $f_1^* = R, b_1^* = R$, or $f_1^* = W, b_1^* = B$—all possible remaining choices for front and back sides. Since this same left–right and front–back "independence" holds for the other cubes, we see that the puzzle can be broken into two disjoint problems.

Decomposition Principle

1. Pick one pair of opposite faces on each cube for the left and right sides of the pile so that these two sides of the pile will have one face of each color; and
2. Pick a different pair of opposite faces on each cube for the front and back sides of the pile so that these two sides will have one face of each.

Now we are ready to present our graph model (due to F. de Carteblanche). Actually we use a multigraph (in which multiple edges and loops are allowed). Make one vertex for each of the four colors. For each pair of opposite faces on cube i, create an edge with label i joining the two vertices representing the colors of these two opposite faces. For opposite faces $l_1 = B, r_1 = W$ on cube 1, we draw an edge labeled 1 between vertex B and vertex W; for $f_1 = R, b_1 = R$, we draw a loop labeled 1 at vertex R; and for $t_1 = G, u_1 = B$, we draw an edge labeled 1 between vertices G and B. The edges for the other cubes are drawn similarly. Figure 2.22 shows this multigraph for the cubes in Figure 2.20.

We can now restate the puzzle in graph-theoretic terms. By the decomposition principle, we can break the puzzle into a left–right part and a front–back part. We initially consider just the left–right part, that is, find one pair of opposite faces on each cube so that the left and right sides of the pile have one face of each color.

Let us simplify this left-right problem slightly by asking only for a set of four opposite-face pairs, one pair from each cube, such that among this total set of eight

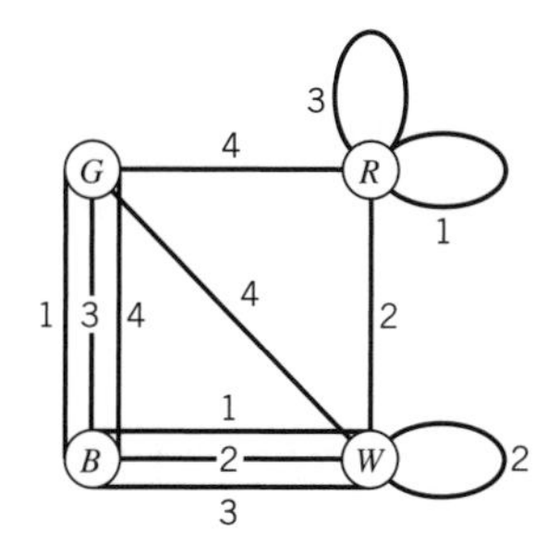

Figure 2.22

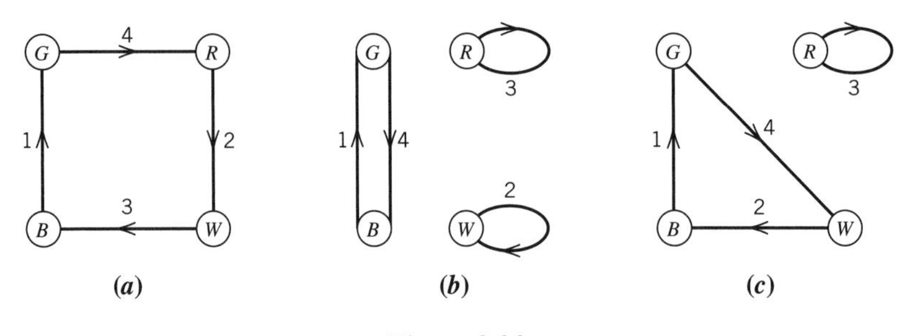

Figure 2.23

faces each color appears twice. Later we will show how to ensure that each color appears once on the left side and once on the right side. Since a color corresponds to a vertex, a cube to an edge number, and a pair of opposite faces to an edge, this simplified left–right problem has the following graph-theoretic restatement: *find four edges, one with each number, such that the family of eight end vertices of these four edges contains each vertex twice.*

This condition on the end vertices of the four edges is equivalent to requiring that the subgraph formed by these four edges has each vertex of degree 2 (a self-loop counts as degree 2 at its vertex). Note that a subgraph with all vertices of degree 2 is just a circuit or collection of disjoint circuits (a self-loop is a circuit of length 1). In an n-vertex multigraph, a set of n edges forming disjoint simple circuits is called a **factor.** Observe that a factor is a natural generalization of a Hamilton circuit. In our Instant Insanity model, let the term **labeled factor** denote a factor in which each edge number appears once. In Figure 2.23 we show three possible labeled factors for the multigraph in Figure 2.22.

The simplified left–right problem in our Instant Insanity graph model reduces to: Find a labeled factor.

We next show how a labeled factor can be transformed into an arrangement of the cubes in which the left and right sides of the pile have one face of each color. We direct the edges in each circuit in a consistent direction, say, in the clockwise direction (see Figure 2.23). Consider the labeled factor in Figure 2.23*a*. As we go around this circuit, we arrange each cube with the color of the trail vertex of an edge as the color on the left side of the cube, and the color of the head vertex on the right side. One can start with any edge on the circuit; we pick the edge labeled 1. This 1-edge in Figure 2.23*a* goes from B to G, and so we arrange cube 1 so that l_1^* (the left face of cube 1 in the pile) $= B$ and $r_1^* = G$ (see Figure 2.24). Following the 1-edge on the circuit (in the clockwise orientation of the circuit), we next encounter a 4-edge from G to R. Accordingly we arrange cube 4 so that $l_4^* = G$ and $r_4^* = R$. Next comes a 2-edge from R to W followed by a 3-edge from W to B. So we arrange cubes 2 and 3 with $l_2^* = R, r_2^* = W$ and $l_3^* = W, r_3^* = B$. Figure 2.24 shows the left and right sides (only) of the pile made by the four cubes arranged as just described.

This process assures that each color appears once on each side, since each vertex (color) is at the head of one edge and at the tail of one edge. If we had chosen the

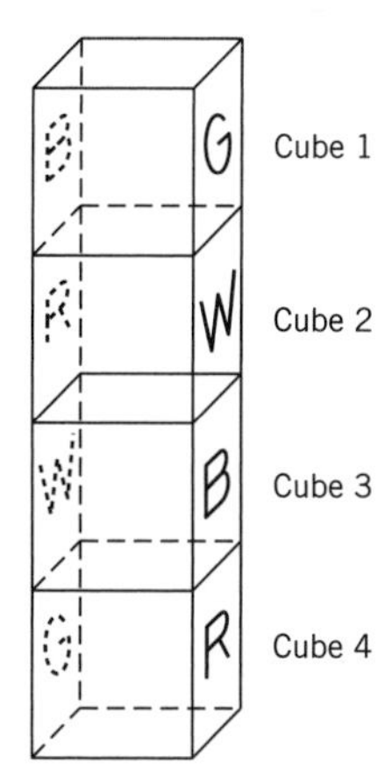

Figure 2.24

labeled factor in Figure 2.23*b*, we would have used the clockwise traversal procedure for all three circuits, yielding $l_1^* = B, r_1^* = G, l_4^* = G, r_4^* = B$ for one circuit, and $l_3^* = r_3^* = R$ and $l_2^* = r_2^* = W$ for the two self-loops.

We are now ready to solve the Instant Insanity puzzle. Restating the decomposition principle in terms of labeled factors:

Graph-theoretic Formulation of Instant Insanity

1. Find two edge-disjoint labeled factors in the graph of the Instant Insanity puzzle, one for left–right sides and one for front–back sides; and then
2. Use the clockwise traversal procedure to determine the left–right and front–back arrangements of each cube.

It is not hard to find two disjoint labeled factors by inspection. The three labeled factors in Figure 2.23 all use the same 1-edge (between B and G). So no two of these three factors are disjoint. The easiest approach is to find one labeled factor, delete its edges, and look for a second labeled factor. If none is found, start with a different labeled factor. For example, suppose we use the labeled factor in Figure 2.23*a* as our first factor. After deleting its edges from the graph in Figure 2.22, we easily find a second labeled factor. Such a second labeled factor is shown in Figure 2.25 (the reader should be able to find a second factor).

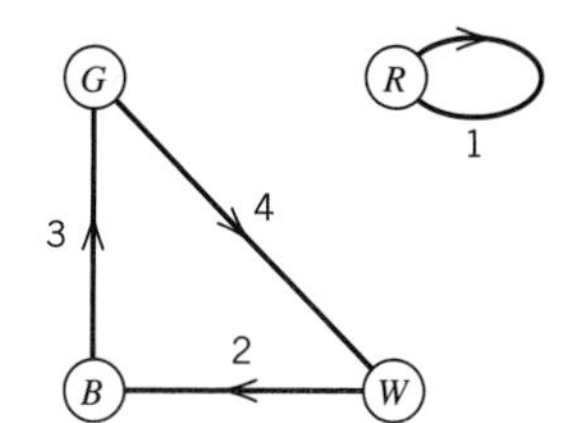

Figure 2.25

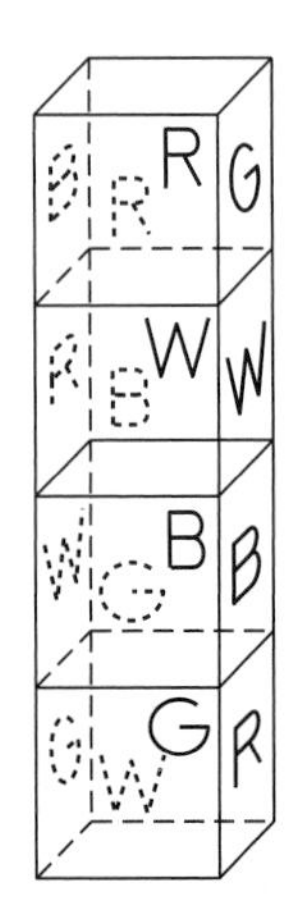

Figure 2.26

We use the factor from Figure 2.23*a* to arrange the left and right sides, as shown in Figure 2.24. Next we rotate each cube about the centers of its left and right faces to arrange the front and back faces according to a clockwise traversal of the circuits in Figure 2.25. Starting with the 2-edge, we make $f_2^* = W$, $b_2^* = B$, $f_3^* = B$, $b_3^* = G$, $f_4^* = G$, $b_4^* = W$ and $f_1^* = b_1^* = R$. Figure 2.26 shows the resulting solution of the Instant Insanity puzzle. Now go buy or borrow a set of Instant Insanity cubes and show your friends that you learned something really useful from this book!

SUPPLEMENT EXERCISES

1. Find all other labeled factors of the multigraph in Figure 2.22 besides the ones in Figure 2.23 and Figure 2.25. Give an argument in the process to show that there can be no other labeled factors.
2. Find all Instant Insanity solutions (a disjoint pair of labeled factors) to the game with associated graph (a).

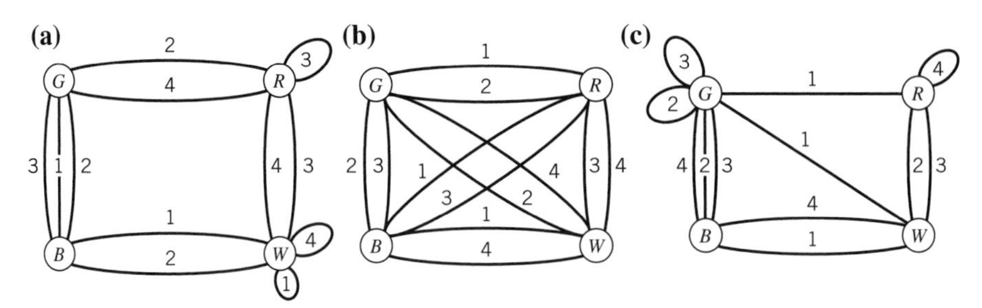

3. Find all Instant Insanity solutions to the game with the associated graph (b) shown above.
4. Show carefully that Instant Insanity graph (c) shown above does not possess a solution, that is, a pair of disjoint labeled factors.

5. **(a)** If a graph G has a Hamiltonian circuit, must it also have a factor? Prove true or give a counterexample.

(b) Repeat part (a) for the case of G having an Euler cycle.

6. Find a factor in the following graphs, if possible:

(a)

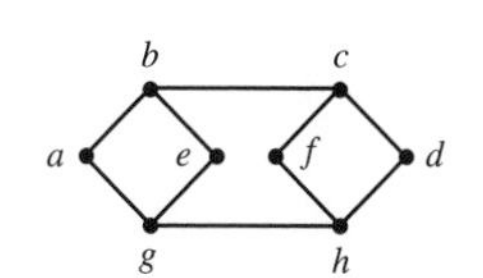

(b)

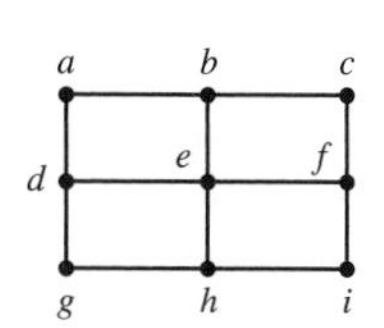

(c)

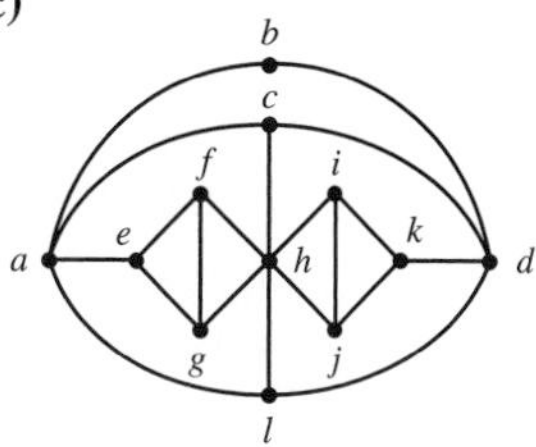

7. Show that the definition of a factor in a graph implies that each vertex is incident to exactly two edges (or one self-loop) in a factor.

CHAPTER 3
TREES AND SEARCHING

3.1 PROPERTIES OF TREES

The most widely used special type of graph is a **tree.** There are two ways to define trees. In undirected graphs, a tree is a connected graph with no circuits. Alternatively, one can define a tree as a graph with a designated vertex called a **root** such that there is a unique path from the root to any other vertex in the tree. This second definition applies to directed as well as undirected graphs.

Intuitively, a tree looks like a tree. See the examples of trees in Figure 3.1. The vertex labeled a is a root for each of the trees in Figure 3.1. The equivalence of the above two definitions is proved in Exercise 5. To illustrate the equivalence, observe that in the tree in Figure 3.1a, the addition of an edge $(\vec{g, h})$ would create a circuit (when edge directions are ignored) and would simultaneously create a second path from root a to h via g.

Trees are a remarkably powerful tool for organizing information and search procedures. In this chapter, we survey some of the diverse settings in which trees can be used to represent and analyze search procedures. These settings include solving puzzles (Section 3.2), solving the "traveling salesperson" problem (Section 3.3), and sorting lists (Section 3.4). In this first section, we present some basic properties of trees and introduce some convenient terminology for working with trees. We prove several useful counting formulas about trees and illustrate some uses of these formulas.

Observe that if a tree is an undirected graph (with no directed edges), then any vertex can be the root. For example, the tree in Figure 3.1b is drawn so that a appears to be a root, but the tree in Figure 3.1c, which is a redrawing of Figure 3.1b, has no single vertex that is a natural root—that is, any vertex can be the root.

In most of this chapter we will be using trees with directed edges. Following common terminology, we call a directed tree a **rooted tree.** A rooted tree T has a unique root, for if vertices a and b were both roots of T, then there would be paths from a to b and from b to a forming a circuit. An undirected tree is unrooted in the sense that it has no one particular root. An undirected tree can be made into a rooted tree by choosing one vertex as the root and then directing all edges away from the root. For example, to root the undirected tree in Figure 3.1b at vertex a, we would simply direct all the edges from left to right.

The standard way to draw a rooted tree T is to place the root a at the top of the figure. Then the vertices adjacent from a are placed one level below a, and so on, as

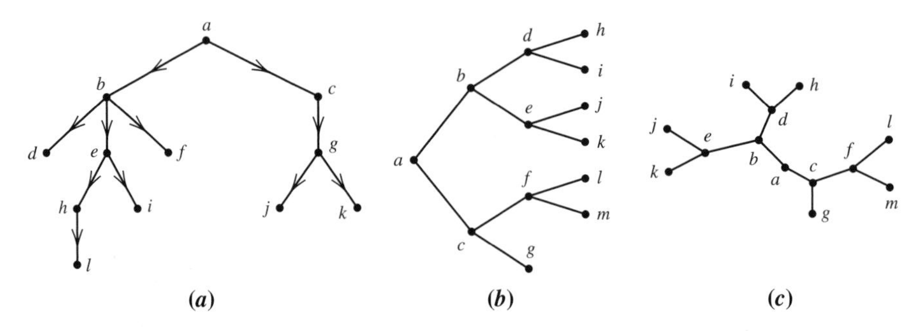

Figure 3.1

in Figure 3.1*a*. We say that the root a is at level 0, vertices b and c in that tree are at level 1, vertices d, e, f, and g in that tree are at level 2, and so forth. The **level number** of a vertex x in T is the length of the (unique) path from the root a to x.

For any vertex x in a rooted tree T, except the root, the **parent** of x is the vertex y with an edge $(\vec{y,x})$ into x (the unique edge directed into x). The **children** of x are vertices z with an edge directed from x to z. Children have level numbers one greater than x. Two vertices with the same parent are **siblings.** The parent–child relationship extends to ancestors and descendants of a vertex. In Figure 3.1*a*, vertex e has b as its parent, h and i as its children, d and f as its siblings, a as its other ancestor, and l as its other descendant. Observe that each vertex x in a tree T is the root of the subtree of x and its descendants. For easy reference, there is a glossary of tree-related terminology at the end of this text.

Theorem 1

A tree with n vertices has $n - 1$ edges.

Proof

Assume that the tree is rooted; if undirected, make it rooted as described above. We can pair off a vertex x with the unique incoming edge $(\vec{x,y})$ from its parent y. Since each vertex except the root has such a unique incoming edge, there are $n - 1$ nonroot vertices and hence $n - 1$ edges. ◆

Vertices of T with no children are called **leaves** of T. Vertices with children are called **internal** vertices of T. If every internal vertex of a rooted tree has m children, we call T an ***m*-ary tree.** If $m = 2$, T is a **binary tree.**

Theorem 2

Let T be an m-ary tree with n vertices, of which i vertices are internal. Then, $n = mi + 1$.

Proof

Each vertex in a tree, other than the root, is the child of a unique vertex (its parent). Each of the i internal vertices has m children, and so there are a total of mi children. Adding the one nonchild vertex, the root, we have $n = mi + 1$. ◆

Corollary

Let T be an m-ary tree with n vertices, consisting of i internal vertices and l leaves. If we know one of n, i, or l, then the other two parameters are given by the following formulas:

(a) Given i, then $l = (m - 1)i + 1$ and $n = mi + 1$.

(b) Given l, then $i = (l - 1)/(m - 1)$ and $n = (ml - 1)/(m - 1)$.

(c) Given n, then $i = (n - 1)/m$ and $l = [(m - 1)n + 1]/m$.

The proof of the corollary's formulas follow directly from $n = mi + 1$ (Theorem 2) and the fact that $l + i = n$. Details are left as an exercise.

Example 1

If 56 people sign up for a tennis tournament, how many matches will be played in the tournament?

The tournament proceeds in a binary-tree-like fashion. The entrants are leaves and the matches are the internal vertices. See Figure 3.2. Given $l = 56$ and $m = 2$, we determine i from part (b) of the corollary: $i = (l - 1)/(m - 1) = (56 - 1)/(2 - 1) = 55$ matches. ■

Example 2

Suppose a telephone chain is set up among 100 parents to warn of a school closing. It is activated by a designated parent who calls a chosen set of three parents. Each of these three parents calls given sets of three other parents, and so on. How many parents will have to make calls? Repeat the problem for a telephone tree of 200 parents.

Such a telephone chain is a rooted tree with 100 vertices. An edge corresponds to a call and an internal vertex corresponds to a parent who makes a call. Since we know $n = 100$ and that the tree is ternary (3-ary), part (c) of the corollary can be used to determine i, the number of callers: $i = (n - 1)/m = (100 - 1)/3 = 33$.

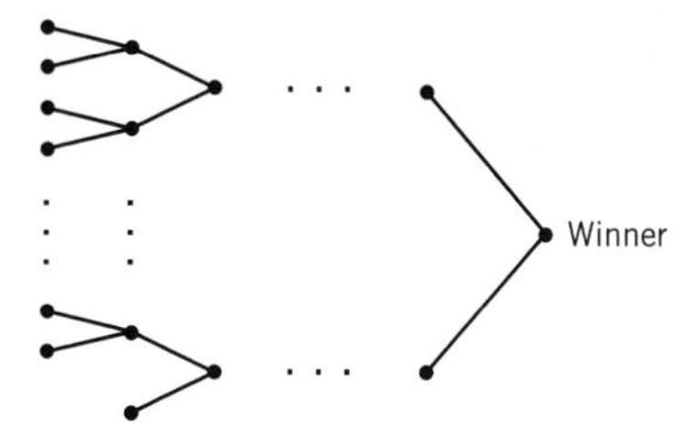

Figure 3.2

When we repeat the computation for an organization of 200 people, we get $i = (200 - 1)/3 = 66\frac{1}{3}$ internal vertices. By Theorem 2, a ternary tree must have a number of vertices n equal to $3i + 1$, for some i—that is, $n \equiv 1 \pmod 3$. But $200 \equiv 2 \pmod 3$, and so we do not have a true ternary tree. Either 199 or 202 parents would give a ternary tree. As a practical matter, with 200 people there will be 66 parents who each make 3 calls and one parent who makes just one call (one internal vertex with one child). ■

The **height** of a rooted tree is the length of the longest path from the root or, equivalently, the largest level number of any vertex. A rooted tree of height h is called **balanced** if all leaves are at levels h and $h - 1$. Balanced trees are "good" trees. The telephone chain tree in Example 2 should be balanced to get the message to everyone as quickly as possible. A tennis tournament's tree should be balanced to be fair; otherwise some players could reach the finals by playing several fewer matches than other players. Making an m-ary tree balanced will minimize its height (Exercise 12). The tree in Figure 3.1b, with a as root, is a balanced binary tree of height 3.

Theorem 3

Let T be an m-ary tree of height h with l leaves. Then:

(a) $l \leq m^h$, and if all leaves are at height h, $l = m^h$.

(b) $h \geq \lceil \log_m l \rceil$, and if the tree is balanced, $h = \lceil \log_m l \rceil$.

Proof

The expression $\lceil r \rceil$ denotes that the smallest integer $\geq r$; that is, $\lceil r \rceil$ rounds r up to the next integer.

(a) An m-ary tree of height 1 has $m^1 = m$ leaves (children of the root). Now we use induction on h to show that an m-ary tree of height h has at most m^h leaves, with $l = m^h$ if all leaves are at level h. An m-ary tree of height h can be broken into m subtrees rooted at the m children of the root. See Figure 3.3. These m subtrees have height at most $h - 1$. By induction, each of these subtrees has at most m^{h-1} leaves, and if all leaves are at height $h - 1$ in the subtrees, each has exactly m^{h-1}. The m subtrees combined have at most $m \times m^{h-1} = m^h$ leaves, and if all leaves are at height h, there are exactly m^h leaves.

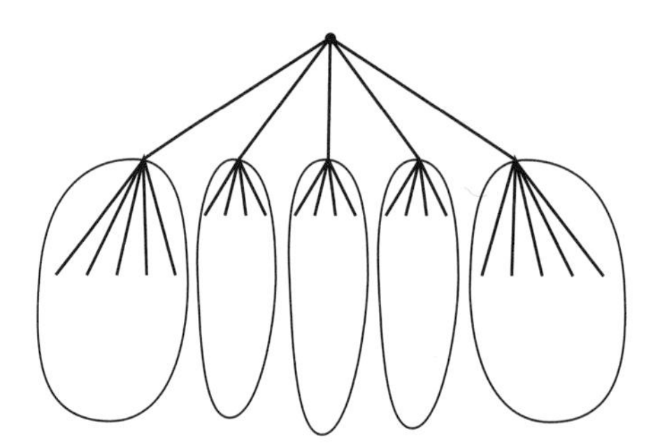

Figure 3.3

(b) Taking the logarithm base m on both sides of the inequality $l \leq m^h$ yields $\log_m l \leq h$. Since h is an integer, we have $\lceil \log_m l \rceil \leq h$. If the tree is balanced with height h, then the largest possible value for l is $l = m^h$ (if all leaves are at level h), and the smallest possible value is $l = (m^{h-1} - 1) + m$ (with m leaves at level h and the rest at level $h - 1$). So $m^{h-1} < l \leq m^h$. Taking logarithms on both sides yields $h - 1 < \log_m l \leq h$, or $h = \lceil \log_m l \rceil$. ◆

The most common use of trees is in searching. The following two sequential testing examples, one a basic computer science problem and the other a logical puzzle, illustrate the use of trees in searching.

Example 3

Let us reexamine the dictionary look-up problem discussed in Example 2 of Section 1.1. We want to identify an unknown word (number) *X* by comparing it to words in a set (dictionary) to which *X* belongs. This time our comparison test will be a three-way branch (less than, equal to, greater than). The test procedure can be represented by a binary, or almost binary, tree. If *X* were one of the first 14 letters of the alphabet, then Figure 3.4 is such a binary search tree. Each vertex is labeled with the letter tested at that stage in the procedure. The procedure starts by testing *X* against *H*. The left edge from a vertex is taken when *X* is less than the letter and the right edge when *X* is greater. Such a tree may have one internal vertex with just one child if the number of vertices is even (as is the case for vertex *N* in Figure 3.4). This tree is built by making the middle letter in the list (in this case, *G* or *H*) the root. The left child of the root is the middle letter in the left subtree (in this case, *D*), and so on.

To minimize the number of tests needed to recognize any *X*, that is, the height of the search tree, we should make the tree balanced. Suppose that *X* were known to belong to a set of n "words." What is the maximum number of tests that would be needed to recognize *X*? By the corollary to Theorem 2, a binary search tree with n vertices has

$$l = \frac{(2-1)n+1}{2} = \frac{n+1}{2}$$

leaves. Then the maximum number of tests needed to recognize X is the height of a balanced $\frac{1}{2}(n+1)$-leaf search tree. By Theorem 3, $h = \lceil log_2[(n+1)/2] \rceil = \lceil log_2(n+1) \rceil - 1$. ■

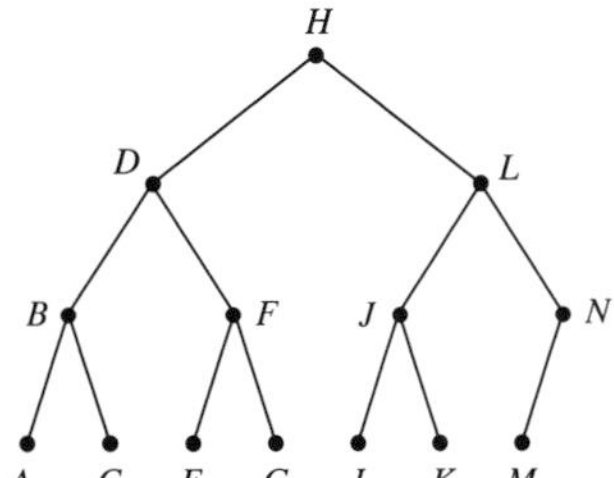

Figure 3.4

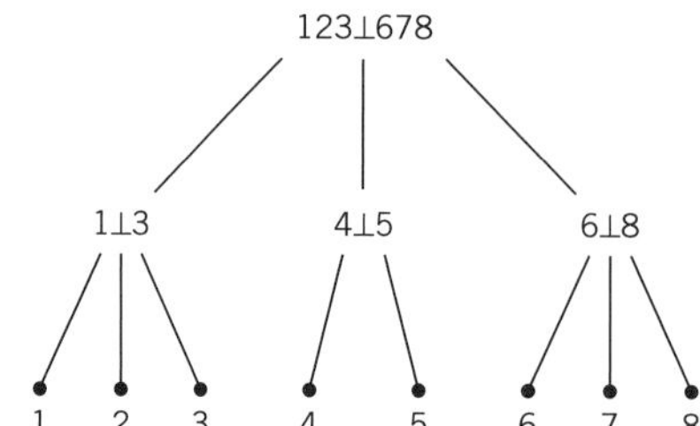

Figure 3.5

Example 4

A well-known logical puzzle has n coins, one of which is counterfeit—too light or too heavy—and a balance to compare the weight of any two sets of coins (the balance can tip to the right, to the left, or be even). For a given value of n, we seek a procedure for finding the counterfeit coin in a minimum number of weighings. Sometimes one is told whether the counterfeit coin is too light or too heavy. If we are told that the fake coin is too light, how many weighings are needed for n coins?

Our testing procedure will form a tree in which the root is the first test, the other internal vertices are the other tests, and the leaves are the solutions, that is, which coin is counterfeit. See the testing procedure in Figure 3.5 for eight coins. The coins are numbered 1 through 8, and the left edge is followed when the left set of coins in a test is lighter, the middle edge when both sets have the same weight, and the right edge when the right set of coins is lighter. Note that when weighing 4 and 5 (and already knowing that 1, 2, 3, 6, 7, 8 are not the light coin), the balance cannot be even.

The test tree is ternary, and with n coins there will be n leaves—that is, n different possibilities of which coin is counterfeit. Theorem 3 tells us that the test tree must have height at least $\lceil \log_3 n \rceil$ to contain n leaves. For the light counterfeit coin problem, this bound can be achieved by successively dividing the current subset known to have the fake coin into three almost equal piles and comparing two of the piles of equal size, as in Figure 3.5.

If the counterfeit coin could be either too light or too heavy, then the problem is harder (see Exercise 29). A particular coin will appear at 2 leaves in the test tree, once when the coin is determined to be too light and another time when too heavy. ■

Determining the number of leaves and height of more complex search trees is a major concern in the field of computer science called analysis of algorithms. In Section 3.4, we determine the number of leaves and height of search trees that arise when sorting a list of n items. Recurrence relations for the number of leaves and height of some other search trees are discussed in Section 7.2 in the enumeration part of this book. We conclude this section with a formula for the number of different undirected trees on n labeled vertices. Let the labels be the numbers 1 through n. For example, there are three different labeled trees on three labels. Each 3-label tree is a path of two edges with the difference being which of the three labels is the one middle vertex: 1–2–3, 1–3–2, and 2–1–3 (switching the position of the two leaves does not produce a different tree). While the formula was first proved by Cayley, we present a simpler proof due to Prufer.

Theorem 4 **(Cayley, 1889)**

There are n^{n-2} different undirected trees on n labels.

Proof

Observe that n^{n-2} is the number of sequences of the n labels of length $n-2$. We now construct a one-to-one correspondence between trees on n labels and $(n-2)$-length sequences of the n labels. Recall that for simplicity, we let the labels be the numbers $1, 2, \ldots, n$.

For any tree on n numbers, we form a sequence $(s_1, s_2, \ldots, s_{n-2})$ of length $n-2$ as follows. Let l_1 be the leaf in the tree with the smallest number and let s_1 be the number of the one vertex adjacent to it. For the tree in Figure 3.6, the leaf with the smallest number is 1 and the number of its neighboring vertex is 6. So $s_1 = 6$. We delete leaf l_1 from the tree and repeat this process. For the tree in Figure 3.6, l_2, the smallest numbered leaf in the tree after 1 is deleted, is 4 and its neighbor is 2. So $s_2 = 2$. Continuing we have $l_3 = 5$ and $s_3 = 2$, then $l_4 = 2$ and $s_4 = 3$, then $l_5 = 6$ and $s_5 = 3$, and then $l_6 = 7$ and $s_6 = 3$. We stop when the remaining tree has been reduced to two leaves joined by an edge. The 6-label sequence for the 8-label tree in Figure 3.6 is thus (6, 2, 2, 3, 3, 3). Such sequences are called *Prufer sequences.*

Next we show that any such $(n-2)$-length sequence of n items defines a unique n-item tree. We simply reverse the procedure in the preceding paragraph used to build the sequence. Observe that leaves (vertices of degree 1) will never appear in the sequence. The first number of the sequence is the neighbor of the smallest numbered leaf. From what we just observed, this smallest numbered leaf is the smallest number that does not appear in the sequence. For the sequence (6, 2, 2, 3, 3, 3), 1 is the smallest number not in the sequence and so 1 is the leaf with 6 as its neighbor.

Now we set the smallest leaf (label 1) aside (its position in the tree—a leaf adjacent to the first item in the sequence—is determined) and we consider the first item (item 6) as a leaf that will be adjacent to some item in the remaining sequence. We then repeat the process of identifying the smallest leaf in the remaining $(n-1)$-label tree specified by the remaining $(n-3)$-label sequence. For the remaining sequence (2, 2, 3, 3, 3), label 4 is the smallest of the remaining numbers (label 1 has been deleted) not in the sequence and so item 4 is a leaf and it is adjacent to label 2, the first number of the remaining sequence. Continuing in the reduced sequence (2, 3, 3, 3), label 5 is the smallest leaf of the remaining numbers and it is adjacent to label 2. Now label 2

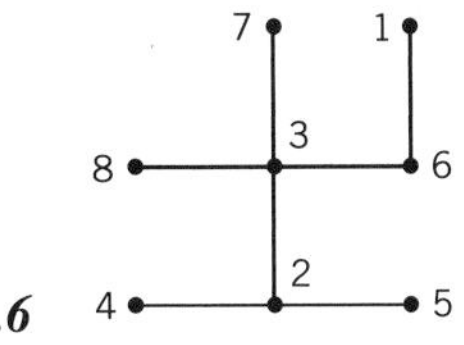

Figure 3.6

becomes a potential "leaf" with respect to the remaining sequence (3, 3, 3). Note that the available leaves are currently labels 2, 6, 7, 8. So label 2, the smallest available leaf, is adjacent to item 3. In the sequence (3,3), label 6 is adjacent to label 3. Finally label 7 is adjacent to label 3. There remain labels 3 and 8 and they must be adjacent to each other.

The preceding construction of a labeled tree from the given Prufer sequence can be applied to any Prufer sequence. Thus the correspondence between n-label trees and sequences of length $n-2$ is one-to-one and the theorem is proved. ◆

3.1 EXERCISES

Summary of Exercises Exercises 3–19 present theory about trees. Exercises 20–29 involve various modeling problems with trees.

1. Draw all nonisomorphic trees with:

(a) Four vertices **(b)** Five vertices **(c)** Six vertices

2. Suppose a connected graph has 20 edges. What is the maximum possible number of vertices?

3. Show that all trees are 2-colorable.

4. Show that all trees are planar.

5. Show that an undirected connected graph G is a tree (i.e., has a unique path from root to each vertex) if any one of the following conditions hold:

(a) G has n vertices and $n-1$ edges.

(b) G has fewer edges than vertices.

(c) Removal of any edge disconnects G.

6. Reprove Theorem 1 by using the fact that trees are planar (Exercise 4) and Euler's formula (Theorem 2 in Section 1.4).

7. Show that any tree with more than 1 vertex has at least 2 vertices of degree 1.

8. Prove the following parts of the corollary to Theorem 2: **(a)** Part a. **(b)** Part b. **(c)** Part c.

9. Reprove that $l \leq m^h$ in an m-ary tree of height h by counting the maximum possible number of choices at each internal vertex when building a path from the root to a leaf.

10. What is the maximum number of vertices (internal and leaves) in an m-ary tree of height h?

11. Show that the fraction of internal vertices in an m-ary tree is about $1/m$.

12. For a given h, show that the height of an m-ary tree with k leaves is minimized when the tree is balanced.

13. What is the size of the largest and smallest numbers of vertices of degree 1 possible in an n-vertex tree, for $n > 2$?

14. Show that a graph is a tree if it has no circuits but the addition of any edge (between two existing vertices) always creates a circuit.

15. Show that the size of the largest independent set (defined in Example 5 of Section 1.1) in an n-vertex tree is at least $n/2$.

16. A *forest* is an unconnected graph that is a disjoint union of trees. If G is an n-vertex forest of t trees, how many edges does it have?

17. Show that the sum of the level numbers of all l leaves in a binary tree is at least $l\lceil \log_2 l \rceil$, and hence the average leaf level is at least $\lceil \log_2 l \rceil$.

18. Let T be an undirected tree. If the choice of vertex x to be the root yields a rooted tree of minimal height, then x is call a *center* of T. Show that any undirected tree has at most two centers.

19. Show that the chromatic polynomial of an n-vertex tree is $k(k-1)^{n-1}$ (see Section 2.3).

20. Any m-ary tree, $m \geq 3$, can be "converted" into a binary tree by the following substitution.

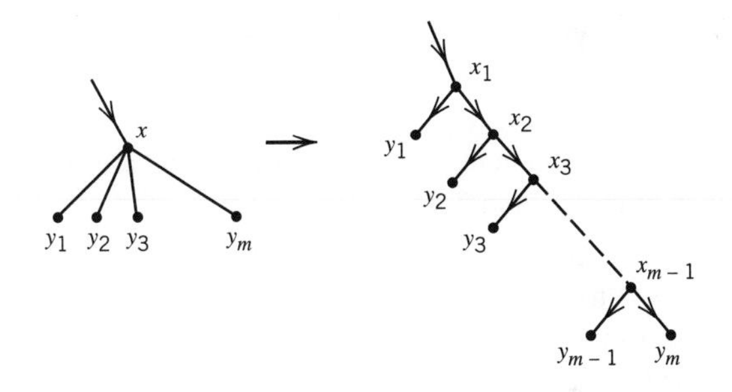

(a) Perform this conversion for the tree in Figure 3.5.

(b) If an m-ary tree has height h, what is the maximum possible height after conversion?

21. Consider the problem of summing n numbers by adding together various pairs of numbers and/or partial sums, for example, $\{[(3+1)+(2+5)]+9\}$.

(a) Represent this addition process with a tree. What will internal vertices represent?

(b) What is the smallest possible height of an "addition tree" for summing 100 numbers?

22. What type of search procedure is represented by the search tree below?

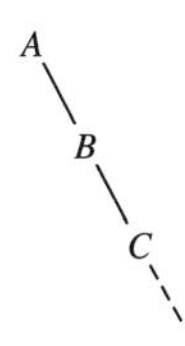

23. A tree can be used to represent a binary code; a left branch is a 0 and a right branch a 1. The path to a letter (vertex) is its binary code. To avoid confusion, one sometimes requires that the initial digits of one letter's code cannot be the code of another letter (e.g., if K is encoded as 0101, then no letter can be encoded as 0 or 01 or 010). Under this requirement, which vertices in a tree represent letters? How many letters can be encoded using n-digit binary sequences?

24. Suppose that each player in a tennis tournament (like the binary-tree tournament in Example 1) brings a new can of tennis balls. One can is used in each match and the other can is taken by the match's winner along to the next round. Use this fact to show that a tennis tournament with n entrants has $n - 1$ matches.

25. Consider a tennis tournament T (with the tree structure illustrated in Figure 3.2) with 32 entrants.

(a) How many players are eliminated (lose) in the first two rounds of matches?

(b) Suppose that the losers in the first two rounds of the tournament qualify for a losers' tournament T'. How many players are eliminated in the first two rounds of matches in T'? Note that all tennis tournaments are balanced trees; the number of people playing matches in each round after the first round is a power of 2.

(c) Suppose that the people who lose in the first two rounds of T' qualify for another losers' tournament T''. How many players are eliminated in the first two rounds of matches in T''?

(d) Suppose that the people who lose in the first two rounds of T'' qualify for another losers' tournament and so on until finally there is just one grand loser (the last tournament has two people). How many losers' tournaments are required to determine this grand loser?

26. Suppose that a chain letter is started by someone in the first week of the year. Each recipient of the chain letter mails copies on to five other people in the next week. After six weeks, how much money in postage (34¢ a letter) has been spent on these chain letters?

27. **(a)** Repeat Example 3 assuming now that only a two-way branch (less than, greater than, or equal to) is available. Draw a balanced search tree for the first 13 letters and determine the height of an n-letter search.

(b) Suppose a two-way branching search tree for letters A, B, C, D, E is to take advantage of the following letter frequencies: A 20%, B 20%, C 30%, D 10%, E 20%. Build a two-way tree that minimizes the average number of tests required to identify a letter.

28. **(a)** Repeat Example 4 for 20 coins with at most one too light.

(b) Prove by induction that 3^n or fewer coins with one too light can be tested in at most n weighings to find out which one is too light.

29. Suppose we have four coins and *possibly* one coin is either too light or too heavy (all four might be true).

(a) Show how to determine which of the nine possible situations holds with just two weighings, if given one additional coin known to be true.

(b) Show that two weighings are not sufficient without the extra true coin.

30. In the proof of Theorem 4, we showed that a Prufer sequence $(s_1, s_2, \ldots, s_{n-2})$ uniquely described a tree on n items. Construct the trees with the following Prufer sequences.

(a) (4, 5, 6, 2) **(b)** (2, 8, 8, 3, 5, 4) **(c)** (3, 3, 3, 3, 3, 3)

3.2 SEARCH TREES AND SPANNING TREES

Trees provide a natural framework for finding solutions to problems that involve a sequence of choices, whether hunting through a graph for a special vertex or finding one's way out of a maze or searching for the cheapest solution to a vehicle routing problem. Most of the problems in the two preceding chapters—isomorphism, Hamilton circuits, minimal colorings, and placing police on street corners—require tree-based searching for computerized solutions. By letting the sequential choices be internal vertices in a rooted tree and the solutions and "dead ends" be the leaves, we can organize our search for possible solutions. Whether searching in a graph or in a maze or through all solutions to an optimization problem, the foremost concern in the search procedure is that it be exhaustive, that is, guaranteed to check all possibilities.

In this section we present applications of tree searches that involve games rather than operations research applications. Most operations research tree enumeration problems involve very large trees and use special tree "pruning" algorithms. As an example of such applications, we solve a small Traveling Salesperson problem in Section 3.3. The next chapter, "Network Algorithms," discusses three important optimization algorithms that implicitly use trees to search through graphs.

We start with searching in a graph. In many applications, graph algorithms are needed to test whether a graph has a certain property, such as connectedness or planarity, or to count all occurrences of a given structure, such as circuits or complete subgraphs. The algorithms usually employ a spanning tree in searching among vertices and edges of a graph to check for these properties or structures. A **spanning tree** of a graph G is a subgraph of G that is a tree containing all vertices of G. Spanning trees can be constructed either by depth-first (backtrack) search or by breadth-first search.

To build a depth-first spanning tree, we pick some vertex as the root and begin building a path from the root composed of edges of the graph. The path, continues until it cannot go any further without repeating a vertex already in the tree. The vertex where this path must stop is a leaf. We now backtrack to the parent of this leaf and try to build a path from the parent in another direction. When all possible paths from this parent y and its other children have been built, we backtrack to the parent of y, and so on, until we come back to the root and have checked all other possible paths from the root.

To build a breadth-first spanning tree, we pick some vertex x as the root and put all edges leaving x (along with the vertices at the ends of these edges) in the tree. Then we successively add to the tree the edges leaving the vertices adjacent from x, unless such an edge goes to a vertex already in the tree. We continue this process in a level-by-level fashion.

It is important to note that if the graph is not connected, then no spanning tree exists. We thus have the following result.

Algorithm to Test Whether an Undirected Graph Is Connected

Use a depth-first or breadth-first search to try to construct a spanning tree. If all vertices of the graph are reached in the search, a spanning tree is obtained and the graph is connected. If the search does not reach all vertices, the graph is not connected.

A formal proof that depth-first (breadth-first) searching does search all vertices in a connected graph is left as an exercise.

An *adjacency matrix* of an (undirected) graph is a (0,1)-matrix with a 1 in entry (i, j) if vertex x_i and vertex x_j are adjacent; entry (i, j) is 0 otherwise. Adjacency matrices were introduced in Supplement I of Chapter 1.

Example 1: Testing for Connectedness

Is the undirected graph G whose adjacency matrix is given in Figure 3.7a connected?

Let us perform a depth-first search of G starting with x_1 as the root. At each successive vertex, we pick the next edge on the tree to be the edge going to the lowest numbered vertex not already in the tree. So from x_1 we go to x_2.

	x_1	x_2	x_3	x_4	x_5	x_6	x_7	x_8
x_1	0	1	1	1	0	1	0	0
x_2	1	0	1	0	1	0	1	0
x_3	1	1	0	0	0	0	0	0
x_4	1	0	0	0	0	0	1	0
x_5	0	1	0	0	0	0	1	0
x_6	1	0	0	0	0	0	0	0
x_7	0	1	0	1	1	0	0	1
x_8	0	0	0	0	0	0	1	0

(*a*)

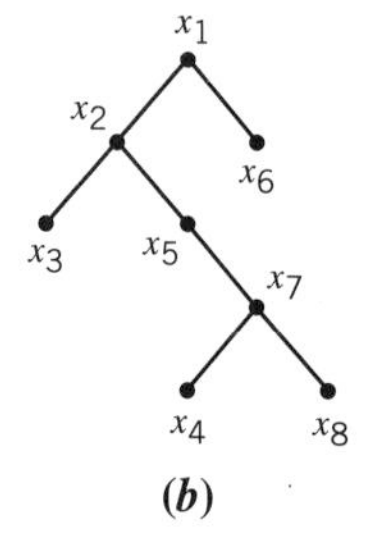

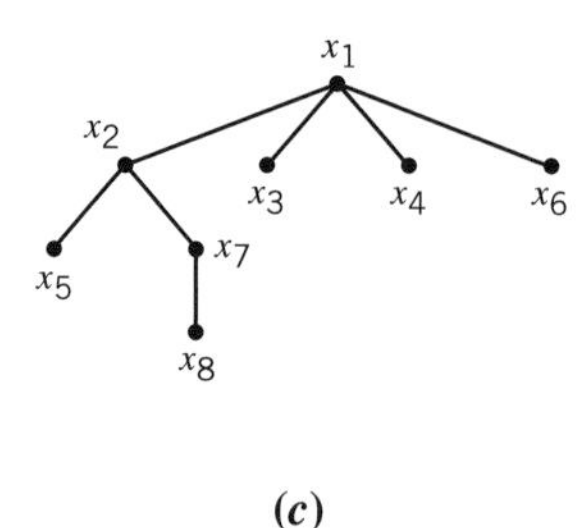

Figure 3.7 (*b*) (*c*)

From x_2 we go to x_3. Since x_3 is not adjacent to any other vertex besides x_1 and x_2 (which are already in the tree), we backtrack from x_3 to x_2 and continue the search from x_2, going to x_5, then to x_7, and then to x_4. At x_4 we backtrack to x_7 and go to x_8. From x_8, we must backtrack all the way back to x_1. From x_1, we go to x_6. This finishes the search—all vertices have been visited. The spanning tree obtained is shown in Figure 3.7*b*.

The result of a breadth-first search is shown in Figure 3.7*c*. ■

The computation time required to make a depth-first search of a graph is proportional to the number of edges in the graph (each edge in the spanning tree is traversed twice, and edges that cannot be used are tried just twice, once from each end vertex). If an undirected graph is not connected, then we can find its components (connected pieces) by applying a depth-first (or breadth-first) search at any vertex to find one component; then apply this search starting at a vertex not on the previous tree to find another component; and continue finding additional components until no unused vertices are left.

We note one important property of a breadth-first search. *A breadth-first spanning tree consists of shortest paths from the root to every other vertex in the graph.* A proof of this claim is left to Exercise 8. If we only wanted to find a shortest path from the root to a particular vertex x in a graph, then the breadth-first search could stop as soon as x was reached (without trying to construct a full spanning tree).

Now we apply the techniques of depth-first and breadth-first search to games. In a maze, the vertices will be the intersection points of paths. In puzzles, the vertices will be the different configurations of the puzzle and the edges will be the possible moves.

Example 2: Traversing a Maze

Consider the maze in Figure 3.8. We start at the location marked with an *S* and seek to reach the end marked with an *E*.

We use a depth-first search. For mazes, there is a convenient rule of thumb (whose verification is left as an exercise) for constructing a depth-first search: stick to the right wall in the maze. When we come to a dead end, we follow the right wall to the end wall, along the end wall, and then backtrack along the left wall (now the right wall as we leave the dead end). When we come to a previously visited corner, we put an artificial (wiggly) dead-end wall to stop us from actually reaching that corner (as at *S* in Figure 3.8). In the maze in Figure 3.8 we use solid lines to indicate the (forward) pathbuilding and dashed lines for backtracking. Because the maze is easily searched directly, we have not drawn the search tree for this problem (in which *S* would be the root, other corners internal vertices, and dead ends and *E* the leaves). ■

The other common method of tree enumeration, called **breadth-first search,** is to determine all edges leaving the root, that is, all possible children of the root; then determine all edges leaving these children; and so on. This procedure fans out uniformly from the root. Again, no vertex can be repeated. Because a breadth-first

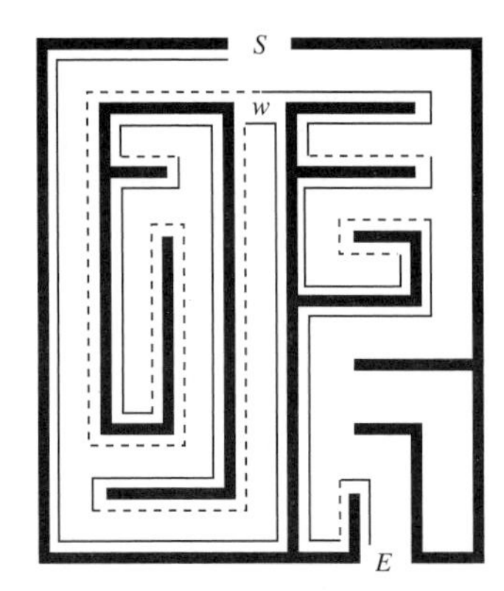

Figure 3.8

tree incorporates vertices into the tree as soon as possible, a breadth-first tree contains shortest paths from the root to each other vertex.

If the tree of possible paths is large, then the breadth-first method quickly becomes unwieldy. The depth-first method that traces only one path at a time is much easier to use by hand or to program. Further, in cases where we need to find only one of the possible solutions, it pays to go searching all the way down a path for a solution rather than to take a long time building a large number of partial paths, only one of which in the end will actually be used. On the other hand, when we want a solution involving a shortest path or when there may be very long dead-end paths (while solution paths tend to be relatively short), then the breadth-first method is better. All the network optimization algorithms in the next chapter use breadth-first searches.

Example 3: Pitcher-Pouring Puzzle

Suppose we are given three pitchers of water, of sizes 10 quarts, 7 quarts, and 4 quarts. Initially the 10-quart pitcher is full and the other two empty. We can pour water from one pitcher into another, *pouring until the receiving pitcher is full or the pouring pitcher is empty*. Is there a way to pour among pitchers to obtain exactly 2 quarts in the 7- or 4-quart pitcher? If so, find a minimal sequence of pourings to get 2 quarts in the 7-quart or 4-quart pitcher.

The positions, or vertices, in this enumeration problem are ordered triples (a, b, c), the amounts in the 10-, 7-, and 4-quart pitchers, respectively. Actually, it suffices to record only (b, c), the 7- and 4-quart pitcher amounts, since $a = 10 - b - c$. A directed edge corresponds to pouring water from one pitcher to another. Let us draw the tree on a b, c-coordinate grid as shown in Figure 3.9a. The grid is bounded by $b = 7, c = 4, b + c = 10$. Pouring between the 10- and 7-quart pitchers will be a horizontal edge, between 10- and 4-quart pitchers a vertical edge, and between 7- and 4-quart pitchers a diagonal edge with slope -1. The beginning of the same tree is shown in the standard form in Figure 3.9b.

The root of this search tree is (0, 0). Since we want a minimal sequence of pourings (a shortest path in the search tree), we will use a breadth-first search. From

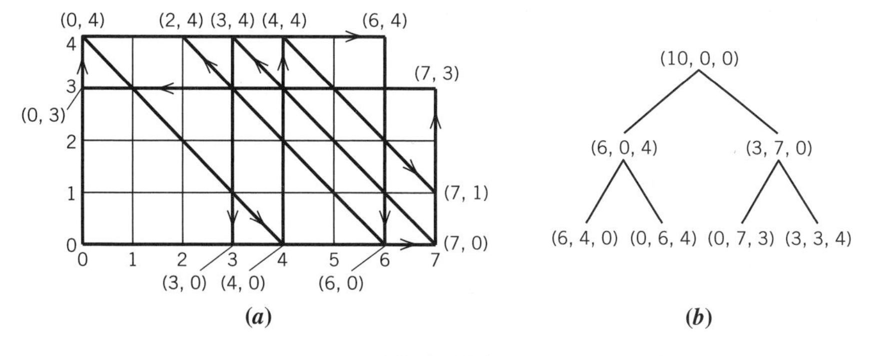

Figure 3.9

the root, we can get to positions (7, 0) and (0, 4). From (7, 0), we can get to new positions (7, 3) and (3, 4), and from (0, 4), we can get to new positions (6, 4) and (4, 0). The tree built thus far is shown in Figure 3.9*b*.

From (7, 3), the only new position is (0, 3), and from (3, 4), the only new position is (3, 0). From (6, 4), the only new position is (6, 0), and from (4, 0), the only new position is (4, 4). We have now checked all paths of length 3. The only new moves now are from (4, 4) to (7, 1) and from (6, 0) to (2, 4). But (2, 4) has 2 quarts in the 7-quart pitcher. So (0, 0) to (0, 4) to (6, 4) to (6, 0) to (2, 4) is a sequence of pourings to obtain 2 quarts. ∎

Example 4: Jealous Wives Puzzle

Three jealous wives and their husbands come to a river. The party must cross the river (from near shore to far shore) in a boat that can hold at most two people. Find a sequence of boat trips that will get the six people across the river without ever letting any husband be alone (without his wife) in the presence of another wife.

Let the wives be represented by the letters A, B, and C, and their respective husbands by a, b, and c. The positions will be the possible partitions of people when the boat is at one of the two shores, and an edge will correspond to a crossing of the river by the boat. We denote a position with the names of the people currently on the near shore plus a star (*) if the boat is on the near shore. An edge is labeled with the direction of the boat and the people in the boat. At each position, we must check that the people will not violate the "jealousy" conditions of an unaccompanied husband in contact with another wife (or vice versa). Figure 3.10 shows one path of feasible positions that gets everyone across the river. There are other similar paths possible; for example, we could start with Bb or Cc instead of Aa, or start with ab or ac or bc, and so forth. ∎

Suppose we have built a tree to provide the framework for searching or organizing information. Searching for a particular vertex or processing information in the tree normally involves a depth-first type of traversal of the spanning tree. However, there are several times during a traversal when internal vertices can be checked. A **preorder**

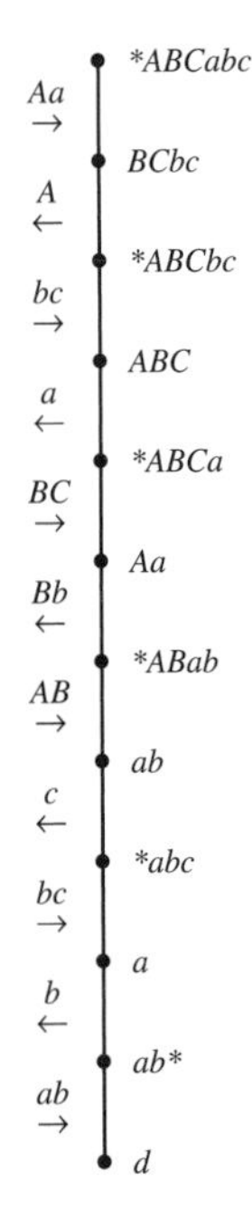

Figure 3.10

traversal of a tree is a depth-first search that examines an internal vertex when the vertex is first encountered in the search. A **postorder traversal** examines an internal vertex when last encountered (before the search backtracks away from the vertex and its subtree). If a tree is binary, we can define an **inorder traversal** that checks an internal vertex in between the traversal of its left and right subtrees. Figures 3.11*a* and 3.11*b* display numberings of the vertices of the tree in Figure 3.11*a* according to preorder and postorder traversals, respectively. (We assume the left child is visited before the right child at each internal vertex.)

An important property of preorder and postorder traversals is that in a preorder traversal a vertex precedes its children and all its other descendants, whereas in a postorder traversal a vertex follows its children and all its other descendants.

We now give examples of each type of traversal. In the binary search tree example of a dictionary look-up (Example 2 in Section 3.1), the alphabetical order of the vertices corresponds to inorder traversal. In searching through a graph for a vertex

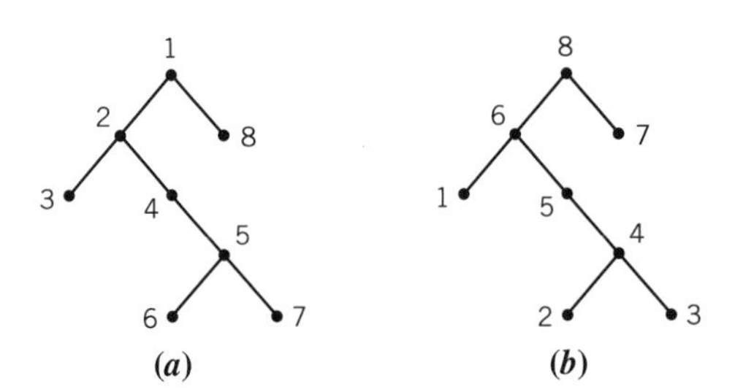

Figure 3.11

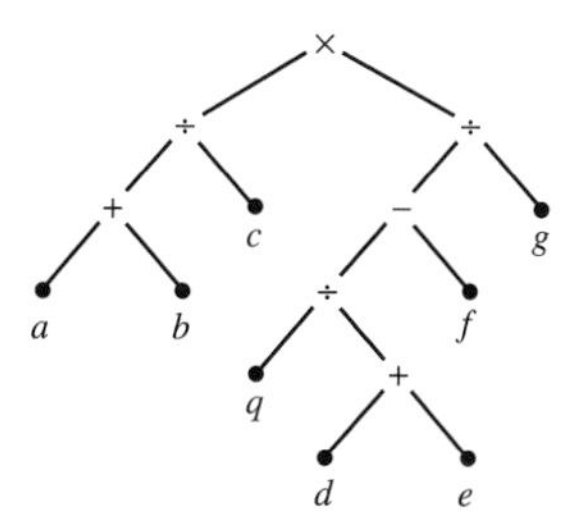

Figure 3.12

with a specified label, we should test each new vertex as it is encountered on a depth-first search to see if it has the desired label. There is no advantage to postponing such testing. Thus, in searching for a special vertex, vertices should be examined according to a preorder traversal.

We demonstrate the need for postorder traversal with an arithmetic tree. The arithmetic expression

$$((a+b)\div c)\times(((q\div(d+e))-f)\div g)$$

can be decomposed into a binary tree as shown in Figure 3.12. The internal vertices are arithmetic operations and the leaves variables. To evaluate this arithmetic expression, we need to execute the operations specified by internal vertices according to a postorder traversal of the arithmetic tree, that is, we cannot perform an operation until the subexpressions represented by the operation vertex's two subtrees are evaluated.

3.2 EXERCISES

Summary of Exercises The first six exercises involve building spanning trees and the next five involve properties of spanning trees. Exercises 12–23 are based on the puzzle examples in this section. Exercises 24–25 involve traversals.

1. Find depth-first spanning trees for each of these graphs;
   (a) K_8 (a complete graph on 8 vertices).
   (b) The graph in Figure 2.5 in Section 2.2.
   (c) The graph in Figure 2.10 in Section 2.3.

2. Find breadth-first spanning trees of each of the graphs in Exercise 1.
3. Find all spanning trees (up to isomorphism) in the following graphs:
   (a) Figure 2.4 **(b)** K_4 **(c)**

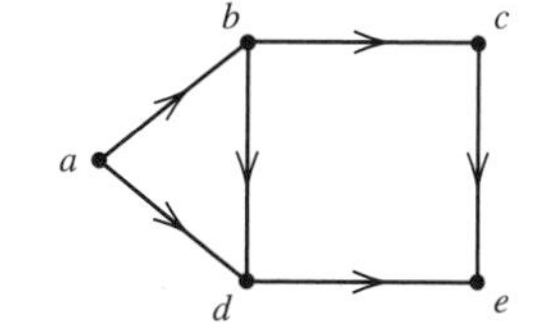

4. Test the graph whose adjacency matrix is given below to see if it is connected.

	x_1	x_2	x_3	x_4	x_5	x_6	x_7	x_8
x_1	0	0	1	0	0	1	0	1
x_2	0	0	1	0	1	0	1	0
x_3	1	1	0	1	0	0	0	0
x_4	0	0	1	0	1	1	1	0
x_5	0	1	0	1	0	0	0	1
x_6	1	0	0	1	0	0	0	0
x_7	0	1	0	1	0	0	0	1
x_8	1	0	0	0	1	0	1	0

5. Consider an undirected graph with 25 vertices $x_2, x_3, \ldots, x_{26}$ with edges (x_i, x_j) if and only if integers i and j have a common divisor. How many components does this graph have? Find a spanning tree for each component.

6. Show that in an n-vertex graph, a set of $n - 1$ edges that form no circuits is a spanning tree.

7. Show that a connected undirected graph with just one spanning tree is a tree.

8. Show that a breadth-first spanning tree contains shortest paths from the root to every other vertex.

9. (a) Prove that a depth-first search reaches all vertices in an undirected connected graph.

(b) Repeat part (a) for breadth-first search.

10. (a) Show that the height of any depth-first spanning tree of a graph G starting from a given root a must be at least as large as the height of a breadth-first spanning tree of G with root a.

(b) By starting with different roots, it might be possible for a depth-first spanning tree of G to have a smaller height than a breadth-first spanning tree of G. Find a graph in which this can happen.

11. A *cutset* is a set S of edges in a connected graph G whose removal disconnects G, but no proper subset of S disconnects G. Show that any cutset of G has at least one edge in common with any spanning tree of G.

12. (a) Use a depth-first search to find your way through this maze from S to E.

(b) How many ways through this maze (without cycles or backing up) are there?

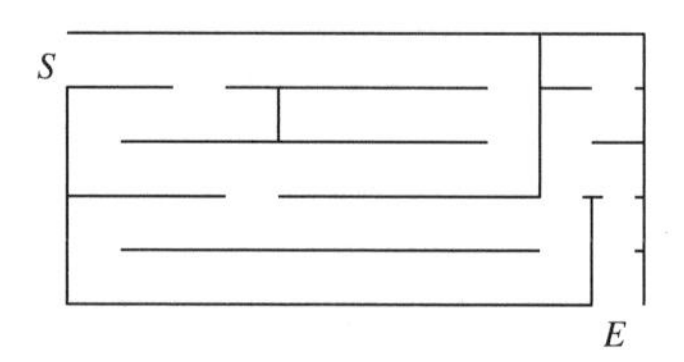

13. Repeat Example 3 using a depth-first search.

14. Perform a breadth-first search of the graph in Figure 3.8 to find the shortest path from S to E.

15. Find another (actually shorter) way in Example 3 to get 2 quarts in one pitcher.

16. **(a)** Repeat Example 3 with pitchers of size 8, 5, and 3 with an objective of 4 quarts in one pitcher.

(b) Repeat Example 3 with pitchers of sizes 12, 8, and 5, and an objective of 7 quarts in one pitcher.

17. Show that the stick-to-the-right-hand-wall rule will always get one out of a maze.

18. Use a depth-first search in Example 3 to show that any amount between 0 and 10 quarts can be obtained in one of the pitchers.

19. Suppose there are a dog, a goat, and a bag of tin cans to be transported across a river in a ferry that can carry only one of these three items at once (along with a ferry driver). If the dog and goat cannot be left alone on a shore (when the driver is not present) nor can the goat and tin cans be left alone, then find a scheme for getting all across the river.

20. **(a)** Repeat Example 4 this time with four jealous wives and their husbands, if possible.

(b) Do part (a) with a three-person boat.

21. **(a)** Suppose three missionaries and three cannibals must cross a river in a two-person boat. Show how this can be done so that at no time do the cannibals outnumber the missionaries on either side (unless there are no missionaries on a shore).

(b) Repeat part (a) with the additional condition that only one of the cannibals can row (all the missionaries can row also).

22. The following is a variation of the pitcher-pouring problem. An hourglass is a timepiece in which a specified period of time will elapse as sand in the top half empties through the narrow opening to the bottom half. If an hourglass is designed to take 10 minutes for the sand to flow down and after, say, 7 minutes the glass is turned upside down, then the sand flow will be reversed and it will take 7 more minutes for the sand to empty. Suppose you have two hourglasses, a 4-minute hourglass and a 7-minute hourglass. Describe a scheme for starting them and turning them upside down so as to measure out a 9-minute period of time. (*Hint:* in the beginning, you start one or both hourglasses; then a decision must be made whenever an hourglass empties whether to restart it and whether to start or invert the other hourglass.)

23. A group of four people have to cross a bridge that will accommodate only two people at a time. It is night and there is one flashlight. Any party who crosses, either one or two people, must have the flashlight with them. (The flashlight must be walked back and forth, it cannot be thrown.) Find a way for all four to get to the other side of the bridge in 17 minutes, given that person A takes 1 minute to cross the bridge, B takes 2 minutes, C takes 5 minutes, D takes 10 minutes. When two people cross, they must go at the speed of the slower person.

24. List the vertices in order of a preorder traversal and a postorder traversal of:

(a) Figure 3.1a **(b)** Figure 3.1b **(c)** Figure 3.4

25. Generalize the arithmetic tree in Figure 3.12 to include unary operations such as inverses or sin( ). Give the tree for the following expression: $\sin(((a + ((b \times c)^{-1} + ((a + d) \times e))) - (c + e)) + (a - b)^{-1})$

26. Let T be a spanning tree in a connected undirected graph G. Show that when any non-tree edge is added to T, a unique circuit results.

27. Prove that any spanning tree T' of a graph G can be converted into any other spanning tree T'' of G by a sequence of spanning trees $T_1, T_2, \ldots, T_m$ where $T'' = T_1$, $T'' = T_m$, and T_k is obtained from T_{k-1} by removing one edge of T_{k-1} that was in T' and adding one edge in T' (m is the number of edges by which T' and T'' differ). (*Hint*: Use Exercise 26 and induction on m.)

28. Suppose that during a preorder traversal of a binary tree T, we write down a 1 for each internal vertex and a 0 for each leaf in the traversal, building a sequence of 1s and 0s. If T has n leaves, the sequence will have n 0s and $n - 1$ 1s. We call this sequence the *characteristic sequence of T*. (Such a sequence determines a unique tree.)

(a) Find the binary tree with the characteristic sequence 110100110100100.

(b) Prove that the last two digits in any characteristic sequence are 0s (assuming $n \geq 2$).

(c) Prove that a binary sequence with n 0s and $n - 1$ 1s, for some n, is a characteristic sequence of some binary tree if and only if the first k digits of the sequence contain at least as many 1s as 0s, for $1 \leq k \leq 2n - 2$.

29. **(a)** Write a program for building a depth-first spanning tree of a graph whose adjacency matrix is given.

(b) Repeat part (a) for a breadth-first spanning tree.

30. Use a breadth-first inverted spanning tree T (edges directed from children to parents) to build an Euler cycle in a directed graph possessing an Euler circuit as follows. Starting at the root of T, trace out a path such that at any vertex x we choose only the edge in T to the parent of x (in T) if there are no other unused edges leaving x at this stage. Verify the correctness of the algorithm. Program this algorithm.

3.3 THE TRAVELING SALESPERSON PROBLEM

In this section, we illustrate the use of trees in graph optimization problems of operations research. The traveling salesperson problem seeks to minimize the cost of the route for a salesperson to visit a set of cities and return to home. One seeks a minimal-cost Hamilton circuit in a complete graph having an associated cost matrix C. Entry

C_{ij} in C is the cost of using the edge from the ith vertex to the jth vertex. The traveling salesperson problem arises in many different guises in operations research. One example is planning the movement of an automated drill press making holes at specified locations on printed circuit boards.

We present two approaches to solving the traveling salesperson problem. Both approaches use trees, but in two very different ways. The first approach is to systematically consider all possible ways to build Hamilton circuits in search of the cheapest one. It uses a "branch and bound" method to limit the number of different vertices in the search tree that must be inspected in search of a minimal solution. Each vertex in the tree will represent a partial Hamilton circuit, and each leaf a complete Hamilton circuit. Note that a complete graph on n vertices has $(n-1)!$ different Hamilton circuits, for example, a 50-city (vertex) problem has $49! \approx 6 \times 10^{62}$ possible circuits. The "branch and bound" method reduces substantially the number of circuits that must be checked, although not enough to make it practical to solve large traveling salesperson problems.

This leads to another strategy, our second approach, which is to construct a near-minimal solution using an algorithm that is quite fast. The huge amount of time required to find exact solutions to problems that involve enumeration of very large trees, such as the traveling salesperson problem, has led researchers to concentrate on heuristic, near-optimal algorithms for such problems. Finding a minimal coloring of any arbitrary graph or testing for isomorphism are other problems we have seen that have underlying search trees with similarly huge numbers of possible solutions. Near-minimal coloring algorithms have received much attention, but unfortunately, a "near-isomorphism" is usually of no value.

We consider a small traveling salesperson problem with four vertices x_1, x_2, x_3, x_4. Let the cost matrix for this problem be the matrix in Figure 3.13. Entry c_{ij} is the cost of going from vertex (city) i to vertex j. (Note that we do not require $c_{ij} = c_{ji}$.) The infinite costs on the main diagonal indicate that we cannot use these entries. A Hamilton circuit will use four entries in the cost matrix C, one in each row and in each column, such that no proper subset of entries (edges) forms a subcircuit. This latter constraint means that if we choose, say, entry c_{23}, then we cannot also use c_{32}, for these two entries form a subcircuit of length 2. Similarly, if entries c_{23} and c_{31} are used, then c_{12} cannot be used.

We first show how to obtain a lower bound for the cost of this traveling salesperson problem. Since every solution must contain an entry in the first row, the edges of a minimal tour will not change if we subtract a constant value from the first row of the cost matrix (of course, the cost of a minimal tour will change by this constant). We subtract as large a number as possible from the first row without making any entry in

	To	1	2	3	4
From	1	∞	3	9	7
	2	3	∞	6	5
	3	5	6	∞	6
	4	9	7	4	∞

Figure 3.13

	To	1	2	3	4
From	1	∞	0	6	4
	2	0	∞	3	2
	3	0	1	∞	1
	4	5	3	0	∞

Figure 3.14

the row negative; that is, we subtract the value of the smallest entry in row 1, namely 3. Do this for the other three rows also. We display the altered cost matrix in Figure 3.14.

All rows in Figure 3.14 now contain a 0 entry. After subtracting a total of $3+3+5+4=15$ from the different rows, a minimal tour using the cost data in Figure 3.15 will cost 15 less than a minimal tour using original cost data in Figure 3.14. Still, the edges of a minimal tour for the altered problem are the same edges that form a minimal tour for the original problem.

In a similar fashion, we can subtract a constant from any column without changing the set of edges of a minimal tour. Since we will want to avoid making any entries negative, we consider subtracting a constant only from columns with all current entries positive. The only such column in Figure 3.14 is column 4, whose smallest value is 1. So we subtract 1 from the last column in Figure 3.14 to get the matrix in Figure 3.15. Every row and column in Figure 3.15 now contains a 0 entry.

The cost of a minimal tour using Figure 3.15 has been reduced by a total of $15+1=16$ from the original cost using Figure 3.13. We can use this reduction of cost to obtain a lower bound on the cost of a minimal tour: A minimal tour using the costs in Figure 3.15 must trivially cost at least 0, and hence a minimal tour using. Figure 3.13 must cost at least 16. *In general, the lower bound for the traveling salesperson problem equals the sum of the constants subtracted from the rows and columns of the original cost matrix to obtain a new cost matrix with a 0 entry in each row and column.*

Now we are ready for the branching part of the "branch and bound" method. We look at an entry in Figure 3.15 that is equal to 0. Say c_{12}. Either we use c_{12} or we do not use c_{12}. We "branch" on this choice. In the case that we do not use c_{12}, we represent the no-c_{12} choice by setting $c_{12}=\infty$. The smallest value in row 1 of the altered Figure 3.15 is now $c_{14}=3$, and so we can subtract this amount from row 1. Similarly we can subtract 1 from column 2. If we do not use entry c_{12}, we obtain the new cost matrix in Figure 3.16. Hence, any tour for the original problem (Figure 3.13) that does not use c_{12} must cost at least $16+(3+1)=20$.

	To	1	2	3	4
From	1	∞	0	6	3
	2	0	∞	3	1
	3	0	1	∞	0
	4	5	3	0	∞

Lower bound = 16

Figure 3.15

	To	1	2	3	4
From	1	∞	∞	3	0
	2	0	∞	3	1
	3	0	0	∞	0
	4	5	2	0	∞

Do not use c_{12}

Lower bound = 20

Figure 3.16

If we use c_{12} to build a tour for Figure 3.15, then the rest of the tour cannot use another entry in row 1 or column 2; also entry c_{21} must be set equal to ∞ (to avoid a subcircuit of length 2). The new smallest value of the reduced matrix in row 2 is 1, and so we subtract 1 from row 2 to obtain the cost matrix in Figure 3.17. The lower bound for a tour in Figure 3.17 using c_{12} is now $16+1=17$.

Since our lower bound of 17 is less than the lower bound of 20 when we do not use c_{12}, we continue our consideration of tours using c_{12} by determining bounds associated with whether or not to use a second entry with value 0. We will continue to extend these partial tours using c_{12} as long as the lower bound is ≤ 20. If the lower bound were to exceed 20, then we would have to go back and consider partial tours not using c_{12}. Our tree of choices for this problem will be a binary tree whose internal vertices represent choices of the form: use entry c_{ij} or do not use c_{ij} (see Figure 3.19). At any stage, as long as the lower bounds for partial tours using c_{ij} are less than the lower bound for tours not using c_{ij}, we do not need to look at the subtree of possible tours not using c_{ij}.

We extend a tour using c_{12} by considering another entry with value 0. This next 0 entry need not connect with c_{12}, that is, need not be of the form c_{j1} or c_{2j}, but for simplicity we shall pick an entry in row 2. The only 0 entry in row 2 of Figure 3.17 is c_{24}.

Again we have the choice of using c_{24} or not using c_{24}. Not using c_{24} will increase the lower bound by 2 (the smallest entry in row 2 of Figure 3.17 after we set $c_{24}=\infty$; column 4 still has a 0). Using c_{24} will not increase the lower bound, and so we further extend the partial tour using c_{24} along with c_{12}. Again we delete row 2 and column 4 in Figure 3.17 and set $c_{41}=\infty$ (to block the subcircuit $x_1{-}x_2{-}x_4{-}x_1$). Figure 3.18 shows the new remaining cost matrix (all rows and columns still have a 0 entry).

Now the way to finish the tour is clear: use entries c_{43} and c_{31} for a complete tour $x_1{-}x_2{-}x_4{-}x_3{-}x_1$. We actually have no choice since not using either of c_{43} or c_{31} forces us to use an ∞ (which represents a forbidden edge). Since $c_{43}=c_{31}=0$, this tour has a cost equal to the lower bound of 17 in Figure 3.18. Further, this tour must be minimal since its cost equals our lower bound. Recheking the tour's cost with the original cost matrix in Figure 3.13, we have $c_{12}+c_{24}+c_{43}+c_{31}=3+5+4+5=17$.

	To	1	3	4
From	2	∞	2	0
	3	0	∞	0
	4	5	0	∞

Use c_{12}

Lower bound = 17

Figure 3.17

		To 1	3
From	3	0	∞
	4	∞	0

Use c_{12} and c_{24}

Lower bound = 17

Figure 3.18

We summarize the preceding reasoning with the decision tree in Figure 3.19 summarizing choices we made and their lower bounds (L.B.).

One general point should be made about how to take best advantage of the branch-and-bound technique. At each stage, we should pick as the next entry on which to branch (use or do not use the entry), the 0 entry whose removal maximizes the increase in the lower bound. In Figure 3.15, a check of all 0 entries reveals that not using entry c_{43} will raise the lower bound by $3+3=6$ (3 is the new smallest value in row 4 and in column 3). So c_{43} would theoretically have been a better entry than c_{12} to use for the first branching, since the greater lower bound for the subtree of tours not using c_{43} makes it less likely that we would ever have to check possible tours in that subtree.

Now consider the 6×6 cost matrix in Figure 3.20. We present the first few branchings in Figure 3.21 for this cost matrix, and circle the 0 entry on which we branch at each stage.

The branch-and-bound method is used not only to solve optimization problems. It is also a central tool in artificial intelligence. Computers that play chess make a tree of possible future moves and then assign some sort of "value" to the resulting chessboard situations. A move that leads to a bad situation gets a very high value, and future moves from this situation are not pursued. At more promising situations, all reasonable next moves are considered and their resulting situations evaluated. The more powerful the computer, the more moves into the future it can examine, although the procedure for evaluating positions is just as important as the machine's speed.

Next we present a quicker algorithm for obtaining good (near-minimal) tours, when the costs are symmetric (i.e., $c_{ij} = c_{ji}$) and the costs satisfy the triangle inequality—$c_{ik} \le c_{ij} + c_{jk}$. These two assumptions are satisfied in most traveling

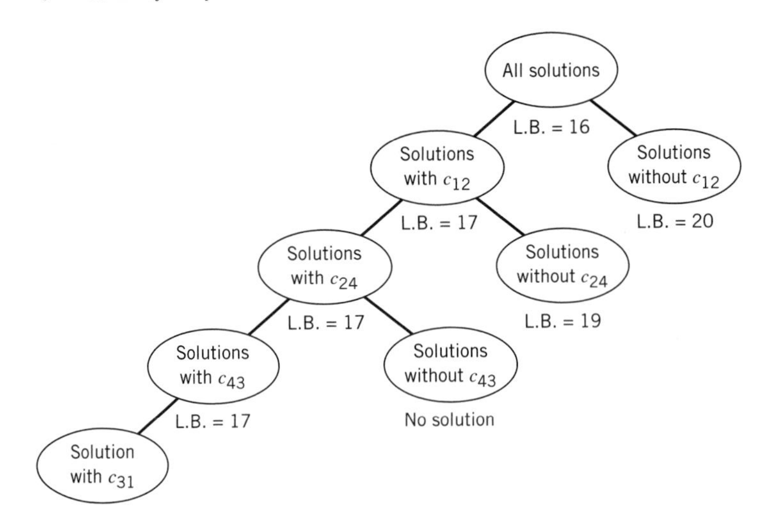

Figure 3.19

From \ To	1	2	3	4	5	6
1	∞	3	4	2	8	3
2	5	∞	3	4	4	5
3	4	1	∞	5	3	4
4	4	2	6	∞	4	5
5	3	3	3	5	∞	4
6	7	4	5	6	7	∞

Figure 3.20

salesperson problems. After describing the algorithm and giving an example of its use, we will prove that at worst the approximate algorithm's tour is always less than twice the cost of a true minimal tour. A bound of twice the true minimum may sound bad, but in many cases where a "ballpark" figure is needed (and the exact minimal tour can be computed later if needed), such a bound is quite acceptable. Furthermore, in practice our approximate algorithm finds a tour that is close to the true minimum. This algorithm uses a successive nearest-neighbor strategy.

Approximate Traveling Salesperson Tour Construction

1. Pick any vertex as a starting circuit C_1 consisting of 1 vertex.

2. Given the k-vertex circuit C_k, $k \geq 1$, find the vertex z_k not on C_k that is closest to a vertex, call it y_k, on C_k.

3. Let C_{k+1} be the $k+1$-vertex circuit obtained by inserting z_k immediately in front of y_k in C_k.

4. Repeat steps 2 and 3 until a Hamilton circuit (containing all vertices) is formed.

Example 1: Approximate Traveling Salesperson Tour

Let us apply the preceding algorithm to the 6-vertex traveling salesperson problem whose cost matrix is given in Figure 3.22. Name the vertices $x_1, x_2, x_3, x_4, x_5, x_6$. We will start with x_1 as C_1. Vertex x_4 is closest to x_1, and so C_2 is $x_1{-}x_4{-}x_1$. Vertex x_3 is the vertex not in C_2 closest to a vertex in C_2, namely, closest to x_4; thus $C_3 = x_1{-}x_3{-}x_4{-}x_1$. There are now two vertices, x_2 and x_6, 3 units from vertices in C_3. Suppose we pick x_2. It is inserted before x_3 to obtain $C_4 = x_1{-}x_2{-}x_3{-}x_4{-}x_1$. Vertex x_6 is still 3 units from x_1, and so we insert x_6 before x_1, obtaining $C_5 = x_1{-}x_2{-}x_3{-}x_4{-}x_6{-}x_1$. Finally, x_5 is within 4 units of x_3 and x_6. Inserting x_5 before x_6, we obtain our near-minimal tour $C_6 = x_1{-}x_2{-}x_3{-}x_4{-}x_5{-}x_6{-}x_1$, whose cost we compute to be 19.

In this case, the length of this tour obtained with the Approximate Tour Construction algorithm is quite close to the minimum (which happens to be 18). The length of the approximate tour typically depends on the starting vertex. If we suspected that our approximate tour was not that close to an optimal length, then by trying other vertices as the starting vertex (that forms C_1) and applying the algorithm, we will get other near-minimal tours. Taking the shortest of this set of tours generated by the Approximate Tour Construction would give us an improved estimate for the true minimal tour. ■

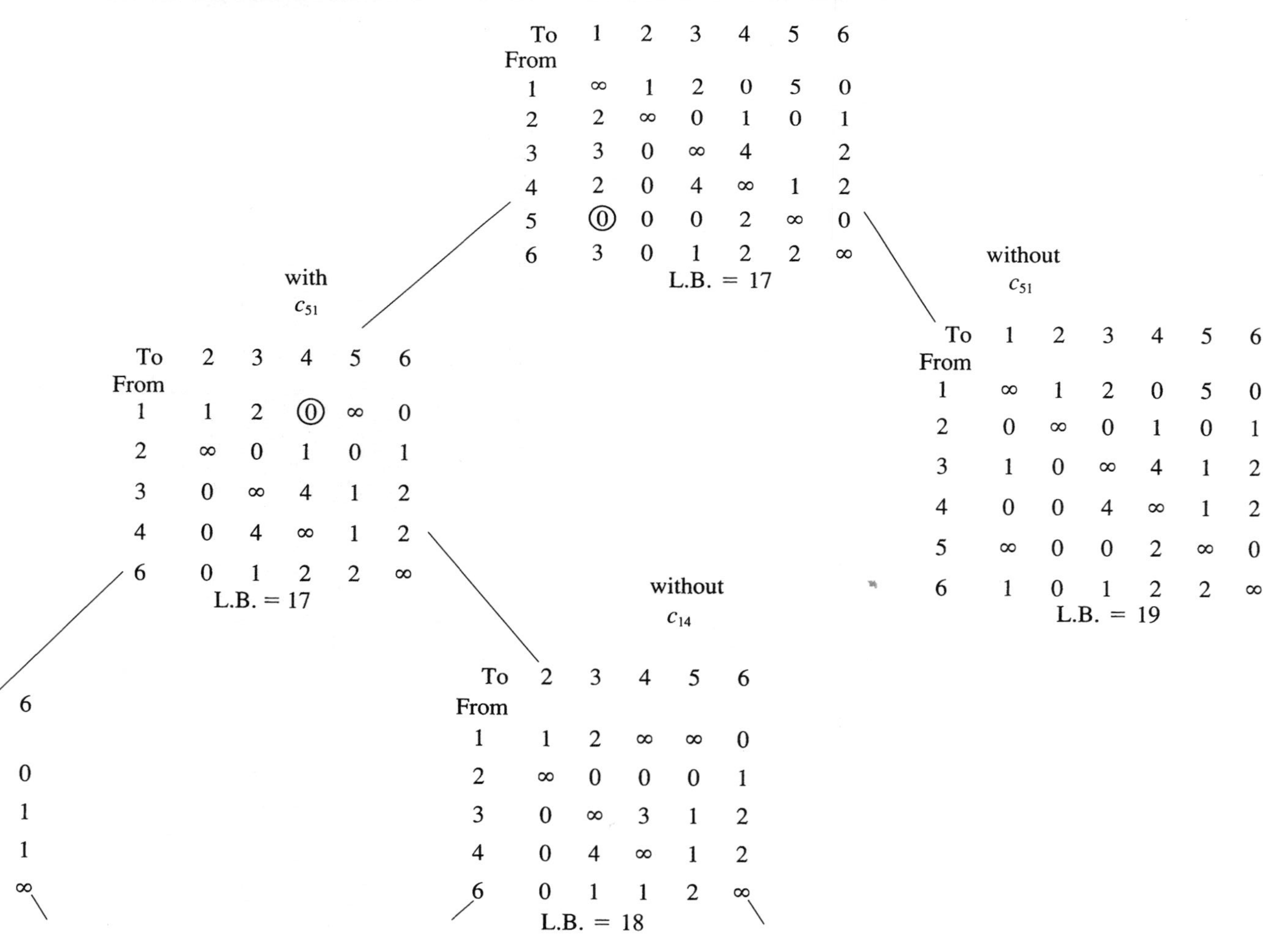

Figure 3.21 Tree of traveling salesperson partial solutions

	To	1	2	3	4	5	6
From	1	∞	3	3	2	7	3
	2	3	∞	3	4	5	5
	3	3	3	∞	1	4	4
	4	2	4	1	∞	5	5
	5	7	5	4	5	∞	4
	6	3	5	4	5	4	∞

Figure 3.22

Theorem

The cost of the tour generated by the approximate tour construction is less than twice the cost of the minimal traveling salesperson tour.

Proof *Optional*

Suppose we are successively building the k-vertex circuits C_k according to the approximate tour construction. Let S_k be a subset of edges in the true minimal tour C^* that connect the vertices in C_k to the other vertices in $G(S_k$ is described precisely below). When $k = 1$, we let S_1 consist of C^*-e^*, where e^* is the costliest edge in C^*. Since C_1 is a single vertex and C^*-e^* is a Hamilton path, $S_1 = C^*-e^*$ will connect C_1 with the rest of G.

For concreteness, see the 8-vertex graph in Figure 3.23a, where C_1 is the vertex x_6 and $x_1 = C^*-e^*$ is the Hamilton path of solid edges. If $z_1 = x_3$ and so C_2 is $x_6-x_3-x_6$, then we set $S_2 = S_1-(x_5, x_6)$ (C_2 is shown by dashed edges and S_2 by solid edges in Figure 3.23b). In general, we will obtain S_{k+1} from S_k by removing the first edge on the path in S_k from C_k to the new vertex z_k being added to C_k. In Figure 3.23c, suppose that $z_2 = x_8$ and $y_2 = x_6$. So x_8 is inserted into C_2 between x_3 and x_6 to obtain $C_3 = x_6-x_8-x_3-x_6$. Edge (x_6, x_7) is removed from S_2 to get S_3, since (x_6, x_7) is the first edge on the path in S_2 from C_2 (specifically x_6) to x_8.

By the shortest edge rule for picking $z_2 = x_8$ and $y_2 = x_6$, we know that c_{68} [the cost of edge (x_6, x_8)] is the smallest cost among all edges between $C_2(= x_6-x_3-x_6)$ and the rest of G. Thus $c_{68} \le c_{67}$ [where (x_6, x_7) was the edge removed from S_2 to get S_3]. Using this fact and the triangle inequality, we shall now prove that inserting x_8 into C_2 to get C_3 has a net increase in cost $\le 2c_{67}$. The increase of C_3 over C_2 is $c_{68} + c_{38} - c_{36}$, since edges (x_6, x_8) and (x_3, x_8) replace (x_3, x_6). But by the triangle inequality, $c_{38} \le c_{36} + c_{68}$, or equivalently $c_{38}-c_{36} \le c_{68}$. Thus $c_{68} + (c_{38}-c_{36}) \le c_{68} + c_{68} \le 2c_{67}$, as claimed.

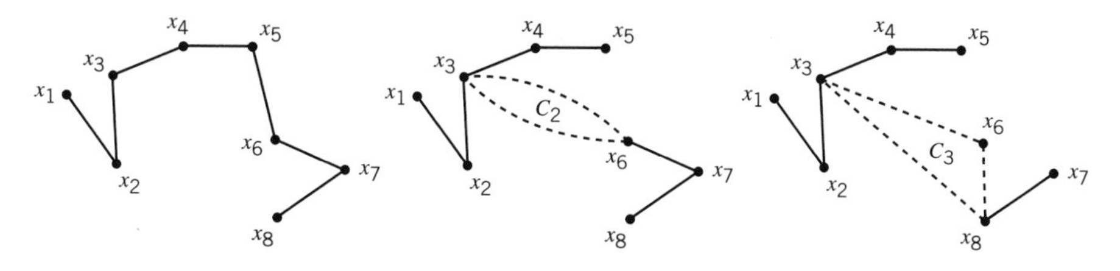

Figure 3.23

This same argument can be applied when we insert each successive z_k into C_k to prove that the increase of C_{k+1} over C_k is at most twice the cost of the edge dropped from S_k to get S_{k+1}. Starting from C_1 and repeating this bound on the insertion cost, we see that C_n, the final near-optimal Hamilton circuit, is bounded by twice the sum of the costs of the edges in $S_1 = C^* - e^*$, which is less than twice the cost of the minimal traveling salesperson tour C^*. ◆

3.3 EXERCISES

1. Subtract the value of each nondiagonal entry in Figure 3.13 from 10. Now solve the traveling salesperson problem for this new cost matrix.
2. Solve the traveling salesperson problem for the cost matrix in Figure 3.20.
3. Write a computer program to solve a six-city traveling salesperson problem and use it for the cost matrix in Figure 3.22.
4. Use ad hoc arguments to show that the cost of a minimal tour for the cost matrix in Figure 3.22 is 18.
5. Every month a plastics plant must make batches of five different types of plastic toys. There is a conversion cost c_{ij} in switching from the production of toy i to toy j, shown in the matrix. Find a sequence of toy production (to be followed for many months) that minimizes the sum of the monthly conversion costs.

To / From	T_1	T_2	T_3	T_4	T_5
T_1	—	3	2	4	3
T_2	4	—	4	5	5
T_3	5	3	—	4	4
T_4	3	5	1	—	6
T_5	5	4	2	3	—

6. The *assignment problem* is a matching problem with n people and n jobs and a cost matrix with entry c_{ij} representing the "cost" of assigning person i to job j. The goal is a one-to-one matching of people to jobs that minimizes the sum of the costs. Set all diagonal entries in Figure 3.13 equal to 5 and solve this 4×4 assignment problem by a branch-and-bound approach. How does an assignment problem differ from a traveling salesperson problem?
7. Use the approximate algorithm to find approximate traveling salesperson tours for the cost matrices in (use just entries above the main diagonal):

   (a) Figure 3.13 **(b)** Figure 3.20 **(c)** Exercise 5

8. Consider the following rule for building approximate traveling salesperson tours. Starting with a single-vertex tour T_1, successively add the vertex whose insertion into T_k to form T_{k+1} minimizes the increase in cost, that is, if x, is inserted between

x_k and x_{k+1} then $c_{kr} + c_{r(k+1)} - c_{k(k+1)}$ should be minimal over all choices of x_r and x_k.

(a) Can you prove an upper bound on this method similar to the one found for the approximate algorithm (making the same assumptions)?

(b) Apply this method to the cost matrix in (i) Figure 3.20 (using just entries above the main diagonal) and (ii) Figure 3.22.

9. Find a 3×3 cost matrix for which two different initial lower bounds can be obtained (with different sets of 0 entries) by subtracting from the rows and columns in different orders.

10. Make up a 5×5 cost matrix for which the Approximate Tour Construction finds:

(a) An optimal tour.

(b) A fairly costly tour (at least 50% over the true minimum).

3.4 TREE ANALYSIS OF SORTING ALGORITHMS

One of the basic combinatorial procedures in computer science is sorting a set of items. Items are sorted according to their own numerical value or the value of some associated variable. To simplify our discussion, we will assume that all items have distinct numerical values.

Some of the sorting procedures we mention will use binary trees explicitly. All the procedures use binary testing trees implicitly, that is, the procedures use a sequence of tests that compare various pairs of items. A binary-testing tree model of sorting has one important consequence. The tree must have at least $n!$ leaves if there are n items, since there are $n!$ possible permutations, or rearrangements, of the set. By Theorem 3 in Section 3.1, the height of a binary testing tree must be at least $\lceil \log_2 n! \rceil$, which is approximately $n \log_2 n$. Thus we obtain the theorem.

Theorem

In the worst case, the number of binary comparisons required to sort n items is at least $O(n \log_2 n)$.

The theorem refers to the worst case, because certain sorting algorithms will sort some sets very quickly (corresponding to short paths in the binary testing tree). A function $g(n)$ is said to be of order $O(f(n))$ if, for large values of n, $g(n) \leq cf(n)$, for some constant c. One can also show that the average number of binary comparisons required to sort n items is at best of order $O(n \log_2 n)$ (see Exercise 3).

The best known sorting algorithm is called **bubble sort,** so named because small items move up the list the way bubbles rise in a liquid. It is compactly written as:

```
FOR m ← 2 TO n DO
    FOR i← n STEP −1 TO m DO
        IF A_i ≤ A_{i−1}, THEN interchange items A_i and A_{i−1}
```

This procedure will always require $(n-1)+(n-2)+\cdots+1 = \frac{1}{2}n(n-1)$ binary comparisons. Thus this procedure requires $O(n^2)$ comparisons, as opposed to the theoretical bound of $O(n \log_2 n)$ comparisons. To see how much faster $O(n^2)$ grows than $O(n \log_2 n)$, observe that for $n = 50$, $n^2 = 2500$, and $n \log_2 n \approx 300$; and for $n = 500$, $n^2 = 250{,}000$, and $n \log_2 n \approx 4500$.

The simplest-to-state sorting procedure that achieves the $O(n \log_2 n)$ bound is **merge sort.** It recursively subdivides the original list and successive sublists in half (or as close to half as possible) until each sublist consists of one item. Then it successively merges the sublists in order. The subdivision process is naturally represented as a balanced binary tree and the merging as a reflected image of the tree.

Example 1: Merge Sort

Sort the list 5, 4, 0, 9, 2, 6, 7, 1, 3, 8 using a merge sort.

The subdivision tree and subsequent ordered merges are shown in Figure 3.24. ■

To analyze the number of binary comparisons in a merge sort, we make the simplifying assumption that $n = 2^r$, for some integer r. Then the subdivision tree will have sublists of size 2^{r-1} at the level-1 vertices. In general, there will be 2^{r-k} items in the sublists at level k. At level r, there are leaves each with one item.

In the merging tree, first pairs of leaves are ordered at each vertex on level $r-1$; this requires one binary comparison (to see which leaf item goes first). See Figure 3.24. In general, at each vertex on level k we merge the two ordered sublists (of 2^{r-k-1} items) of the 2 children into an ordered sublist of 2^{r-k} items; this merging will require $2^{r-k}-1$ binary comparisons (verification of this number is left as an exercise). Finally, the two ordered sublists of 2^{r-1} items at level 1 are merged at the root. The number of binary comparisons at all the vertices on level k is $2^k(2^{r-k}-1)$ since there are 2^k different vertices on level k. Summing over all levels, we compute the total number of binary comparisons

$$\begin{aligned}
\sum_{k=0}^{r-1} 2^k(2^{r-k}-1) &= \sum_{k=0}^{r-1}(2^r - 2^k) \\
&= \sum_{k=0}^{r-1} 2^r - \sum_{k=0}^{r-1} 2^k \\
&= r2^r - (2^r - 1) = (\log_2 n)n - (n-1)
\end{aligned}$$

since $n = 2^r$ and hence $r = \log_2 n$. Thus the number of binary comparisons in a merge sort is $O(n \log_2 n)$. However, extra computer time is required to implement the initial subdivision process. This extra work also requires only $O(n \log_2 n)$ steps.

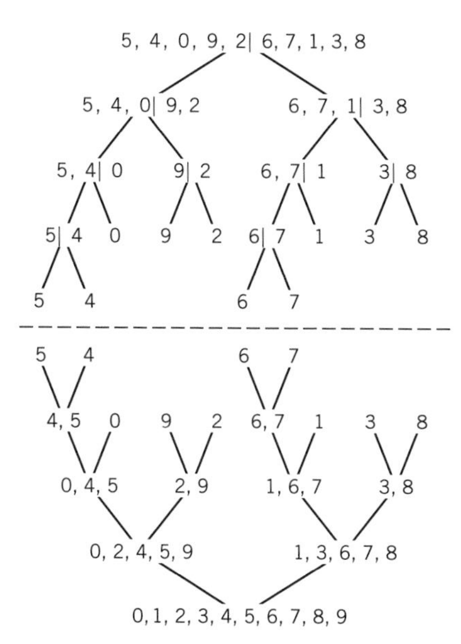

Figure 3.24

A closely related sorting procedure is called **QUIK sort.** In QUIK sort, one takes the first item A_1 in the list L and uses it to divide the rest of the list into sublists L_1 and L_2 such that all items in L_1 are less than A_1 and all items in L_2 are greater than A_1. This subdivision would require $n-1$ comparisons (of each item against A_1). Item A_1 is then put at the end of sublist L_1. Next the same subdivision procedure is used with the first item in each of the two sublists (L_1 is divided into L_{11} and L_{12}), and so on. Finally all sublists have only one item, but the sublists are ordered and by concatenating the sublists in order we obtain a sorted list.

This procedure is simpler to implement than a merge sort, but it may not divide the lists evenly (for example, A_1 could be the largest item so that L_1 is empty). In the worst case, QUIK sort can be shown to require as many comparisons as a bubble sort and hence is an $O(n^2)$ algorithm. But when QUIK sort is used with many randomly arranged lists, the average number of comparisons is only $O(n \log_2 n)$.

We conclude this section with a more complicated tree-based sorting procedure called **heap sort.** A **heap** is a binary or almost-binary (some internal vertices may have just one child) tree with each internal vertex's value being numerically greater than the values of its children. Figure 3.25*a* shows a heap involving the numbers 0

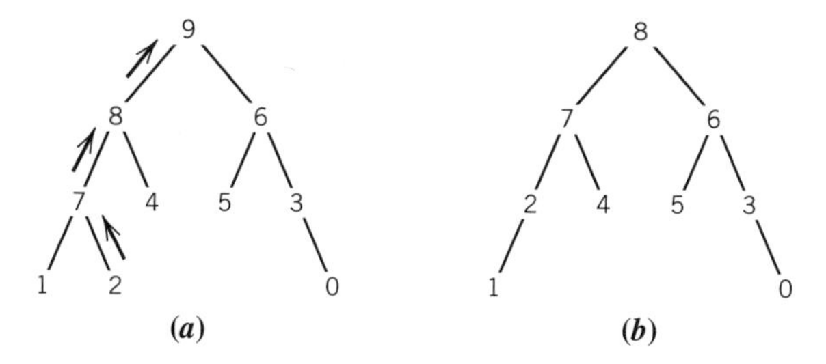

Figure 3.25

through 9. The root of the heap will have the largest value in the set. If we remove the root of the heap (and place it at the end of the ordered list we are building), then we can reestablish a heap by making the larger child of the root the new root and recursively replacing each internal vertex moved up with the larger of its two children. The arrows in Figure 3.25*a* show how vertices move up when the root 9 is removed. The new heap is shown in Figure 3.25*b*. Now we again remove the root of the heap (and make it the next-to-last item in the sorted list). Repeat this procedure until the heap is emptied and the sorted list complete. The one important missing step is creating the initial heap. This problem and further discussion of heap sort are left for the exercises.

3.4 EXERCISES

Summary of Exercises The first exercises are based on the sorting methods presented in this section. The remaining exercises present other sorting schemes.

1. Apply merge sort and QUIK sort to the following sequences:
   (a) 6, 9, 5, 0, 3, 1, 8, 4, 2, 7
   (b) 15, 27, 4, −7, 9, 13, 8, 28, 12, 20, −80
2. Show that at most $n - 1$ comparisons are needed to merge two sorted sublists into a single sorted list of n items.
3. Use the result of Exercise 13 in Section 3.1 to show that the average number of binary comparisons required to sort n items is at least $O(n \log_2 n)$.
4. **(a)** Find a way to represent QUIK's sublist divisions with a binary tree.
   (b) Draw this tree for a QUIK sort of the list in Exercise 1(a).
5. Give an example of an n-item list for which QUIK sort requires $\frac{1}{2}n(n-1)$ comparisons.
6. **(a)** Describe how to build an initial heap from an unordered list of n items (the initial heap should be a balanced tree).
   (b) Use your method in part (a) to make a heap for the lists in Exercise 1.
   (c) Apply heap sort to the heaps in part (b).
7. Show that heap sort requires $O(n \log_2 n)$ comparisons to sort n items (this includes the initial construction of a heap).
8. Modify the following sorting methods to allow for repeated (two or more equal) items:
   (a) Bubble sort **(b)** Merge sort **(c)** QUIK sort **(d)** Heap sort
9. Write a computer implementation of (use a recursive language such as PASCAL):
   (a) Merge sort **(b)** QUIK sort **(c)** Heap sort

10. Consider the following sorting scheme, which we call *tree sort*. We build a "dictionary look-up" binary tree recursively. The first item in the list to be sorted is the root. The second item is a left or right child of the root, depending on whether it numerically precedes or follows the root. We continue to add each successive item as a leaf to this growing tree. For example, if a list begins 8, 4, 11, 6,..., the tree after four items would be

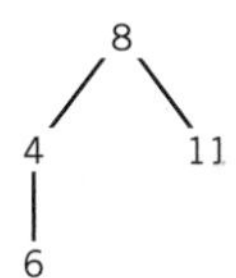

When all items have been incorporated into the tree, the sorted order is obtained by an inorder traversal of the tree (inorder traversals were defined in Section 3.2).

(a) Build the tree for the list in Exercise 1(a).

(b) Compare the tree in part (a) with the tree in Exercise 4(b) for this list.

(c) Generalize part (b) to show that tree sort is equivalent to QUIK sort.

11. Consider the following sorting scheme for the list $A_1, A_2, \ldots, A_n (n = 2^r)$. First do a sort (one comparison) of A_i and $A_{i+n/2}$; call this the ith ordered pair, for $i = 1, 2, \ldots, n/2$. Next do a merge sort of the ith and the $(i + n/4)$th ordered pairs; call this the ith 4-tuple. Next do a merge sort of the ith and the $(i + n/8)$th ordered 4-tuples. Continue until there is just one sorted list of all n items.

(a) Apply this sorting method to the list 10, 12, 7, 0, 5, 8, 11, 15, 1, 6, 3, 9, 13, 4, 2, 14.

(b) How many comparisons does this sorting method require for a list of n numbers?

3.5 SUMMARY AND REFERENCES

This chapter examined a variety of search and data organization problems. The common graph-theoretic tool for all these problems was trees. Section 3.1 presented basic properties and terminology of trees. Section 3.2 introduced spanning trees and demonstrated the uses of depth-first and breadth-first search. Section 3.3 gave tree-based branch-and-bound and heuristic approaches to a famous optimization problem, the traveling salesperson problem. Section 3.4 looked at the decision trees underlying sorting algorithms.

The first paper implicitly using trees was Kirchhoff's 1847 fundamental paper about electrical networks. Cayley was the first person to use the term *tree* in an 1857 formula for counting ordered trees (Theorem 4 in Section 3.1). Searching methods have been around for years (see Lucas [3]), but a systematic development of this subject came only in recent years with the advent of digital computers. There are several good computer science books about searching, sorting, and graph algorithms. See Kruse [1] or Sedgewick [4]. For a good survey of the Traveling Salesperson Problem, see Lawler et al. [2].

1. R. Kruse, *Data Structures and Program Design in C++,* Prentice-Hall, Upper Saddle River, NJ, 1998.

2. E. Lawler, J. Lenstra, A. Rinnooy Kan, and D. Shmoys, *The Traveling Salesman Problem,* John Wiley, New York, 1985.

3. E. Lucas, *Recreations Mathematiques,* Gauthier-Villars, Paris, 1891.

4. R. Sedgewick, *Algorithms in C,* 3rd. ed. Addison-Wesley, Reading MA, 1997.

CHAPTER 4
NETWORK ALGORITHMS

4.1 SHORTEST PATHS

In this chapter we present algorithms for the solution of three important network optimization problems: shortest paths, minimal spanning trees, and maximal flows. By a **network** we mean a graph with a positive integer $k(e)$ assigned to each edge e. This integer will typically represent the "length" of an edge, in units such as miles, or represent "capacity" of an edge, in units such as megawatts or gallons per minute. The optimization problems we shall discuss arise in hundreds of different guises in management science and system analysis settings. Thus good systematic procedures for their solution are essential. In the case of network flows, we shall see that the flow optimization algorithm can also be used to prove several well-known combinatorial theorems.

We begin with an algorithm for a relatively simple problem, finding a shortest path in a network from point a to point z. We do not say *the* shortest path because, in general, there may be more than one shortest path from a to z. For the rest of this section, let us assume that all networks are *undirected* and *connected.*

Let us immediately eliminate one possible shortest path algorithm: Determine the lengths of all paths from a to z, and choose a shortest one. The computer is fast, but not that fast—such enumeration is already infeasible for most networks with 20 vertices. So when we find a shortest path, we must be able to prove it is shortest without explicitly comparing it with all other a–z paths. Although the problem is now starting to sound difficult, there still is a straightforward algorithmic solution.

The algorithm we present is due to Dijkstra. This algorithm gives shortest paths from a given vertex a to all other vertices. Let $k(e)$ denote the length of edge e. Let the variable m be a "distance counter." For increasing values of m, the algorithm labels vertices whose minimal distance from vertex a is m. The first label of a vertex x will be the previous vertex on the shortest path from a to x. The second label of x will be the length of the shortest path from a to x.

Shortest Path Algorithm

1. Set $m = 1$ and label vertex a with $(-, 0)$ (the "$-$" represents a blank).

2. Check each edge $e = (p, q)$ from some labeled vertex p to some unlabeled vertex q. Suppose p's labels are $[r, d(p)]$. If $d(p) + k(e) = m$, label q with (p, m).

3. If all vertices are not yet labeled, increment m by 1 and go to Step 2. Otherwise go to Step 4. If we are only interested in a shortest path to z, then we go to Step 4 when z is labeled.

4. For any vertex y, a shortest path from a to y has length $d(y)$, the second label of y. Such a path may be found by backtracking from y (using the first labels) as described below.

Observe that instead of concentrating on the distances to specific vertices, this algorithm solves the questions: How far can we get in 1 unit, how far in 2 units, in 3 units, ..., in m units, ...? Formal verification of this algorithm requires an induction proof (based on the number of labeled vertices).

The key idea is that to find a shortest path from a to any other vertex we must first find shortest paths from a to the "intervening" vertices. If $P_k = (s_1, s_2, \dots, s_k)$ is a shortest path from $s_1 = a$ to s_k then $P_k = P_{k-1} + (s_{k-1}, s_k)$, where $P_{k-1} = (s_1, s_2, \dots, s_{k-1})$ is a shortest path to s_{k-1}. Similarly $P_{k-1} = P_{k-2} + (s_{k-2}, s_{k-1})$, and so on.

To record a shortest path to s_k all we need to store (as the first part of a label in the above algorithm) is the name of the next-to-last vertex on P_k namely, s_{k-1}. Preceding s_{k-1} on the shortest path from a is s_{k-2}, the next-to-last vertex on P_{k-1}. By continuing this backtracking process, we can recover all of P_k.

The algorithm given above has two significant inefficiencies. First, if all sums $d(p) + k(e)$ in Step 2 have values of at least $m' > m$, then the distance counter m should be increased immediately to m'. Second, one does not need to repeat the computation of the expressions $d(p) + k(e)$ in Step 2 every time m increases. Instead, each unlabeled vertex q can be given a "temporary" label $(p, d^*(q))$, where $d^*(q)$ equals the minimum of the $d(p) + k(e)$, for those labeled vertices p with an edge $e = (p, q)$ to q. Thus, $d^*(q)$ represents the current shortest path length to q using labeled vertices. Now q's temporary label needs to be updated only when a newly labeled vertex p is adjacent to q. Details of this and other improvements in the form of Dijkstra's algorithm given here can be found in Ahuja et al. [1].

Example 1: Shortest Path

A newly married couple, upon finding that they are incompatible, want to find a shortest path from point N (Niagara Falls) to point R (Reno) in the road network shown in Figure 4.1. We apply the shortest path algorithm. First N is labeled $(-, 0)$. For $m = 1$, no new labeling can be done [we check edges $(N, b), (N, d)$, and (N, f)]. For $m = 2$, $d(N) + k(N, b) = 0 + 2 = 2$, and we label b with $(N, 2)$. For $m = 3, 4$, no new labeling can be done. For $m = 5$, $d(b) + k(b, c) = 2 + 3 = 5$. So we label c with $(b, 5)$. We continue to obtain the labeling shown in Figure 4.1. Backtracking from R, we find the shortest path to be *N–b–c–d–h–k–j–m–R* with length 24. ∎

If we want simultaneously to find the shortest distances between all pairs of vertices (without directly finding all the associated shortest paths), we can use the

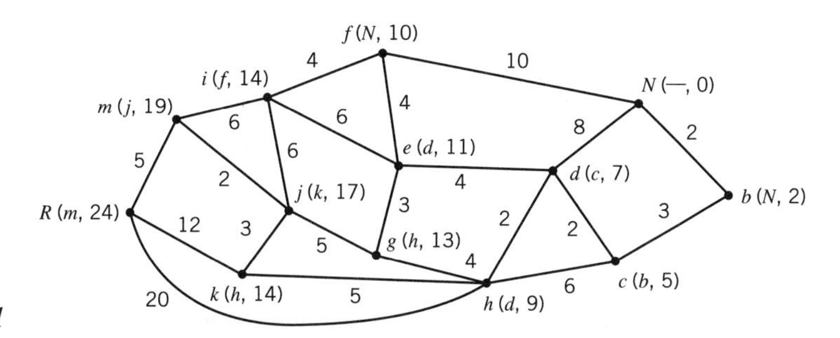

Figure 4.1

following simple algorithm due to Floyd. Let matrix D have entry $d_{ij} = \infty$ (or a very large number) if there is no edge from the ith vertex to the jth vertex; otherwise d_{ij} is the length of the edge from x_i, to x_j. Then Floyd's algorithm is most easily stated with the following deceptively simple computer program:

```
FOR k←1 TO n DO
   FOR i←1 TO n DO
      FOR j←1 TO n DO
          IF dik + dkj < dij THEN dij ← dik + dkj;
```

When finished, d_{ij} will be the shortest distance from the ith vertex to the jth vertex.

4.1 EXERCISES

Summary of Exercises The first six exercises involve shortest path calculations. The remaining exercises discuss associated theory (fairly easy theory), and the last exercise asks for a program.

1. Use the shortest path algorithm to find the shortest path between vertex c and vertex m in Figure 4.1.
2. Find the shortest path between the following pairs of vertices in the network in Figure 4.2 in Section 4.2.

   (a) a and y **(b)** d and r **(c)** e and g

3. The network below shows the paths to success from L (log cabin) to W (White House). The first number on an edge is the time (number of years) it takes to traverse the edge; the second is the number of enemies you make in taking that edge. Use the shortest path algorithm to answer the following:

   (a) Find the quickest path to success (from L to W).

   (b) Find the quickest path to success that avoids points c, g, and k.

   (c) Find the path to success (from L to W) that minimizes the total number of enemies you make.

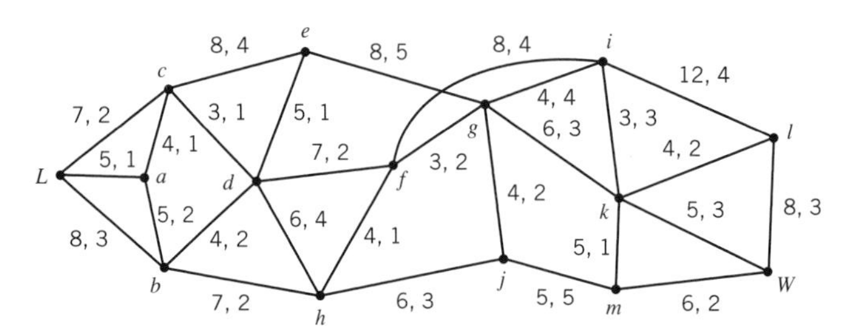

(d) Alter the order in which edges are checked in Step 2 of the shortest path algorithm to get another minimal-enemies path. How many such minimal-enemies paths are there?

4. With reference to Exercise 3, let the roughness index R of a path to success be $R = T + 2E$, where T is the time to get from L to W and E is the total number of enemies made.

(a) Find the smoothest (least rough) path to success.

(b) Find the smoothest path to success that includes edge (f, i); this edge can be traversed in either direction.

5. Ignore the numbers on the edges in Exercise 3 and use the shortest path algorithm to find the following shortest (fewest edges) paths:

(a) Shortest path from L to W.

(b) Shortest path from L to W including vertex d.

(c) Shortest path from L to W including both vertex e and vertex m.

6. Suppose that the edges in Exercise 3 are directed according to the alphabetical order of the endpoints, where L precedes a and W follows m [so edge $(r \vec{,} s)$ goes from r to s if r precedes s in the alphabet]. Find the quickest path from L to W in this directed network.

7. Prove by induction on m that Dijkstra's shortest path algorithm finds the shortest path from a to every other vertex in the network.

8. Prove that Floyd's algorithm finds the shortest path between all pairs of vertices.

9. Make up an example to show that Dijkstra's algorithm fails if negative edge lengths are allowed.

10. **(a)** Show that if the edges are properly ordered (and the edges are checked in this order in Step 2), the shortest path algorithm will produce any given shortest path from a to z when more than one shortest path exists.

(b) Order the edges of the figure in Exercise 3 so that in Exercise 3(c), the "shortest" path found will be $(L, a, c, d, f, g, k, m, W)$.

11. Show that in the shortest path algorithm the edges used in Step 2 to label new vertices form a spanning tree.

12. The *transitive closure* of a directed graph G is obtained by adding to G an edge $(x_i \vec{,} x_j)$ for each nonadjacent pair x_i, x_j with a directed path from to x_i to x_j. Let $d_{ij} = 1$ if $(x_i \vec{,} x_j)$ is an edge in G and $= 0$ otherwise. Replace the IF statement in

Floyd's algorithm with

$$\texttt{IF } d_{ik} \cdot d_{kj} > d_{ij} \texttt{ THEN } d_{ij} \leftarrow 1$$

Show that this revised Floyd's algorithm finds the transitive closure of G.

13. Write a computer program implementing Dijkstra's shortest path algorithm.

4.2 MINIMAL SPANNING TREES

A **minimal spanning tree** in a network is a spanning tree whose sum of edge lengths $k(e)$ is as small as possible. Minimal spanning trees arise in a variety of important commercial settings, such as finding a minimal-length fiber-optic network to link a given set of sites. This problem appears to be harder than finding a shortest path between two given vertices, but with the proper algorithm it is actually easier to solve by hand than the shortest path problem. The reason for the simple solution is that there are straightforward "greedy" algorithms that can build a minimal spanning tree by successively picking a shortest available edge.

We present two greedy algorithms for building a minimal spanning tree. Let n denote the number of vertices in the network.

Kruskal's Algorithm

Repeat the following step until the set T has $n - 1$ edges (initially T is empty): Add to T the shortest edge that does not form a circuit with edges already in T.

Prim's Algorithm

Repeat the following step until tree T has $n - 1$ edges: Add to T the shortest edge between a vertex in T and a vertex not in T (initially pick any edge of shortest length).

In both algorithms, when there is a tie for the shortest edge to be added, any of the tied edges may be chosen. A proof of the validity of Kruskal's algorithm is given in Exercise 12.

Example 1: Minimal Spanning Tree

We seek a minimal spanning tree for the network in Figure 4.2. Both algorithms start with a shortest edge. There are three edges of length 1: (a, f), (l, q), and (r, w). Suppose we pick (a, f). If we follow Prim's algorithm, the next edge we would add is (a, b) of length 2, then (f, g) of length 4, (g, l), then (l, q), then (l, m), and so forth. The next-to-last addition would be either (m, n) or (o, t) both of length 5 [suppose we choose (m, n)], and either one would be followed by (n, o). The final tree is indicated with darkened lines in Figure 4.2.

On the other hand, if we follow Kruskal's algorithm, we first include all three edges of length 1: (a, f), (l, q), (r, w). Next we would add all the edges of length 2: (a, b), (e, j), (g, l), (h, i), (l, m), (p, u), (s, x), (x, y). Next we would add almost all the edges of length 3: (c, h), (d, e), (k, l), (k, p), (q, v), (r, s), (v, w), but not (w, x) unless (r, s) were omitted [if both were present we would get a circuit containing these

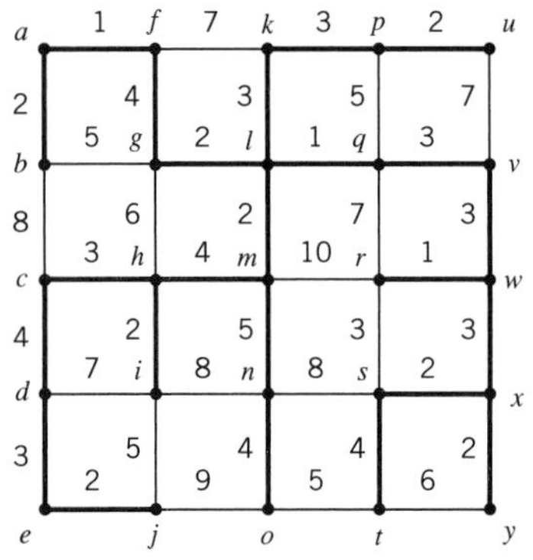

Figure 4.2

two edges together with edges (r, w) and (s, x)]. Next we would add all the edges of length 4 and finally either (m, n), or (o, t) to obtain the same minimal spanning tree(s) produced by Prim's algorithm. This similarity is no coincidence (see Exercise 14). ■

The difficult part in the minimal spanning tree problem is proving the minimality of the two algorithms. We give the proof for Prim's algorithm and leave Kruskal's as an exercise (Exercise 12).

Theorem

Prim's algorithm yields a minimal spanning tree.

Proof

For simplicity, assume that the edges all have different lengths. Let T' be a minimal spanning tree chosen to have as many edges as possible in common with the tree T^* constructed by Prim's algorithm.

If $T^* \neq T'$, let $e_k = (a, b)$ be the first edge chosen by Prim's algorithm that is not in T'. This means that the subtree T_{k-1}, composed of the first $k - 1$ edges chosen by Prim's algorithm, is part of the minimal tree T'. In Figure 4.3, edges of T_{k-1} are in bold, e_k's edge is dashed, and the other edges of T' are drawn normally. Since e_k is not in the minimal spanning tree T', there is a path, call it P, in T' that connects a to b. At least one of the edges of P must not be in T_{k-1}, for otherwise $P \cup e_k$ would form a circuit in the tree T_k formed by the first k edges in Prim's algorithm. *Let* e^* be the first edge along P (starting from a) that is not in T_{k-1} (see Figure 4.3). Note that one end vertex of e^* is in T_{k-1}.

If e^* is shorter than e_k, then on the kth iteration, Prim's algorithm would have incorporated e^* not e_k. If e^* is greater than e_k, we remove e^* from T' and replace it

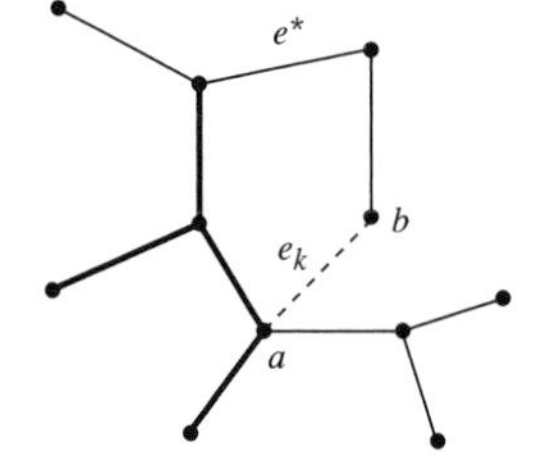

Figure 4.3

with e_k. The new T' is still a spanning tree (see Exercise 10 for details), but its length is shorter—this contradicts the minimality of the original T'. ◆

4.2 EXERCISES

Summary of Exercises The first five exercises involve building minimal spanning trees. Exercises 6–15 are variations and theory questions about the two minimal spanning tree algorithms.

1. The network below shows the paths to success from L (log cabin) to W (White House). The first number is the cost of building a freeway along the edge and the second is the number of trees (in thousands) that would have to be cut down.

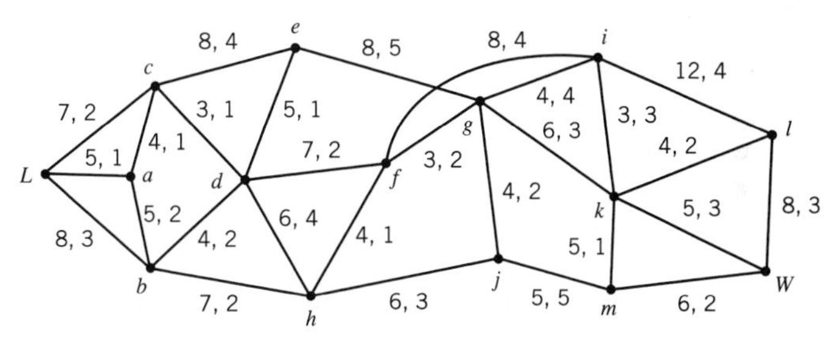

   (a) Use Kruskal's algorithm to find a minimal-cost set of freeways connecting all the vertices together.

   (b) Court action by conservationists rules out use of edges (c, e), (d, f), and (k, W). Now find the minimal-cost set of freeways.

   (c) Find two nonadjacent vertices such that the tree in part (a) does not contain the cheapest path between them.

   (d) Use Prim's algorithm to find a set of connecting freeways that minimizes the number of trees cut down.

2. With reference to Exercise 1, suppose that the set of freeways must include city c or city d (but not necessarily both) and all the other cities. Find the minimal-cost set of freeways.

3. Find a minimal spanning tree for the network in Figure 4.1 using:

   (a) Prim's algorithm (b) Kruskal's algorithm

4. With reference to Exercise 1, suppose that the governor's summer home is along edge (f, i). Find a minimal-cost set of freeways such that (f, i) is in that set.

5. Find a *maximal* spanning tree (whose sum of edge lengths is maximal) for the network in Figure 4.2.

6. If each edge has a different cost, show that the minimal spanning tree is unique.

7. Modify Kruskal's algorithm so that it finds a minimal spanning tree that contains a prescribed edge. Prove that your modification works.

8. Modify Prim's algorithm so that it finds a maximal spanning tree.

9. In the proof of the Theorem, show that:

(a) $e_k = (a, b)$ has one end vertex (a or b) in the tree T_{k-1}.

(b) e^* has one of its end vertices in the tree T_{k-1}.

10. Show that the new T' (mentioned in the last sentence of the proof of the Theorem) with e^* replaced by e_k is a tree, that is, that T' is connected and circuit-free.

11. Let T be the spanning tree found by Prim's algorithm in an undirected, connected network N.

(a) Prove that T contains all edges of shortest length in N unless such edges include a circuit.

(b) Prove that if $e^* = (a, b)$ is any edge of N not in T and if P is the unique path in T from a to b, then for each edge e in P, $k(e) \leq k(e^*)$.

(c) Prove that part (b) characterizes a minimal spanning tree, that is, any spanning tree T (not just those formed by Prim's algorithm) is a minimal spanning tree if and only if part (b) is always true for T.

12. Prove that Kruskal's algorithm gives a minimal spanning tree.

13. Construct an undirected, connected network with 8 vertices and 15 edges that has a minimal spanning tree containing the shortest path between every pair of vertices.

14. Suppose that the edges of the undirected, connected network N are ordered and that in both Prim's and Kruskal's algorithms, when there is a tie for the next edge to be added, the smaller indexed edge is chosen.

(a) Prove that the edges can be ordered so that Prim's algorithm will yield any given minimal spanning tree.

(b) Prove that the edges can be ordered so that Kruskal's algorithm will yield any given minimal spanning tree.

(c) Prove that with ordered edges, both algorithms give the same tree.

15. Given an undirected, connected n-vertex graph G with lengths assigned to each edge, we form a graph G_N whose vertices correspond to minimal spanning trees of G with two vertices v_1, v_2 adjacent if the corresponding minimal spanning trees T_1, T_2 differ by one edge, that is, $T_1 = T_2 - e' + e''$ (for some e', e'').

(a) Produce an 8-vertex network G such that G_N is a chordless 4-circuit.

(b) Prove that if T_1 and T_2 are minimal spanning trees which differ by k edges, that is $|T_1 \cap T_2| = n - k$, then in G_N there is a path of length k between the corresponding vertices.

16. Write a computer program to implement (as efficiently as possible):

(a) Kruskal's algorithm **(b)** Prim's algorithm

4.3 NETWORK FLOWS

In this section we interpret the integer $k(e)$ associated with edge e in a network as a capacity. We seek to maximize a "flow" from vertex a to vertex z such that the flow in each edge does not exceed that edge's capacity. Many transport problems are of this general form, for example, maximizing the flow of oil from Houston to New York through a large pipeline network (here the capacity of an edge represents the capacity in barrels per minute of a section of pipeline), or maximizing the number of telephone calls possible between New York and Los Angeles through the wires in a telephone network. It is convenient to assume initially that *all networks are directed.*

We define an a–z **flow** $f(e)$ in a directed network N to be an integer-valued function $f(e)$ defined on each edge e—$f(e)$ is the flow in e— together with a **source** vertex a and a **sink** vertex z satisfying the following conditions. In(x) and Out(x) denote the sets of edges directed into and out from vertex x, respectively.

(**a**) $0 \le f(e) \le k(e)$

(**b**) For $x \neq a$ or z, $\displaystyle\sum_{e\in\mathrm{In}(x)} f(e) = \sum_{e\in\mathrm{Out}(x)} f(e).$

To simplify our analysis, we assume that the flow goes from a to z, never in the reverse direction:

(**c**) $f(e) = 0$ if $e \in \mathrm{In}(a)$ or $e \in \mathrm{Out}(z)$.

A sample flow is shown in Figure 4.4; the capacity and flow in each edge e are written beside the edge: $k(e), f(e)$. The assumption of integer capacities and flows is not restrictive, that is, the units we count could be thousandths of an ounce rather than barrels. The requirement that there be a single supply vertex, the source a, and a single demand vertex, the sink z, also turns out not to be a restriction. The reason is that we can recast network problems with multiple sources and multiple sinks into a single-source, single-sink network, as illustrated in the following example.

Example 1: Flow Networks with Supplies and Demands

Consider the network of solid edges in Figure 4.5 with supplies and demands. Vertex b can supply up to 60 units of flow, and vertices c and d can each supply 40 units. Vertices h, i, and j have flow demands of 50, 40, and 40 units, respectively. Can we

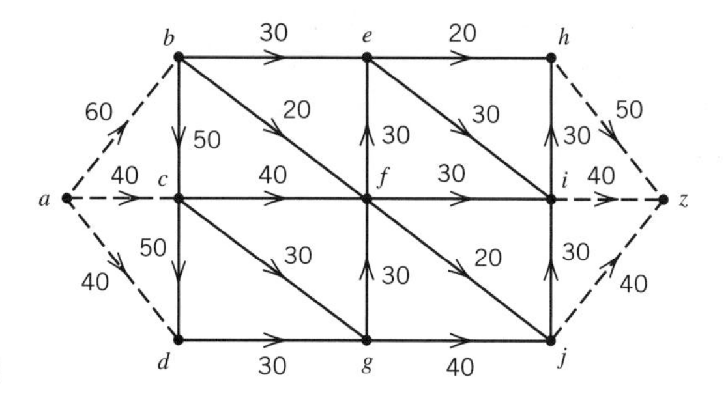

Figure 4.4

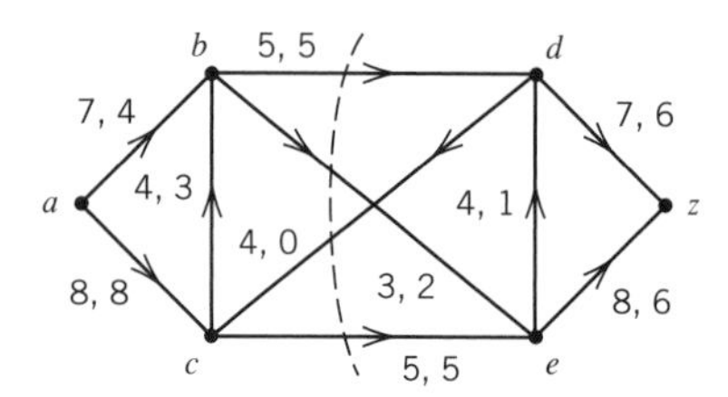

Figure 4.5

meet all the demands? The sources could be oil refineries and the sinks oil-truck distribution centers, or the sources factories and the sinks warehouses.

We model this network problem with a standard one-source, one-sink network as follows. Make b, c, and d regular nonsource vertices. Add a new source vertex a and edges $(a \vec{,} b)$, $(a \vec{,} c)$, and $(a \vec{,} d)$ having capacities 60, 40, and 40, respectively. This construction simulates the role of the capacitated sources b, c, d. Next make h, i, and j regular nonsink vertices and add a new sink vertex z and edges $(h \vec{,} z)$, $(i \vec{,} z)$, and $(j \vec{,} z)$, having capacities 50, 40, and 40, respectively.

A flow satisfying the original demands is equivalent to a flow in the new network that saturates the edges coming into z, that is, a flow of value 130. Such a flow exists if no a–z cut has capacity less than 130. ■

Let $(P, \overline{P})$ denote the set of all edges $(x \vec{,} y)$ with vertex $x \in P$ and $y \in \overline{P})$ (where $\overline{P}$ denotes the complement of P). We call such a set $(P, \overline{P})$ a **cut.** The cut $(\{a, b, c\}, \{d, e, z\})$ in Figure 4.5 consists of the edges $(b \vec{,} d)$, $(b \vec{,} e)$, $(c \vec{,} e)$; the edge $(d \vec{,} c)$ is not in the cut because it goes from $\overline{P}$ to P. This cut is represented in Figure 4.5 by a dashed line that separates the vertices in $P = \{a, b, c\}$ from the vertices in $\overline{P}$; the edges in the cut are the edges crossing the dashed line from left to right. We call $(P, \overline{P})$ an **a–z cut** if $a \in P$ and $z \in \overline{P}$.

Let $f(e)$ be an a–z flow in the network N and let P be a subset of vertices in N not containing a or z. Summing together the conservation-of-flow equations in condition (b) for each $x \in P$, we obtain

$$\sum_{x \in \mathrm{P}} \sum_{e \in \mathrm{In}(x)} f(e) = \sum_{x \in \mathrm{P}} \sum_{e \in \mathrm{Out}(x)} f(e)$$

Certain $f(e)$ terms occur on both sides of the preceding equality; namely, edges from one vertex in P to another vertex in P. After eliminating such $f(e)$ from both sides, the left side becomes $\sum_{e \in (\overline{P}, P)} f(e)$ and the right side $\sum_{e \in (P, \overline{P})} f(e)$. Thus we have:

(**b**$'$) For each vertex subset P not containing a or z,

$$\sum_{e \in (\bar{P}, P)} f(e) = \sum_{e \in (P, \overline{P})} f(e)$$

That is, the flow into P equals the flow out of P. The following intuitive result is readily verified with (b').

Theorem 1

For any a–z flow $f(e)$ in a network N, the flow out of a equals the flow into z.

Proof

Assume temporarily that N contains no edge $(\vec{a, z})$. Let P be all vertices in N except a and z. So $\overline{P} = \{a, z\}$.

Remember that the only flow into P from $\{a, z\}$ must be from a, since condition c, forbids flow from z. Similarly by condition c, all flow out of P must go to z. Then $(P, \overline{P})$ consists of edges going into z (from P) and $(\overline{P}, P)$ consists of edges going out of a (to P). Thus

$$\text{flow out of } a = \sum_{e \in (\overline{P}, P)} f(e) \overset{\text{by}(b')}{=} \sum_{e \in (P, \overline{P})} f(e) = \text{flow into } z$$

The flow equality still holds if there is flow in an edge $(\vec{a, z})$. ◆

The **value** of the a–z flow $f(e)$, denoted $|f|$, equals the sum of the flow in edges coming out of a, or equivalently by Theorem 1, the flow into z. Let us consider the question of how large $|f|$ can be. One upper bound is the sum of the capacities of the edges leaving a, since

$$|f| = \sum_{e \in \mathrm{Out}(a)} f(e) \leq \sum_{e \in \mathrm{Out}(a)} k(e)$$

Similarly, the sum of the capacities of the edges entering z is an upper bound for $|f|$. Intuitively, $|f|$ is bounded by the sum of the capacities of any set of edges that cut all flow from a to z.

We define the **capacity** $k(P, \overline{P})$ of the cut $(P, \overline{P})$ to be

$$k(P, \overline{P}) = \sum_{e \in (P, \overline{P})} k(e)$$

The capacity of the a–z cut $(P, \overline{P})$, where $P = \{a, b, c\}$, in Figure 4.5 is 13. This tells us that no a–z flow in the network in Figure 4.5 can have a value greater than 13. This motivates the following theorem.

Theorem 2

For any a–z flow f and any a–z cut $(P, \overline{P})$ in a network N, $|f| \leq k(P, \overline{P})$.

Proof

Informally, $|f|$ should equal the total flow from P to $\overline{P}$ [which is bounded by $k(P, \overline{P})$], but we cannot prove an equality of this type using condition (b') without first modifying N. We cannot use (b') in N because condition (b') requires that P not contain the source a.

Expand the network N by adding a new source vertex a' with an edge $e' = (\vec{a', a})$ of immense capacity. Assign a flow value of $|f|$ to e', yielding a valid a'–z flow in the expanded network (see Figure 4.6). In effect, the old source a now gets its flow from the "super source" a'. Note that a' will be part of $\overline{P}$. Observe that new source a' is playing a role similar to the source a, we added to the multiple-source network in Example 1.

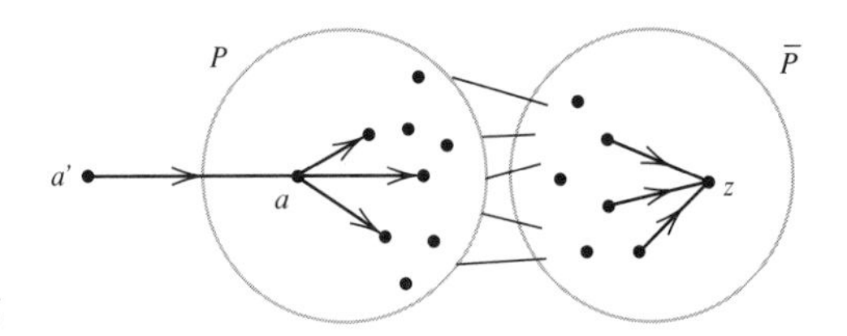

Figure 4.6

In the new network, we can apply condition (b') to P. It says that the flow into P, which is at least $|f|$ (other flow could come into P along edges from $\overline{P}$), equals the flow out of P Thus

$$|f| \le \sum_{e\in(\overline{P},P)} f(e) \overset{\text{by}(b')}{=} \sum_{e\in(P,\overline{P})} f(e) \le \sum_{e\in(P,\overline{P})} k(e) = k(P,\overline{P}) \ \blacklozenge \qquad (*)$$

Corollary 2a

For any a–z flow f and any a–z cut $(P, \overline{P})$ in a network N, $|f| = k(P, \overline{P})$ if and only if:

(i) For each edge $e \in (\overline{P}, P)$, $f(e) = 0$.

(ii) For each edge $e \in (P, \overline{P})$, $f(e) = k(e)$.

Further, when $|f| = k(P, \overline{P})$, f is a maximal flow and $(P, \overline{P})$ is an a–z cut of minimal capacity.

Proof

Consider the two inequalities in (*) of the preceding proof. The first inequality is an equality—the flow from $\overline{P}$ into P (in the expanded network) equals $|f|$—if condition (i) holds; otherwise the flow into P is greater than $|f|$. The second inequality is an equality—the flow out of P equals $k(P, \overline{P})$—if condition (ii) holds; otherwise it is less. Thus equality holds in (*) if and only if conditions (i) and (ii) are both true. The last sentence in the corollary follows directly from Theorem 2. ◆

Now that we have developed all the concepts needed to present our flow maximizing algorithm, we next discuss an intuitive but faulty technique that can sometimes be used as a shortcut in place of the correct algorithm. After the fault in the shortcut is exposed, the algorithm can be more easily understood and appreciated.

All normal flows can be decomposed into a sum of **unit-flow paths** from a to z, for short, ***a–z* unit flows** (abnormal flows that cannot be so decomposed are discussed in Exercise 23). For example, in a telephone network, the flow from New York to Los Angeles can be decomposed into individual telephone calls. Similarly, flow of oil in a pipeline network can be decomposed into the paths of each individual petroleum molecule. Formally, an a–z unit flow f_L along a–z path L is defined as $f_L(e) = 1$ if e is in L and $= 0$ if e is not in L.

This suggests a way to build a maximal flow. We build up the flow as much as possible by successively adding a–z unit flows together, always being sure not to exceed any edge's capacity. An additional unit flow can use only **unsaturated** edges, edges where the present flow does not equal the capacity. So we must build paths consisting of unsaturated edges. We define the **slack** $s(e)$ of edge e in flow f by

$$s(e) = k(e) - f(e)$$

If s is the minimum slack among edges in the a–z unit flow f_L, then we can use an additional flow along L of $sf_L = f_L + f_L + f_L + \cdots + f_L$ (s times).

If $f_1, f_2, \ldots, f_m$ are a–z unit flows, then $f = f_1 + f_2 + \cdots + f_m$ will satisfy conditions (b) and (c) in the definition of a flow (given at the beginning of this section) since $f_1, f_2, \ldots, f_m$ satisfy these conditions. If in addition, φ satisfies condition (a)—$f(e) \leq k(e)$, for all e—then f is a valid flow.

Example 2: Building a Flow with Flow Paths

Let us use the method just outlined to build a maximal a–z flow for the network in Figure 4.7*a*. Note, as an upper bound, that the value of a flow cannot exceed 10, the capacity of edges going out of a.

We start with no flow, that is, $f(e) = 0$, for all e. Now we find some path from a to z, for example, the a–z path $L_1 = a$–b–d–z. The minimum slack on L_1 is 3 (at the start, the slack of each edge is just its capacity). So to our initial zero flow, we add the flow $3f_{L_1}$. All edges except $(b\vec{,}d)$ are still unsaturated.

Suppose that we next find the a–z path $L_2 = a$–c–e–z, also with minimum slack 3. Our current flow is $3f_{L_1} + 3f_{L_2}$, as shown in Figure 4.7*b*. The path $L_3 = a$–b–e–z with minimum slack 2 can be used to get the augmenting flow $2f_{L_3}$. Figure 4.7*c* shows the remaining unsaturated edges with their slacks.

The only a–z path in Figure 4.7*c* is $L_4 = a$–c–d–z with minimum slack 1. After adding the flow f_{L_4} (see Figure 4.7*d*), we can get no further than from a to c. Observe that we have saturated the edges in the a–z cut $(P, \overline{P})$, where $P = \{a, c\}$. The value of the final flow, 9, equals the capacity $k(P, \overline{P})$ of this cut, and so by Corollary 2a the flow must be maximal. ∎

Example 3: Faulty Flow Building

Suppose the network in Figure 4.7*a* is redrawn as in Figure 4.7*e*, with the positions of d and e switched.

Let us again choose augmenting a–z flow paths across the top and bottom of the network, now having sizes 5 and 1, respectively. Since edges $(a\vec{,}b)$ and $(c\vec{,}d)$, are saturated by these flow paths, the only possible a–z path along unsaturated edges is $L_5 = a$–c–e–z with minimum slack 1 [the minimum occurring in edge (e, z)]. After adding the a–z unit flow f_{L_5}, we get the flow f_0 shown in Figure 4.7*e*. The cut $(P_0, \overline{P}_0)$, where $P_0 = \{a, c, e\}$, is saturated, and so no more augmenting a–z unit flows exist. Yet $|f_0| = 7$ and $k(P_0, \overline{P}_0) = 12$. Remember that a flow of size 9 was obtained for this same network in Example 2! What has happened?

We now see that an arbitrary sequence of successive augmenting a–z unit flows need not inevitably yield a maximal flow. We are also faced with a flow f_0 and a saturated a–z cut $(P_0, \overline{P}_0)$ such that $|f_0| < k(P_0, \overline{P}_0)$. Corollary 2a implies that there

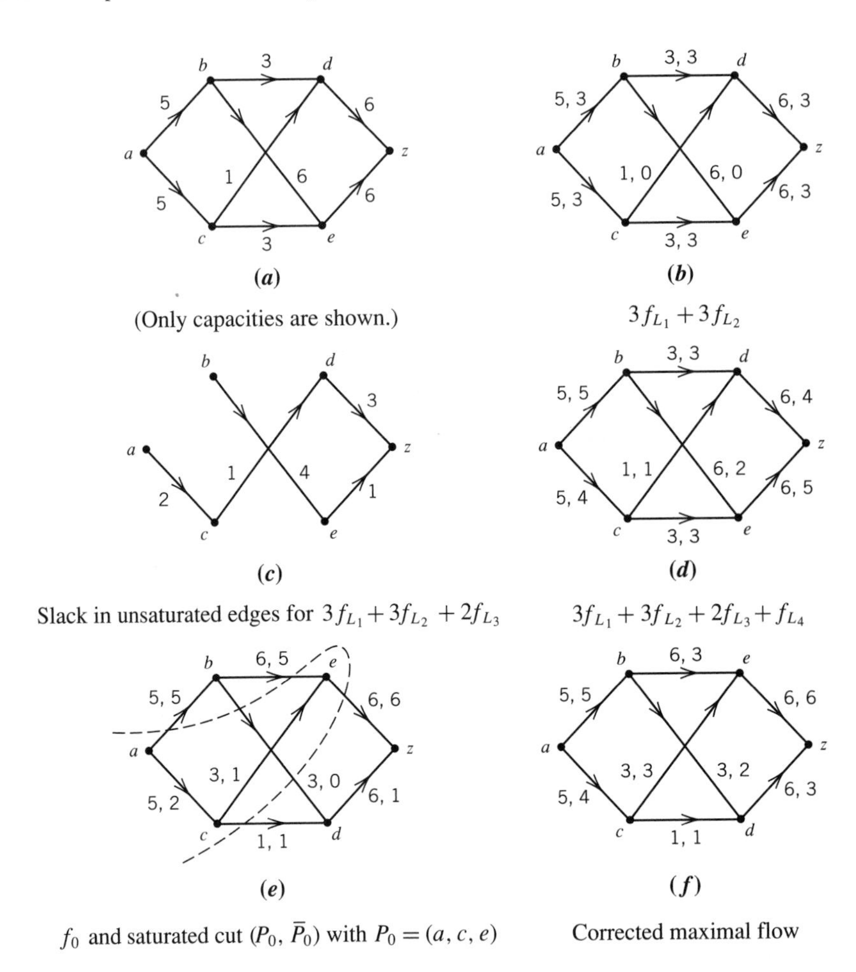

Figure 4.7

must be some flow in an edge $e \in (\overline{P}_0, P_0)$. Looking at Figure 4.7*e*, we see that the flow path $L' = a$–b–e–z crosses the a–z cut $(P_0, \overline{P}_0)$ backwards (from $\overline{P}$ to P_0) on the edge $(\vec{b, e})$; or equivalently, L' crosses the cut forward twice, on edge $(\vec{a, b})$ and again on $(\vec{e, z})$. Thus the 5 units of flow along L' use up 10 units of capacity in the cut, whence $k(P_0, \overline{P}_0)$ is 5 units greater than $|f_0|$.

The reason that the sequence of augmenting flow paths in this example did not lead to a maximal flow can be explained intuitively as follows. By sending 5 units of flow along L' (see Figure 4.7*e*), we have routed all the flow passing through b on to e and none of it to d. Then only 1 unit of flow passing through c can be routed on to e and then along edge $(\vec{e, z})$. But much of the flow through c must go to e, since the capacity of $(\vec{c, d})$ is only 1. In sum, the initial 5-unit flow along a–b–e–z was a "mistake" because some of the capacity in edge $(\vec{e, z})$ should have been "reserved" for flow from c. ■

How can we avoid or correct such mistakes? If we understood where the mistakes were made, we could change some of the flow paths and try a new sequence of augmenting flow paths. However, we are likely to make other mistakes in subsequent constructions. Indeed, there may be certain networks in which it is impossible not to make such a mistake, no matter what sequence of flow paths is used. In terms of cuts, we may always end with a saturated a–z cut that one of our flow paths crosses twice.

Fortunately, there is a procedure to correct "mistakes" and thereby further increase the flow. The method will not look for edges that must be "reserved" for later flow paths, as suggested above (that is too hard a problem). Rather, it looks for flow that is going the wrong way (backwards) across an a–z cut. Then it finds a way to decrease the backward flow without changing the forward flow across the cut, the result being more total flow across the cut (and through the network). The following example presents the idea behind this procedure.

Example 3: (continued)

The edge $(b\vec{,}e)$ contains 5 units of flow that is going backward across the saturated cut $(P_0, \overline{P}_0)$, where $P_0 = \{a, c, e\}$. By how much can the flow in $(b\vec{,}e)$ be reduced?

Condition (b) of a flow—flow in equals flow out—requires that a reduction of the flow into e from b must be compensated by an increase to e from elsewhere in P_0 (if the compensating flow comes from $\overline{P}_0$ we would have a new backward flow). Such an increase must in the end come from a. Thus, we need an a–e flow path in P_0. The only such path is $K_1 = a$–c–e. Similarly, a reduction of the flow out of b to e must be compensated by an increase in flow out of b to somewhere else in $\overline{P}_0$. So we need a b–z flow path in $\overline{P}_0$. The only such path is $K_2 = b$–d–e.

The minimum slack along K_1 is 2 and the minimum slack along K_2 is 3. Then we can decrease the flow in $(b\vec{,}e)$ by 2 while increasing the flow in K_1 and K_2 by 2. Figure 4.7f shows the resulting maximal flow. Note that this new maximal flow is different from the maximal flow in Figure 4.7d, although both saturate the a–z cut $(\{a, c\}, \{b, d, e, z\})$. ■

A **chain** in a directed graph is a sequence of edges that forms a path when the direction of the edges is ignored. A **unit-flow chain** from a to z along the a–z chain K is a "flow" f_K with a value of 1 in each edge of K forwardly directed, a value of -1 in each edge of K backwardly directed, and a value of 0 elsewhere. Note that f_K is not really a flow because it can assume a negative value on some edges. However, if f is a flow that already has positive values in each backwardly directed edge of K and has slack in each forwardly directed edge of K, then $f + f_K$ is a valid flow.

The flow correction made in the continuation of Example 3 consisted of a (2-unit) a–z flow chain along chain $K = a$–c–e–b–d–z. When no backwardly directed edges occur in a flow chain, then it is simply a flow path. We shall see that any sequence of augmenting a–z flow chains can be extended to a maximal flow. Flow chains are the appropriate generalization of flow paths that both build additional flow and simultaneously correct possible "mistakes."

We now present a flow chain algorithm to increase the value of a flow. The algorithm is designed so that if it fails to obtain an augmenting a–z flow chain, it will produce a saturated a–z cut whose capacity equals the value $|f|$ of the current flow f. Thus by Corollary 2a, f would be maximal.

The algorithm recursively tries to build augmenting flow chains from a to all vertices in a manner reminiscent of the shortest path algorithm in Section 4.1. The algorithm assigns two labels to a vertex q: $(p^{\pm}, \Delta(q))$, where p is the previous vertex on a flow chain from a to q, the superscript of p is $+$ if the last edge of the chain is $(\vec{p, q})$ and is $-$ if the last edge is $(\vec{q, p})$ and $\Delta(q)$ is the amount of additional flow that can be sent from a to q. That is, $\Delta(q)$ is the minimum slack among the edges of the chain from a to q. On a backwardly directed edge e, the slack is the amount of flow that can be removed, namely $f(e)$. A $+$ superscript on q means flow is being added to edge $(\vec{p, q})$; a $-$ superscript means flow is being subtracted from edge $(\vec{q, p})$.

Augmenting Flow Algorithm

1. Give vertex a the labels $(-, \infty)$.
2. Call the vertex being scanned p with second label $\Delta(p)$. Initially, $p = a$.
   (a) Check each incoming edge $e = (\vec{q, p})$. If $f(e) > 0$ and q is unlabeled, then label q with $[p^-, \Delta(q)]$, where $\Delta(q) = \min[\Delta(p), f(e)]$.
   (b) Check each outgoing edge $e = (\vec{p, q})$. If $s(e) = k(e) - f(e) > 0$ and q is unlabeled, then label q with $[p^+, \Delta(q)]$, where $\Delta(q) = \min[\Delta(p), s(e)]$.
3. If z has been labeled, go to Step 4. Otherwise choose another labeled vertex to be scanned (which was not previously scanned) and go to Step 2. If there are no more labeled vertices to scan, let P be the set of labeled vertices, and now $(P, \overline{P})$ is a saturated a–z cut. Moreover, $|f| = k(P, \overline{P})$, and thus f is maximal.
4. Find an a–z chain K of slack edges by backtracking from z as in the shortest path algorithm. Then an a–z flow chain f_K along K of $\Delta(z)$ units is the desired augmenting flow. Increase the flow in the edges of K by $\Delta(z)$ units (decrease flow if edge is backward directed in K).

Like the shortest path algorithm, the flow algorithm extends partial-flow chains from currently labeled vertices to adjacent unlabeled vertices. Before we prove that repeated application of our algorithm always leads to a maximal flow, let us give some examples.

Example 3: (continued)

Let us apply our augmenting flow algorithm to the flow in Figure 4.7*e*.

Vertex a is labeled $(-, \infty)$ by Step 1 of the algorithm. Next we apply Step 2b at a (Step 2a does not apply at a since the definition of a flow does not allow flow into the source a). The edge $(\vec{a, b})$ is saturated, but $(\vec{a, c})$ has slack $5 - 2 = 3$. So we label c $(a^+, 3)$. At c, Step 2a finds no incoming flow from an unlabeled vertex, but Step 2b finds slack in edge $(\vec{c, e})$ going to unlabeled vertex e. We label e $(c^+, 2)$

[2 is the minimum of $\Delta(c)$, the extra flow we can get to c, and the slack in edge $(c \vec{,} e)$]. At e, Step 2a finds a positive flow entering on edge $(b \vec{,} e)$ from unlabeled vertex b. We label $b\ (e^-, 2)$. From b, we label $d\ (b^+, 2)$, and from d we label $z\ (d^+, 2)$. Sink z is now labeled and so Step 4 tells us we can get $\Delta(z) = 2$ more units of flow from a to z. Backtracking with the labels, the flow chain K (in backwards order) is z–d–b–e–c–a. The new flow $f_0 + 2f_K$ is shown in Figure 4.7*f*. ∎

Example 4: Using Augmenting Flow Algorithm

Consider the network shown in Figure 4.8*a*. We seek a maximal flow from a to z.

If a maximal flow were being found by a computer, it would have to start with a zero flow. When solving a flow problem by hand, we can speed the process by starting with a (nonzero) flow obtained by inspection (in small networks we can often obtain a maximal flow by inspection). Let the initial flow be $f = 4f_{K_1} + 4f_{K_2} + 5f_{K_3}$, where $K_1 = a$–b–e–z, $K_2 = a$–c–d–f–z, and $K_3 = a$–d–f–z. See Figure 4.8*a*. The reader may see that we sent too much flow from c to d. Some of it should have gone directly from c to f.

We now apply the labeling algorithm to the flow f. Label vertex $a\ (-, \infty)$. Scanning edges at a, there cannot be incoming edges to the source with flow [by part (b) of the definition of a flow], but there are three outgoing edges: edge $(a \vec{,} b)$ has slack $s(a \vec{,} b) = 2 > 0$ and b is unlabeled, so we label $b\ (a^+, 2)$, where 2 is the minimum of $\Delta(a)$ $(= \infty)$ and $s(a \vec{,} b)$; edge $(a \vec{,} c)$ has no slack; edge $(a \vec{,} d)$ has slack $s(a \vec{,} d) = 2$ and d is unlabeled, so we label it $(a^+, 2)$.

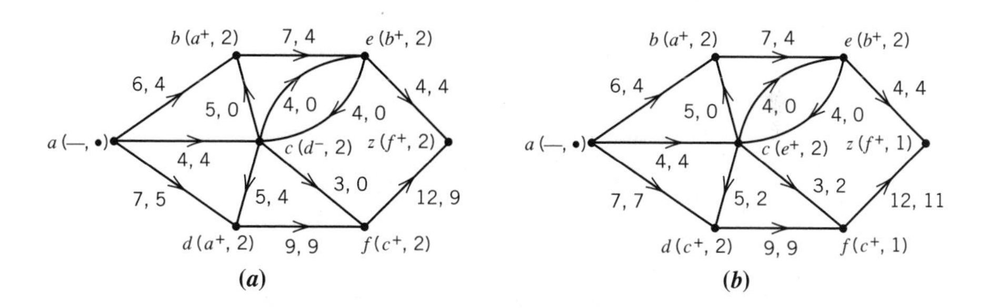

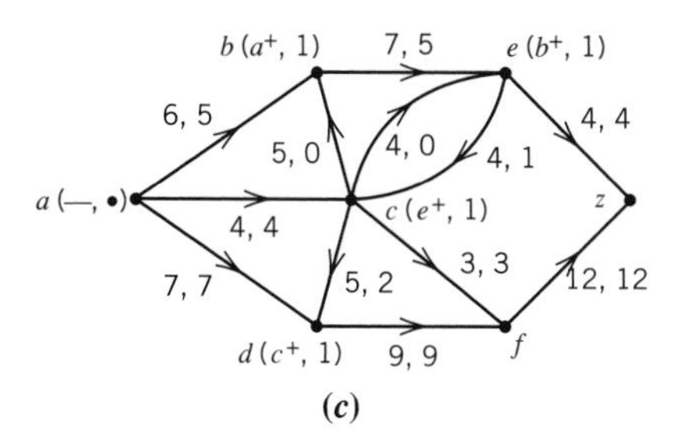

Figure 4.8

Next we scan b (we will scan vertices in the order that they were labeled). There are two incoming edges at b; edge $(a\vec{,}b)$ has $f(a\vec{,}b) = 4 > 0$ but a is already labeled; edge $(c\vec{,}b)$ has no flow. There is one outgoing edge at b: edge $(b\vec{,}e)$ has slack $s(b\vec{,}e) = 3$ and e is unlabeled, so we label e $(b^+, 2)$, where 2 is min $[\Delta(b), s(b\vec{,}e)]$.

Next we scan at d. There are two incoming edges at d: edge $(a\vec{,}d)$ comes from a labeled vertex; edge $(c\vec{,}d)$ has $f(c\vec{,}d) = 4$ and c is unlabeled, so using Step 2a we label c $(d^-, 2)$, where $2 = \min[\Delta(d), f(c\vec{,}d)]$. There is one outgoing edge at d: edge $(d\vec{,}f)$ is saturated.

No labeling can be done from e. At c we label f with $(c^+, 2)$. At f we label z with $(f^+, 2)$.

Since z is now labeled, the labeling procedure terminates. We can send 2 $[= \Delta(z)]$ units in the augmenting a–z flow chain f_{K_4}, where $K_4 = a\text{–}d\text{–}c\text{–}f\text{–}z$, found by the backtracking procedure. Recall that since edge $(c\vec{,}d)$ is backwardly directed in K_4, the flow chain f_{K_4} subtracts 2 units from the current flow in edge $(c\vec{,}d)$. The new flow $f' = f + 2f_{K_4}$ is shown in Figure 4.8*b*.

We eliminate all the current labels and restart the labeling algorithm from scratch. Label vertex $a(-, \infty)$. From a we label b with $(a^+, 2)$. At b we label e with $(b^+, 2)$. At e we label c with $(e^+, 2)$. At c we label d with $(c^+, 2)$ and f with $(c^+, 1)$. At d we can make no new labels. At f we label z with $(f^+, 1)$. The augmenting a–z flow chain is f_{K_5}, with $K_5 = a\text{–}b\text{–}e\text{–}c\text{–}f\text{–}z$, using forward edge $(e\vec{,}c)$. Our new flow is $f'' = f + 2f_{K_4} + f_{K_5}$, shown in Figure 4.8*c*.

The reader may have observed that the flow f'' is maximal. The incoming edges at z are now saturated. Let us again apply the augmenting flow algorithm to flow f''. At a we label b; at b we label e; at e we label c; and at c we label d. No more vertices can be labeled. Let P be the set of labeled vertices. Then $(P, \overline{P})$ is the saturated a–z cut specified by the algorithm with $|f''| = 16 = k(P, \overline{P})$. ■

In applying our algorithm, we must always check incoming edges in Step 2a for the possibility of minus labeling of vertices, even though this labeling is very infrequent. Recall that the minus labeling corresponds to correcting a mistaken flow assignment. A permissible shortcut would be to use only positive labeling (as in the faulty procedure discussed earlier) until no new flow paths can be found, and then apply the full algorithm to hunt for "mistakes." The use of this shortcut serves to increase the importance of having a rigorous proof that, when repeatedly applied to any given flow, our algorithm will yield a maximal flow.

Theorem 3

For any given a–z flow f, a finite number of applications of the augmenting flow algorithm yields a maximal flow. Moreover, if P is the set of vertices labeled during the final (unsuccessful) application of the algorithm, then $(P, \overline{P})$ is a minimal a–z cut set.

Proof

There are two main parts to the proof. First, if f is the current flow and f_K is the augmenting a–z unit flow chain (along chain K) found by the algorithm with

$m = \Delta(z)$, then we must show that the new flow $f + mf_K$ is indeed a legal flow in the network. Both f and f_K satisfy flow conditions (b) and (c) and are integer-valued. Hence $f + mf_K$ also satisfies (b) and (c) and is integer-valued. The labeling algorithm is designed so that $m = \Delta(z)$ is the minimum slack (of the appropriate kind) along chain K and hence $f + mf_K$ satisfies flow condition (a): $0 \leq f(e) + mf_K(e) \leq k(e)$.

Since m is a positive integer, each new flow is larger by an integral amount. The capacities and the number of edges are finite, and so the algorithm must eventually halt—fail to label z. Let P be the set of labeled vertices when the algorithm halts. Clearly $(P, \overline{P})$ is an a–z cut, since a is labeled and z is not. Observe that there cannot be an unsaturated edge from a labeled vertex p to an unlabeled vertex q, or else at p we could have labeled q in Step 2b. Similarly, there cannot be a flow in an edge from an unlabeled vertex q to a labeled vertex p, or else again at p we could have labeled q in Step 2a. Thus, both conditions of Corollary 2a are satisfied. Hence the value of the final flow equals $k(P, \overline{P})$ and is maximal. Also $(P, \overline{P})$ is a minimal a–z cut. ◆

Corollary 3a Max Flow–Min Cut Theorem

In any directed flow network, the value of a maximal a–z flow is equal to the capacity of a minimal a–z cut.

Let us now indicate how flows in a directed network can be used to model a large variety of extensions in directed and undirected networks. In Example 1, we showed how to model a problem with several supplies and demands by a network with one source and one sink.

Example 5: Undirected Networks

Suppose the undirected network in Figure 4.9 represents a network of telephone lines (the capacity of an edge is the number of calls the line can handle). We wish to know the maximal number of calls that the network can simultaneously carry between locations a and z. That is, we seek the value of a maximal flow in this network.

To make the network directed, we can replace each undirected edge (x, y) by the two edges $(\vec{x, y})$ and $(\vec{y, x})$, each with the same capacity as (x, y). An equivalent approach is to allow a directed flow in undirected edges. If $e = (x, y)$, $f(e)$ would be a number with an "arrow" indicating whether the flow goes from x to y or from y to x. Step 2 of the flow algorithm is modified as follows: When checking edges at a

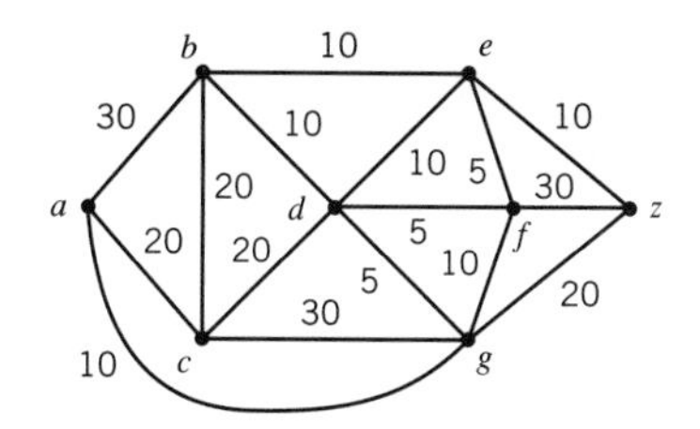

Figure 4.9

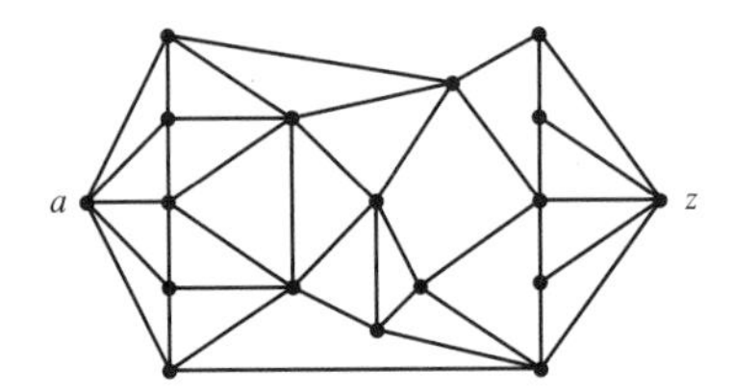

Figure 4.10

labeled vertex p, edges with a flow directed inward are treated like incoming edges and edges with no flow or flow away from p are treated like outgoing edges. ■

Example 6: Edge-Disjoint Paths in a Graph

We are going to send messengers from a to z in the graph shown in Figure 4.10. Because certain edges (roads) may be blocked, we require each messenger to use different edges. How many messengers can be sent? That is, we want to know the number of edge-disjoint paths.

We convert this path problem into a network flow problem by assigning unit capacities to each edge. One could think of the flow as "flow messengers," and the unit capacities mean that at most one messenger can use any edge. The number of edge-disjoint paths (number of messengers) is thus equal to the value of a maximal flow in this undirected network. (See Example 5 for flows in undirected networks.)

Observe that we have implicitly shown that a maximal a–z flow problem for a unit-capacity network is equivalent to finding the maximal number of edge-disjoint a–z paths in the associated graph (where edge capacities are ignored). This equivalence can be extended using multigraphs to all networks by replacing each k-capacity edge (x, y) with k multiple unit-capacity edges and now proceeding with the above conversion. ■

Example 7: Dynamic Network Flows

We want to know how many autos can be shipped from location a to location z in four days through the network in Figure 4.11a. We assume that each edge $(\vec{x, y})$ is the route of a train that leaves location x daily for a nonstop run to location y. The first number associated with an edge is the capacity of the trains (number of autos). The second number is the number of days the trip takes. Autos may be left temporarily at any location in the network.

We turn this dynamic problem into a static a–z flow problem by adding the dimension of time to our network: Each vertex x is replaced by five vertices x_0, x_1, x_2, x_3, x_4, where the subscript refers to the ith day in the four-day shipping time (the starting day is day 0). For each original edge $(\vec{x, y})$, which takes k days to traverse, we make edges $(\vec{x_0, y_k})$, $(\vec{x_1, y_{k+1}})$, $\ldots$, $(\vec{x_{4-k}, y_4})$, each with the same capacity as $(\vec{x, y})$. For each vertex x, we make four edges of the form $(\vec{x_i, x_{i+1}})$ with very large (in effect, infinite) capacity; these edges correspond to the provision that permits autos to be left temporarily at any vertex from one day to the next. See Figure 4.11b.

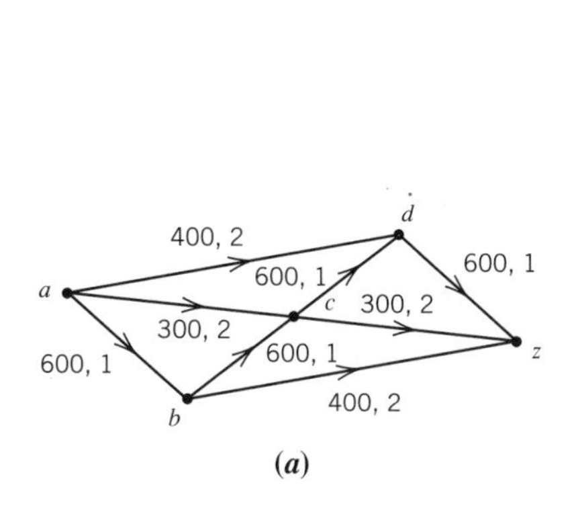

(*a*)

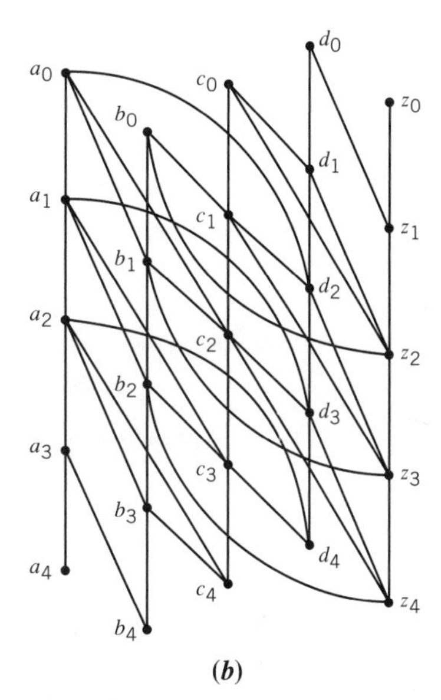
(*b*)
(edges directed downward and to right)

Figure 4.11

A maximal flow in the new network gives the maximal dynamic flow in the original network. Note that for vertices other than a and z, the range of the subscripts of usable vertices will be at most 1 to 3. Nondaily trains could easily be incorporated into the model. ■

It is hoped that the preceding examples have impressed the reader with the versatility of our basic static a–z flow model. More examples are to be found in the exercises.

4.3 EXERCISES

Summary of Exercises Exercises 1–17 mimic or extend the flow computations and models in Examples 3–7. Exercises 18–38 develop the theory of network flows (later exercises are very challenging). Exercises 39–42 present programming projects.

In the following problems, unless directed otherwise, the reader should route most of the flow by inspection and then, when a near-optimal flow is obtained, use the augmenting flow algorithm to get further flow and afterwards a minimal cut.

1. Apply the augmenting flow algorithm to the flow in Figure 4.5.

2. Find a maximal a–z flow and minimal capacity a–z cut in the following networks:

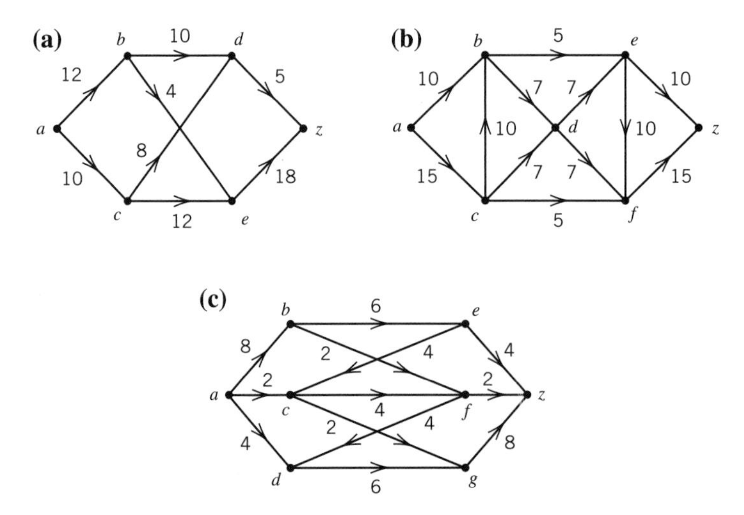

3. Find a maximal flow from a to z in the network of Figure 4.9 (using the associated directed network). Also give a minimal capacity a–z cut.
4. Treat the first number assigned to each edge in Exercise 3 of Section 4.1 as a capacity and let the edges be directed by the alphabetical order of their endpoints with L preceding a and W following n [e.g., edge (f, i) goes from f to i].
   (a) Find a maximal L–W flow and a minimal L–W cut in this network.
   (b) By inspection, find a different maximal L–W flow.
   (c) Make a misrouted flow that yields a saturated L–W cut whose capacity is greater than the flow (similar to the situation in Figure 4.7*e*). Now apply the algorithm.
   (d) In addition, let the second number of each edge be a lower bound on the flow in that edge. Try to find an L–W flow satisfying both constraints.
5. Delete vertices p through y in Figure 4.2 in Section 4.2 and treat the remaining edge numbers as capacities (edges are still undirected). Use the undirected version of the flow algorithm suggested in Example 5.
   (a) Find a maximal flow from b to j and a minimal b–j cut.
   (b) Build a b–j flow that saturates edge (k, l) (in either direction) and then apply the algorithm.
   (c) Treat the edge numbers as lower bounds and modify the algorithm to find a minimal b–j flow, using the whole network (a through y). Remember that there is no flow into b or out of j.
6. Is there a flow meeting the demands in Figure 4.4?
7. Suppose vertices b, c, d in Figure 4.4 have *unlimited* supplies. How much flow can be sent to the set $\{h, i, j\}$? Explain your model.
8. Vertices b, c, d have supplies 30, 20, 10, respectively, and vertices j, k have demands of 30 and 25 in this network.

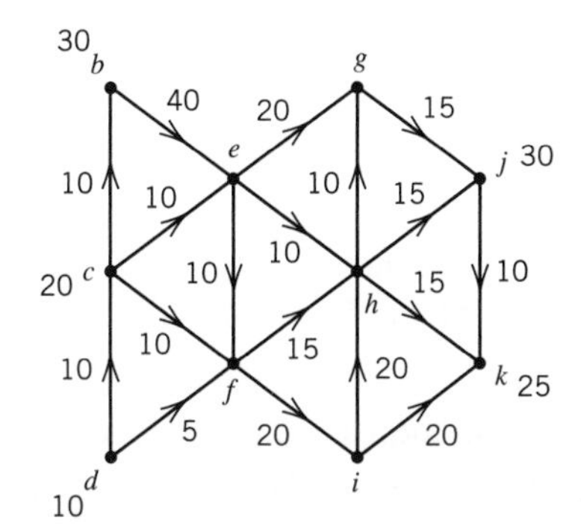

(a) Find a flow satisfying the demands, if possible.

(b) Reverse the direction of edge (h, g) and repeat part (a).

9. Solve the messenger problem in Example 6.

10. Suppose that up to three messengers can use each edge in Example 6. Now how many messengers can be sent? Is the answer with such a modification always just three times the answer to the original problem?

11. **(a)** Ignoring the numbers of the edges in Exercise 3 of Section 4.1, what is the size of the largest set of edge-disjoint paths from L to W?

(b) What is the size of the largest set of paths from L to W such that no edge is used by more than five paths?

12. Suppose that no more than five units of flow can go through each intermediate vertex b, c, d, in Figure 4.8a. Now find a maximal flow in this revised network and associated minimal a–z cut.

13. Suppose that no more than 20 units of flow can go through each intermediate vertex b, c, d, e, f, g in Figure 4.4. Now find the maximal flow in this revised network.

14. What is the size of a largest set of vertex-disjoint paths from a to z in Figure 4.10?

15. Solve the dynamic flow problem in Example 7.

16. In Example 7, suppose that the trains do not run every day. Let trains depart from a and c on Monday, Wednesday, and Friday, and from b and d on Tuesday, Thursday, and Saturday. In one week, Monday through Sunday, how many cars can be sent from a to z in that network?

17. Suppose that it takes one day to traverse each edge in Figure 4.5. How many units can be moved from a to z in five days in this network?

18. In the proof of Theorem 3, show that $f^0 = mf_K$ satisfies the second part of condition (b).

19. **(a)** Prove that if a directed network contains edges $(\overrightarrow{x, y})$ and $(\overrightarrow{y, x})$ for some x, y, then the augmenting flow algorithm would never make assignments that would have flow occurring simultaneously in both edges.

(b) Could edges $(\vec{x, y})$ and $(\vec{y, x})$ both get flow if the augmenting flow algorithm in Step 2 checked outgoing edges before incoming edges?

20. **(a)** Restate the augmenting flow algorithm so that the labeling starts at z and "works back" to a.

(b) Restate the augmenting flow algorithm for undirected networks (see Example 5). Sketch a proof of this algorithm.

21. Show that the set of edges used to label vertices in Steps 2a and 2b of the augmenting flow algorithm form a tree rooted at a.

22. **(a)** Give a weakened replacement for condition (b) in the definition of an a–z flow. The new condition should still ensure that the net flow is from a to z.

(b) Suppose condition (b) is eliminated. Can there exist maximal flows that violate condition (b)? Prove or give a counterexample.

23. Build a flow in the network in Figure 4.8*a* with the prescribed properties:

(a) Its value $|f|$ is 0 but not all edges have 0 flow.

(b) Its value is 2 but it cannot be decomposed into a sum of a–z flow paths.

24. **(a)** Show that for a flow f in a (directed or undirected) network, if f is *circuit-free,* that is, there is no set of edges with flow that form a (directed) circuit, then f can be decomposed into a sum of a–z flow paths. (*Hint:* Prove by induction on the value of f.)

(b) Use the proof to get a decomposition algorithm for any such f.

(c) Conclude that any flow f can be decomposed into $|f|$ a–z flow paths plus a set of circuits.

25. **(a)** Prove that starting from a circuit-free flow (see Exercise 24), perhaps a zero flow, the maximal flow generated by the augmenting flow algorithm is circuit-free and hence [by Exercise 24(b)] can be decomposed into a sum of a–z flow paths.

(b) Use part (a) and Exercise 24(b) to find the routings of a maximal set of phone calls in Example 5.

26. How many times is an edge checked in Step 2 to perform one complete iteration of the augmenting flow algorithm (resulting in increased flow or a min-cut)?

27. Show that if a flow is decomposed into unit-flow paths and if each unit-flow path crosses a given saturated cut once, then the flow is maximal.

28. A *cut-set* in an undirected graph G is a set S of edges whose removal disconnects G but no proper subset of S disconnects G. Prove that in an undirected flow network, every cut-set that separates a and z is an a–z cut and every minimal a–z cut is a cut-set.

29. Let G be a connected, undirected graph and a, b be any two vertices in G.

(a) Show that there are k edge-disjoint paths between a and b if and only if every a–b cut has at least k edges.

(b) Show that there are k vertex-disjoint paths between a and b if and only if every set of vertices disconnecting a from b has at least k vertices.

30. As mentioned in Example 6, we can model a flow network N (directed or undirected) by another multigraph flow network N' in which each edge has unit capacity; N' has the same vertices as N and for each edge e in N there are $k(e)$ edges in N' paralleling e. Since a flow in N' takes 0 or 1 values on the edges, we can drop the capacities in N' to get a multigraph G'. A flow in N' is just a subset of edges (with flow) in G'.

(a) Characterize the subsets of edges in G' that correspond to an a–z flow in N'.

(b) Restate the augmenting flow algorithm in terms of G'.

(c) Using the multigraph model, prove that any a–z flow f in N' contains $|f|$ a–z flow paths.

(d) Using the multigraph model and assuming the result in Exercise 29(a), prove Corollary 3a (max flow–min cut theorem).

31. Suppose the numbers on the edges in a directed network represent lower bounds for the flow. State a decreasing flow algorithm. Sketch a proof of this algorithm and deduce the counterpart of Corollary 3a.

32. **(a)** Explain how the algorithm in Exercise 31 can be applied to a bipartite graph to find a minimal set of edges incident with every vertex in the graph.

(b) Find such an edge set for the bipartite graph in Figure 4.12 of Section 4.4.

33. Suppose we have upper and lower bounds $k_1(e)$ and $k_2(e)$, respectively, on the flow in each edge e in a directed network and that we are given a feasible flow (satisfying these constraints).

(a) Modify the augmenting flow algorithm so that it can be used to construct a maximal flow from a given feasible flow.

(b) Prove that the maximal flow has value equal to the minimum of $k_1(P, \overline{P}) - k_2(\overline{P}, P)$, among all a–z cuts $(P, \overline{P})$, where $k_1(S, \overline{S})$ and $k_2(S, \overline{S})$ are the sums of the upper and lower bounds of edges in a cut $(S, \overline{S})$.

34. Consider a directed network with supplies and demands as in Example 1. Let $z(P)$ be the total demand of vertices in set P and $a(P)$ be the total supply of vertices in P.

(a) Prove that the demands can be met if and only if for all sets P, $z(P) - a(P) \leq k(\overline{P}, P)$. (*Hint:* Generalize the reasoning in Example 1.)

(b) Prove that the supplies can all be used if and only if for all P, $a(P) - z(P) \leq k(P, \overline{P})$.

35. Suppose that the edges in Figure 4.11*a* in Example 7 were undirected. How would we construct a static flow model to simulate this dynamic flow problem so that a maximal static flow in the new network would correspond to a maximal dynamic flow? Are there any difficulties?

36. Suppose an undirected flow network is a planar graph and a (at the left side) and z (at the right side) are both on the unbounded region surrounding the network. Draw edges extending infinitely to the left from a and to the right from z; give them infinite capacity. This divides the unbounded region into two unbounded regions, an upper and a lower unbounded region. Now form the dual network (see Section 1.4) of this planar network with each dual edge assigned the capacity of the original edge it crosses.

(a) Show that a path from the upper unbounded region's vertex to the lower unbounded region's vertex in the dual network corresponds to an a–z cut in the original network. Thus, a shortest such path in the dual network is a minimal a–z cut in the original network.

(b) Draw the dual network for the network in Figure 4.9 and find a shortest path (using the algorithm in Section 4.1) corresponding to a minimal cut in the original network.

37. Suppose an undirected flow network N is a planar graph and a (at the left side) and z (at the right side) are both on the unbounded region surrounding the network. Consider the following flow-path building heuristic. Starting from a, build an a–z flow path by choosing at each vertex x the first unsaturated edge in clockwise order starting from the edge used to enter x.

(a) Apply this heuristic to the network in Figure 4.9 to find a maximal flow.

(b) Show that repeated use of this heuristic yields a maximal flow in N.

38. Prove that repeated use of the augmenting flow algorithm yields a maximal flow in a finite number of applications for networks with irrational capacities provided that vertices are ordered (indexed) and that the next vertex scanned in Step 2 of the algorithm is the labeled vertex with lowest index.

39. Write a program to find maximal flows in directed networks (the networks are input data).

40. Write a program to find the maximal number of paths in an undirected graph between two given vertices such that:

(a) The paths are edge disjoint.

(b) The paths are vertex disjoint.

41. Write a program that, when given a network with an a–z flow of value k, will extract from the flow k unit-flow paths from a to z.

42. Write a program to find, for a given pair of vertices a and b in a given connected graph, a minimal set of vertices whose removal disconnects a from b (see Exercise 29).

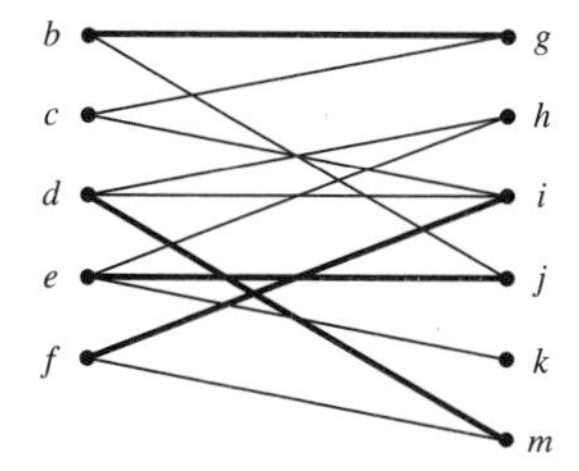

Figure 4.12

4.4 ALGORITHMIC MATCHING

In this section, we apply network flows to the theory of matchings. Recall that a **bipartite graph** $G = (X, Y, E)$ is an undirected graph with two specified vertex sets X and Y and with all edges of the form (x, y), $x \in X$, $y \in Y$. See Figure 4.12. Bipartite graphs are a natural model for matching problems. We let X and Y be the two sets to be matched and edges (x, y) represent pairs of elements that may be matched together.

A **matching** in a bipartite graph is a set of **independent edges** (with no common endpoints). The thicker edges in Figure 4.12 constitute a matching. An ***X*-matching** is a matching involving all vertices in X. A **maximal matching** is a matching of the largest possible size. As with network flows, we cannot always obtain a maximal matching in a bipartite graph by simply adding more edges to a nonmaximal matching. The matching indicated in Figure 4.12 cannot be so increased, even though it is not maximal.

A typical matching problem involves pairing off compatible boys and girls at a dance or the one-to-one assignment of workers to jobs for which they are trained. A closely related problem is to find a **set of distinct representatives** for a collection of subsets. We need to pick one element from each subset without using any element twice. In the bipartite graph model, we make one X-vertex for each subset, one Y-vertex for each element, and an edge (x, y) whenever element y is in subset x. Now an X-matching picks a distinct representative element for each subset. Conversely, any matching problem can be modeled as a set-of-distinct-representatives problem.

We employ a modification of the trick in Example 1 of Section 4.3 to turn a matching problem into a network flow problem. Associate a supply of 1 at each X-vertex and a demand of 1 at each Y-vertex. The capacities of the edges from X to Y can be any large positive integers, but it is convenient to assume that these capacities are ∞. We also assume that edges are directed from X to Y. Now we apply the technique in Example 1 of Section 4.3 with source a connected by a unit-capacity edge to each X-vertex and sink z connected by a unit capacity edge from each Y-vertex.

We call such a network a **matching network**. See Figure 4.13. The X-Y edges used in an a–z flow constitute a matching. A maximal flow is a maximal matching. A flow saturating all edges from source a corresponds to an X-matching.

Example 1: Maximal Matching

Suppose the bipartite graph in Figure 4.12 represents possible pairings of boys b, c, d, e, f with girls g, h, i, j, k, m. A tentative matching indicated by thicker edges in

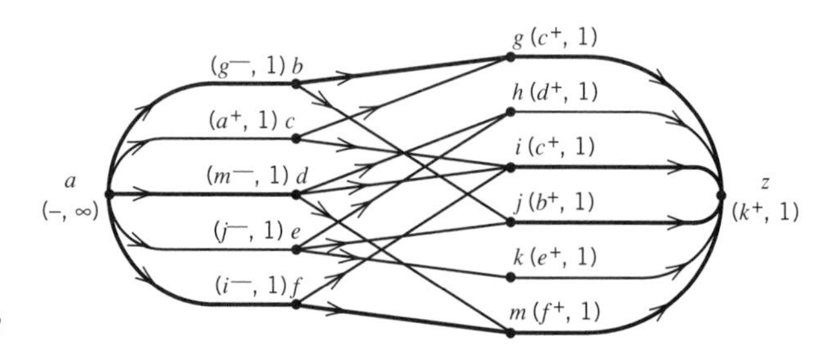

Figure 4.13

Figure 4.12 was made. Although this matching cannot be extended, we still wonder whether a complete X-matching is possible.

As has happened before in flow problems, we have made a "mistake" in this matching and now need to make some reassignments. We convert this matching into the corresponding flow in the associated matching network: darkened edges in Figure 4.13 have a flow of 1, other edges have a flow of 0. Now we apply the augmenting flow algorithm. See Figure 4.13.

From a, the only outgoing edge that is not saturated is $(\vec{a, c})$, since c is the only X-vertex not involved in the initial matching. We label c $(a^+, 1)$. From c we can label the two Y-vertices, g and i, adjacent to c with the label $(c^+, 1)$. At g the one outgoing edge—to z—is saturated. Then there must be an incoming edge to g with positive flow, namely $(\vec{b, g})$, coming from the unlabeled X-vertex b. We label b $(g^-, 1)$. By similar reasoning, from i we label f $(i^-, 1)$. At b and f we can label Y-vertices not currently matched to b and f. We label j $(b^+, 1)$ and m $(f^+, 1)$.

Summarizing our labeling thus far, we are trying to find a match for c (the only X-vertex currently unmatched). We have possible matches of c with g and i, but to make either match, we must sever the current match of g with b or of i with f and rematch b or f with other Y-vertices—in this network, b can be rematched with j or f rematched with m. Continuing this reasoning, to rematch b with j requires us to delete j's current match with e and find a new match for e. Or to rematch f with m requires us to delete m's current match with d and find a new match for d. Now either d can be matched with h (currently unmatched) or e can be matched with k (also currently unmatched). The fact that h and k are unmatched is reflected in the fact that from either h or k one can label z. See the labels in Figure 4.13.

If we label z $(k^+, 1)$, the augmenting flow chain prescribed by the algorithm is a–c–g–b–j–e–k–z. Our augmenting flow chain specifies that we add edge $(\vec{c, g})$ to our matching, remove edge $(\vec{b, g})$, add edge $(\vec{b, j})$, remove edge $(\vec{e, j})$, and add edge $(\vec{e, k})$. The new flow corresponds to the matching b–j, c–g, d–m, e–k, f–i. Another application of the algorithm would label only a. The set of edges leaving a now forms a (saturated) minimal a–z cut—a sign that we have an X-matching. ∎

Observe that the action of the augmenting flow algorithm in matching problems can be described as follows: Starting from an unmatched vertex x_1 in X, we go on a nonmatching edge (an edge not currently used in the matching) to a matched Y-vertex y_1; then we move back from y_1 along a matching edge to a matched X-vertex x_2;

then we go on a nonmatching edge from x_2 to a matched vertex y_2; and so on until a nonmatching edge (x_k, y_k) goes from a matched X-vertex x_k to an unmatched Y-vertex y_k. In short, starting from an unmatched X-vertex, we create an odd-length alternating path L of nonmatching and matching edges in search of an unmatched Y-vertex. Given such a path, we get a new, larger set of matching edges by interchanging the roles of matching and nonmatching edges on L.

We have seen that in bipartite graphs, a matching is analogous to a flow in a network. What then is the bipartite graph counterpart to an a–z cut? Actually it is convenient to restrict our attention to finite-capacity a–z cuts. The corresponding concept is an **edge cover**, a set S of vertices such that every edge has a vertex of S as an endpoint. Recall that edge covers were encountered in a street surveillance problem in Example 4 of Section 1.1.

Lemma

Let $G = (X, Y, E)$ be a bipartite graph and let N be the matching network associated with G. For any subsets (possibly empty) $A \subseteq X$ and $B \subseteq Y$, $S = A \cup B$ is an edge cover if and only if $(P, \overline{P}))$ is a finite capacity a–z cut in N, where $P = a \cup (X - A) \cup B$. In terms of P, $S = (\overline{P} \cap X) \cup (P \cap Y)$. Further, $|S| = k(P, \overline{P})$.

Proof

A finite capacity a–z cut cannot contain an edge between X and Y (whose capacity is ∞). Then a finite-capacity cut $(P, \overline{P})$ must consist of edges of the form $(\vec{a, x})$, $x \in A$ and $(\vec{y, z})$, $y \in B$ for some sets $A \subseteq X$ and $B \subseteq Y$. These edges block all flow (and thus are an a–z cut) if and only if any X–Y edge starts at some $x \in A$ or ends at some $y \in B$, that is, if and only if $S = A \cup B$ is an edge cover. In terms of A and B, $P = a \cup (X - A) \cup B$, and in terms of P, $A = \overline{P} \cap X$ and $B = P \cap Y$. See Figure 4.14. Also, $|S| = k(P, \overline{P})$, since the edges in $(P, \overline{P})$ all have unit capacity. ◆

We now prove two famous theorems about matching in bipartite graphs.

Theorem 1 **(König-Egevary, 1931)**

In a bipartite graph $G = (X, Y, E)$, the size of a maximal matching equals the size of a minimal edge cover.

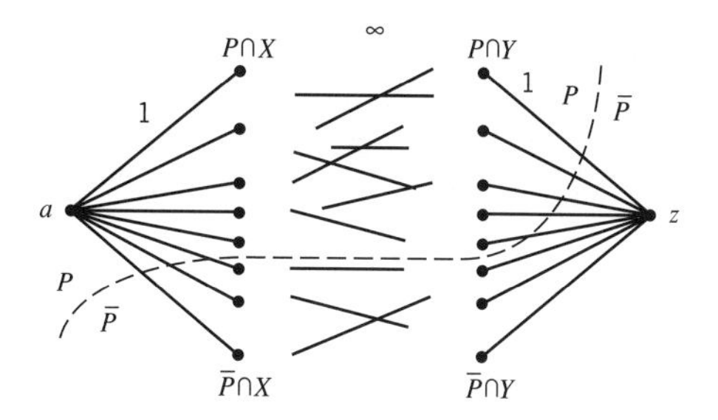

Figure 4.14

Matchings correspond to flows in the associated matching network and, by the lemma, edge covers correspond to (finite capacity) a–z cuts. Theorem 1 is simply a bipartite-graph restatement of the max flow–min cut theorem (Corollary 3a in Section 4.2).

In the next theorem, the **range** $R(A)$ of A denotes the set of vertices adjacent to a vertex in A.

Theorem 2 **(Hall's Marriage Theorem, 1935)**

A bipartite graph $G = (X, Y, E)$ has an X-matching if and only if for each subset $A \subseteq X$, $|R(A)| \geq |A|$.

Proof

The condition is necessary, for if $|R(A)| < |A|$, a matching of the vertices in A is clearly impossible, and so an X-matching is impossible.

Next we show that if $|R(A)| \leq |A|$, for all $A \subseteq X$, then G has a X-matching. By Theorem 1, an X-matching exists if and only if each edge cover S has size $|S| \geq |X|$. If S contains only X-vertices—that is, $S = X$—then the result is immediate. We must show that S cannot be made smaller by dropping some X-vertices and in their place using a smaller number of Y-vertices. If A is the set of X-vertices not in S, then $S \cap Y$ must contain the vertices in $R(A)$ to cover the edges between A and $R(A)$. Then we have

$$\begin{aligned} |S| = |S \cap X| + |S \cap Y| &= |X| - |A| + |S \cap Y| \\ &\geq |X| - |A| + |R(A)| \end{aligned}$$

Since $|R(A)| \geq |A|$, then $|X| - |A| + |R(A)| \geq |X|$, and so $|S| \geq |X|$, proving that there is a matching of size X. ◆

Theorems 1 and 2 are the starting point for a large family of matching theorems (for example, see Exercises 16 and 21).

Example 2

Country A would like to spy on all meetings in its territory between diplomats from Country X and from Country Y. We know which pairs of diplomats are likely to meet. We can make a bipartite graph expressing this likely-to-meet relationship. Country A cannot afford to assign spies to every X diplomat or to every Y diplomat. Instead it wants to find a minimal set S of diplomats such that every possible meeting would involve a diplomat of S.

Country A hires a graduate of this course, who immediately sees that this minimal spying problem is really a minimal edge cover problem in disguise. The problem can thus be solved by using the augmenting flow algorithm to find a minimal a–z cut in the associate matching network and, from the a–z cut, obtain a minimal edge cover using the lemma. ■

Example 3

Suppose there are n people and n jobs, each person is qualified for k jobs, and for each job there are k qualified people. Is it possible to assign each person to a (different) job they can do?

We model this problem using a bipartite graph $G = (X, Y, E)$ with X-vertices for people and Y-vertices for jobs. Note that each vertex will have degree k. The question is now, Is there an X-matching? From Theorem 2, we can deduce a yes answer as follows: Let A be any subset of X. Since each vertex has degree k, there will be $k|A|$ edges leaving A. Since at most k of these edges can go to any one vertex in Y, it follows that $R(A)$ has at least $|A|$ vertices. Then by Theorem 2, there is an X-matching. ■

We close this chapter with an application of matching networks to the sports problem of determining late in a season whether a particular team still has a mathematical chance of being conference champion. The network model and associated analysis presented here are due to Schwartz [4].

Example 4: Elimination from Contention

Suppose we know how many games each of the teams in a sports conference has won to date and how many games remain to be played between each pair of teams. We want to know if there exists a scenario under which a specified team could finish the season with the most wins in the conference.

Let us consider a concrete example with four teams, the Bears, the Lions, the Tigers, and the Vampires with the following records of wins and games to play. See Figure 4.15. Suppose we wonder if it is possible for the Bears (currently with fewest wins) to end the season with the most wins. The problem is that, even assuming the Bears win the rest of their games to finish with 23 wins, the outcomes of the games between the other teams may always result in one of those teams finishing with more wins than the Bears. Thus, the challenge is to find an answer to the question:

> Does there exist a way to assign the wins in the games among other teams so that each of those teams finishes with at most 23 wins?

We describe how to construct a matching network model to answer this question. Figure 4.16 gives the network for the data in Figure 4.15. The X-vertices will represent the other teams besides the Bears. In Figure 4.16 the X-vertices will be L (for Lions),

Team	*Wins to date*	*Games to play*	*with Bears*	*with Lions*	*with Tigers*	*with Vampires*
Bears (t_1)	16	7	—	2	2	3
Lions (t_2)	22	7	2	—	3	2
Tigers (t_3)	20	8	2	3	—	3
Vampires (t_4)	19	8	3	2	3	—

Figure 4.15

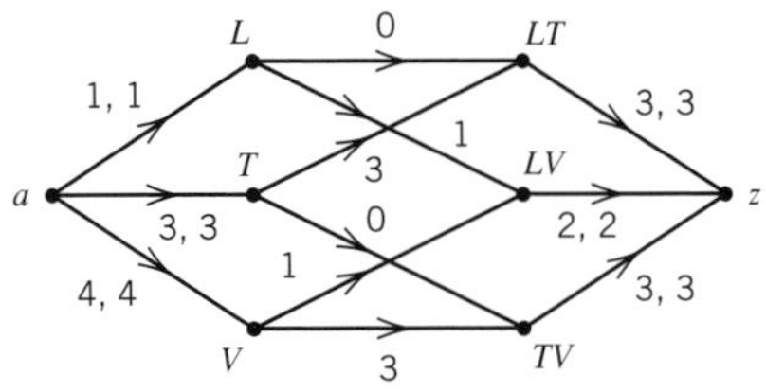

Figure 4.16 (middle edges have capacity ∞)

T (for Tigers), and V (for Vampires). The Y-vertices will represent all different pairs of other teams. In this case, LT, LV, and TV.

The size of the flow in the edge from source a to X-vertex t_i will stand for the number of games won by the i-th team t_i in the rest of the season. The capacity of the edge $(\vec{a, t_i})$ will be $23 - w_i$, where w_i denotes the number of wins to date by t_i. Observe that if team t_i has won w_i games to date, and if it wins at most $23 - w_i$ games in the rest of the season, then t_i cannot finish with more than 23 wins, enabling the Bears to be the champion, or possibly co-champion (assuming the Bears win all remaining games).

The flow in the X–Y edge $(\vec{t, (t_i, t_j)})$ will represent the number of wins of t_i over t_j in their remaining games. We can let the capacity of all X–Y edges be ∞. The flow in the edge from the Y-vertex (t_i, t_j) to the sink z represents the number of games that will be played between teams t_i and t_j. The capacity of $(\vec{(t_i, t_j), z})$ is the number of scheduled remaining games between t_i and t_j. Conservation of flow at vertex t_i requires that flow in (the number of games won in the rest of the season) = flow out (the number of future wins over various other teams). Conservation of flow at the vertex (t_i, t_j) requires that flow in (the wins of t_i over t_j and of t_j over t_i) = flow out (the number of games these two teams will play).

To have a flow that represents a possible real scenario, we require that the edges from Y to z all be saturated—the number of games played between t_i and t_j is the actual number of games scheduled between them. If a flow exists in this network that saturates the Y–z edges, it produces a scenario in which the Bears ends the season with the most wins. Figure 4.16 presents a flow saturating the Y–z edges showing how the Bears could end up with the most wins (actually all four teams would tie for first place). ■

Optional We now extend the preceding sports network model to give a necessary and sufficient condition for a particular team t^* to be able to be conference champion. For a subset S of teams, let $w(A)$ denote the total number of wins to date by all teams in S and let $r(S)$ be the total number of games remaining to be played between all pairs of teams in S. Observe that for any subset S of teams, the total number of wins at the end of the season by teams in S will be at least $w(S) + r(S)$ (because when two teams in S play each other, one of them must win). Let W^* be the final number of wins by t^* if t^* wins all remaining games. One constraint that guarantees that t^* cannot finish with the most wins is if the average number of wins by any subset S not containing t^* exceeds W^*, that is,

$$\text{for any subset } S \text{ not containing } t^*, \ \{w(S) + r(S)\}/|S| > W^* \qquad (*)$$

We shall show that constraint (*) is sufficient as well as necessary to characterize when team t^* is eliminated. If the min-cut is the set of Y–z edges, then there is a flow saturating all Y–z edges and specifying a scenario in which t^* has the most wins (as described in the example above). Suppose there is an a–z cut $(P, \overline{P})$ of smaller capacity. We now show that (*) is satisfied by some subset S.

$(P, \overline{P})$ cannot contain any of the infinite-capacity edges from X to Y, and so it consists solely of a–X edges and Y–z edges. Recall that edge $(\vec{a, t_i})$ has capacity $W^* - w_i$. Then if $A = \overline{P} \cap X$, the a–X edges in $(P, \overline{P})$ have capacity $|A|W^* - w(A)$. One can show that if t_i and t_j are in $\overline{P}$, then (t_i, t_j) must be in $\overline{P}$ (details are left as an exercise). Then the Y–z edges in $(P, \overline{P})$ have capacity $\Sigma_{(t_i,t_j)\not\subset A} r_{ij}$, where r_{ij} denotes the capacity of edge $(\vec{(t_i, t_j), z})$ (= the number of games remaining to be played between t_i and t_j) and we sum over all (t_i, t_j) pairs for which at least one of t_i or t_j is not in A. So $k(P, \overline{P}) = [|A|W^* - w(A)] + \Sigma_{(t_i,t_j)\not\subset A} r_{ij}$.

Assuming $k(P, \overline{P})$ is less than Σr_{ij}, the capacity of all the edges from Y to z, then

$$|A|\,W^* - w(A) + \sum_{(t_i,t_j)\not\subset A} r_{ij} < \sum r_{ij}$$

$$\Rightarrow |A|\,W^* - w(A) < \sum r_{ij} - \sum_{(t_i,t_j)\not\subset A} r_{ij} = \sum_{(t_i,t_j)\subset A} r_{ij} = r(A)$$

$$\Rightarrow |A|W^* < w(A) + r(A) \Rightarrow W^* < \{w(A) + r(A)\}/|A|$$

This last inequality is (*), as claimed, with the subset being A. The teams in A average more wins than t^*, and hence t^* cannot be conference champion.

4.4 EXERCISES

Summary of Exercises Exercises 1–12 involve matching networks. Exercises 13–22 develop theory.

1. Bill is liked by Ann, Diana, and Lolita; Fred is liked by Bobbie, Carol, and Lolita; George is liked by Ann, Bobbie, and Lolita; John is liked by Carol and Lolita; and Larry is liked by Diana and Lolita. We want to pair each girl with a boy she likes. (Make girls the vertices on the left side.)
   - **(a)** Set up the associated matching network and maximize its flow to solve this problem.
   - **(b)** Make a flow corresponding to a partial pairing that has Bill with Diana and Fred with Carol along with two other matches (chosen by the reader). Now apply the flow augmenting algorithm to increase the matching to a complete matching.

2. Suppose there are five committees: committee A's members are a, c, e; committee B's members are b, c; committee C's members are a, b, d; committee D's members are d, e, f; and committee E's members are e, f. We wish to have each committee send a different representative to a convention.

(a) Set up the associated matching network and maximize its flow to solve this problem.

(b) Make a flow corresponding to the partial assignment: A sends e, B sends b, C sends a, and D sends f. Now apply the flow-augmenting algorithm to increase the matching.

3. Let us repeat Exercise 1, but this time Bill gets a total of 5 dates, Fred 4 dates, George 3, John 5, and Larry 3, while Ann gets 4, Bobbie 3, Diana 5, Carol 4, and Lolita 4. Compatible pairs may have any number of dates together. Model with a network to find a possible set of pairings.

4. Let us repeat Exercise 1, but this time we want to pair each boy twice (with two different girls) and each girl twice. Find the pairings using an appropriate network flow model.

5. Suppose that there are 6 universities and each will produce 5 mathematics Ph.D.s this year, and there are 5 colleges that will be hiring 7, 7, 6, 6, 5 math Ph.D.s, respectively. No college will hire more than one Ph.D. from any given university. Will all the Ph.D.s get a job? Explain.

6. In Example 4, is it possible for the Vampires to be champions (or co-champions) if they win all remaining games? Build the appropriate network model.

7. In the following table of remaining games, is it possible for the Bears to be champions (or co-champions) if they win all remaining games? Build the appropriate network model.

Team	*Wins to date*	*Games to play*	*with Bears*	*with Lions*	*with Tigers*	*with Vampires*
Bears	26	14	—	4	4	6
Lions	34	5	4	—	0	1
Tigers	32	8	4	0	—	4
Vampires	29	11	6	1	4	—

8. In the following table of remaining games, is it possible for the Bears to be champions (or co-champions) if they win all remaining games? Build the appropriate network model.

Team	*Wins to date*	*Games to play*	*with Bears*	*with Lions*	*with Tigers*	*with Vampires*
Bears	20	8	—	1	2	5
Lions	25	3	1	—	1	1
Tigers	23	5	2	1	—	2
Vampires	20	8	5	1	2	—

9. There are n boys and n girls in a computer dating service. The computer has made nm pairings so that each boy dates m different girls and each girl dates m different boys ($m < n$).

(a) Show that it is always possible to schedule the nm dates over m nights, that is, the pairings may be partitioned into m sets of complete pairings.

(b) Show that in part (a), no matter how the first k complete pairings are selected $(0 < k < m)$, the partition can always be completed.

10. We want to construct an $n \times m$ matrix whose entries will be nonnegative integers such that the sum of the entries in row i is r_i, and the sum of the entries in column j is c_j. Clearly the sum of the r_is must equal the sum of the c_js.

(a) What other constraints (if any) should be imposed on the r_is and c_js to assure such a matrix exists?

(b) Construct such a 5×6 matrix with row sums 20, 40, 10, 13, 25 and column sums all equal to 18.

11. We have a group of people and each is a member of a subset of committees. In addition, each person graduated from (exactly) one of three different universities. In this extension of the "distinct representatives" problem, we seek a unique person to represent each committee with the added constraint that a third of the representatives must have graduated from each of the three universities. Describe how build a network to model this problem. (Assume that m, the number of committees, is a multiple of 3.)

12. (Due to Bacharach) We are given an n by n matrix with numerical entries that all have one digit to the right of the decimal point (e.g., 13.3). We want to round the entries to whole numbers in a fashion so that the sum of the rounded entries in each column (and row) is a rounded value of the original column (row) sum; e.g., if the first column has entries 2.5, 6.4, 5.7 summing to 16.6, then the sum of the rounded values of these three entries would have to be 16 or 17.

(a) Describe how to build a matching-type network flow model of this problem. Let X-vertices represent the columns and Y-vertices represent the rows. Note that this model will have lower as well as upper bounds for each edge.

(b) Find the required rounding, if possible, using the network model developed in (a) for the following matrix:

$$\begin{matrix} 4.5 & 7.5 & 2.5 \\ 6.8 & 4.3 & 5.7 \\ 3.6 & 1.6 & 4.3 \end{matrix}$$

13. Give necessary and sufficient conditions for the existence of a circuit or a set of vertex-disjoint circuits that pass through each vertex once in a directed graph $G = (V, E)$. [*Hint:* Make a bipartite graph $G' = (X, Y, E)$ with X and Y copies of V, and for each edge $(v_1 \vec{,} v_2)$ in G, G' has an edge (x_1, y_2); restate the problem.]

14. Prove that a bipartite graph $G = (X, Y, E)$ has a matching of size t if and only if for all A in X, $|R(A)| \geq |A| + t - |X| = t - |X - A|$. (*Hint:* Add $|X| - t$ new vertices to Y and join each new Y-vertex to each X-vertex.)

15. Show that every bipartite-graph matching problem can be modeled as a set-of-distinct-representatives problem.

16. In the analysis following Example 4, show that if t_i and t_j are in $\overline{P}$, then vertex (t_i, t_j) is in $\overline{P}$.

17. Prove Theorem 2 using Exercise 34 in Section 4.3, which applies directly to the bipartite network, not the augmented *a–z* matching network.

18. Prove Theorem 2 for complete *Y*-matchings (without simply interchanging the roles of *X* and *Y*). By symmetry, the same condition is required, but the set *A* in the reproof is chosen differently from the *A* in the text's proof.

19. Let $\delta(G) = \max_{A \subseteq X}(|A| - |R(A)|$. δ(G) is called the *deficiency* of the bipartite graph $G = (X, Y, E)$ and gives the worst violation of the condition in Theorem 2. Note that $\delta(G) \geq 0$ because $A = \emptyset$ is considered a subset of *X*.

(a) Use Exercise 16 to prove that a maximal matching of *G* has size $|X| - \delta(G)$.

(b) Given a maximal matching of size $t = |X| - \delta(G)$ [assume $\delta(G) > 0$], describe how the associated minimal edge covering of Theorem 1 can be used to find an *A* such that $|A| - |R(A)| = \delta(G)$.

20. **(a)** Show that the size of the largest independent set of vertices (mutually non-adjacent vertices) in $G = (X, Y, E)$ is equal to $|Y| + \delta(G)$ (see Exercise 19). Describe how to find such an independent set.

(b) Use part (a) to find such an independent set in Figure 1.3.

21. (Due to J. Hopcroft) Suppose each vertex of a bipartite graph *G* has degree 2^r, for some *r*. Partition the edges of *G* into circuits, and delete every other edge in each circuit. Repeat this process on the new graph (where each vertex now has degreee 2^{r-1}), and continue repeating until a graph is obtained with each vertex of degree 1. The edges in this final graph constitute a matching of the vertices of *G*. Show that such a partition of edges into circuits exists in each successive graph and can be found in a number of steps proportional to the number of edges in the current graph.

22. Suppose we are given a partial matching to an arbitrary graph (such a matching is a set of edges with no common endpoints).

(a) Prove that a generalization of the interchange method along a path alternating between nonmatching and matching edges (described following Example 1) can be used to increase the size of the partial matching until it is a maximal matching.

(b) Randomly pick a partial matching for the graph in Figure 4.9 and use this method to get a maximal matching.

23. Write a computer program to find a maximal matching and a minimal edge cover in a bipartite graph (the graph is input data).

4.5 SUMMARY AND REFERENCES

In this chapter we presented algorithms for three basic network optimization problems: shortest path, minimal spanning trees, and maximal flow. Principal emphasis was placed on a thorough discussion of maximal flows. We showed how these flows could be applied to a wide variety of other network problems. In Section 4.4 we used flow models to develop a combinatorial theory of matching. All the material about flows in this chapter is discussed in greater detail in the pioneering work *Flows in Networks* by Ford and Fulkerson [2] and in *Network Flow Theory* by Ahuja et al. [1]. We have omitted discussion of the speed of these algorithms. The interested reader is referred to Ahuja et al. [1] for efficient implementations of the shortest path and maximal flow algorithms. Modern maximal flow algorithms require less than $O(n^3)$ operations for an n-vertex network.

It is often natural in network flow problems to have costs associated with edges so that when many possible maximal flows exist, one can ask for a least-cost maximal flow. Such problems are called **trans-shipment** and **transportation** problems. Similarly, in a matching problem with many solutions (X-matchings), one can ask for a least-cost matching. Such a problem is called an **assignment** problem. Efficient algorithms exist for all these minimization problems (see [1], [2], or [3]). Furthermore, any flow optimization problem, with or without the above-mentioned minimization, is a problem of optimizing a linear function of the edge flows subject to linear equalities and inequalities, such as the flow constraints (a), (b), and (c) of Section 4.3. Such a constrained linear optimization problem is called a **linear program**. Linear programming is a principal tool of operations research, and good algorithms exist for solving linear programs. However, it is much more efficient to solve network problems with the network-specific algorithms presented in this chapter.

1. R. Ahuja, T. Magnanti, and J. Orlin, *Network Flow Theory, Algorithms and Applications,* Prentice-Hall, Englewood Cliffs, NJ, 1993.
2. L. Ford and D. Fulkerson, *Flows in Networks,* Princeton University Press, Princeton, NJ, 1962.
3. F. Hillier and G. Lieberman, *Introduction to Operations Research,* Holden Day, San Francisco, 1988.
4. B. Schwartz, "Possible winners in partially completed tournaments," *SIAM Review* 8(1966), 302–308.

PART TWO
ENUMERATION

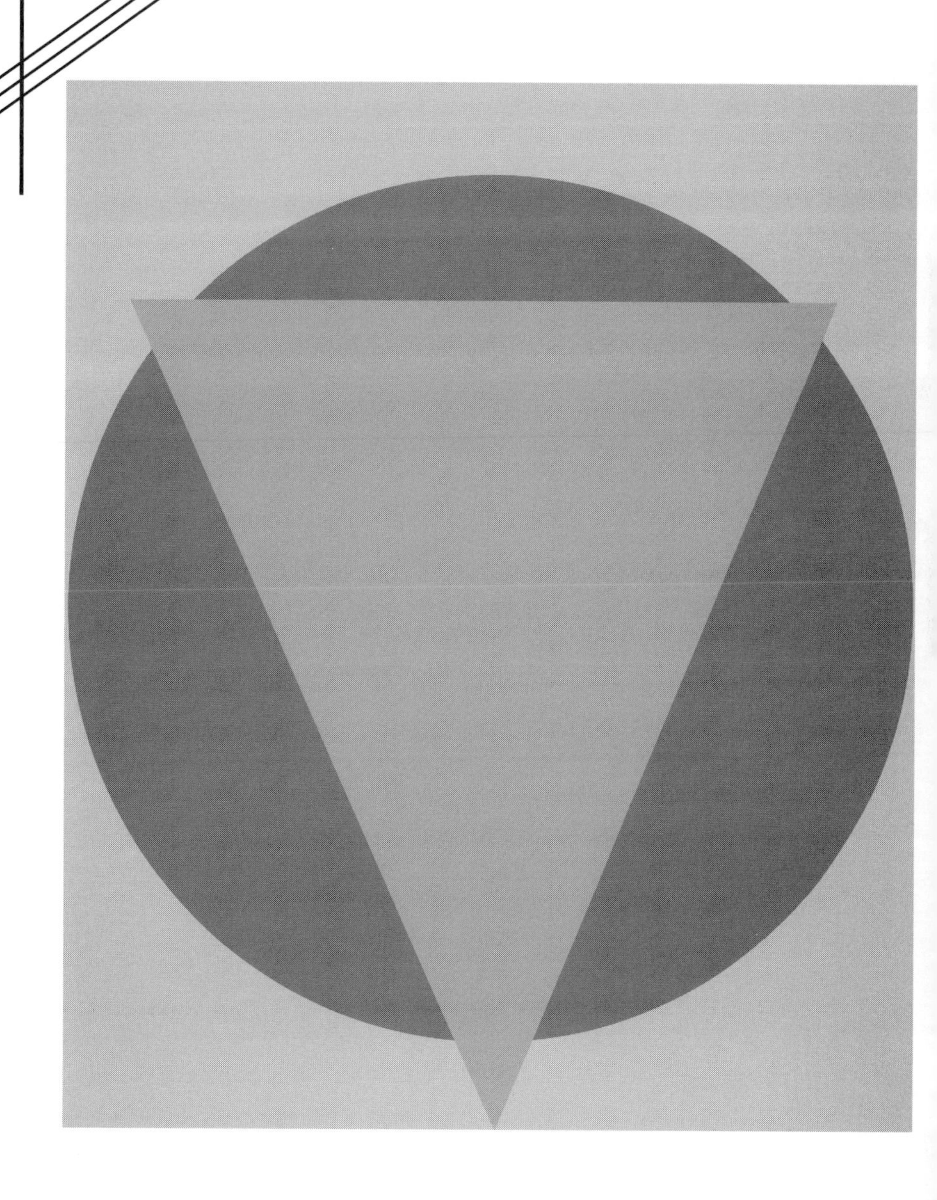

CHAPTER 5
GENERAL COUNTING METHODS FOR ARRANGEMENTS AND SELECTIONS

5.1 TWO BASIC COUNTING PRINCIPLES

In this chapter we discuss counting problems for which no specific theory exists. We present a few basic formulas, involving permutations and combinations, most of which the reader has seen before. Then we examine a variety of word problems involving counting and show how they can be broken down into sums and products of simple numerical factors. Having read through the examples as passive readers, students next must assume the active role of devising solutions on their own to the exercises. The first exercises at the end of each section are similar to the examples discussed in the section. The later exercises, however, have little in common with the examples except that they require the same general types of logical reasoning, clever insights, and mathematical modeling. Facility with these three basic skills in problem solving, as much as an inventory of special techniques, is the key to success in most combinatorial applications. These skills are very helpful in constructing computer programs and in mastering many theoretical topics of computer science. In sum, for many students, this is the most challenging and most valuable chapter in this book.

Subsequent chapters in the enumeration part of this book develop specialized theory and techniques that considerably simplify the solution of specific classes of counting problems. The usefulness of this theory is particularly evident after one has grappled with counting problems in this chapter using only first principles. Unfortunately, there is no such theory to assist in the solution of most real-world counting problems.

In order to motivate our problem solving, we mention applications to probability and statistics, computer science, operations research, and other disciplines. However, the details of the counting problems arising from such applications often appear tedious if one is not actively working in the area of application. So instead, we will base many of the worked-out examples (and exercises) on recreational problems, such as poker probabilities. The solution of these problems requires the same mathematical skills used in more substantive applications.

This section starts with two elementary but fundamental counting principles whose simplicity masks both their power and the ease with which they can be misused.

The Addition Principle

If there are r_1 different objects in the first set, r_2 different objects in the second set, ..., and r_m different objects in the mth set, and *if the different sets are disjoint,* then the number of ways to select an object from one of the m sets is $r_1 + r_2 + \cdots + r_m$.

The Multiplication Principle

Suppose a procedure can be broken into m successive (ordered) stages, with r_1 different outcomes in the first stage, r_2 different outcomes in the second stage, ..., and r_m different outcomes in the mth stage. If the number of outcomes at each stage is independent of the choices in previous stages and *if the composite outcomes are all distinct,* then the total procedure has $r_1 \times r_2 \times \cdots \times r_m$ different composite outcomes.

Remember that the addition principle requires disjoint sets of objects and the multiplication principle requires that the procedure break into ordered stages and that the composite outcomes be distinct. The validity of these two principles follows directly from the definitions of addition and multiplication of integers. That is, the sum $a + b$ is the number of items resulting when a set of a items is added to a set of b items; and the product $a \times b$ is the number of sequences (A,B), when A can be any of a items and B can be any of b items. The two principles are standard m-way extensions of these two binary operations.

Example 1: Adding Sets of Students

Professor Mindthumper has 40 students in an algebra class and 40 students in a geometry class. How many different students are in these two classes?

By the addition principle, the answer is 80 students, provided that no students are in both classes. Suppose 10 students are in both classes. To obtain disjoint sets of students, we categorize students as just in algebra, just in geometry, and in both classes. Since 10 students are in both classes, then $40 - 10 = 30$ algebra students are just in algebra. Similarly, 30 students are just in geometry. Now we can safely use the addition principle to sum the numbers of students in these three disjoint sets. The total number of students is $30 + 30 + 10 = 70$. ■

Example 2: Rolling Dice

Two dice are rolled, one green and one red.

(a) How many different outcomes of this procedure are there?

(b) What is the probability that there are no doubles (not the same value on both dice)?

(a) There are six outcomes of a single die. So, by the multiplication principle, there are $6 \times 6 = 36$ outcomes of the procedure.

(b) We calculate the probability of no doubles using the probability formula for an event: the subset of outcomes producing the desired event divided by all possible outcomes (see Appendix A.3 for details). The denominator in the probability fraction is 36, the total number of outcomes [determined in part (a)]. To get the numerator—the number of outcomes with no doubles—we use the multiplication principle. Suppose that the roll of the red die is the first stage of the procedure. This first stage has six possible outcomes. Once the red die is rolled, then there are five permissible values for the green die, independent of the particular value of the red die. So there are $6 \times 5 = 30$ outcomes with no doubles, and the probability of no doubles is $30/36 = 5/6$. ■

Note that for us to be able to apply the multiplication principle to the second part of Example 2, the constraint "no doubles" must be recast as "the value of the green die is different from the value of the red die." Finding the appropriate rephrasing of a constraint so that the addition and multiplication principles can be applied is the challenging "art" of combinatorial problem-solving. The number of no-doubles in Example 2(b) could also be answered by subtracting the 6 outcomes of doubles from all 36 outcomes, yielding $36 - 6 = 30$ outcomes with no doubles.

Example 3: Arranging Books

There are five different Spanish books, six different French books, and eight different Transylvanian books. How many ways are there to pick an (unordered) pair of two books not both in the same language?

If one Spanish and one French book are chosen, the multiplication principle says that the selection can be done in $5 \times 6 = 30$ ways; if one Spanish and one Transylvanian book, $5 \times 8 = 40$ ways; and if one French and one Transylvanian book, $6 \times 8 = 48$ ways. These three types of selections are disjoint, and so by the addition principle there are all together $30 + 40 + 48 = 118$ ways in all. ■

The preceding example typifies a basic way of thinking in combinatorial problem solving: Always try first to break a problem into a moderate number of manageable subproblems. There may be cleverer ways to solve the problem, but if we can reduce the original problem to a set of subproblems with which we are familiar, then we are less likely to make a mistake.

Example 4: Sequences of Letters

How many ways are there to form a three-letter sequence using the letters a, b, c, d, e, f: **(a)** with repetition of letters allowed? **(b)** without repetition of any letter? **(c)** without repetition and containing the letter e? **(d)** with repetition and containing e?

(a) With repetition, we have six choices for each successive letter in the sequence. So by the multiplication principle there are $6 \times 6 \times 6 = 216$ three-letter sequences with repetition.

(b) Without repetition, there are six choices for the first letter. For the second letter, there are five choices, the five remaining letters (no matter what the first choice was). Similarly for the third letter, there are four choices. Thus there are $6 \times 5 \times 4 = 120$ three-letter sequences without repetition.

(c) It is often helpful to make a diagram displaying the positions in a sequence, even a sequence with just three letters:

$$\_\ \_\ \_$$

Such a diagram helps focus on choices involving the positions. If the sequence must contain e, then there are three choices for which position in the sequence is e, as shown in the following three diagrams:

$$\underline{e}\ \_\ \_ \qquad \_\ \underline{e}\ \_ \qquad \_\ \_\ \underline{e}$$

In each diagram, there are 5 choices for which of the other 5 letters (excluding e) goes in the first remaining position and 4 choices for which of the remaining 4 letters goes in the other position. Thus there are $3 \times 5 \times 4 = 60$ three-letter sequences with e. Another approach to solving this problem is given in the exercises.

(d) Let us try the approach used in part (c) when repetition is allowed. As before, there are three choices for e's position. For any of these choices for e's position, there are $6 \times 6 = 36$ choices for the other two positions, since e and the other letters can appear more than once. But the answer of $3 \times 36 = 108$ is not correct.

The multiplication principle has been violated because the outcomes are not distinct. Consider the sequence

$$\underline{e}\ \underline{c}\ \underline{e}$$

It was generated two times in our procedure: once when e was put in the first position followed by $c\ e$ as one of the 36 choices for the latter two positions, and a second time when e was put in the last position with $e\ c$ in the first two positions.

We must use an approach for breaking the problem into parts that ensures distinct outcomes. Let us decompose the problem into disjoint cases based on where the first e in the sequence occurs. First suppose the first e is in the first position:

$$\underline{e}\ \_\ \_$$

Then there are six choices (including e) for the second and for the third positions—6×6 ways.

Next suppose the first e is in the second position:

$$\underset{\text{no } e}{\_}\ \underline{e}\ \_$$

Then there are five choices for the first position (cannot be e) and six choices for the last position—5×6 ways. Finally, let the first (and only) e be in the last position:

$$\underset{\text{no } e}{\_}\ \underset{\text{no } e}{\_}\ \underline{e}$$

There are five choices each for the first two positions—5×5 ways. The correct answer is thus $(6 \times 6) + (5 \times 6) + (5 \times 5) = 91$. ■

The hardest part about solving most counting problems is finding a structure in the problem that allows it to be broken into subcases or stages. In other words, the difficulty is in "getting started." At the same time, one must be sure that the decomposition into cases or stages generates outcomes that are all distinct.

Many counting problems require their own special insights. For such problems, knowing the solution to problem A is typically of little help in solving problem B. Learning how to construct solutions to such counting problems is what a combinatorics course is about. This skill cannot be acquired in reading textbook examples. It is only gained by working many problems oneself.

The following example illustrates this type of special insight in combinatorial problem solving.

Example 5: Nonempty Collections

How many nonempty different collections can be formed from five (identical) apples and eight (identical) oranges?

Readers with some experience in combinatorial problem solving may want to break the problem into subcases based on the number of objects in the collections. Any one of these subcases can be counted quite easily, but there are 13 possible subcases; that is, collections can have 1 or 2 or ... up to 13 pieces of fruit.

In counting different possibilities, we must concentrate on what makes one collection different from another collection. The answer is, the number of apples and/or the number of oranges will be different in different collections. Then we can characterize any collection by a pair of integers (a, o), where a is the number of apples and o is the number of oranges.

Now the number of collections is easy to count. There are 6 possible values for a (including 0): 0, 1, 2, 3, 4, 5, and 9 possible values for o. Together there are $6 \times 9 = 54$ different collections. (Note that we multiply 6 and 9, not add them, because a collection combines any number of apples and any number of oranges; we add if we want to count the ways to get some amount of apples or some amount of oranges, but not both.)

Since the problem asked for nonempty collections and one of the possibilities allowed was (0,0), the desired answer is $54 - 1 = 53$. ■

Here is one piece of advice to consider when you are stuck and cannot get started with a problem. Try writing down in a systematic fashion some (a dozen or so) of the possible outcomes you want to enumerate. In listing outcomes, you should start to see a pattern emerge. Think of your list as being part of one particular subcase. Then ask yourself, how many outcomes would your list need to include to complete that subcase? Next ask what other (hopefully similar) subcases need to be counted. Once a first subcase has been successfully analyzed, then other subcases are usually easier.

We summarize the two key facets we have encountered in combinatorial problem-solving. First, we must find a way to recast the constraints in a problem so that some combination of the addition and multiplication principles can be applied. Second, to use these principles we must break the problem into pieces or stages, and be sure that the outcomes in the different pieces are distinct.

5.1 EXERCISES

Summary of Exercises The first eight exercises are straightforward. Then the exercises become more challenging and require analysis that will be different for each problem. For the harder problems, readers must devise their own method of solution rather than mimic a method used in one of the text's examples. All exercises should be read carefully two times to avoid misinterpretation.

The word "between" is always used in the inclusive sense; that is, "integers between 0 and 50" means $0, 1, 2, \ldots, 49, 50$. The *probability* of a particular outcome is the number of such particular outcomes divided by the number of all outcomes. See Appendix A.3 for more about probability.

1. **(a)** How many ways are there to pick a sequence of two different letters of the alphabet that appear in the word BOAT? In MATHEMATICS?

(b) How many ways are there to pick first a vowel and then a consonant from BOAT? From MATHEMATICS?

2. **(a)** How many integers are there between 0 and 50 (inclusive)?

(b) How many of these integers are divisible by 2?

(c) How many (unordered) pairs of these integers are there whose difference is 5?

3. A store carries 8 styles of pants. For each style, there are 10 different possible waist sizes, 6 pants lengths, and 4 color choices. How many different types of pants could the store have?

4. How many different sequences of heads and tails are possible if a coin is flipped 100 times? Using the fact that $2^{10} = 1024 \approx 1000 = 10^3$, give your answer in terms of an (approximate) power of 10.

5. How many four-letter "words" (sequence of any four letters with repetition) are there? How many with no repeated letters?

6. How many ways are there to pick a man and a woman who are not husband and wife from a group of n married couples?

7. Given eight different English books, seven different French books, and five different German books:

(a) How many ways are there to select one book?

(b) How many ways are there to select three books, one of each language?

(c) How many ways are there to make a row of three books in which exactly one language is missing (the order of the three books makes a difference)?

8. There are four different roads between town A and town B, three different roads between town B and town C, and two different roads between town A and town C.

(a) How many different routes are there from A to C altogether?

(b) How many different routes are there from A to C and back (any road can be used once in each direction)?

(c) How many different routes are there from A to C and back in part (b) that visit B at least once?

(d) How many different routes are there from A to C and back in part (b) that do not use any road twice?

9. How many ways are there to pick 2 different cards from a standard 52-card deck such that:

(a) The first card is an Ace and the second card is not a Queen?

(b) The first card is a spade and the second card is not a Queen? (*Hint:* Watch out for the Queen of spades.)

10. How many nonempty collections of letters can be formed from three As and five Bs?

11. How many ways are there to roll two dice to yield a sum divisible by 3?

12. How many four-letter "words" (sequences of letters with repetition) are there in which the first and last letter are vowels? In which vowels appear only (if at all) as the first and last letter?

13. (a) How many different five-digit numbers are there (leading zeros, e.g., 00174, not allowed)?

(b) How many even five-digit numbers are there?

(c) How many five-digit numbers are there with exactly one 3?

(d) How many five-digit numbers are there that are the same when the order of their digits is inverted (e.g., 15251)?

14. How many different numbers can be formed by various arrangements of the six digits 1, 1, 1, 1, 2, 3?

15. What is the probability that the top two cards in a shuffled deck do not form a pair?

16. **(a)** How many different outcomes are possible when a pair of dice, one red and one white, are rolled two successive times?

(b) What is the probability that each die shows the same value on the second roll as on the first roll?

(c) What is the probability that the sum of the two dice is the same on both rolls?

(d) What is the probability that the sum of the two dice is greater on the second roll?

17. A rumor is spread randomly among a group of 10 people by successively having one person call someone, who calls someone, and so on. A person can pass the rumor on to anyone except the individual who just called.

(a) By how many different paths can a rumor travel through the group in three calls? In n calls?

(b) What is the probability that if A starts the rumor, A receives the third call?

(c) What is the probability that if A does not start the rumor, A receives the third call?

18. **(a)** How many different license plates involving three letters and three digits are there if the three letters appear together either at the beginning or end of the license?

(b) How many license plates involving one, two, or three letters and one, two, or three digits are there if the letters must appear in a consecutive grouping?

19. Re-solve the problem in Example 4 of counting the number of three-letter sequences without repetition using a, b, c, d, e, f that have an e by first counting the number with no e.

20. What is the probability that the sum of two randomly chosen integers between 20 and 40 inclusive is even (the possibility of the two integers being equal is allowed)?

21. How many three-letter sequences without repeated letters can be made using a, b, c, d, e, f in which either e or f (or both) is used?

22. How many ternary (0, 1, 2) sequences of length 10 are there without any pair of consecutive digits the same?

23. How many integers between 1000 and 10,000 are there with (make sure to avoid sequences of digits with leading 0s):

(a) Distinct digits?

(b) Repetition of digits allowed but with no 2 or 4?

(c) Distinct digits and at least one of 2 and 4 must appear?

24. How many 12-digit decimal sequences are there that start and end with a sequence of at least two 3s?

25. How many sequences of length 5 can be formed using the digits 0, 1, 2, ..., 9 with the property that exactly two of the 10 digits appear, e.g., 05550?

26. What is the probability that an integer between 1 and 10,000 has exactly one 8 and one 9?

27. How many different five-letter sequences can be made using the letters A, B, C, D with repetition such that the sequence does not include the word BAD—that is, sequences such as A<u>BAD</u>D are excluded.

28. **(a)** How many election outcomes are possible with 20 people each voting for one of seven candidates (the outcome includes not just the totals but also who voted for each candidate)?

(b) How many election outcomes are possible if only one person votes for candidate A and only one person votes for candidate D?

29. There are 15 different apples and 10 different pears. How many ways are there for Jack to pick an apple or a pear and then for Jill to pick an apple and a pear?

30. How many times is the digit 5 written when listing all numbers from 1 to 100,000?

31. How many times is "25" written when listing all numbers from 1 to 100,000? (This is an extension of the previous exercise.)

32. What is the probability that if one letter is chosen at random from the word RECURRENCE and one letter is chosen from RELATION, the two letters are the same?

33. How many four-digit numbers are there formed from the digits 1, 2, 3, 4, 5 (with possible repetition) that are divisible by 4?

34. How many nonempty collections of letters can be formed from n As, n Bs, n Cs and n Ds?

35. There are 50 cards numbered from 1 to 50. Two different cards are chosen at random. What is the probability that one number is twice the other number?

36. If two different integers between 1 and 100 inclusive are chosen at random, what is the probability that the difference of the two numbers is 15?

37. If three distinct dice are rolled, what is the probability that the highest value is twice the smallest value?

38. How many different numbers can be formed by the product of two or more of the numbers 3, 4, 4, 5, 5, 6, 7, 7, 7?

39. There are 10 different people at a party. How many ways are there to pair them off into a collection of 5 pairings?

40. A chain letter is sent to five people in the first week of the year. The next week each person who received a letter sends letters to five new people, and so on. How many people have received letters in the first five weeks?

41. How many ways are there to place two identical rooks in a common row or column of an 8×8 chessboard? an $n \times m$ chessboard?

42. How many ways are there to place two identical kings on an 8×8 chessboard so that the kings are not in adjacent squares? on an $n \times m$ chessboard?

43. How many ways are there to place two identical queens on an 8×8 chessboard so that the queens are not in a common row, column, or diagonal?

44. How many different positive integers can be obtained as a sum of two or more of the numbers 1, 3, 5, 10, 20, 50, 82?

45. How many ways are there for a man to invite some (nonempty) subset of his 10 friends to dinner?

46. How many different rectangles can be drawn on an 8×8 chessboard (the rectangles could have sides of length 1 through 8; two rectangles are different if they contain different subsets of individual squares)?

47. How many ways are there to place a red checker and a black checker on two black squares of a checkerboard so that the red checker can jump over the black checker? (A checker jumps on the diagonal from in front to behind.)

48. Use induction to verify formally:

(a) The addition principle.

(b) The multiplication principle.

49. On the real line, place n white pegs at positions $1, 2, \ldots, n$ and n blue pegs at positions $-1, -2, \ldots, -n$ (0 is open). Whites move only to the left, blues to the right. When beside an open position, a peg may move one unit to occupy that position (provided it is in the required direction). If a peg of one color is in front of a peg of the other color that is followed by an open position (in the required direction), a peg may jump two units to the open position (the jumped peg is not removed). By a sequence of these two types of moves (not necessarily alternating between white and blue pegs), one seeks to get the positions of the white and blue pegs interchanged. (See the article on this game in *Mathematics Teacher,* January 1982.)

(a) Play this game for $n = 3$ and $n = 4$.

(b) Use a combinatorial argument to show that, in general, $n^2 + 2n$ moves (unit steps and jumps) are required to complete the game.

5.2 SIMPLE ARRANGEMENTS AND SELECTIONS

A **permutation** of n distinct objects is an arrangement, or ordering, of the n objects. An ***r*-permutation** of n distinct objects is an arrangement using r of the n objects. An ***r*-combination** of n distinct objects is an unordered selection, or *subset,* of r out of the n objects. We use $P(n, r)$ and $C(n, r)$ to denote the number of r-permutations and r-combinations, respectively, of a set of n objects. From the multiplication principle we obtain

$$P(n,2) = n(n-1), \qquad P(n,3) = n(n-1)(n-2),$$
$$P(n,n) = n(n-1)(n-2) \times \cdots \times 3 \times 2 \times 1$$

In enumerating all permutations of n objects, we have n choices for the first position in the arrangement, $n-1$ choices (the $n-1$ remaining objects) for the second position, ..., and finally one choice for the last position. Using the notation $n! = n(n-1)(n-2)\cdots \times 3 \times 2 \times 1$ ($n!$ is said "n factorial"), we have the formulas

$$P(n, n) = n!$$

and

$$P(n, r) = n(n-1)(n-2) \times \cdots \times [n-(r-1)] = \frac{n!}{(n-r)!}$$

Our formula for $P(n, r)$ can be used to derive a formula for $C(n, r)$. All r-permutations of n objects can be generated by first picking any r-combination of the n objects and then arranging these r objects in any order. Thus $P(n, r) = C(n, r) \times P(r, r)$, and solving for $C(n, r)$ we have

$$C(n, r) = \frac{P(n, r)}{P(r, r)} = \frac{n!/(n-r)!}{r!} = \frac{n!}{r!(n-r)!}$$

The numbers $C(n, r)$ are frequently called **binomial coefficients** because of their role in the binomial expansion $(x+y)^n$. We study the binomial expansion and identities involving binomial coefficients in Section 5.5. It is common practice to write the expression $C(n, r)$ as $\binom{n}{r}$ and to say "n choose r" [we usually write $C(n, r)$ in this book]. Note that $C(n, r) = C(n, n-r)$, since the number of ways to pick a subset of r out of n distinct objects equals the number of ways to throw away a subset of $n-r$.

The rest of this section is a continuation of the combinatorial problem-solving examples in the previous section. Along with the introduction of permutations and combinations, these examples involve more complicated situations. With many new constraints to keep straight, readers may find themselves becoming confused about even basic issues such as when to add and when to multiply. One must think very carefully about each step in the analysis of a counting problem.

Example 1: Ranking Wizards

How many ways are there to rank n candidates for the job of chief wizard? If the ranking is made at random (each ranking is equally likely), what is the probability that the fifth candidate, Gandalf, is in second place?

A ranking is simply an arrangement, or permutation, of the n candidates. So there are $n!$ rankings. Intuitively, a random ranking should randomly position Gandalf, and so we would expect that he has probability $1/n$ of being second, or being in any given position. Equivalently, each of the n candidates should have the same probability of being in second place. Formally, we calculate the probability of Gandalf being second using the probability formula for an event: the subset of outcomes producing the desired event divided by all possible outcomes (see Appendix A.3 for details).

$$\text{Prob(Gandalf second)} = \frac{\text{no. of rankings with Gandalf second}}{\text{total no. of rankings}}$$

First put Gandalf in second place—1 way—and then put the remaining wizards in the remaining $n - 1$ places—$(n - 1)!$ ways. So Gandalf is second in $(n - 1)!$ rankings, and Prob(Gandalf second) $= (n - 1)!/n! = 1/n$, as expected. ■

There are two important lessons to take from this example.

1. There are two ways to analyze arrangement problems. We have the option of either picking which item goes in the first position, then which item goes in the second position, and so on, or picking which position to choose for the first item, which position to choose for the second item, and so on. In the Gandalf-second problem, we pick which positions for the items.

2. To count arrangements where a particular position is fixed—in the preceding example, that Gandalf is in second place—we do not count the ways to make Gandalf second (there is only one way to put Gandalf in second place) but rather count all the ways to arrange the remaining wizards in the remaining positions. We count what has *not* been constrained.

Example 2: Arrangements with Repeated Letters

How many ways are there to arrange the seven letters in the word SYSTEMS? In how many of these arrangements do the 3 Ss appear consecutively?

The difficulty of multiple Ss can be avoided by picking the positions that the other four letters E, M, T, Y will occupy in the seven-letter arrangement, and then the 3 Ss will fill the remaining three positions in one way. There are 7 possible positions for E, 6 for M, 5 for T, and 4 for Y. Thus there are $P(7, 4) = 7!/3! = 840$ arrangements. (A general formula for counting arrangements with repeated objects is given in the next section.)

Here again we counted arrangements in terms of where successive items should be placed rather than what should go in successive positions.

Next we consider the case where the 3 Ss appear consecutively, that is, the 3 Ss are all side-by-side. The "trick" for handling this new consecutivity constraint is to realize that when the 3 Ss are grouped together they now become a single composite letter. So the problem reduces to arranging the 5 distinct letters, Y, T, E, M and SSS (treated as a single letter), which can be done in $5! = 120$ ways.

Another way to look at this problem is to think of temporarily setting aside two of the Ss, arranging the 5 remaining letters, Y, T, E, M, S, in 5! ways, and then in each resulting arrangement inserting beside the S the other 2 Ss. ■

Example 3: Binary Sequences

How many different 8-digit binary sequences are there with six 1s and two 0s?

What distinguishes one 8-digit binary sequence with six 1s from another such sequence? Obviously, the answer is the positions of the six 1s (or the two 0s). Though this problem initially reads as an arrangement problem, what must be counted is the different possible placements of the six 1s, that is, different possible *subsets* of six of the eight positions in the binary sequence. (A sequence with 1s in positions 1, 2, 3, 4, 5, 6 is the same as a sequence with 1s in positions 6, 5, 2, 3, 4, 1—the order does not matter, just the collection of positions involved.) So the answer is $C(8, 6) = 28$.

We could alternatively have focused on picking a subset of two of the eight positions for 0s. ■

Example 4: Poker Probabilities

(a) How many 5-card hands (subsets) can be formed from a standard 52-card deck?

(b) If a 5-card hand is chosen at random, what is the probability of obtaining a flush (all 5 cards in the hand are in the same suit)?

(c) What is the probability of obtaining 3, but not 4, Aces?

(a) A 5-card hand is a subset of 5 cards chosen from the 52 cards in a deck, and so there are $C(52, 5) = 52!/(47!5!) = 2{,}598{,}960$ different 5-card hands.

(b) To find the probability of a flush, we need to find the number of 5-card subsets with all cards of the same suit. There are 4 suits, and a subset of 5 cards from the 13 cards in a given suit can be chosen in $C(13, 5) = 13!/(5!8!) = 1287$ ways. So there are $4 \times 1287 = 5148$ flushes, and

$$\text{Prob(5-card hand is a flush)} = \frac{5148}{2{,}598{,}960} = 0.00198 (\approx 0.2\%)$$

(c) To count the number of hands with exactly 3 Aces, we must pick 3 of the 4 Aces—done in $C(4, 3) = 4$ ways—and then fill out the hand with 2 cards chosen from the 48 non-Ace cards—done in $C(48, 2) = 1128$ ways. So there are $4 \times 1128 = 4512$ hands with exactly 3 Aces, and

$$\text{Prob(5-card hand has exactly 3 Aces)} = \frac{4512}{2{,}598{,}960} = .00174 \quad ■$$

Note that to count hands with 3 Aces in Example 4, we implicitly used the multiplication principle to multiply the ways to pick 3 Aces times the ways to fill out the hand with 2 non-Ace cards. However, a hand is an unordered collection, and by ordering it into two parts, we might generate two outcomes that are really the same set, violating the distinctness condition of the multiplication principle. In this problem, hands could safely be decomposed into an Aces part and a non-Aces part because the numbers of cards in the two parts were different. The next example illustrates this disjointness difficulty more fully.

Example 5: Forming Committees

A committee of k people is to be chosen from a set of 7 women and 4 men. How many ways are there to form the committee if:

(a) The committee consists of 3 women and 2 men?

(b) The committee can be any positive size but must have equal numbers of women and men?

(c) The committee has 4 people and one of them must be Mr. Baggins?

(d) The committee has 4 people and at least 2 are women?

(e) The committee has 4 people, two of each sex, and Mr. and Mrs. Baggins cannot both be on the committee?

(a) Applying the multiplication principle to disjoint collections, we can count all committees of 3 women and 2 men by composing all subsets of 3 women with all subsets of 2 men, $C(7, 3) \times C(4, 2) = 35 \times 6 = 210$ ways.

(b) To count the possible subsets of women and men of equal size on the committee, we must know definite sizes of the subsets. That is, we must break the problem into the four disjoint subcases of: 1 woman and 1 man, 2 each, 3 each, and 4 each (there are only 4 men available). So the total number is the sum of the possibilities for these four subcases, $[C(7, 1) \times C(4, 1)] + [C(7, 2) \times C(4, 2)] + [C(7, 3) \times C(4, 3)] + [C(7, 4) \times C(4, 4)] = 7 \times 4 + 21 \times 6 + 35 \times 4 + 35 \times 1 = 329$.

(c) If Mr. Baggins must be on the committee, this simply means that the problem reduces to picking 3 other people on the 4-person committee from the remaining 10 people (7 women and 3 men). So far we have made all selections from the set of men or the set of women. Now the other three people must be chosen from the set of all remaining people, and so the answer is $C(10, 3) = 120$. (It is easy to get in a mindset where one uses the same sets in the current problem that one used in the previous problem, here the set of women and the set of men—problem solvers must always be alert to this trap.)

(d) One approach is to pick 2 women first, $C(7, 2) = 21$ ways, and then pick any 2 of the remaining set of 9 people (5 women and 4 men). However, counting all committees in this fashion counts some outcomes more than once, since any woman in one of these committees could be chosen as either one of the first 2 women or one of the 2 remaining people. For example, if W_i denotes the ith woman and M_i the ith man, then (W_1, W_3) composed with the 2 remaining people (W_2, M_3) yields the same set as (W_1, W_2) composed with (W_3, M_3).

A correct solution to this problem must use a subcase approach, as in part (b). That is, break the problem into three subcases that specify exactly how many women and how many men are on the committee: 2 women and 2 men, 3 women and 1 man, and 4 women. The answer is thus $[C(7, 2) \times C(4, 2)] + [C(7, 3) \times C(4, 1)] + C(7, 4) = 21 \times 6 + 35 \times 4 + 35 = 301$.

(e) We need to recast the condition "Mr. and Mrs. Baggins cannot both be on the committee" into several subcases in which we know exactly which of Mr.

and Mrs. Baggins is on the committee. Note that the possibility of neither Mr. nor Mrs. Baggins' being on the committee satisfies the given condition. There are three subcases to consider.

The first subcase is, Mrs. Baggins is on the committee and Mr. Baggins is not. Then 1 more woman must be chosen from the remaining 6 women and 2 more men must be chosen from the remaining 3 men (Mr. Baggins is excluded). This can be done in $C(6, 1) \times C(3, 2) = 6 \times 3 = 18$ ways. The other two cases are: Mrs. Baggins is off and Mr. Baggins is on, which by a similar argument yields $C(6, 2) \times C(3, 1) = 15 \times 3 = 45$ ways; and neither is on the committee, $C(6, 2) \times C(3, 2) = 15 \times 3 = 45$ ways. The total answer is $18 + 45 + 45 = 108$.

An easier solution to this problem can be obtained by taking a complementary approach. We can consider all $C(7, 2) \times C(4, 2)$ 2-women–2-men committees and then subtract the forbidden committees that contain both Bagginses. The forbidden committees are formed by picking one more woman and one more man to join Mr. and Mrs. Baggins—done in $C(6, 1) \times C(3, 1)$ ways. This approach yields a simpler formula for the same answer: $21 \times 6 - 6 \times 3 = 108$. ■

The mistake of counting the same outcome twice, which arose in part (d) of Example 5, arises in many guises. The following principle should help the reader avoid this problem.

The Set Composition Principle

Suppose a set of distinct objects is being enumerated using the multiplication principle, multiplying the number of ways to form some *first part* of the set by the number of ways to form a *second part* (for a given first part). The members of the first and second parts must be disjoint. Another way to express this condition is: Given any set S thus constructed, one must be able to tell uniquely which elements of S are in the first part of S and which elements are in the second part.

In Example 5(a), the part W of 3 women and the part M of 2 men are disjoint, and so by the set composition principle the number of committees with 3 women and 2 men is the size of W times the size of M. In the "committee with at least two women" problem in Example 5(d), the method of choosing two of the seven women first and then picking any remaining two people (women or men) violates the Set Composition Principle because the members of the two parts are not disjoint—a woman could be in either the first or second part. For example, in the committee $\{W_1, W_2, W_3, M_3\}$ it is impossible to say which two women are chosen in the first part. The scheme of picking two women and then any two remaining people generates the committee $\{W_1, W_2, W_3, M_3\}$ in $C(3, 2)$ different ways, namely, all compositions of: (i) two of women W_1, W_2, W_3; with (ii) M_3 plus the remaining woman in W_1, W_2, W_3 not chosen in (i).

In Example 2 about arrangements of the letters of SYSTEMS, we dealt with the constraint that certain letters had to be grouped together in the arrangement. In Example 2, it was three consecutive Ss. In the following continuation of that example, we consider the constraint of requiring a particular letter to appear somewhere before another letter in the arrangement. We also show how to grapple with the two constraints combined. ■

Example 2: (Continued) Arrangements with Repetitions

How many arrangements of the seven letters in the word SYSTEMS have the E occurring somewhere before the M? How many arrangements have the E somewhere before the M and the three Ss grouped consecutively?

The key to the constraint of E being somewhere before the M (not necessarily immediately before the M) is to focus on the pair of positions where E and M will go. Thus, we start by picking the pair of the 7 positions in an arrangement where the E and M will go—$C(7, 2) = 21$ ways—and then we put E in the first of this pair of positions and M in the second one. Now we fill in the 5 other positions in the arrangement by picking a position for the Y and the T—$P(5, 2) = 5 \times 4 = 20$ ways—and then putting the 3 Ss in the 3 remaining positions. The answer is thus $21 \times 20 = 420$.

While it may sound scary to deal with the two constraints at once, it often turns out to be less hard than expected if one handles the constraints in the right order. If we first pick the pair of the positions for the E and the M, things get messy for the consecutivity constraint because the different placements of the E and M will impact differently the positions in the arrangement where there is enough room to place 3 consecutive Ss.

So we try dealing with the consecutive Ss first. This simply requires that we "glue" the 3 Ss together into one composite letter and consider the reduced problem of arranging 5 letters, Y, T, E, M, (SSS). Now we turn to the other constraint and pick the pair of positions, out of the 5 new positions, for the E and the M—$C(5, 2) =$ 10 ways—with the E going in the first of the two chosen positions. Then we arrange the Y, T and (SSS) in the remaining 3 positions in $3! = 6$ ways. So the answer is $10 \times 6 = 60$. ■

The following type of counting problem arises frequently in quality-control problems.

Example 6: Counting Defective Products

A manufacturing plant produces ovens. At the last stage, an inspector marks the ovens A (acceptable) or U (unacceptable). How many different sequences of 15 As and Us are possible in which the third U appears as the twelfth letter in the sequence?

This problem is a binary sequence problem similar to Example 5 except now the elements are A and U, rather than 0 and 1. If the third U appears at the twelfth letter in the sequence, then the subsequence composed of the first 11 letters must have exactly 2 Us (and 9 As). Following the reasoning in Example 5, there are $C(11, 2) = 55$ possible sequences for the first 11 letters. There is one way to pick the twelfth letter—it is specified to be U. The remaining 3 letters in the sequence can be either A or U—$2^3 = 8$ possibilities. All together, there are $55 \times 1 \times 8 = 440$ sequences. ■

Example 7: Probability of Repeated Digits

What is the probability that a 4-digit campus telephone number has one or more repeated digits?

There are $10^4 = 10{,}000$ different 4-digit phone numbers. We break the problem of counting 4-digit phone numbers with repeated digits into four different cases of repetitions:

(a) All 4 digits are the same.

(b) 3 digits are the same, the other is different.

(c) 2 digits are the same, the other 2 digits are also the same (e.g., 2828).

(d) 2 digits are the same, the other 2 digits are each different (e.g., 5105).

(a) There are 10 numbers formed by repeating one of the 10 digits four times.

(b) One way to decompose the process of generating these case-(b) numbers is as follows (several other decompositions are possible). First pick which digit appears once—10 choices—then where it occurs in the 4-digit number—4 choices—and finally which other digit appears in the other three positions—9 choices. This yields $10 \times 4 \times 9 = 360$ numbers.

(c) First pick which two digits are each to appear twice—$C(10, 2) = 45$ choices—and then how to arrange these 4 digits: pick which two positions are used by, say, the smaller of the digits—$C(4, 2) = 6$ ways. This yields $45 \times 6 = 270$ numbers.

(d) First pick which pair of digits appear once—$C(10, 2) = 45$ choices—then pick a position for, say, the smaller of these two digits and a position for the larger digit—$4 \times 3 = 12$ choices—and finally pick which other digit appears in the remaining two positions—8 choices. This yields $45 \times 12 \times 8 = 4320$ numbers.

In sum, there are $10 + 360 + 270 + 4320 = 4960$ 4-digit phone numbers with a repeated digit. The probability of a repeated digit is thus $4960/10{,}000 = 0.496 \approx 0.5$.

We note that there happens to be a simpler way, using the complementary set, to count phone numbers with repeated digits: count numbers with no repeated digits. These are just the $P(10, 4) = 5040$ four-permutations of the 10 digits. So the remaining repeated-digit numbers amount to $10{,}000 - 5040 = 4960$. ∎

One point of caution about cases (c) and (d) where two different digits both occur once or both occur twice. In case (d), we pick the two digits occurring once as an unordered pair in $C(10, 2)$ ways and arrange those digits (and then pick the digit to go in the remaining two positions) rather than pick a first digit, position it, then pick a second digit, position it (and then pick the digit to go in the remaining two positions)—$10 \times 4 \times 9 \times 3 \times 8$ ways. In this latter (wrong) approach, we cannot tell for a telephone number such as 2529 whether the 5 was chosen first and put in the second position and then the 9 chosen next and put in the fourth position, or whether the 9 was chosen first and put in the fourth position and then the 5 chosen next and put in the second position. The disjointness requirement of the multiplication principle is being violated and each outcome in case (d) would be counted twice.

Example 8: Voter Power

We consider a way of measuring the influence of different players in weighted voting. Suppose that in a 5-person regional council there are 3 representatives from small towns, call them a, b, c, who each cast one vote, and there are 2 representatives from large towns, call them D, E, who each cast two votes. With a total of 7 votes cast, it takes 4 votes (a majority of votes) in favor of legislation to enact it. Suppose that in forming a coalition to vote for some legislation, the people join the coalition in order (an arrangement of the people). The *pivotal* person in a coalition arrangement is the person whose vote brings the number of votes in the coalition up to 4. For example, in the coalition arrangement $bDcaE$, the pivotal person is c. A measure of the "power" of a person p in the council is the fraction of coalition arrangements in which p is the pivotal person. This measure of power is called the *Shapley–Shubik index*.

Determine the Shapley–Shubik index of person a and person D in this council, that is, determine the fraction of all coalition arrangements in which a and D, respectively, are pivotal. (By symmetry, the 1-vote people b and c will have the same index as a, and similarly E will have the same index as D.)

If a is pivotal in a coalition arrangement, then people with (exactly) 3 votes must precede a in the arrangement and people with 3 votes must follow a. Since there are only two other 1-vote people, the 3 votes preceding a must come from one 1-vote person—b or c—and one 2-vote person—D or E. Then the beginning of the coalition can be formed in 2 (choice of 1-vote person) $\times$ 2 (choice of 2-vote person) $\times$ 2 (whether 1-vote or 2-vote person goes first) $=$ 8 ways. The remaining 1-vote and remaining 2-vote person will follow a, with 2 ways to arrange them. In total there are $8 \times 2 = 16$ coalition arrangements in which a is pivotal. There are $5! = 120$ arrangements in all, and so the Shapley–Shubik index of a is $16/120 = 4/30$.

If D is pivotal in a coalition arrangement, there can be people with 2 or 3 votes preceding D. Suppose there are 2 votes before D and 3 votes after D. Either the arrangement starts with 2 of the three 1-vote people—arranged 3×2 ways—then D, followed by the other 1-vote person and E in either order—2 ways—or the arrangement starts with E, then D, followed by an arrangement of the three 1-vote people—$3!$ ways. In total, there are $(3 \times 2 \times 2) + 3! = 18$ arrangements with 2 votes before D and 3 votes after D. By interchanging the people before D with the people after D in these arrangements, we obtain the arrangements with of 3 votes before D and 2 votes after D. So there are 18 of the latter arrangements. In total, there are $18 + 18 = 36$ arrangements in which D is pivotal, and so D's Shapley–Shubik index is $36/120 = 9/30$.

Observe that a 2-vote person has an index $2\frac{1}{4}$ times the size of a 1-vote person. ■

We close this section by noting that for large values of n, $n!$ can be approximated by the number $s_n = \sqrt{2\pi n}(n/e)^n$, where e is Euler's constant ($e = 2.718 \ldots$). This approximation is due to Stirling and its derivation is given in most advanced calculus texts (see Buck [1]). The error $|n! - s_n|$ increases as n increases, but the relative error $|n! - s_n|/s_n$ is always less than $1/11n$. The following table gives some sample values of $n!$, s_n, and $|n! - s_n|/s_n$.

n	$n!$	s_n	$\lvert n! - s_n \rvert / s_n$
1	1	.922	.085
2	2	1.919	.042
5	120	118.02	.017
10	3,628,800	3,598,600	.008
20	2.433×10^{18}	2.423×10^{18}	.004
100	9.333×10^{157}	9.328×10^{157}	.0005

5.2 EXERCISES

Summary of Exercises As in the previous section, most of these exercises require individual analysis, different for each problem. Remember to read problems carefully to avoid misinterpretation. Pay special attention to whether a problem involves arrangements or subsets. Note that the problems assume that people are distinct objects (no identical people).

1. How many ways are there to arrange the cards in a 52-card deck?

2. How many different 10-letter "words" (sequences) are there with no repeated letters formed from the 26-letter alphabet?

3. How many ways are there to distribute nine different books among 15 children if no child gets more than one book?

4. How many ways are there to seat five different boys and five different girls along one side of a long table with 10 seats? How many ways if boys and girls alternate seats?

5. How many arrangements are there of the 8 letters in the word VISITING?

6. How many ways are there to pick a subset of 4 different letters from the 26-letter alphabet?

7. How many ways are there to pick a 5-person basketball team from 10 possible players? How many teams if the weakest player and the strongest player must be on the team?

8. There are 10 white balls and 5 red balls in an urn. How many different ways are there to select a subset of 5 balls, assuming the 15 balls are different? What is the probability that the selection has 3 whites and 2 reds?

9. If a coin is flipped 10 times, what is the probability of 8 or more heads?

10. What is the probability that an arrangement of a, b, c, e, f, g begins and ends with a vowel?

11. Given 5 distinct pairs of gloves, 10 distinct gloves in all, how many ways are there to distribute two gloves to each of 5 sisters:

(a) If the two gloves someone receives might both be for the left hand or right hand?

(b) If each sister gets one left-hand glove and one right-hand glove?

12. How many ways are there to partition 14 people into:

(a) Three groups of sizes 3, 5 and 6?

(b) Two (unordered) groups of size 7?

13. How many 5-letter sequences (formed from the 26 letters in the alphabet, with repetition allowed) contain exactly one A and exactly two Bs?

14. **(a)** On a 10-question test, how many ways are there to answer exactly eight questions correctly?

(b) Repeat part (a) with the requirement that the first or second question, but not both, are answered correctly.

(c) Repeat part (a) in the case that three of the first five questions are answered correctly.

15. How many n-digit ternary (0, 1, 2) sequences with exactly nine 0s are there?

16. What is the probability that a five-card poker hand has the following?

(a) Four Aces

(b) Four of a kind

(c) Two pairs (not four of a kind or a full house)

(d) A full house (three of a kind and a pair)

(e) A straight (a set of five consecutive values)

(f) No pairs (possibly a straight or flush)

17. How many ways can a committee be formed from four men and six women with:

(a) At least two men and at least twice as many women as men?

(b) Between 3 and 5 people, and Ms. Wonder is excluded?

(c) Five people, and not all of the three O'Hara sisters can be on the committee?

(d) Four members, at least two of whom are women, and Mr. and Mrs. Baggins cannot both be chosen?

18. There are eight applicants for the job of dog catcher and three different judges who each rank the applicants. Applicants are chosen if and only if they appear in the top three in all three rankings.

(a) How many ways can the three judges produce their three rankings?

(b) What is the probability of Mr. Dickens, one of the applicants, being chosen in a random set of three rankings?

19. There are six different French books, eight different Russian books, and five different Spanish books. How many ways are there to arrange the books in a row on a shelf with all books of the same language grouped together?

20. If 13 players are each dealt 4 cards from a 52-card deck, what is the probability that each player gets one card of each suit?

21. How many 10-letter (sequences) are there using 5 different vowels and 5 different consonants (chosen from the 21 possible consonants)? What is the probability that one of these words has no consecutive pair of consonants?

22. What is the probability that a random 9-digit Social Security number has at least one repeated digit?

23. What is the probability that an arrangement of a, b, c, d, e, f has

(a) a and b side-by-side?

(b) a occurring somewhere before b?

24. How many ways are there to pair off 10 women at a dance with 10 out of 20 available men?

25. How many triangles are formed by (assuming no three lines cross at a point):

(a) Pieces of n nonparallel lines; for example, the four lines below form four triangles: *acd*, *abf*, *efd*, and *ebc*?

(b) Pieces of n lines, m of which are parallel and the others mutually nonparallel?

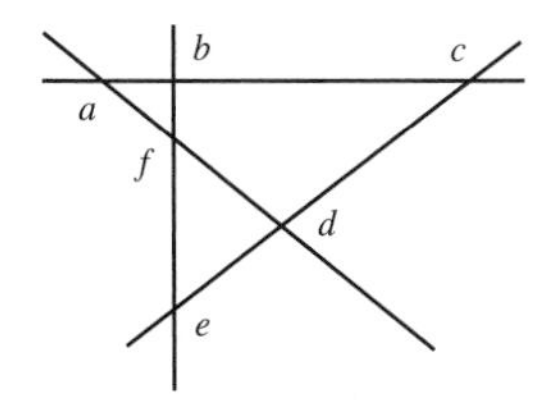

26. There are three women and five men who will split up into two four-person teams. How many ways are there to do this so that there is (at least) one woman on each team?

27. A man has n friends and invites a different subset of four of them to his house every night for a year (365 nights). How large must n be?

28. How many arrangements of JUPITER are there with the vowels occurring in alphabetic order?

29. Determine the Shapley–Shubik index of a 1-vote person and a 2-vote person in the councils with the following make-up:

(a) Three 1-vote people and one 2-vote person.

(b) Two 1-vote people and two 2-vote people (majority is 4).

(c) Four 1-vote people and three 2-vote people (majority is 6).

(d) Four 1-vote people and four 2-vote people (majority is 7).

30. How many 7-card hands dealt are there with 3 pairs (each of a different kind, plus a seventh card of a different kind)?

31. How many arrangements of the 26 letters of the alphabet in which:

(a) a occurs before b?

(b) a occurs before b and c occurs before d?

(c) the 5 vowels appear in alphabetical order?

32. A student must answer 5 out of 10 questions on a test, including at least 2 of the first 5 questions. How many subsets of 5 questions can be answered?

33. How many 6-letter sequences are there with at least 3 vowels (A, E, I, O, U)? No repetitions are allowed.

34. How many arrangements of 1, 1, 1, 1, 2, 3, 3 are there with the 2 not beside either 3?

35. How many arrangements of INSTRUCTOR are there in which there are exactly two consonants between successive pairs of vowels?

36. How many "words" can be formed by rearranging INQUISITIVE so that U does not immediately follow Q?

37. Suppose a subset of 60 different days of the year are selected at random in a lottery. What is the probability that there are 5 days from each month in the subset? (For simplicity, assume a year has 12 months with 30 days each.)

38. Suppose a subset of 8 different days of the year is selected at random. What is the probability that each day is from a different month? (For simplicity, assume a year has 12 months with 30 days each).

39. What is the probability of randomly choosing a permutation of the 10 digits 0, 1, 2, ..., 9 in which:

(a) An odd digit is in the first position and 1, 2, 3, 4, or 5 is in the last position?

(b) 5 is not in the first position and 9 is not in the last position?

40. **(a)** What is the probability that a five-card hand has at least one card of each suit?

(b) Repeat for a six-card hand.

41. What is the probability that a five-card hand has

(a) At least one of each of the four values: Ace, King, Queen, and Jack?

(b) The same number of hearts and spades?

42. How many n-digit decimal (0, 1, 2) sequences are there with k 1s?

43. What fraction of all arrangements of INSTRUCTOR have:

(a) Three consecutive vowels?

(b) Two consecutive vowels?

44. If one quarter of all 3-subsets of the integers 1, 2, ..., m contain the integer 5, determine m.

45. How many ways are there to form an (unordered) collection of 4 pairs of 2 people chosen from a group of 30 people?

46. What fraction of all arrangements of GRACEFUL have:

(a) F and G appearing side by side?

(b) No pair of consecutive vowels?

47. How many arrangements of SYSTEMATIC are there in which each S is followed by a vowel (this includes Y)?

48. How many arrangements of MATHEMATICS are there in which each consonant is adjacent to a vowel?

49. **(a)** What is the probability that k is the smallest integer in a subset of four different numbers chosen from 1 through 20 $(1 \le k \le 17)$?

(b) What is the probability that k is the second smallest?

50. What is the probability that the difference between the largest and the smallest numbers is k in a subset of four different numbers chosen from 1 through 20 $(3 \le k \le 19)$?

51. How many ways are there to place eight identical black pieces and eight identical white pieces on an 8×8 chessboard?

52. How many 8-letter arrangements can be formed from the 26 letters of the alphabet (without repetition) that include at most 3 of the 5 vowels and in which the vowels appear in alphabetical order? (*Hint*: break into cases.)

53. What is the probability that 2 (or more) people in a random group of 25 people have a common birthday? (This is the famous *Birthday Paradox Problem.*)

54. How many ways are there to arrange n (distinct) people so that: (i) Mr. and Mrs. Smith are side by side; and (ii) Mrs. Tucker is k positions away from the Smiths (i.e., $k-1$ people between Mrs. Tucker and the Smiths)?

55. A basketball team has 5 players, 3 in "forward" positions (this includes the "center") and 2 in "guard" positions. How many ways are there to pick a team if there are 6 forwards, 4 guards, and 2 people who can play forward or guard?

56. A family has two boys and three girls to send to private schools. There are five boys' schools, eight girls' schools, and three coed schools. If each child goes to a different school, how many different *subsets* of five schools can the family choose for their children?

57. Given a collection of $2n$ objects, n identical and the other n all distinct, how many different subcollections of n objects are there?

58. A batch of 50 different automatic typewriters contains exactly 10 defective machines. What is the probability of finding:

(a) At least 1 defective machine in a random group of 5 machines?

(b) At least 2 defective machines in a random group of 10 machines?

(c) The first defective machine to be the kth machine taken apart for inspection in a random sequence of machines?

(d) The last defective machine to be the kth machine taken apart?

59. Ten fish are caught in a lake, marked, and then returned to the lake. Two days later 20 fish are again caught, 2 of which have been marked.

(a) Find the probability of 2 of the 20 fish being marked if the lake has k fish (assuming the fish are caught at random).

(b) What value of k maximizes the probability?

60. How many different Hamilton paths are there in each of the following graphs?

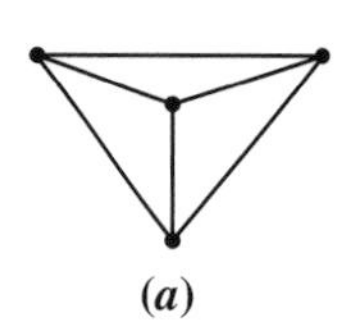

(*a*)

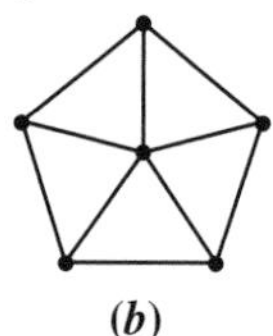

(*b*)

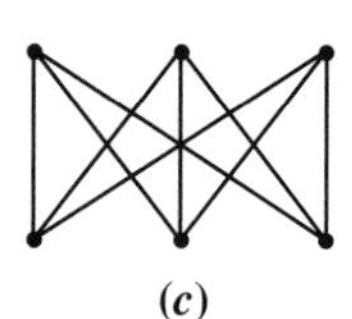

(*c*)

61. How many different 4-colorings do each of the graphs (a) and (b) in Exercise 60 have?

62. For the given map of roads between city A and city B,

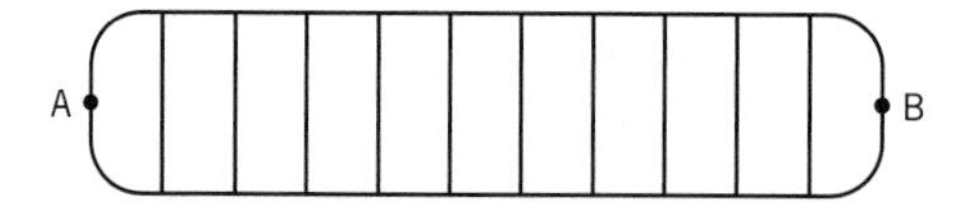

(a) How many routes are there from A to B that do not repeat any road (a road is a line segment between two intersections)?

(b) How many ways are there for two people to go from A to B without both ever traversing the same road in the same direction?

63. What is the probability that a random five-card hand has:

(a) Exactly one pair (no three of a kind or two pairs)?

(b) One pair or more (three of a kind, two pairs, four of a kind, full house)?

(c) The cards dealt in order of decreasing value?

(d) At least one spade, at least one heart, no diamonds or clubs, and the values of the spades are all greater than the values of the hearts?

64. How many ways are there to pick a group of n people from 100 people (each of a different height) and then pick a second group of m other people such that all people in the first group are taller than the people in the second group?

65. How many subsets of three different integers between 1 and 90 inclusive are there whose sum is:

(a) An even number?

(b) Divisible by 3?

(c) Divisible by 4?

66. What is the probability that two different letters selected from MISSISSIPPI are the same? That three letters are all different?

67. If a coin is tossed eight times, what is the probability of getting:

(a) Exactly four heads in a row?

(b) At least four heads in a row?

68. How many ways are there for a woman to invite different subsets of three of her five friends on three successive days? How many ways if she has n friends?

69. How many triangles can be formed by joining different sets of three corners of an octagon? How many triangles if no pair of adjacent corners is permitted?

70. How many arrangements of five 0s and ten 1s are there with no pair of consecutive 0s?

71. **(a)** How many points of intersection are formed by the chords of an n-gon (assuming no three of these lines cross at one point)?

(b) Into how many line segments are the lines in part (a) cut by the intersection points?

(c) Use Euler's formula $\mathbf{r} = \mathbf{e} - \mathbf{v} + 2$ and parts (a) and (b) to determine the number of regions formed by the chords of an n-gon.

72. Given 14 positive integers, 12 of which are even, and 16 negative integers, 11 of which are even, how many ways are there to pick 12 numbers from this collection of 30 integers such that 6 of the 12 numbers are positive and 6 of the 12 numbers are even?

73. How many triangles are formed by (assuming no three lines cross at a point):

(a) Pieces of three chords of a convex 10-gon such that the triangles are wholly within the 10-gon (a corner of the 10-gon cannot be a corner of any of these triangles)?

(b) Pieces of three chords or outside edges of a convex n-gon?

74. How many arrangements of the integers $1, 2, \ldots, n$ are there such that each integer differs by one (except the first integer) from some integer to the left of it in the arrangement?

75. A man has seven friends. How many ways are there to invite a different subset of three of these friends for a dinner on seven successive nights such that each pair of friends are together at just one dinner?

5.3 ARRANGEMENTS AND SELECTIONS WITH REPETITIONS

In this section we discuss arrangements and selections with repetition—arrangements of a collection with repeated objects, such as the collection *b, a, n, a, n, a,* and selections when an object can be chosen more than once, such as ordering six hot dogs chosen from three varieties. We motivate the formulas for these counting problems with the two examples just mentioned.

Example 1: Arrangements of *banana*

How many arrangements are there of the six letters *b, a, n, a, n, a*?

Consider a possible arrangement:

$$\underline{n}\ \underline{a}\ \underline{b}\ \underline{n}\ \underline{a}\ \underline{a}$$

This problem is solved by an extension of reasoning used to solve the problem of counting 8-digit binary sequences with six 1s (Example 3 in Section 5.2). The key is to focus on the subset of positions where the *a*s go and the subset of positions where the *n*s go. For example, the above arrangement is characterized by having *a*s in positions 2, 5, 6, *n*s in positions 1, 4, and *b* in position 3. We count the arrangements by first choosing the subset of three positions in the arrangement where the *a*s will go—$C(6, 3) = 20$ ways—then the subset of two positions (out of the remaining three) where the *n*s will go—$C(3, 2) = 3$ ways—and finally the last remaining position gets the *b*—1 way. Thus there are $C(6, 3) \times C(3, 2) \times C(1, 1) = 20 \times 3 \times 1 = 60$ arrangements. ∎

Theorem 1

If there are n objects, with r_1 of type 1, r_2 of type 2, ..., and r_m of type m, where $r_1 + r_2 + \cdots + r_m = n$, then the number of arrangements of these n objects, denoted $P(n; r_1, r_2, \ldots, r_m)$, is

$$P(n; r_1, r_2, \ldots, r_m) = \binom{n}{r_1}\binom{n - r_1}{r_2}\binom{n - r_1 - r_2}{r_3} \cdots \binom{n - r_1 - r_2 \cdots - r_{m-1}}{r_m}$$

$$= \frac{n!}{r_1! r_2! \ldots r_m!} \qquad (*)$$

Proof 1

First pick r_1 positions for the first types, then r_2 of the remaining positions for the second types, and so on. A mathematically precise proof of the "etc." part requires induction (see Exercise 33). The second line of formula (*) is just a simplification that results from canceling factorials in the binomial coefficients; for example,

$$P(6; 3, 2, 1) = \binom{6}{3}\binom{3}{2}\binom{1}{1} = \frac{6!}{3!3!} \times \frac{3!}{2!1!} \times \frac{1!}{1!} = \frac{6!}{3!2!1!} \quad \blacklozenge$$

Proof 2

This proof is similar to our derivation of $C(n, r)$ through the equation $P(n, r) = C(n, r) \times P(r, r)$. Suppose that for each type, the r_i objects of type i are given subscripts numbered 1, 2,..., r_i to make each object distinct. Then there are $n!$ arrangements of the n distinct objects. Let us enumerate these $n!$ arrangements of distinct objects by enumerating all $P(n; r_1, r_2, \ldots, r_m)$ patterns (without subscripts) of the objects, and then for each pattern placing the subscripts in all possible ways. For example, the pattern *baanna* can have subscripts on *a*s placed in the 3! ways:

$$\begin{array}{ccc} ba_1a_2nna_3 & ba_2a_1nna_3 & ba_3a_1nna_2 \\ ba_3a_2nna_1 & ba_1a_3nna_2 & ba_2a_3nna_1 \end{array}$$

For each of these 3! ways to subscript the *a*s, there are 2! ways to subscript the *n*s. Thus, in general a pattern will have $r_1!$ ways to subscript the r_1 objects of type 1,

$r_2!$ ways for type 2, and $r_m!$ ways for type m. Then

$$n! = P(n; r_1, r_2, \ldots, r_m) r_1! r_2! \ldots r_m!$$

or

$$P(n; r_1, r_2, \ldots, r_m) = \frac{n!}{r_1! r_2! \ldots r_m!} \quad \blacklozenge$$

Example 2: Ordering Hot Dogs

How many different ways are there to select six hot dogs from three varieties of hot dog?

To solve such a selection-with-repetition problem, we recast it as an arrangement problems as follows. Suppose the three varieties are regular dog, chili dog, and super dog. Let a selection be written down on an order form (when a person places this order) in the following fashion:

Regular	*Chili*	*Super*
x	$xxxx$	x

Each x represents a hot dog. The request shown on the form above is one regular, four chili, and one super. Since the hot dog chef knows that the sequence of dogs on the form is regular, chili, super, the request can simply be written as $x|xxxx|x$ without column headings.

Any selection of r hot dogs will consist of some sequence of r xs and two |s. Conversely, any sequence of r xs and two |s represents a unique selection: the xs before the first | count the number of regular dogs; the xs between the two |s count chilis; and the final xs count supers. So there is a one-to-one correspondence between orders and such sequences. Counting the number of sequences of six xs and two |s is an arrangement-with-repetition problem whose answer is $P(8; 6, 2) = \binom{8}{6}\binom{2}{2} = \frac{8!}{6!2!}$. As discussed in Example 3 of Section 5.2, counting such sequences of xs and |s is simply a matter of picking the subset of positions where the xs (or the |s) go—again, $\binom{8}{6}$ ways. ■

Theorem 2

The number of selections with repetition of r objects chosen from n types of objects is $C(r + n - 1, r)$.

Proof

We make an "order form" for a selection just as in Example 2, with an x for each object selected. As before, the xs before the first | count the number of the first type of object, the xs between the first and second |s count the number of the second type, ..., and the xs after the $(n - 1)$-st | count the number of the nth type ($n - 1$ slashes are needed to separate n types). The number of sequences with r xs and $n - 1$ |s is $C(r + (n - 1), r)$. ◆

Example 3: Grouping Classes

Nine students, three from Ms. A's class, three from Mr. B's class, and three from Ms. C's class, have bought a block of nine seats for their school's homecoming game. If three seats are randomly selected for each class from the nine seats in a row, what is the probability that the three A students, three B students, and three C students will each get a block of three consecutive seats?

In probability problems, we seek the number of favorable outcomes divided by all outcomes. The first question is, what is the set of all outcomes in this problem? They are the $P(9; 3, 3, 3) = 9!/3!3!3! = 1680$ ways to arrange three As, three Bs and three Cs in the row of nine seats.

If the three students of each class are to sit together, then we want to count outcomes that are arrangements of the three blocks, AAA, BBB, and CCC. Thus, instead of nine letters, we really are working now with three composite letters. There are $3! = 6$ ways to arrange these three composite letters. So the probability that each class sits together is $6/1680$. ■

Example 4: Sequencing Genes

The genetic code of organisms is stored in DNA molecules as a long string of four nucleotides: A (adenine), C (cytosine), G (guanine), and T (thymine). Short strings of DNA can be "sequenced"—the sequence of letters determined—by various modern biotech methods. Although the DNA sequence for a single gene typically has hundreds or thousands of letters, there exist special enzymes that will split a long string into short fragments (which can be sequenced) by breaking the string immediately following each appearance of a particular letter.

Suppose a C-enzyme (which splits after each appearance of C) breaks a 20-letter string into 8 fragments, which are identified to be: AC, AC, AAATC, C, C, C, TATA, TGGC. Note that each fragment, except the last one on the string, must end with a C. How many different strings could have given rise to this set of fragments?

Since the fragment TATA does not end with a C, it must go at the end of the string. The other seven fragments can occur in any order. These fragments consist of three Cs, two ACs, one AAATC, and one TGGC. Treat each fragment as a letter, similar to the reasoning in Example 3. Then we must arrange 7 letters, three of one type, two of a second type, and one each of two other types. There are thus $P(7; 3, 2, 1, 1) = 420$ possible arrangements of the fragments to form a 20-letter string. ■

If we use an A-enzyme to break the same string into fragments and look at all the possible arrangements of these fragments and then do the same with a G-enzyme and a T-enzyme, there will normally be only one string that appears in all four sets of arrangements. This will be the true original string. More sophisticated variations on this approach are used to determine the DNA sequence of entire genes.

Example 5: Sequences with Varying Repetitions

How many ways are there to form a sequence of 10 letters from 4 *a*s, 4 *b*s, 4 *c*s, and 4 *d*s if each letter must appear at least twice?

To apply Theorem 1 we need to know exactly how many *a*s, *b*s, *c*s, and *d*s will be in the arrangement. Thus we have to break this problem into a set of subproblems that each involves sequences with given numbers of *a*s, *b*s, *c*s, and *d*s. There are two categories of letter frequencies that sum to 10 with each letter appearing two or more times. The first category is four appearances of one letter and two appearances of each other letter. The second is three appearances of two letters and two appearances of the other two letters.

In the first category, there are four cases for choosing which letter occurs four times and $P(10; 4, 2, 2, 2) = 18{,}900$ ways to arrange four of one letter and two of the three others. In the second category, there are $C(4, 2) = 6$ cases for choosing which two of the four letters occur three times and $P(10; 3, 3, 2, 2) = 25{,}200$ ways to arrange three of two letters and two of the two others. The final answer is $4 \times 18{,}900 + 6 \times 25{,}200 = 226{,}800$ ways. ■

Example 6: Selecting Doughnuts

How many ways are there to fill a box with a dozen doughnuts chosen from five different varieties with the requirement that at least one doughnut of each variety is picked?

We solve this problem by following the typical procedure of people making such a doughnut selection. They would first pick one doughnut of each variety and then pick the remaining seven doughnuts any way they pleased. There is no choice (only one way) in picking one doughnut of each type. The choice occurs in picking the remaining seven doughnuts from the five types. Think of placing one doughnut of each type in a box and then covering them with a sheet of waxed paper, and then choosing the remaining seven doughnuts. The variety in outcomes involves only the seven doughnuts chosen to go on top of the waxed paper. So the answer by Theorem 2 is $C(7+5-1, 7) = 330$. (Note that the set composition principle does not apply here, because the objects are not all distinct.) ■

Example 7: Selections with Lower and Upper Bounds

How many ways are there to pick a collection of exactly 10 balls from a pile of red balls, blue balls, and purple balls if there must be at least 5 red balls? If at most 5 red balls?

The first problem is similar to Example 6. We put 5 red balls in a box, cover them with waxed paper, and then select 5 more balls arbitrarily (possibly including more red balls). The 5 balls above the waxed paper can be chosen in $C(5+3-1, 5) = 21$ ways.

There is no direct way to count selections when there is an upper bound of the number of repetitions of some object. To handle the constraint of at most 5 red balls, we count the complementary set. Of all $C(10+3-1, 10) = 66$ ways to pick 10 balls from the three colors without restriction, there are $C(4+3-1, 4) = 15$ ways to choose a collection with at least 6 red balls (put 6 red balls in a box, cover with waxed paper, and then arbitrarily choose 4 more balls). So there are $66 - 15 = 51$ ways to choose 10 balls without more than 5 red balls. ∎

Example 8: Arrangements with Restricted Positions

We return to Example 1 about arrangements of the letters in *banana,* but now with some constraints of the sort encountered in Section 5.2. How many arrangements are there of the letters *b, a, n, a, n, a* such that:

(a) The *b* is followed (immediately) by an *a*: We use the method for counting arrangements with consecutive letters introduced in Example 2 of Section 5.2: that is, we glue the *b* and one of the *a*s together to form the multiletter *ba*. Now we want to count all arrangements of the 5 "letters": *ba, a, a, n, n*. By Theorem 1, there are $5!/1!2!2! = 30$ arrangements.

(b) The pattern *bnn* never occurs: We solve this problem by counting all arrangements of *b, a, n, a, n, a* without constraint and then subtracting off the arrangements with the pattern *bnn*. Arrangements with this pattern are handled the same way as the pattern *ba* in part (a). We treat *bnn* as a single multiletter and now count arrangements of the 4 letters bnn, a, a, a—$4!/1!3! = 4$ ways. Subtracting the 4 forbidden arrangements from all $6!/1!3!2! = 60$ arrangements of *b, a, n, a, n, a* yields the answer, $60 - 4 = 56$.

(c) The *b* occurs before any of the *a*s (not necessarily immediately before an *a*): In other words, the relative order of the *b* and 3 *a*s is b–a–a–a. We use the method of handling relative order introduced in the continuation of Example 2 in Section 5.2. We pick a subset of 4 positions (for these 4 letters) from the 6 positions in an arrangement—C(6, 4) = 15 ways. We put the *b, a, a, a* in these 4 positions in the required relative order—1 way. Next we fill the 2 remaining positions in the arrangement with the 2 *n*s—1 way. So the final answer is $15 \times 1 \times 1 = 15$. ∎

5.3 EXERCISES

Summary of Exercises Most of these problems are not too difficult variations on the section's examples. Be careful to distinguish between arrangements and subsets problems.

1. How many ways are there to roll a die six times and obtain a sequence of outcomes with one 1, three 5s, and two 6s?

2. How many ways are there to arrange the letters in MISSISSIPPI?

3. **(a)** How many 6-digit numbers can be formed with the digits 3, 5, and 7?

(b) What fraction of the numbers in part (a) have two 3s, two 5s, and two 7s?

4. How many ways are there to invite one of three different friends over for dinner on six successive nights such that no friend is invited more than three times?

5. How many ways are there to pick a collection of 10 coins from piles of pennies, nickels, dimes, and quarters?

6. If three identical dice are rolled, how many different outcomes can be recorded?

7. How many ways are there to select a committee of 15 politicians chosen from a room full of indistinguishable Democrats, indistinguishable Republicans, and indistinguishable Independents if every party must have at least two members on the committee? If, in addition, no group may have a majority of the committee members?

8. Ten different people walk into a delicatessen to buy a sandwich. Four always order tuna fish, two always order chicken, two always order roast beef, and two order any of the three types of sandwich.

(a) How many different sequences of sandwiches are possible?

(b) How many different (unordered) collections of sandwiches are possible?

9. How many ways are there to pick a selection of coins from \$1 worth of identical pennies, \$1 worth of identical nickels, and \$1 worth of identical dimes if:

(a) You select a total of 10 coins?

(b) You select a total of 15 coins?

10. How many ways are there to pick 10 balls from large piles of (identical) red, white, and blue balls plus 1 pink ball, 1 lavender ball, and 1 tan ball?

11. How many numbers greater than 3,000,000 can be formed by arrangements of 1, 2, 2, 4, 6, 6, 6?

12. How many different rth-order partial derivatives does $f(x_1, x_2, \ldots, x_n)$ have?

13. How many 8-digit sequences are there involving exactly six different digits?

14. How many 9-digit numbers are there with twice as many different odd digits involved as different even digits (e.g., 945222123 with 9, 3, 5, 1 odd and 2, 4 even).

15. How many arrangements are possible with five letters chosen from MISSISSIPPI?

16. How many ways are there to select an unordered group of eight numbers between 1 and 25 inclusive with repetition? In what fraction of these ways is the sum of these numbers even?

17. How many arrangements of letters in REPETITION are there with the first E occurring before the first T?

18. In an international track competition, there are 5 United States athletes, 4 Russian athletes, 3 French athletes, and 1 German athlete. How many rankings of the 13 athletes are there when:

(a) Only nationality is counted?

(b) Only nationality is counted and all the U.S. athletes finish ahead of all the Russian athletes?

19. How many arrangements of the letters in MATHEMATICS are there in which TH appear together but the TH is not immediately followed by an E (not THE)?

20. How many arrangements of the letters in PEPPERMILL are there with:

(a) The M appearing to the left of all the vowels?

(b) The first P appearing before the first L?

21. How many arrangements of the letters in MISSISSIPPI in which

(a) The M is followed (immediately) by an I?

(b) The M is beside an I, that is, an I is just before or just after the M. Possibly there is an I both just before and just after the M—special care is required to make sure you count arrangements with the pattern IMI correctly.

22. How many ways are there first to pick a subset of r people from 50 people (each of a different height), and next to pick a second subset of s people such that everyone in the first subset is shorter than everyone in the second subset (explain your answer carefully).

23. How many ways are there to split a group of $2n$ αs, $2n$ βs, and $2n$ γs in half (into two groups of $3n$ letters)? (*Note:* The halves are unordered; there is no first half.)

24. How many ways are there to place nine different rings on the four fingers of your right hand (excluding the thumb) if:

(a) The order of rings on a finger does not matter?

(b) The order of rings on a finger is considered? (Hint: *tricky*.)

25. Show that $\Sigma P(10; k_1, k_2, k_3) = 3^{10}$, where k_1, k_2, k_3 are nonnegative integers ranging over all possible triples such that $k_1 + k_2 + k_3 = 10$.

26. How many arrangements of six 0s, five 1s, and four 2s are there in which:

(a) The first 0 precedes the first 1?

(b) The first 0 precedes the first 1, which precedes the first 2?

27. How many arrangements are there of $4n$ letters, four of each of n types of letters, in which each letter is beside a similar letter?

28. How many ways are there for 10 people to have five simultaneous telephone conversations?

29. When a coin is flipped n times, what is the probability that:

(a) The first head comes after exactly m tails?

(b) The ith head comes after exactly m tails?

30. How many arrangements are there of TINKERER with two but not three consecutive vowels?

31. How many arrangements are there of seven *a*s, eight *b*s, three *c*s, and six *d*s with no occurrence of the consecutive pairs *ca* or *cc*?

32. How many arrangements of 5 αs, 5 βs, and 5 γs are there with at least one β and at least one γ between each successive pair of αs?

33. **(a)** Use induction to give a rigorous proof of Theorem 1 (Proof 1).

(b) Use induction to prove that:

$$\binom{n}{r_1}\binom{n-r_1}{r_2}\binom{n-r_1-r_2}{r_3}\cdot\cdots\cdot\binom{n-r_1-r_2\cdot\cdots\cdot-r_{m-1}}{r_m}$$
$$= \frac{n!}{r_1!r_2!\cdot\cdots\cdot r_m!}$$

5.4 DISTRIBUTIONS

Generally a distribution problem is equivalent to an arrangement or selection problem with repetition. Specialized distribution problems must be broken up into subcases that can be counted in terms of simple permutations and combinations (with and without repetition). General guidelines for modeling distribution problems are:

Distributions of distinct objects are equivalent to arrangements

and

Distributions of identical objects are equivalent to selections.

Basic Models for Distributions

Distinct Objects The process of distributing r distinct objects into n different boxes is equivalent to putting the distinct objects in a row and stamping one of the n different box names on each object. The resulting sequence of box names is an arrangement of length r formed from n items (box names) with repetition. Thus there are $n \times n \times \cdots \times n$ (r ns) $= n^r$ distributions of the r distinct objects. If r_i objects must go in box i, $1 \le i \le n$, then there are $P(r; r_1, r_2, \ldots, r_n)$ distributions.

Identical Objects The process of distributing r identical objects into n different boxes is equivalent to choosing an (unordered) subset of r box names with repetition from among the n choices of boxes. Thus there are $C(r+n-1, r) = (r+n-1)!/r!(n-1)!$ distributions of the r identical objects.

Example 1: Assigning Diplomats

How many ways are there to assign 100 different diplomats to five different continents? How many ways if 20 diplomats must be assigned to each continent?

According to the model for distributions of distinct objects, this assignment equals the number of sequences of length 100 involving the five continental destinations—5^{100} such sequences (think of the diplomats lined up in a row holding attaché cases stamped with their destination). The constraint that 20 diplomats go to each continent means that each continent name should appear 20 times in the sequence. This can be done $P(100; 20, 20, 20, 20, 20) = 100!/(20!)^5$ ways.

Example 2: Bridge Hands

In bridge, the 52 cards of a standard card deck are randomly dealt 13 apiece to players North, East, South, and West. What is the probability that West has all 13 spades? That each player has one Ace?

There are $P(52; 13, 13, 13, 13)$ distributions of the 52 cards into four different 13-card hands, using the same reasoning as in the preceding example. Distributions in which West gets all the spades may be counted as the ways to distribute to West all the spades—1 way—times the ways to distribute the 39 non-spade cards among the 3 other hands—$P(39; 13, 13, 13)$ ways. So the probability that West has all the spades is

$$\frac{39!}{(13!)^3} \bigg/ \frac{52!}{(13!)^4} = 1 \bigg/ \frac{52!}{13!39!} = 1 \bigg/ \binom{52}{13}$$

The simple form of this answer can be directly obtained by considering West's possible hands alone (ignoring the other three hands). A random deal gives West one of the $C(52, 13)$ possible 13-card hands. Thus, the unique hand of 13 spades has probability $1/C(52, 13)$.

To count the ways in which each player gets one Ace, we divide the distribution up into an Ace part—4! ways to arrange the 4 Aces among the four players—and a non-Ace part—$P(48; 12, 12, 12, 12)$ ways to distribute the remaining 48 non-Ace cards, 12 to each player. So the probability that each player gets an Ace is

$$\frac{4!48!}{(12!)^4} \bigg/ \frac{52!}{(13!)^4} = \frac{13!^4}{12!^4} \times \frac{4!48!}{52!} = 13^4 \bigg/ \binom{52}{4} = 0.105 \ \blacksquare$$

Example 3: Distributing Candy

How many ways are there to distribute 20 (identical) sticks of red licorice and 15 (identical) sticks of black licorice among five children?

Using the identical objects model for distributions, we see that the ways to distribute 20 identical sticks of red licorice among five children is equal to the ways to select a collection of 20 names (destinations) from a set of 5 different names with repetition. This can be done $C(20+5-1, 20) = 10{,}626$ ways. By the same modeling argument, the 15 identical sticks of black licorice can be distributed in $C(15+5-1, 15) = 3876$ ways. The distributions of red and of black licorice are disjoint procedures. So the number of ways to distribute red and black licorice is $10{,}626 \times 3876 = 41{,}186{,}376$. ■

Example 4: Distributing a Combination of Identical and Distinct Objects

How many ways are there to distribute four identical oranges and six distinct apples (each a different variety) into five distinct boxes? In what fraction of these distributions does each box get exactly two objects?

There are $C(4+5-1, 4) = 70$ ways to distribute four identical oranges into five distinct boxes and $5^6 = 15{,}625$ ways to put six distinct apples in five distinct boxes. These two processes are disjoint, and so there are $70 \times 15{,}625 = 1{,}093{,}750$ ways to distribute the four identical and six distinct fruits.

The additional constraint of two objects in each box substantially complicates matters. To count constrained distributions of distinct objects, we must know exactly how many of the distinct objects should go in each box. But in this problem, the number of distinct objects that can go in a box depends on how many identical objects are in the box. So we deal with the (identical) oranges first. There are three possible categories of distributions of the four oranges (without exceeding two in any box).

Case 1 Two (identical) oranges in each of two boxes and no oranges in the other three boxes. The two boxes to get the pair of oranges can be chosen in $C(5, 2) = 10$ ways, and the six (distinct) apples can then be distributed in those three other boxes, two to a box, in $P(6; 2, 2, 2) = 90$ ways. So Case 1 has $10 \times 90 = 900$ possible distributions.

Case 2 Two (identical) oranges in one box, one orange in each of two other boxes, and the remaining two boxes empty. The one box with two oranges can be chosen $C(5, 1) = 5$ ways, the two boxes with one orange can be chosen in $C(4, 2) = 6$ ways [or combining these two steps, we can think of arranging the numbers 2, 1, 1, 0, 0, among the five boxes in $P(5; 1, 2, 2) = 30$ ways]. Then the two boxes still empty will each get two apples and the two boxes with one orange will get one apple. Thus, six distinct apples can then be distributed in $P(6; 2, 2, 1, 1) = 180$ ways, for a total of $5 \times 6 \times 180 = 5400$ ways.

Case 3 One orange each in four of the five boxes. This can be done in $C(5, 4) = 5$ ways, and then the apples can be distributed in $P(6; 2, 1, 1, 1, 1) = 360$ ways, for a total of $5 \times 360 = 1800$ ways.

Summing these cases, there are $900 + 5400 + 1800 = 8100$ distributions with two objects in each box. The fraction of such distributions among all ways to distribute the four oranges and six apples is $8100/1{,}093{,}750 = .0074$—a lot smaller than one might guess. (If all 10 fruits were distinct, the fraction with two in each box would be about 50 percent larger. Why?) ■

Example 5: Distributing Balls

Show that the number of ways to distribute r identical balls into n distinct boxes with at least one ball in each box is $C(r-1, n-1)$. With at least r_1 balls in the first box, at least r_2 balls in the second box, . . . , and at least r_n balls in the nth box, the number is $C(r - r_1 - r_2 - \cdots - r_n + n - 1, n - 1)$.

The requirement of at least one ball in each box can be incorporated into an equivalent selection-with-repetition model (as in the doughnut selection problem in Section 5.3). One could form such a collection of box destinations by putting one label for each of the n boxes on a tray, covering these labels with a piece of waxed paper, and then picking the remaining $r - n$ labels in all possible ways.

Alternatively, we can handle this constraint directly in terms of the boxes as follows. First put one ball in each box and then put a false bottom in the boxes to conceal the ball in each box. Now it remains to count the ways to distribute without restriction the remaining $r - n$ balls into the n boxes.

With either approach, the answer is

$$C((r-n)+n-1, (r-n)) = \frac{[(r-n)+n-1]!}{(r-n)!(n-1)!} = C(r-1, n-1) \text{ ways}$$

In the case where at least r_i balls must be in the ith box, we first put r_i balls in the ith box, and then distribute the remaining $r - r_1 - r_2 - \cdots - r_n$ balls in any way into the n boxes. This can be done in

$$\begin{aligned} &C((r - r_1 - r_2 - \cdots - r_n) + n - 1, (r - r_1 - r_2 - \cdots - r_n)) \\ &= C((r - r_1 - r_2 - \cdots - r_n) + n - 1, n - 1) \text{ ways} \quad \blacksquare \end{aligned}$$

The next counting problem will play a critical role in building generating functions in the next chapter.

Example 6: Integer Solutions

How many integer solutions are there to the equation $x_1 + x_2 + x_3 + x_4 = 12$, with $x_i \geq 0$? How many solutions with $x_i \geq 1$? How many solutions with $x_1 \geq 2, x_2 \geq 2, x_3 \geq 4, x_4 \geq 0$?

By an integer solution to this equation, we mean an ordered set of integer values for the x_is summing to 12, such as $x_1 = 2, x_2 = 3, x_3 = 3, x_4 = 4$. We can model this problem as a distribution-of-identical-objects problem or as a selection-with-repetition problem. Let x_i represent the number of (identical) objects in box i or the number of objects of type i chosen. The integer on the right side of the equation is the number of objects to be distributed or selected. Using either of these models, we see that the number of integer solutions is $C(12 + 4 - 1, 12) = 455$.

Solutions with $x_i \geq 1$ correspond in these models to putting at least one object in each box or choosing at least one object of each type. Solutions with $x_1 \geq 2, x_2 \geq 2,$

$x_3 \geq 4, x_4 \geq 0$ correspond to putting at least two objects in the first box, at least two in the second, at least four in the third, and any number in the fourth (or equivalently in the selection-with-repetition model). Formulas for these two types of distribution problems were given in Example 5. The respective answers are $C(12-1, 4-1) = 165$ and $C((12-2-2-4)+4-1, 4-1) = C(7,3) = 35$. ■

Equations with integer-valued variables are called *diophantine equations*. They are named after the Greek mathematician Diophantus, who studied them 2250 years ago.

We have now encountered three equivalent forms for selection with repetition problems.

Equivalent Forms for Selection with Repetition

1. The number of ways to select r objects with repetition from n different types of objects.
2. The number of ways to distribute r identical objects into n distinct boxes.
3. The number of nonnegative integer solutions to $x_1 + x_2 + \cdots + x_n = r$.

It is important that the reader be able to restate a problem given in one of the foregoing settings in the other two. Many students find version 2 the most convenient way to look at such problems because a distribution is easiest to picture on paper (or in one's head). Indeed, the original argument with order forms for hot dogs used to derive our formula for selection with repetition was really a distribution model. Version 3 is the most general (abstract) form of the problem. It is the form needed in Chapter 6 to build generating functions.

We next present three problems that we solve by recasting as problems involving distributions of identical or distinct objects.

Example 7: Ingredients for a Wife's Brew

A warlock goes to a store with five dollars to buy ingredients for his wife's Witch's Brew. The store sells bat tails for 25¢ apiece, lizard claws for 25¢ apiece, newt eyes for 25¢ apiece, and calf blood for \$1 a pint bottle. How many different purchases (subsets) of ingredients will these dollars buy?

The first step is to make our unit of money 25¢ (a quarter). So the warlock has 20 units to spend, with blood costing four units and the other three items each costing one unit. An integer-solutions-of-equation model of this problem is

$$T + S + E + 4B = 20, \qquad T, S, E, B \geq 0$$

The simplest way to handle the fact that B has a coefficient of 4 is first to specify exactly how much blood is bought. If one pint is bought, $B = 1$, then we have $T + S + E = 16$, an equation with $C(16 + 3 - 1, 16) = 153$ nonnegative integer solutions. In general, if $B = i$, then we have $T + S + E = 20 - 4i$, an equation with $C((20 - 4i) + 3 - 1, 20 - 4i) = C(22 - 4i, 2)$ solutions, for $i = 0, 1, 2, 3, 4, 5$. Summing these possibilities, we obtain

$$\binom{22}{2} + \binom{18}{2} + \binom{14}{2} + \binom{10}{2} + \binom{6}{2} + \binom{2}{2} = 536 \text{ different purchases} \quad \blacksquare$$

The output from electronic sensors is often a binary sequence of 0s and 1s. When a certain pattern of 0s and 1s is seen frequently in a sensor's output, there is a natural tendency to be concerned that the sensor may be malfunctioning. The next example looks at the probability of one particular pattern of binary sequences.

Example 8: Binary Patterns

What fraction of binary sequences of length 10 consists of a (positive) number of 1s, followed by a number of 0s, followed by a number of 1s, followed by a number of 0s? An example of such a sequence is 1110111000.

There are $2^{10} = 1024$ 10-bit binary sequences. We model the problem of counting this special type of 10-bit binary sequences as a distribution problem as follows. We create four distinct boxes, the first box for the initial set of 1s, the second box for the following set of 0s, and so on. We have 10 identical markers (call them xs) to distribute into the four boxes. Each box must have at least one marker, since each subsequence of 0s or of 1s must be nonempty. By Example 5 the number of ways to distribute 10 xs into 4 boxes with no box empty is $C(10 - 1, 4 - 1) = 84$. So there are 84 such binary sequences and thus $84/1024 \approx 8\%$ of all 10-bit binary sequences are of this special type. ■

Example 9: Nonconsecutive Vowels

How many arrangements of the letters a, e, i, o, u, x, x, x, x, x, x, x, x (eight xs) are there if no two vowels can be consecutive?

First we count the ways to fill in the xs before, between, and after the five vowel positions. We require at least one x between successive vowels. Think of arrangements of 8 xs and 5 vs. The situation can be modeled by creating boxes before, between, and after the vs and distributing the xs into the six resulting boxes with at least one x in the middle four boxes (to assure no two vowels are consecutive). The following pattern illustrates one possible distribution of the xs:

$$\underset{\text{box 1}}{\underline{\quad}}\ V\ \underset{\text{box 2}}{\underline{x\,x}}\ V\ \underset{\text{box 3}}{\underline{x}}\ V\ \underset{\text{box 4}}{\underline{x}}\ V\ \underset{\text{box 5}}{\underline{x\,x\,x}}\ V\ \underset{\text{box 6}}{\underline{x}}$$

Initially we put one x in boxes 2, 3, 4, 5. Then we distribute the remaining four xs into the six boxes without constraint—$C(4+6-1, 4) = 126$ ways. This counts all possible patterns of xs and vs (the positions of the vs are forced by the way we choose the xs).

Next in each pattern, we arrange a, e, i, o, u in $5! = 120$ ways among the five positions with vs. In total, there are $120 \times 126 = 15{,}120$ arrangements. ■

We close this section with a table summarizing the different basic types of counting problems we have encountered in this chapter.

Ways to Arrange, Select, or Distribute *r* Objects from *n* Items or into *n* Boxes

	Arrangement (ordered outcome) or *Distribution of distinct objects*	*Combination (unordered outcome)* or *Distribution of identical objects*
No repetition	$P(n, r)$	$C(n, r)$
Unlimited repetition	n^r	$C(n+r-1, r)$
Restricted repetition	$P(n; r_1, r_2, \ldots r_m)$	—

5.4 EXERCISES

Summary of Exercises Be careful to distinguish whether a problem involves distinct or identical objects. Exercises 27–30 involve problem restatement (no actual numerical answers are sought). Exercise 47 presents an important alternative approach for analyzing nonconsecutivity problems.

1. How many ways are there to distribute 40 identical jelly beans among four children:

(a) Without restrictions?

(b) With each child getting 10 beans?

(c) With each child getting at least 1 bean?

2. How many ways are there to distribute 20 different toys among five children?

(a) Without restrictions?

(b) If two children get 7 toys and three children get 2 toys?

(c) With each child getting 4 toys?

3. In a bridge deal, what is the probability that:

(a) West has four spades, three hearts, three diamonds, and three clubs?

(b) North and South have five spades, West has two spades, and East has one spade?

(c) One player has all the Aces?

(d) All players have a (4, 3, 3, 3) division of suits?

4. How many ways are there to distribute five (identical) apples, six oranges, and seven pears among three different people:

(a) Without restriction?

(b) With each person getting at least one pear?

5. How many ways are there to distribute 18 chocolate doughnuts, 12 cinnamon doughnuts, and 14 powdered sugar doughnuts among four school principals if each principal demands at least 2 doughnuts of each kind?

6. How many distributions of 24 different objects into three different boxes are there with twice as many objects in one box as in the other two combined?

7. How many ways are there to arrange the letters in VISITING with no pair of consecutive Is?

8. How many ways are there to arrange the 26 letters of the alphabet so that no pair of vowels appear consecutively (Y is considered a consonant)?

9. If you flip a coin 20 times and get 14 heads and 6 tails, what is the probability that there is no pair of consecutive tails?

10. How many integer solutions are there to $x_1 + x_2 + x_3 + x_4 + x_5 = 28$ with:

(a) $x_i \geq 0$ **(b)** $x_i > 0$ **(c)** $x_i \geq i (i = 1, 2, 3, 4, 5)$

11. How many integer solutions are there to $x_1 + x_2 + x_3 = 0$ with $x_i \geq -5$?

12. How many positive integer solutions are there to $x_1 + x_2 + x_3 + x_4 < 100$?

13. How many ways can a deck of 52 cards be broken up into a collection of unordered piles of sizes:

(a) Four piles of 13 cards?

(b) Three piles of 8 cards and four piles of 7 cards?

14. How many ways are there to distribute k balls into n distinct boxes ($k < n$) with at most one ball in any box if:

(a) The balls are distinct?

(b) The balls are identical?

15. How many ways are there to distribute three different teddy bears and nine identical lollipops to four children:

(a) Without restriction?

(b) With no child getting two or more teddy bears?

(c) With each child getting three "goodies"?

16. Suppose that 30 different computer games and 20 different toys are to be distributed among 3 different bags of Christmas presents. The first bag is to have 20 of the computer games. The second bag is to have 15 toys. The third bag is to have 15 presents, any mixture of games and toys. How many ways are there to distribute these 50 presents among the 3 bags?

17. Suppose a coin is tossed 12 times and there are 3 heads and 9 tails. How many such sequences are there in which there are at least 5 tails in a row?

18. How many binary sequences of length 20 are there that:

(a) Start with a run of 0s, that is, a consecutive sequence of (at least) one 0; then a run of 1s; then a run of 0s; then a run of 1s; and finally finishing with a run of 0s?

(b) Repeat part a with the constraint that each run is of length at least 2.

19. How many binary sequences of length 18 are there that start with a run of 1s, that is, a consecutive sequence of (at least) one 1, then a run of 0s, then a run of 1s, and then a run of 0s, and such that one run of 1s has length at least 8.

20. What fraction of all arrangements of EFFLORESCENCE has consecutive Cs and consecutive Fs but no consecutive Es?

21. How many arrangements of MISSISSIPPI are there with no pair of consecutive Ss?

22. How many ways are there to distribute 15 identical objects into four different boxes if the number of objects in box 4 must be a multiple of 3?

23. If n distinct objects are distributed randomly into n distinct boxes, what is the probability that:

(a) No box is empty?

(b) Exactly one box is empty?

(c) Exactly two boxes are empty?

24. How many ways are there to distribute eight balls into six boxes with the first two boxes collectively having *at most* four balls if:

(a) The balls are identical?

(b) The balls are distinct?

25. In Example 4, if all 10 pieces of fruit were distinct, what would be the fraction of outcomes with two pieces of fruit in each box? Why is this fraction greater when the pieces of fruit are distinct?

26. **(a)** How many ways are there to make 35 cents change in 1952 pennies, 1959 pennies, and 1964 nickels?

(b) In 1952 pennies, 1959 pennies, 1964 nickels, and 1971 quarters?

27. State an equivalent distribution version of each of the following arrangement problems:

(a) Arrangements of 8 letters chosen from piles of as, bs, and cs.

(b) Arrangements of 2 as, 3 bs, 4 cs.

(c) Arrangements of 10 letters chosen from piles of as, bs, cs, and ds with the same number of as and bs.

(d) Arrangements of 4 letters chosen from 2 as, 3 bs, and 4 cs.

28. State an equivalent arrangement version of each of the following distribution problems:

(a) Distributions of n distinct objects into n distinct boxes.

(b) Distributions of 15 distinct objects into five distinct boxes with 3 objects in each box.

(c) Distributions of 15 distinct objects into five boxes with i objects in the ith box, $i = 1, 2, 3, 4, 5$.

(d) Distributions of 12 distinct objects into three distinct boxes with at most 3 objects in box 1 and at most 5 objects in box 2.

29. State an equivalent distribution version and an equivalent integer-solution-of-an-equation version of the following selection problems.

(a) Selections of six ice cream cones from 31 flavors.

(b) Selections of five marbles from a group of five reds, four blues, and two pinks.

(c) Selections of 12 apples from four types of apples with at least 2 apples of each type.

(d) Selections of 20 jelly beans from four different types with an even number of each type and not more than 8 of any one type.

30. State an equivalent selection version and an equivalent integer-solution-of-an-equation version of the following distribution problems:

(a) Distributions of 30 black chips into five distinct boxes.

(b) Distributions of 18 red balls into six distinct boxes with at least 2 balls in each box.

(c) Distributions of 20 markers into four distinct boxes with the same number of markers in the first and second boxes.

31. How many election outcomes are possible (numbers of votes for different candidates) if there are three candidates and 30 voters? If, in addition, some candidate receives a majority of the votes?

32. How many election outcomes in the race for class president are there if there are five candidates and 40 students in the class and

(a) Every candidate receives at least two votes?

(b) One candidate receives at most one vote and all the others receive at least two votes?

(c) No candidate receives a majority of the votes?

(d) Exactly three candidates tie for the most votes?

33. How many numbers between 0 and 10,000 have a sum of digits:

(a) Equal to 7? **(b)** Less than or equal to 7? **(c)** Equal to 13?

34. How many integer solutions are there to the equation $x_1 + x_2 + x_3 + x_4 \leq 15$ with $x_i \geq -10$?

35. How many nonnegative integer solutions are there to the equation $2x_1 + 2x_2 + x_3 + x_4 = 12$?

36. How many nonnegative integer solutions are there to the pair of equations $x_1 + x_2 + \cdots + x_6 = 20$ and $x_1 + x_2 + x_3 = 7$?

37. How many nonnegative integer solutions are there to the inequalities $x_1 + x_2 + \cdots + x_6 \leq 20$ and $x_1 + x_2 + x_3 \leq 7$?

38. How many nonnegative integer solutions are there to $x_1 + x_2 + \cdots + x_5 = 20$:

(a) With $x_i \leq 10$? **(b)** With $x_i \leq 8$? **(c)** With $x_1 = 2x_2$?

39. How many ways are there to split four red, five blue, and seven black balls among:

(a) Two boxes without restriction?

(b) Two boxes with no box empty?

40. How many ways are there to arrange the letters in UNIVERSALLY so that the 4 vowels appear in two clusters of 2 consecutive letters with at least two consonants between the two clusters? An example of such an arrangement is LNUILYSVAER.

41. How many ways are arrange the letters in UNIVERSALLY so that no two vowels occur consecutively and also the consonants appear in alphabetical order?

42. How many 8-letter arrangements can be formed from the 26 letters of the alphabet (without repetition) that include at most 3 of the 5 vowels and in which the vowels are nonconsecutive?

43. How many arrangements of letters in INSTITUTIONAL have all of the following properties:

(a) No consecutive Ts,

(b) The 2 Ns are consecutive, and

(c) Vowels in alphabetical order?

44. How many arrangements of the letters in INSTRUCTOR have all of the following properties:

(a) The vowels appearing in alphabetical order,

(b) At least 2 consonants between each vowel, and

(c) Begin or end with the 2 Ts (the Ts are consecutive)?

45. How many arrangements of letters in STATISTICS have all of the following properties:

(a) No consecutive Ss,

(b) Vowels in alphabetical order, and

(c) The 3 Ts are consecutive (appear as 3 Ts in a row)?

46. **(a)** How many arrangements are there of REVISITED with vowels not in increasing order—that is, an I before one (or both) of the Es?

(b) How many arrangements with no consecutive Es and no consecutive Is?

(c) How many arrangements with vowels not in increasing order and no consecutive Es and no consecutive Is?

47. This exercise presents an alternative approach to counting arrangements with a certain type of object not occurring consecutively. Take Exercise 7, which seeks arrangements of VISITING with no consecutive Is. Arrange the 5 consonants in 5! ways and then pick a subset of 3 positions (with no repetition) for Is from among the 6 locations before, between, and after the 5 consonants—done in $C(6, 3)$ ways.

   (a) Rework Example 9 using this approach.

   (b) Use this approach to solve Exercise 8.

48. Among all arrangements of WISCONSIN without any pair of consecutive vowels, what fraction have W adjacent to an I?

49. How many bridge deals are there in which North and South get all the spades?

50. What is the probability in a bridge deal that each player gets at least three honors (an honor is an Ace, or King, or Queen, or Jack)?

51. How many ways are there to distribute 15 distinct oranges into three different boxes with at most 8 oranges in a box?

52. How many ways are there to distribute 20 toys to m children such that the first two children get the same number of toys if:

   (a) The toys are identical? **(b)** The toys are distinct?

53. How many ways are there to distribute 25 different presents to four people (including the boss) at an office party so that the boss receives exactly twice as many presents as the second most popular person?

54. How many ways are there to distribute r identical balls into n distinct boxes with exactly m boxes empty?

55. How many subsets of six integers chosen (without repetition) from $1, 2, \ldots, 20$ are there with no consecutive integers (e.g., if 5 is in the subset, then 4 and 6 cannot be in it)?

56. How many arrangements are there of eight αs, six βs, and seven γs in which each α is beside (on at least one side) another α? (*Hint:* Watch out for two clusters of αs occurring consecutively.)

57. How many arrangements are there with n 0s and m 1s, with k *runs* of 0s? [*A run* is a consecutive set (1 or more) of the same digit; e.g., $\underline{000}111\underline{0}1\underline{00}$ has three (underlined) runs of 0s.]

58. What is the probability that a random arrangement of a deck of 52 cards has exactly k runs of hearts (see Exercise 57 for a definition of a run)?

59. How many binary sequences of length n are there that contain exactly m occurrences of the pattern 01?

60. How many ways are there to distribute 20 distinct flags onto 12 distinct flagpoles if:

   (a) In arranging flags on a flagpole, the order of flags from the ground up makes a difference?

   (b) No flagpole is empty and the order on each flagpole is counted?

61. How many ways are there to distribute r identical balls into n distinct boxes with the first m boxes collectively containing at least s balls?

62. How many ways are there to deal four cards to each of 13 different players so that exactly 11 players have a card of each suit?

5.5 BINOMIAL IDENTITIES

In this section we show why the numbers $C(n, r)$ are called binomial coefficients. Then we present some identities involving binomial coefficients. We use three techniques to verify the identities: combinatorial selection models, "block walking," and the binomial expansion. The study of binomial identities is itself a major subfield of combinatorial mathematics and we will necessarily just scratch the surface of this topic.

Consider the polynomial expression $(a+x)^3$. Instead of multiplying $(a+x)$ by $(a+x)$ and the product by $(a+x)$ again, let us formally multiply the three factors term by term:

$$(a+x)(a+x)(a+x) = aaa + aax + axa + axx + xaa + xax + xxa + xxx$$

Collecting similar terms, we reduce the right-hand side of this expansion to

$$a^3 + 3a^2x + 3ax^2 + x^3 \tag{1}$$

The formal expansion of $(a+x)^3$ was obtained by systematically forming all products of a term in the first factor, a or x, times a term in the second factor times a term in the third factor. There are two choices for each term in such a product, and so there are the $2 \times 2 \times 2 = 8$ formal products obtained above. If we were expanding $(a+x)^{10}$, we would obtain $2^{10} = 1024$ different formal products.

Now we ask the question, how many of the formal products in the expansion of $(a+x)^3$ contain k xs and $(3-k)$ as? This question is equivalent to asking for the coefficient of $a^{3-k}x^k$ in (1). Since all possible formal products of as and xs are formed, and since formal products are just three-letter sequences of as and xs, we are simply asking for the number of all three-letter sequences with k xs and $(3-k)$ as. The answer is $C(3, k)$ and so the reduced expansion for $(a+x)^3$ can be written as

$$\binom{3}{0}a^3 + \binom{3}{1}a^2x + \binom{3}{2}ax^2 + \binom{3}{3}x^3$$

By the same argument, we see that the coefficient of $a^{n-k}x^k$ in $(a+x)^n$ will be equal to the number of n-letter sequences formed by k xs and $(n-k)$ as, that is, $C(n, k)$. If we set $a = 1$, we have the following theorem.

Binomial Theorem

$$(1+x)^n = \binom{n}{0} + \binom{n}{1}x + \binom{n}{2}x^2 + \cdots + \binom{n}{k}x^k + \cdots + \binom{n}{n}x^n$$

Another proof of this expansion, using induction, is given in the Exercises. Just as important as the Binomial Theorem is the equivalence we have established between the number of k-subsets of n objects and the coefficient of x^k in $(1+x)^n$. We exploit this equivalence extensively in the next chapter.

Let us now consider some basic properties of binomial coefficients. The most important identity for binomial coefficients is the symmetry identity

$$\binom{n}{k} = \frac{n!}{k!(n-k)!} = \binom{n}{n-k} \tag{2}$$

In words, this identity says that the number of ways to select a subset of k objects out of a set of n objects is equal to the number of ways to select a group of $n-k$ of the objects to set aside (the objects not in the subset).

The other fundamental identity is

$$\binom{n}{k} = \binom{n-1}{k} + \binom{n-1}{k-1} \tag{3}$$

This identity can be verified algebraically. We will give a combinatorial argument instead. We classify the $C(n, k)$ committees of k people chosen from a set of n people into two categories, depending on whether or not the committee contains a given person P. If P is not part of the committee, there are $C(n-1, k)$ ways to form the committee from the other $n-1$ people. On the other hand, if P is on the committee, the problem reduces to choosing the $k-1$ remaining members of the committee from the other $n-1$ people. This can be done $C(n-1, k-1)$ ways. Thus $C(n, k) = C(n-1, k) + C(n-1, k-1)$.

The following example presents another binomial identity that can be verified algebraically or by a combinatorial argument.

Example 1

Show that

$$\binom{n}{k}\binom{k}{m} = \binom{n}{m}\binom{n-m}{k-m} \tag{4}$$

The left-hand side of Eq. (4) counts the ways to select a group of k people chosen from a set of n people and then to select a subset of m leaders within the group of k people. Equivalently, as counted on the right side, we could first select the subset of m leaders from the set of n people and then select the remaining $k-m$ members of the group from the remaining $n-m$ people. Note the special form of Eq. (4) when $m = 1$:

$$k\binom{n}{k} = n\binom{n-1}{k-1} \quad \text{or} \quad \binom{n}{k} = \frac{n}{k}\binom{n-1}{k-1} \; \blacksquare \tag{5}$$

For a fixed integer n, the values of the binomial coefficients $C(n, k)$ increase as k increases as long as $k \le n/2$. Then the values of $C(n, k)$ decrease as k increases for

$k \geq n/2+1$. To verify this assertion, we observe that the binomial coefficients are increasing if and only if the ratio $C(n, k)/C(n, k-1)$ is greater than 1.

$$\frac{\binom{n}{k}}{\binom{n}{k-1}} = \frac{\dfrac{n!}{k!(n-k)!}}{\dfrac{n!}{(k-1)!\,(n-k+1)!}} = \frac{n-k+1}{k}$$

So the $C(n, k)$s are increasing when $(n-k+1)/k > 1$ or equivalently when $n-k+1 > k$. Solving for k in terms of n, we have $k < (n+1)/2$. This is the same bound on integer values of k as $k \leq n/2$.

Using Eq. (3) and the fact that $C(n, 0) = C(n, n) = 1$ for all nonnegative n, we can recursively build successive rows in the following table of binomial coefficients, called **Pascal's triangle**. Each number in this table, except the first and last numbers in a row, is the sum of the two neighboring numbers in the preceding row.

Table of binomial coefficients: kth number in row n is $\binom{n}{k}$

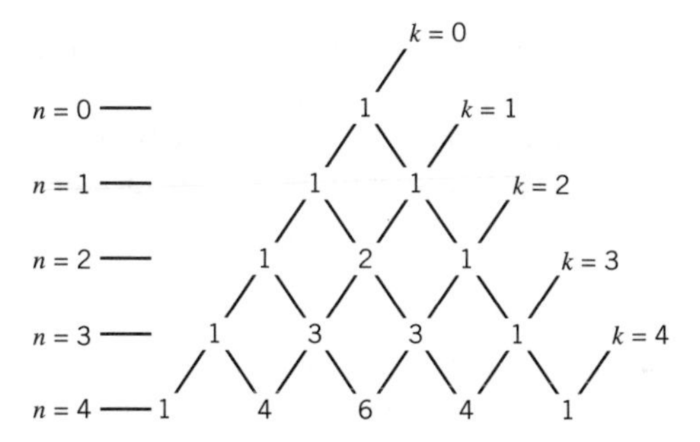

Pascal's triangle has the following nice combinatorial interpretation, due to George Polya. Consider the ways a person can traverse the blocks in the map of streets shown in Figure 5.1. The person begins at the top of the network, at the spot marked (0, 0), and moves down the network (down the page) making a choice at each intersection to go right or left (for simplicity, let "right" be your right as you look at this page, not the right of the person moving down the network). We label each street corner in the network with a pair of numbers (n, k), where n indicates the number of blocks traversed from (0, 0) and k the number of times the person chose the right branch at intersections. Figure 5.1 shows a possible route from the start (0, 0) to the corner (6, 3).

Any route to corner (n, k) can be written as a list of the branches (left or right) chosen at the successive corners on the path from (0, 0) to (n, k). Such a list is just a sequence of k Rs (right branches) and $(n-k)$ Ls (left branches). To go to corner (6, 3) following the route shown in Figure 5.1, we have the sequence of turns LLRRRL.

Let $s(n, k)$ be the number of possible routes from the start (0, 0) to corner (n, k) (moving down in the network). This is the number of sequences of k Rs and $(n-k)$ Ls, and so $s(n, k) = C(n, k)$. Another useful interpretation of binomial coefficients is the committee selection model, used above to verify Eqs. (3) and (4).

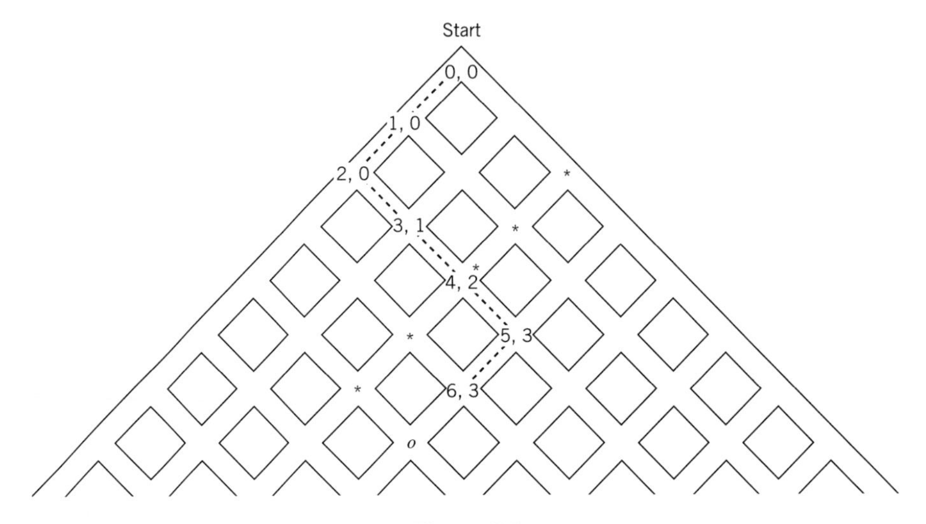

Figure 5.1

Let us show how our "block-walking" model for binomial coefficients can be used to get an alternate proof of identity (3). At the end of a route from the start to corner (n, k), a block walker arrives at (n, k) from either corner $(n-1, k)$ or corner $(n-1, k-1)$. For example, to get to corner (6, 3) in Figure 5.1, the person either goes to corner (5, 3) and branches left to (6, 3), or goes to corner (5, 2) and branches right to (6, 3). Thus, the number of routes from (0, 0) to corner (n, k) equals the number of routes from (0, 0) to $(n-1, k)$ plus the number of routes from (0, 0) to $(n-1, k-1)$. We have verified identity (3): $s(n, k) = s(n-1, k) + s(n-1, k-1)$.

We now list seven well-known binomial identities and verify three of them (the others are left as exercises). These binomial identities are of much practical interest. Expressions involving sums and products of binomial coefficients arise frequently in complicated real-world counting problems. These identities can be used to simplify such expressions.

$$\binom{n}{0} + \binom{n}{1} + \binom{n}{2} + \cdots + \binom{n}{n} = 2^n \tag{6}$$

$$\binom{n}{0} + \binom{n+1}{1} + \binom{n+2}{2} + \cdots + \binom{n+r}{r} = \binom{n+r+1}{r} \tag{7}$$

$$\binom{r}{r} + \binom{r+1}{r} + \binom{r+2}{r} + \cdots + \binom{n}{r} = \binom{n+1}{r+1} \tag{8}$$

$$\binom{n}{0}^2 + \binom{n}{1}^2 + \binom{n}{2}^2 + \cdots + \binom{n}{n}^2 = \binom{2n}{n} \tag{9}$$

$$\sum_{k=0}^{r} \binom{m}{k}\binom{n}{r-k} = \binom{m+n}{r} \tag{10}$$

$$\sum_{k=0}^{m} \binom{m}{k}\binom{n}{r+k} = \binom{m+n}{m+r} \tag{11}$$

$$\sum_{k=s-n}^{m-r} \binom{m-k}{r}\binom{n+k}{s} = \binom{m+n+1}{r+s+1} \tag{12}$$

Here $C(n, r) = 0$ if $0 \leq n < r$. These identities can be explained by the "committee" type of combinatorial argument used for Eqs. (2), (3), and (4) or by "block-walking arguments." We give a committee argument for Eq. (6). Consider the following two ways of counting all subsets (of any size) of a set of n people: (a) summing the number of subsets of size 0, of size 1, of size 2, and so on—this yields the left-hand side of Eq. (6)—and (b) counting all subsets by whether or not the first person is in the subset, whether or not the second person is in the subset, and so on—this yields $2 \times 2 \times 2 \ldots \times 2$ (n times) $= 2^n$, the right-hand side of Eq. (6). A simpler proof of Eq. (6) is given at the end of this section.

In general, a combinatorial argument for proving such identities consists of specifying a selection counted by the term on the right side, partitioning the selection into subcases, and showing that the terms in the summation count the subcases.

The advantage of proofs involving block walking is that we can draw pictures of the "proof." The picture shows that all the routes to a certain corner—this amount is the right-hand side of the identity—can be decomposed in terms of all routes to certain intermediate corners (or blocks)—the sum on the left-hand side of the identity.

Example 2

Verify identity (8) by block-walking and committee-selection arguments.

As an example of this identity, we consider the case where $r = 2$ and $n = 6$. The corners $(k, 2)$, $k = 2, 3, 4, 5, 6$, are marked with a * in Figure 5.1 and corner (7, 3) is marked with an o. Observe that the right branches at each starred corner are the locations of last possible right branches on routes from the start (0, 0) to corner (7, 3). After traversing one of these right branches, there is just one way to continue on to corner (7, 3), by making all remaining branches left branches. In general, if we break all routes from (0, 0) to (n + 1, $r + 1$) into subcases based on the corner where the last right branch is taken, we obtain identity (8).

We restate the block-walking model as a committee selection: If the kth turn is right, this corresponds to selecting the kth person to be on the committee; if the kth turn is left, the kth person is not chosen. We break the ways to pick $r + 1$ members of a committee from $n + 1$ people into cases depending on who is the last person chosen: the $(r + 1)$st, the $(r + 2)$nd, ..., the $(n + 1)$st. If the $(r + k + 1)$st person is the last chosen, then there are $C(r + k, r)$ ways to pick the first r members of the committee. Identity (8) now follows. ■

Example 3

Verify identity (9) by a block-walking argument.

The number of routes from (n, k) to $(2n, n)$ is equal to the number of routes from $(0, 0)$ to $(n, n-k)$, since both trips go a total of n blocks with $n-k$ to the right (and k to the left). So the number of ways to go from $(0, 0)$ to (n, k) and then on to $(2n, n)$ is $C(n, k) \times C(n, n-k)$. By (2), $C(n, n-k) = C(n, k)$, and thus the number of routes from $(0, 0)$ to $(2n, n)$ via (n, k) is $C(n, k)^2$. Summing over all k—that is, over all intermediate corners n blocks from the start—we count all routes from $(0, 0)$ to $(2n, n)$. So this sum equals $C(2n, n)$, and identity (9) follows. ∎

Now we show how binomial identities can be used to evaluate sums whose terms are closely related to binomial coefficients.

Example 4

Evaluate the sum $1 \times 2 \times 3 + 2 \times 3 \times 4 + \cdots + (n-2)(n-1)n$.

The general term in this sum $(k-2)(k-1)k$ is equal to $P(k, 3) = k!/(k-3)!$. Recall that the numbers of r-permutations and of r-selections differ by a factor of $r!$ That is, $C(k, 3) = k!/(k-3)!3! = P(k, 3)/3!$, or $P(k, 3) = 3!C(k, 3)$. So the given sum can be rewritten as

$$3!\binom{3}{3} + 3!\binom{4}{3} + \cdots + 3!\binom{n}{3} = 3!\left(\binom{3}{3} + \binom{4}{3} + \cdots + \binom{n}{3}\right)$$

By identity (8), this sum equals $3!\binom{n+1}{4}$. ∎

Example 5

Evaluate the sum $1^2 + 2^2 + 3^2 + \cdots + n^2$.

A strategy for problems whose general term is not a multiple of $C(n, k)$ or $P(n, k)$ is to decompose the term algebraically into a sum of $P(n, k)$-type terms. In this case, the general term k^2 can be written as $k^2 = k(k-1) + k$. So the given sum can be rewritten as

$$\begin{aligned}
&[(1 \times 0) + 1] + [(2 \times 1) + 2] + [(3 \times 2) + 3] + \cdots + [n(n-1) + n] \\
&\quad = [(2 \times 1) + (3 \times 2) + \cdots + n(n-1)] + (1 + 2 + 3 + \cdots + n) \\
&\quad = \left(2\binom{2}{2} + 2\binom{3}{2} + \cdots + 2\binom{n}{2}\right) + \left(\binom{1}{1} + \binom{2}{1} + \cdots + \binom{n}{1}\right) \\
&\quad = 2\binom{n+1}{3} + \binom{n+1}{2}
\end{aligned}$$

by identity (8).

Note that as part of Example 5, we showed that $1 + 2 + 3 + \cdots + n = C(n+1, 2) = \frac{1}{2}n(n+1)$, a result we verify by induction in Example 1 of Appendix A.2. ∎

There is another simple way to verify identities with binomial coefficients. We start with the binomial expansion in the Binomial Theorem. Then we substitute appropriate values for x. The following identities can be obtained from the binomial expansion: Eq. (6) by setting $x = 1$, Eq. (13), below by setting $x = -1$, and Eq. (14) by differentiating both sides of the binomial expansion and setting $x = 1$.

$$(1+1)^n = 2^n = \binom{n}{0} + \binom{n}{1} + \binom{n}{2} + \cdots + \binom{n}{n} \tag{6}$$

$$(1-1)^n = 0 = \binom{n}{0} - \binom{n}{1} + \binom{n}{2} - \cdots + (-1)^n \binom{n}{n} \tag{13}$$

or

$$\binom{n}{0} + \binom{n}{2} + \cdots = \binom{n}{1} + \binom{n}{3} + \cdots = 2^{n-1} \tag{13'}$$

(since the sum of both sides is 2^n)

$$n(1+x)^{n-1} = 1\binom{n}{1} + 2\binom{n}{2}x + 3\binom{n}{3}x^2 + \cdots + n\binom{n}{n}x^{n-1} \tag{14}$$

and so

$$n(1+1)^{n-1} = n2^{n-1} = 1\binom{n}{1} + 2\binom{n}{2} + 3\binom{n}{3} + \cdots + n\binom{n}{n} \tag{14'}$$

5.5 EXERCISES

Summary of Exercises Combinatorial identities are a well-developed field whose surface was barely scratched in this section. The later problems in this exercise set go well beyond the level of examples worked in this section.

1. **(a)** Verify identity (3) algebraically (writing out the binomial coefficients in factorials).

(b) Verify identity (4) algebraically.

(c) Verify that $C(n, k)/C(n, k-1) = (n+k-1)/k$.

2. Verify the following identities by block walking.

(a) (6) **(b)** (7) **(c)** (10) **(d)** (11)

3. Verify the following identities by a committee selection model.

(a) (7) **(b)** (9) **(c)** (10) **(d)** (11) **(e)** (13)

4. Verify the following identities by mathematical induction [*Hint:* Use Eq. (3)].

(a) (3) **(b)** (5) **(c)** (6) **(d)** (7) **(e)** (13′)

5. Prove the binomial theorem by mathematical induction.

6. Show that identity (7) can be obtained as a special case of Eq. (11).

7. Show that $C(2n, n) + C(2n, n-1) = \frac{1}{2}C(2n+2, n+1)$.

8. If $C(n,\ 3) + C(n+3-1,\ 3) = P(n,\ 3)$, find n.

9. Show by a combinatorial argument that

(a) $\binom{2n}{2} = 2\binom{n}{2} + n^2$

(b) $(n-r)\binom{n+r-1}{r}\binom{n}{r} = n\binom{n+r-1}{2r}\binom{2r}{r}$

10. Show that $C(k+m+n,\ k)C(m+n,\ m) = (k+m+n)!/(k!m!n!)$.

11. (a) Show that $\binom{n}{1} + 6\binom{n}{2} + 6\binom{n}{3} = n^3$

(b) Evaluate $1^3 + 2^3 + 3^3 + \cdots + n^3$.

12. (a) Evaluate $\sum_{k=0}^{n} 12(k+1)k(k-1)$. (*Hint:* Use Example 4.)

(b) Evaluate $\sum_{k=0}^{n} (2+3k)^2$.

(c) Evaluate $\sum_{k=0}^{n} k(n-k)$.

13. (a) Evaluate the sum

$$1 + 2\binom{n}{1} + \cdots (k+1)\binom{n}{k} + \cdots + (n+1)\binom{n}{n}$$

by breaking this sum into two sums, each of which is an identity in this section.

(b) Evaluate the sum

$$\binom{n}{0} + 2\binom{n}{1} + \binom{n}{2} + 2\binom{n}{3} + \cdots$$

14. By setting x equal to the appropriate values in the binomial expansion (or one of its derivatives, etc.) evaluate:

(a) $\sum_{k=0}^{n} (-1)^k \binom{n}{k}$

(b) $\sum_{k=0}^{n} k(k-1)\binom{n}{k}$

(c) $\sum_{k=0}^{n} 2^k \binom{n}{k}$

(d) $\sum_{k=1}^{n} k3^k \binom{n}{k}$

(e) $\sum_{k=1}^{n} (-1)^k k \binom{n}{k}$

(f) $\sum_{k=0}^{n} \frac{1}{k+1}\binom{n}{k}$

(g) $\sum_{k=0}^{n} (2k+1)\binom{n}{k}$

15. Show that $\sum_{k=m}^{n} \binom{k}{r} = \binom{n+1}{r+1} - \binom{m}{r+1}$.

16. Show that $\sum_{k=1}^{n}\binom{m+k-1}{k}=\sum_{k=1}^{m}\binom{n+k-1}{k}$.

17. Show that $\sum_{k=0}^{n-1} P(m+k, m)=\dfrac{P(m+n, m+1)}{(m+1)}$.

18. **(a)** Consider a sequence of $2n$ distinct people in a line at a cashier. Suppose n of the people owe \$1 and n of the people are due a \$1 payment. Show that the number of arrangements in which the cashier never goes into debt (i.e., at every stage at least as many people have paid in \$1 as were paid out \$1) is equal to $\binom{2n}{n}-\binom{2n}{n+1}$. [*Hint:* Use a symmetry argument in a block-walking type of model to demonstrate a one-to-one correspondence between sequences where at some stage the cashier goes (at least) \$1 into debt and all sequences of $2n$ people in which $n+1$ of the people are owed \$1.]

(b) Repeat part (a) with the following constraint: If the first person pays \$1, how many arrangements of the people are possible in which the cashier always has a positive amount of money until the last person in line (who is owed \$1)?

19. Show that $\left[\binom{n}{0}+\binom{n}{1}+\binom{n}{2}+\cdots+\binom{n}{n}\right]^2=\sum_{k=0}^{2n}\binom{2n}{k}$.

20. Find the value of k that maximizes

(a) $\binom{n}{k}$ **(b)** $\binom{2n+k}{n}\binom{2n-k}{n}$.

21. Evaluate $\sum_{k=1}^{n}\binom{n}{k}\binom{n}{k-1}$.

22. **(a)** Prove that $\binom{n}{1}+3\binom{n}{3}+5\binom{n}{5}+\cdots=2\binom{n}{2}+4\binom{n}{4}+6\binom{n}{6}+\cdots$

(b) What is the value of the sum on each side?

23. Evaluate $\binom{n}{0}-2\binom{n}{1}+3\binom{n}{2}+\cdots+(-1)^n(n+1)\binom{n}{n}$.

24. Show that $\sum_{k=0}^{n-1}(-1)^k\binom{n}{k+1}=1$. (*Hint:* Rewrite as a binomial coefficient with k alone in the bottom.)

25. Show that $\sum_{k=0}^{n}\dfrac{(2n)!}{k!^2(n-k)!^2}=\binom{2n}{n}^2$.

26. Give a combinatorial argument to evaluate $\sum_{k=0}^{n}\binom{n}{k}\binom{m}{k}$, $n\le m$.

27. **(a)** For any given k, show that an integer n can be represented as:

$$n=\binom{m_1}{1}+\binom{m_2}{2}+\cdots+\binom{m_k}{k} \quad \text{where } 0<m_1<m_2<\cdots<m_k.$$

(b) Prove that for a given k, the representation in part (a) is unique. [*Hint:* Use the fact that

$$\binom{n}{k} - 1 = \binom{n-1}{k} + \binom{n-2}{k-1} + \cdots + \binom{n-k}{1}.]$$

28. Show that $\sum_{j=0}^{k} \binom{n+k-j-1}{k-j}\binom{m+j-1}{j} = \binom{n+m+k-1}{k}$.

29. Show that $\sum_{k=1}^{m} \frac{(P(m, k)}{P(n, k)} = \frac{1}{\binom{n}{m}} \sum_{k=0}^{2n} \binom{n-k}{n-m} = \frac{n+1}{n-m+1}, m \leq n.$

30. Show that $\sum_{k=0}^{m} \frac{m!(n-k)!}{n!(m-k)!} = \frac{n+1}{n-m+1}, m \leq n.$

31. If $C_{2n} = \frac{1}{n+1}\binom{2n}{n}$, show that $C_{2m} = \sum_{k=1}^{m} C_{2k-2}C_{2m-2k}$.

32. Show that $\sum_{k=0}^{m} \binom{n}{k}\binom{n-k}{m-k} = 2^m \binom{n}{m}, \ m < n.$

33. Consider the problem of three-dimensional block walking. Show by a combinatorial argument that $P(n; i-1, j, k) + P(n; i, j-1, k) + P(n; i, j, k-1) = P(n+1; i, j, k)$.

34. Give a combinatorial argument to show that

$$(x_1 + x_2 + \cdots + x_k)^n = \Sigma P(n; i_1, i_2, \ldots, i_k) x_1^{i_1} x_2^{i_2} \cdots x_k^{i_k}$$

where the sum is over all $i_1 + i_2 + \cdots + i_k = n, \ i_j \geq 0$.

35. Show for sums over all $i_1 + i_2 + \cdots + i_k = n, \ i_j \geq 0$, that

(a) $\Sigma P(n; i_1, i_2, \ldots, i_k) = k^n$.

(b) $\Sigma i_1 i_2 \cdots i_k P(n; i_1, i_2, \ldots, i_k) = P(n, k) k^{n-k}$.

5.6 GENERATING PERMUTATIONS AND COMBINATIONS AND PROGRAMMING PROJECTS

Having spent a whole chapter counting different types of arrangements and selections, it is only fair to say a little about actually enumerating, or listing, these permutations and combinations. In this section, we present algorithms for enumerating all permutations of an n-set and all r-combinations of an n-set. With these algorithms, one can write programs to enumerate all the elements in various combinatorial sets; we do this in the second half of the section.

The following examples illustrate how such algorithms are used in operations research. Suppose that we have eight different products to manufacture in a month

Table 5.1

1. *abc*
2. *acb*
3. *bac*
4. *bca*
5. *cab*
6. *cba*

at a factory. Given the cost of converting from product i to product j, we must find the order of products that minimizes the sum of the conversion costs. The easiest solution to this problem is simply to generate all arrangements and see which one costs least (a "branch and bound" approach to this type of problem is given in Section 3.3).

Suppose that each of 10 workers knows a different subset of 30 skills needed on a project, and we must find all groups of 5 workers who collectively know all 30 skills. We need to check all 5-subsets of the 10 workers to find the ones that satisfy this special condition (this also can be solved by "branch and bound" methods).

There are several different algorithms for enumerating permutations and combinations. We will present algorithms that enumerate these collections in **lexicographic order,** that is, we treat an arrangement of objects as a sequence of "letters" forming a "word" and list these "words" in alphabetical order. This approach requires that the objects be numbered (or otherwise ordered). We write $A < B$ if sequence A is lexicographically smaller than B. A subset is treated similarly, with the "letters" (objects) of each subset arranged in increasing alphabetical order. For example, Table 5.1 lists all permutations of a, b, c in lexicographic order, and Table 5.2 lists all 3-combinations of 1, 2, 3, 4, 5 in lexicographic order.

To list all permutations of an n-set in lexicographic order, we need to be able to determine for any given permutation $A = a_1a_2a_3 \ldots a_n$, what is the next permutation $B = b_1b_2b_3 \ldots b_n$ in lexicographic order. If $a_{n-1} < a_n$, then interchanging the positions of a_{n-1} and a_n in A yields a lexicographically larger sequence, the desired sequence B. For example, 14325 is followed by 14352.

Suppose $a_{n-1} > a_n$. Then interchanging the positions of a_{n-1} and a_n in A yields a lexicographically smaller sequence. Instead, we must turn to a_{n-2}. If $a_{n-2} < a_{n-1}$, then the last three digits of A can be rearranged to yield a lexicographically larger sequence. To increase the sequence as little as possible, position $n - 2$ should have the next larger "digit" (larger than a_{n-2}) chosen from a_{n-1} and a_{n-2}. The remaining two of the original last three "digits" go in positions $n - 1$ and n in increasing order.

Table 5.2

1.	123	6.	145
2.	124	7.	234
3.	125	8.	235
4.	134	9.	245
5.	135	10.	345

For example, 14352 would be followed by 14523, while 14253 would be followed by 14325. If $a_{n-2} > a_{n-1}$ (and $a_{n-1} > a_n$), then we must turn to a_{n-3}.

In general, we start from the right end of the sequence and, moving left, look for the first pair a_i, a_{i+1} such that $a_i < a_{i+1}$ (and $a_{i+1} > a_{i+2} > a_{i+3} > \cdots > a_{n-1} > a_n$). Then the next larger sequence is obtained by placing $a_k = \min\{a_j \mid a_j > a_i \text{ and } j > i\}$ in position i and placing the remaining digits $a_i, a_{i+1}, \ldots, a_n$ (excluding a_k) in positions $i+1$ through n in increasing order.

Algorithm for Next Sequence in Lexicographic Listing of Permutations

```
PROCEDURE NEXTPERMUT(A, n)
Comment—find lexicographically next permutation of
n-element array A
Syntactic Comment—GOTO's are used instead of WHILE
statements for simplicity.
BEGIN
   FOR i←n−1 TO 1 STEP −1 DO
     IF A(i) < A(i+1) THEN GOTO 1);
   Comment—find min A(h) such that A(h) > A(i) and swap
   A(h), A(i)

1) FOR h←n TO i+1 STEP −1 DO
      IF A(i) < A(h) THEN GOTO 2);
2) TEMP ← A(i); A(i) ← A(h); A(h) ← TEMP;
   Comment—complete next arrangement by reversing rest
   of sequence
    FOR h←i+1 TO n DO TEM(h) ← A(h);
    FOR h←i+1 TO n DO A(h) ← TEM(n+j+1−h);
END NEXTPERMUT.
```

Next we show how to generate all r-combinations of an n-set in lexicographic order. For simplicity, let us assume that the n-set is the set of integers 1, 2, …, n. Recall that we list the elements in each r-combination in increasing order (see Table 5.2). The successor $b_1 b_2 \ldots b_r$ to the current r-combination $a_1 a_2 \ldots a_r$ is found by searching from right to left until the first $a_i \neq n - r + i$ is found. For this i, set $b_i = a_i + 1$ and $b_j = b_{j-1} + 1$, for $i < j \leq r$.

Algorithm for Next Subset in Lexicographic Listing of Combinations

```
PROCEDURE NEXTSUBSET (A, n, r)
Comment—given an r-subset A, find the
lexicographically next r-subset from 1, 2, 3,...,n.
```

```
BEGIN
   FOR i ← r TO 1 STEP −1 DO
      IF A(i) < n − r + i THEN GOTO 1);
1) A(i) ← A(i) + 1;
   FOR h ← i + 1 TO r DO A(h) ← A(h − 1) + 1;
END  NEXTSUBSET.
```

Programming Projects

Using the preceding algorithms, we can enumerate most combinatorial sets. Recall from Section 5.3 that selection with repetition can be modeled as a selection without repetition problem and that arrangement with repetition can be modeled as a succession of selection problems.

The following examples give computer programs to list all outcomes in two set enumeration problems from preceding sections.

Example 1: Enumerating Pairings

List all outcomes in Exercise 24 of Section 5.2. How many ways are there to pair off 10 women at a dance with 10 out of the 20 available men?

Let the men be named 1, 2, 3, . . . , 20. Let B be the array with the current subset of 10 men being paired (in 10! ways) with the 10 women. Let $A(i)$ be the man paired with the ith women, $i = 1, 2, \ldots, 10$. Procedures *NEXTPERMUT* and *NEXTSUBSET* are given earlier in this section.

```
BEGIN
   Comment—construct lexicographically first subset of
   10 men
      FOR i ← 1 TO 10 DO B(i) ← i;
    Comment—loop to list all 10-subsets of the 20 men
      FOR i ← 1 TO (20 over 10) DO
         BEGIN
            FOR i ← 1 TO 10 DO A(i) ← B(i);
   Comment—loop to list all arrangements of men in
   subset A with the women
      FOR k ← 1 TO 10! DO
         BEGIN
            print out pairing (i, A(i), i = 1, 2, ..., 10);
            NEXTPERMUT(A, 10)
         END;
      NEXTSUBSET(B, 20, 10);
       END;
END. ■
```

Example 2: Enumerating Arrangements Without Consecutive Is

List all outcomes in Exercise 7 of Section 5.4. How many ways are there to arrange the letters in VISITING with no pair of consecutive Is?

Let *OUTCOME(k)*, $k = 1, 2, \ldots, 8$, be the array where the possible outcomes will be constructed. Let *CONS(k)*, $k = 1, 2, \ldots, 5$, be the permutation of the consonants V, S, T, N, G to be used in *OUTCOME*.

Let *POS(j)*, $j = 1, 2, 3$, be the positions in array *OUTCOME* where Is will occur. Think of the 6 locations before, between, and after the five consonants G, N, S, T, V in VISITING. We pick a subset of 3 of these 6 locations for the 3 Is (this approach to nonconsecutive elements is discussed in Exercise 47 of Section 5.4). Suppose we picked locations 2, 3, 6. Location 2 for the first I is between the first and second consonant, and so the first I goes in position 2. Location 3 for the second I is between the second and third consonant, but also after the first I, and so the second I will be in position 4 (not position 3). Location 6 for the third I comes after all five consonants and after the other two Is, and so the third I goes in position 8. In general, the position of the first I is the first number in the 3-subset. The position of the second I is the second number in the 3-subset plus 1. The position of the third I is the third number in the subset plus 2.

```
BEGIN
   Comment—construct lexicographically first 3-subset
   of 1, 2, 3, 4, 5, 6
      FOR i ← 1 TO 3 DO P(i) ← i;
   Comment—loop on all 3-subsets of 1, 2, 3, 4, 5, 6.
      FOR i ← 1 TO (6 choose 3) DO
        BEGIN
          Comment—convert 3-subset into positions
          for Is:
            FOR k ← 1 TO 3 DO POS(k) ← P(k) + k − 1;
          Comment—construct lexicographically first
          permutation of consonants
            CONS(1) ← 'G';   CONS(2) ← 'N';
            CONS(3) ← 'S';   CONS(4) ← 'T';
            CONS(5) ← 'V';
          Comment—loop on all permutations of the
          five consonants
             FOR j ← 1 TO 5! DO
                BEGIN
          Comment—construct current  OUTCOME
             i1 ← 1;  j1 ← 1;
             FOR k ← 1 TO 8 DO
             IF k= POS(i1) THEN
                BEGIN
                   OUTCOME(k) ← 'I';  i1 ← j1+1
                END;
```

```
                ELSE
                   BEGIN
                      OUTCOME(k) ←  CONS(j1); j1 ← j1+1;
                   END;
                PRINT OUTCOME(i), i=1,2,...,8;
                NEXTPERMUT(CONS, 5);
           END;
        NEXTSUBSET(P, 6, 3);
     END;
END. ■
```

5.6 EXERCISES

Summary of Exercises The first five exercises use the three algorithms presented in this section. The next six exercises involve variants and related properties of these algorithms. The last nine exercises are programming projects.

1. Enumerate all arrangements of 1, 2, 3, 4 in lexicographic order.
2. Enumerate all arrangements of a, c, e, h in lexicographic order.
3. Enumerate all 4-subsets of 1, 2, 3, 4, 5, 6 in lexicographic order.
4. Enumerate all 3-subsets of a, d, h, j in lexicographic order.
5. What are the lexicographically first and last elements in:
   (a) The collection of all arrangements of the integers $1, 2, \ldots, 10$?
   (b) The collection of all 4-subsets of the 26 letters?
6. Prove by induction that the successor rule given in this section correctly finds the next lexicographic element in:
   (a) The collection of all arrangements of an n-set.
   (b) The collection of r-subsets of an n-set.
7. Devise another algorithm for enumerating all arrangements of an n-set.
8. Devise another algorithm for enumerating all r-subsets of an n-set.
9. **(a)** Develop a procedure to compute the location of a given arrangement in the lexicographic order of all arrangements of $1, 2, \ldots, n$.
   (b) What is the location of 314526 in the lexicographic order of $1, 2, \ldots, 6$?
10. **(a)** Develop a procedure to compute the location of a given r-subset in the lexicographic order of all r-subsets of $1, 2, \ldots, n$.
    (b) What is the location of {3,4,7} in the 3-subsets of $1, 2, \ldots, 8$?
11. Devise a method to list all subsets of an n-set in lexicographic order.

Programming Projects

12. The following table gives the cost of converting from job i to job j. In what order should the four jobs be performed in order to minimize the total conversion costs?

From\To	1	2	3	4
1	—	4	5	2
2	6	—	4	8
3	3	5	—	7
4	2	7	6	—

13. The following table tells which people know each of a set of trades. Find all subsets of four people who know all 10 trades.

Person\Trade	1	2	3	4	5	6	7	8	9	10
1	1	1	0	1	0	0	0	1	0	0
2	1	0	1	0	1	0	1	0	0	1
3	0	1	0	0	0	1	0	0	1	1
4	0	0	1	1	1	0	1	1	0	0
5	0	1	0	1	0	0	1	0	1	0
6	1	0	1	0	1	0	0	0	1	0
7	1	0	0	1	0	1	0	0	0	1

14. A salesperson wants to make a tour visiting each of five cities, starting and ending at city 1. Use the intercity cost matrix below to find the cheapest tour.

From\To	1	2	3	4	5
1	—	4	5	3	6
2	7	—	3	4	7
3	3	4	—	3	6
4	2	5	6	—	5
5	3	3	7	4	—

15. Write a program to list all 5-subsets of 1, 2, 3, 4, 5, 6 with repetition.

16. Write a program to list all arrangements of three *a*s, three *b*s, and three *c*s.

17. Write a program to list all distributions of r identical objects into n different boxes with between three and six objects in each box.

18. Write a program to list all integer solutions of $x_1 + x_2 + \cdots + x_r = n$ where:

(a) $x_i \geq k$ **(b)** $0 \leq x_1 \leq x_2 \leq x_3 \leq \cdots \leq x_r$

19. Write a program to list all distributions of r distinct objects into n different boxes with at least m objects in each box.

20. Write a program to list all r-subsets of 1, 2, ..., n with no pair of consecutive integers.

5.7 SUMMARY AND REFERENCES

This chapter introduced the basic formulas and logical reasoning used in arrangement and selection problems. There were four formulas for arrangements and selections with and without repetition, and there were three principles for composing subproblems. But these formulas and principles were just the building blocks for constructing answers to dozens of examples and hundreds of exercises. These problems required a thorough, logical analysis before one could begin to decompose them into tractable subproblems. Such logical analysis of possibilities arises often in computer science, probability, and operations research. It is the basic methodology of discrete mathematics and the most important skill for a student to develop in this course. Note that such logical reasoning is the very essence of mathematical model building.

The $n!$ formula for arrangements was known at least 2500 years ago (see David [2] for more details of the history of combinatorial mathematics). Problems involving binomial coefficients and the binomial expansion were mentioned in a primitive way in Chinese, Hindu, and Arab works 800 years ago. Pascal's triangle appears in a 14th-century work of Shih-Chieh (see text cover). The first appearance of the triangle in the West was 200 years later. However, it was not until the end of the seventeenth century that Jacob Bernoulli gave a careful proof of the binomial theorem. The examination of probabilities related to gambling by Pascal and Fermat around 1650 was the beginning of modern combinatorial mathematics. The formula for selection with repetition was discovered soon afterward in the following context: the probability when flipping a coin that the nth head appears after exactly r tails is $\left(\frac{1}{2}\right)^{r+n} C(n+r-1, r)$ (there are n "boxes" before and between the occurrences of the n heads for placing the r tails). The formula for arrangements with repetition was discovered around 1700 by Leibnitz in connection with the multinomial theorem (see Exercise 34 of Section 5.5). Jacques Bernoulli's *Ars Conjectandi* (1713) was the first book presenting basic combinatorial methods. The first comprehensive textbook on permutation and combination problems was written by Whitworth [5] in 1901.

The problem of systematically enumerating arrangements, selections, and other combinatorial collections was not addressed until the advent of modern digital computers. One of the first people to work on such enumeration in the 1950s was Lehmer [3]. For further information about algorithms for enumerating permutations and combinations, see Skiena [4].

See General References (at end of text) for a list of other introductory texts on enumeration.

1. R. Buck, *Advanced Calculus,* 3rd ed., McGraw-Hill, New York, 1978.

2. F. David, *Games, Gods, and Gambling: A History of Probability and Statistical Ideas,* Dover Press, New York, 1998.
3. D. Lehmer, "The Machine Tools of Combinatorics," in *Applied Combinatorial Mathematics,* E. Beckenbach (ed.), John Wiley & Sons, New York, 1964.
4. S. Skiena, *The Algorithm Design Manual,* Springer-Verlag, New York, 1997.
5. W. Whitworth, *Choice and Chance,* 5th ed. (1901), Hafner Press, New York, 1965.

SUPPLEMENT:
SELECTED SOLUTIONS TO PROBLEMS IN CHAPTER 5

Section 5.1

Exercise 13

(a) How many different five-digit numbers are there (leading zeros, e.g., 00174, not allowed)?

(b) How many even five-digit numbers are there?

(c) How many five-digit numbers are there with exactly one 3?

(d) How many five-digit palindromic numbers (numbers that are the same when the order of their digits is inverted, e.g., 15251) are there?

Answer

(a) All numbers from 10,000 to 99,999 form the 5-digit numbers. Answer is $9 \times 10 \times 10 \times 10 \times 10$.

(b) For a number to be even, the rightmost position must have an even digit (0 or 2 or 4 or 6 or 8). Answer: $9 \times 10 \times 10 \times 10 \times 5$.

(c) If leftmost digit is 3, there are 9^4 numbers. If any other digit is 3, 8×9^3 numbers. Total is $9^4 + 4 \times (8 \times 9^3)$.

(d) A palindromic number must have the same digit in the leftmost and rightmost position and similarly for the second-leftmost and second-rightmost positions. Then the problem becomes one of counting the ways to pick values for 3 leftmost digits (the 2 right digits are then forced by palindromic symmetry)—$9 \times 10 \times 10$ ways.

Exercise 30

How many times is the digit 5 written when listing all numbers from 1 to 100,000?

Answer

It is simpler to answer this question for numbers between 0 and 99,999 (adding 0 and omitting 100,000 makes no difference since neither contains a 5). We thus ask how many times 5 appears in writing 5-digit sequences (5-digit numbers with leading 0s allowed). Consider the problem, How many times does a 5 occur in the third (middle) position in these 5-digit sequences, that is, how many 5-digit sequences are there with a 5 in the third position? The answer is 10^4. For all 5 positions in the sequence, the answer is 5×10^4.

Exercise 33

How many four-digit numbers are there formed from the digits 1, 2, 3, 4, 5 (with possible repetition) that are divisible by 4?

Answer

A number is divisible by 4 if and only if the number formed by its two rightmost digits is divisible by 4. Also to be divisible by 4, the 1s digit (the rightmost digit) must be even, in this problem, 2 or 4. Observe that composing any given 10s digit with the two even 1s digits, such as 32, 34, we get two consecutive even numbers. Exactly one of every two consecutive even numbers is divisible by 4. So to generate a number divisible by 4 using digits 1, 2, 3, 4, 5 there can be any of these digits in the 1000s, the 100s and 10s positions followed by one choice for the 1s position—5^3 ways.

Section 5.2

Exercise 43

What is the probability that an arrangement of INSTRUCTOR has:

(a) Three consecutive vowels?

(b) Two consecutive vowels?

Answer

Total number of arrangements is $C(10, 2) \times C(8, 2) \times 6!$ and so outcomes in parts (a) and (b) will be divided by this amount to get the probabilities of these events.

(a) Replace the sequence of 3 consecutive vowels by a special symbol V. Now we arrange the 8 letters, C, N, R, R, S, T, T, V. We pick a pair (subset of size 2) of positions for the 2 Rs, then a pair of (remaining) positions for the Ts, and then arrange the 4 distinct letters—$C(8, 2) \times C(6, 2) \times 4!$ arrangements. Next there are $3!$ arrangements of the 3 vowels to put in place of V. Answer: $3! \times C(8, 2) \times C(6, 2) \times 4!$

(b) Initially proceed as in part (a), now with V as a double vowel and V' as a single vowel. There are $C(9, 2) \times C(7, 2) \times 5!$ arrangements of C, N, R, R, S, T, T, V, V'. There are 3 choices for the vowel to be V' and 2 arrangements of the remaining vowels for V. In total, $3 \times 2 \times C(9, 2) \times C(7, 2) \times 5!$ arrangements. However, if V is followed (immediately) by V' or if V' is followed by V, we obtain three vowels in a row. Every arrangement with three vowels in a row will be generated in two different ways, e.g., TIOUCRRTNS arises from T(IO)(U)CRRTNS and T(I)(OU)CRRTNS. Subtracting arrangements with 3 vowels in a row [part (a)], we have $\{3 \times 2 \times C(9, 2) \times C(7, 2) \times 5!\} - \{3! \times C(8, 2) \times C(6, 2) \times 4!\}$.

Exercise 49

(a) What is the probability that k is the smallest integer in a subset of four different numbers chosen from 1 through 20 ($1 \leq k \leq 17$)?

(b) What is the probability that k is the second smallest?

Answer

(a) As in most counting problems, the key here is to focus on the numbers that remain to be chosen, not on what one is told must be in the subset. To be concrete, initially assume $k = 8$. For 8 to be the smallest number in the subset, the other three numbers in the subset must be larger than 8—chosen in $C(20 - 8, 3)$ ways. For a general k, the probability is $C(20 - k, 3)/C(20, 4)$.

(b) If 8 is the second smallest number in the subset, there is one smaller number—chosen in $C(8 - 1, 1)$ ways—and two larger numbers—chosen in $C(20 - 8, 2)$ ways. In general, the probability is $C(k - 1, 1) \times C(20 - k, 2)/C(20, 4)$.

Exercise 53

What is the probability that 2 (or more) people in a random group of 25 people have a common birthday?

Answer

The "trick" in this problem is that it is easier to count the probability that no one has a common birthday and subtract this probability from 1. We want the fraction of possible birthday dates for 25 people in which everyone has a different birthday. The denominator, all possibilities of various birthdays for the 25 (different) people, is 365^{25}. The numerator, the possibilities where everyone has a different birthday, is $P(365, 25)$. The desired probability equals $1 - P(365, 25)/365^{25}$. (With a calculator or computer, one determines this probability to be about .57.)

Exercise 57

Given a collection of $2n$ objects, n identical and the other n all distinct, how many different subcollections of n objects are there?

Answer

The trick is to decide which of the n distinct objects will be in the collection. Any subset of distinct objects can be chosen in $C(n, 0) + C(n, 1) + C(n, 2) + \cdots + C(n, n)$ ways, with the remaining elements made up of identical objects—done in 1 way (since the objects are identical). An alternative approach is to say that we have the choice to use or not use each distinct object—2^n outcomes.

Exercise 63

What is the probability that a random five-card hand has:

(a) Exactly one pair (no three of a kind or two pairs)?

(b) One pair or more (three of a kind, two pairs, four of a kind, full house)?

(c) The cards dealt in order of decreasing value?

(d) At least one spade and at least one heart and the values of the spades are all greater than the values of the hearts?

Answer

The denominator is $C(52, 5)$.

(a) There are $C(13, 1) \times C(4, 2)$ ways to pick a pair (2 cards of the same kind). To fill out the rest of the hand, pick one card of a second kind—48 ways—then one card of a third kind—44 ways—and finally one card of a fourth kind—40 ways. The problem is that this rest-of-hand count is ordered; that is, the sequence of choices 7♠, Q♥, 5♦ yields the same rest-of-hand as 5♦, Q♥, 7♠. Divide by 3! to "un-order" this rest-of-hand, yielding the answer: $\{C(13, 1) \times C(4, 2) \times (48 \times 44 \times 40/3!)\}/C(52, 5)$.

Another approach for the rest-of-hand is to pick a subset of 3 other kinds that will appear in the rest-of-hand—$C(12, 3)$ ways—and pick a card of each of these kinds—4^3 ways, yielding $\{C(13, 1) \times C(4, 2) \times C(12, 3) \times 4^3\}/C(52, 5)$.

(b) Determine the probability of each of the 5 cards being of a different kind and subtract this probability from $1 : 1 - C(13, 5) \times 4^5/C(52, 5)$.

(c) Any 5-card hand has a $1/5!$ chance of being dealt in a particular order (increasing or otherwise).

(d) Pick a subset of the 5 kinds (values) that will appear in the hand—$C(13, 5)$ ways. The lowest k $(1 \leq k \leq 4)$ kinds must be hearts and the other kinds spades. It remains only to pick the value of k—4 choices: $4 \times C(13, 5)$. (Wasn't that sneaky!)

Exercise 71

(a) How may points of intersection are formed by the chords of an n-gon (assuming no 3 of these lines cross at one point)?

(b) Into how many line segments are the lines in part (a) cut by the intersection points?

(c) Use Euler's formula $\mathbf{r} = \mathbf{e} - \mathbf{v} + 2$ and parts (a) and (b) to determine the number of regions formed by the chords of an n-gon.

Answer

(a) Every subset of four vertices of the n-gon forms a unique intersection point generated by the intersection of the two chords joining opposite vertices in the subset of four vertices. Thus the answer is $C(n, 4)$.

(b) The number of chords is $C(n, 2) - n$ (all pairs of vertices are connected by chords except for the n edges of the polygon). Each intersection point splits two line segments, increasing the number of line segments by 2. Then the answer to part (b) is $C(n, 2) - n + 2C(n, 4)$.

(c) The total number of edges $\mathbf{e}$ equals the number of line segments—$C(n, 2) - n + 2C(n, 4)$ [from part (b)]—plus the number of edges of the polygon—n edges—yielding $\mathbf{e} = C(n, 2) + 2C(n, 4)$. The number of vertices $\mathbf{v}$ is the number of polygon vertices plus intersection points, $\mathbf{v} = n + C(n, 4)$. By Euler's formula, the number $\mathbf{r}$ of regions (excluding the infinite region) is $\mathbf{r} = \mathbf{e} - \mathbf{v} + \mathbf{1} = [C(n, 2) + 2C(n, 4)] - [n + C(n, 4)] + 1 = C(n, 2) + C(n, 4) - n + 1$.

Exercise 75

A man has seven friends. How many ways are there to invite a different subset of three of these friends for a dinner on seven successive nights such that each pair of friends are together at just one dinner?

Answer

We need to find all possible unordered collections of seven 3-subsets such that each pair of the seven friends appears in exactly one subset. Consider the following table of one such collection of 3-subsets (an X means that the element of that row is in the subset of that column):

		Subsets						
		1	2	3	4	5	6	7
F	A	X	X	X				
r	B	X			X	X		
i	C	X					X	X
e	D		X		X			X
n	E		X			X	X	
d	F			X	X		X	
s	G			X		X		X

Let the first three 3-subsets be the ones involving friend A. There are $\left[\binom{6}{2} \times \binom{4}{2} \times \binom{2}{2}\right]/3! = 15$ collections of 3-subsets involving A—we count all ways to pair off the other 6 friends into 3-subsets containing A, and divide by 3! so that the pairings are not ordered. Let the first subset involve B as well as A. A general collection of three 3-subsets with A has the form (A, B, C′), (A, D′, E′), (A, F′, G′) where C′ is the other friend in the subset with A and B, D′ is the (alphabetically) first of the friends not in the first subset, E′ is the other friend in the subset with A and D′, and F′ is alphabetically earlier than G′. Replacing C, D, E, F, and G by C′, D′, E′, F′, and G′ in the above table, the only remaining choice is whether B or C′ forms 3-subsets with D′, F′ and E′, G′—2 choices. So there are 15×2 collections of seven 3-subsets of the seven friends such that each pair appears in one 3-subset.

Each such collection can be arranged over the seven successive nights in 7! ways. So the answer is $15 \times 2 \times 7!$.

Section 5.3

Exercise 23

How many ways are there to split a group of $2n$ αs $2n$ βs, and $2n$ γs in half (into two groups of $3n$ letters)? (*Note:* The halves are unordered; there is no first half.)

Answer

Count the ways to select $3n$ letters from the 3 types of letters and then subtract outcomes with $2n+1$ or more of one letter—$C(3n+3-1, 3n) - 3 \times C((n-1)+3-1, n-1)$. Since each split forms *two* groups of $3n$ letters, it appears we should divide this count of $3n$-letter groups by 2. However, the split in which each group consists of n letters of each type is *not* double counted. So the answer is $\frac{1}{2}[C(3n+3-1, 3n) - 3 \times C((n-1)+3-1, n-1) - 1] + 1$.

Exercise 26

How many arrangements of six 0s, five 1s, and four 2s are there in which:

(a) The first 0 precedes the first 1?

(b) The first 0 precedes the first 1 which precedes the first 2?

Answer

(a) Position the four 2s among the 15 positions—$C(15, 4)$ ways—then put a 0 in the first of the remaining positions—1 way—and then pick five other positions for the remaining 0s—$C(10, 5)$ ways. Answer: $C(15, 4) \times C(10, 5)$.

(b) Put a 0 in the first position—1 way—then pick five other positions for the remaining 0s—$C(14, 5)$ ways—then put a 1 in the first of the remaining positions—1 way—and then pick four other positions for the remaining 1s—$C(8, 4)$ ways: $C(14, 5) \times C(8, 4)$.

Exercise 27

How many arrangements are there of $4n$ letters, four of each of n types of letters, in which each letter is beside a similar letter?

Answer

There cannot be exactly three consecutive letters the same, because that would leave the fourth letter of that type alone. So there are either two in a row or four in a row. But four in a row is the same as two consecutive two-in-a-rows. So our problem reduces to counting arrangements of $2n$ objects (two-in-a-rows) of n types with 2 of each type. There are $2n!/(2!)^n$ such arrangements.

Exercise 29

When a coin is flipped n times, what is the probability that:

(a) The first head comes after exactly m tails?

(b) The ith head comes after exactly m tails?

Answer

(a) There are 2^n outcomes in all. The sequence of flips begins with m successive tails followed by a head. The sequence can be completed in $2^{n-(m+1)}$ ways. Probability: $2^{n-(m+1)}/2^n = 2^{-(m+1)}$.

(b) If the ith head comes after exactly m tails, then the first $m + (i - 1)$ flips contain m tails and $i - 1$ heads—$C(m + (i - 1), m)$ ways. The remaining flips are unrestricted—$2^{n-(m+i)}$ ways. Probability: $C(m + (i - 1), m)2^{n-(m+i)}/2^n = C(m + i - 1, m)2^{-(m+i)}$.

Exercise 32

How many arrangements of 5 αs, 5 βs and 5 γs are there with at least one β and at least one γ between each successive pair of αs?

Answer

There are three cases:

(a) Exactly one β and exactly one γ between each pair of αs: Between each of the four pairs of αs, the β or the γ can be first—2^4 ways. The fifth β and fifth γ along with the sequence of the rest of the letters can be considered as 3 objects to be arranged—3! ways. Altogether, $2^4 \times 3! = 96$ ways.

(b) Exactly one β between each pair of αs and two γs between some pair of αs (or two βs between some pair of αs and exactly one γ between each pair of αs): There are 4 choices for between which pair of αs the two γs go and 3 ways to arrange the two γs and one β there. There are 2^3 choices for whether the β or the γ goes first between the other 3 pairs of αs and 2 choices for at which end of the arrangement the fifth β goes. Multiplying by 2 for the case of two βs between some pair of αs, we obtain $2 \times (4 \times 3 \times 2^3 \times 2) = 384$ ways.

(c) Two βs between some pair of αs and two γs between some pair of αs: There are two subcases. If the two βs and two γs are between the same pair of αs, there are 4 choices for which pair of αs, $C(4, 2)$ ways to arrange them between this pair of αs, and 2^3 choices for whether the β or the γ goes first between the other 3 pairs of αs. If two βs and two γs are between the different pairs of αs, there are 4×3 ways to pick between which αs the two βs and then between which αs the two γs go, 3^2 ways to arrange the two γs and one β and to arrange the one γ and two βs, and 2^2 choices for whether the β or the γ goes first between the other 2 pairs of αs. Together, $4 \times C(4, 2) \times 2^3 + 4 \times 3 \times 3^2 \times 2^2 = 1056$ ways.

Section 5.4

Exercise 23

If n distinct objects are distributed randomly into n distinct boxes, what is the probability that:

(a) No box is empty?

(b) Exactly one box is empty?

(c) Exactly two boxes are empty?

Answer

(a) $n!/n^n$.

(b) Pick which box is empty—n choices—then which other box gets two objects—$n - 1$ choices—then which two objects go into this box—$C(n, 2)$ choices—and then distribute the remaining $n - 2$ objects into the remaining $n - 2$ boxes, one per box—$(n - 2)!$ ways. Probability: $n \times (n - 1) \times C(n, 2) \times (n - 2)!/n^n$.

(c) Pick which two boxes are empty—$C(n, 2)$ choices. Two cases: Case (i). One box has three objects: pick the box with three objects—$n - 2$ choices—then pick which three objects go in this box—$C(n, 3)$ choices—and then distribute the remaining $n - 3$ objects into the remaining $n - 3$ boxes—$(n - 3)!$ ways. Case (ii). Two boxes which each have two objects: pick the two boxes with two objects—$C(n - 2, 2)$ choices—then pick which two objects go into the first 2-object box and which two objects go into the second 2-object box—$C(n, 2) \times C(n - 2, 2)$ choices—and then distribute the remaining $n - 4$ objects into the remaining $n - 4$ boxes—$(n - 4)!$ ways. The probability is $C(n, 2) \times \{(n - 2) \times C(n, 3) \times (n - 3)! + C(n - 2, 2) \times C(n, 2) \times C(n - 2, 2) \times (n - 4)!\}/n^n$.

Exercise 24

How many ways are there to distribute eight balls into six boxes with the first two boxes collectively having at most four balls if:

(a) The balls are identical?

(b) The balls are distinct?

Answer

Break into 5 cases of 0 or 1 or 2 or 3 or 4 balls in first two boxes.

(a) $\sum_{k=0}^{4} C(k + 2 - 1, k) \times C(8 - k + 4 - 1, 8 - k)$.

(b) $\sum_{k=0}^{4} C(8, k) \times 2^k \times 4^{8-k}$: If k distinct balls are in the first two boxes, pick which k balls—$C(8, k)$ ways—then decide for each ball into which of the first two boxes it goes—2^k ways—and then distribute remaining $8 - k$ distinct balls into the other 4 boxes—4^{8-k} ways.

Exercise 48

Among all arrangements of WISCONSIN without any pair of consecutive vowels, what fraction have W adjacent to an I?

Answer

Arrangements of WISCONSIN without any pair of consecutive vowels: $C[(6 - 2) + 4 - 1, (6 - 2)] = C(7, 4)$ patterns of vowels and consonants without consecutive vowels; 3 ways to distribute I, I, O among 3 vowel positions; $6!/2!2!$ ways to distribute consonants among consonant positions. Denominator in fraction is then $C(7, 4) \times 3 \times 6!/2!2!$

For the numerator, we look at three cases:

(a) W is beside an I that is the first vowel. If W is just before the I (glue W and I together), there are $C((5-2)+4-1, (5-2)) = C(6, 3)$ patterns for distributing the other consonants to assure no consecutive vowels, whereas if W is just after I, there are $C((5-1)+4-1, (5-1)) = C(7, 4)$ patterns. In either case, there are 2 ways to order the other vowels and $5!/2!2!$ ways to arrange the other consonants.

(b) If W is beside an I that is the last vowel, the number of outcomes is the same as in case (a).

(c) If W is beside an I that is the middle vowel, there are $C((5-1)+4-1, (5-1)) = C(7, 4)$ patterns for distributing the other consonants to assure no consecutive vowels whether W is just before or just after the I—2 choices for W. Again the other vowels and other consonants can be placed in $2 \times 5!/2!2!$ ways. We must subtract the arrangements with the subsequence IWI (that is, W is the only consonant between two Is): $C((5-1)+3-1, (5-1)) \times 2 \times 5!/2!2!$ In total, the numerator is $[2 \times C(6, 3) + 4 \times C(7, 4) - C(6, 4)] \times 2 \times 5!/2!2!$

Exercise 55

How many subsets of six integers chosen (without repetition) from 1, 2, ..., 20 are there with no consecutive integers (e.g., if 5 is in the subset, then 4 and 6 cannot be in it)?

Answer

Form a binary sequence of length 20 with six 1s and fourteen 0s to represent which integers are in the subset (a 1 in the ith position means that i is in the subset). In this form, we seek all 20-digit binary sequences with 6 nonconsecutive 1s. $C((14-5)+7-1, (14-5))$.

Exercise 57

How many arrangements are there of n 0s and m 1s with k runs of 0s? A *run* is a consecutive set (1 or more) of the same digit; e.g., 0001110100 has three (underlined) runs of 0s.

Answer

Represent a run no matter what its length as an R. Then we arrange m 1s and k Rs with no consecutive Rs. Using the reasoning in Exercise 55, there are $C(m+1, k)$ arrangements. Next pick how many 0s there are in each run. We select distribute the n 0s into k boxes (runs) with at least one 0 in each box—$C(n-1, k-1)$ ways: $C(m+1, k) \times C(n-1, k-1)$.

Exercise 60

How many ways are there to distribute 20 distinct flags onto 12 distinct flagpoles if:

(a) In arranging flags on a flagpole, the order of flags from the ground up makes a difference?

(b) No flagpole is empty and the order on each flagpole is counted?

Answer

(a) Lay the flagpoles on the ground end-to-end with a slash (|) between flagpoles. Then we need to count all arrangements of the 20 distinct flags and $(12 - 1)$|s—$31!/11!$ ways.

(b) First we distribute 20 identical flags into 12 flagpoles with no flagpole empty in $C(20 - 1, 12 - 1)$ ways. Now as in part (a), lay the flagpoles end-to-end. Then arrange the 20 distinct flags among the 20 places where there are (identical) flags in 20! ways: $C(20 - 1, 12 - 1) \times 20!$

CHAPTER 6
GENERATING FUNCTIONS

6.1 GENERATING FUNCTION MODELS

In this chapter, we introduce the concept of a generating function. Generating functions are developed in this chapter to handle special constraints in selection and arrangement problems with repetition. They are used in Chapters 7, 8, and 9 to solve other combinatorial problems. Generating functions are one of the most abstract problem-solving techniques introduced in this text. But once understood, they are also the easiest way to solve a broad spectrum of combinatorial problems.

Suppose a_r is the number of ways to select r objects in a certain procedure. Then $g(x)$ is a **generating function** for a_r if $g(x)$ has the polynomial expansion

$$g(x) = a_0 + a_1x + a_2x^2 + \cdots + a_rx^r + \cdots + a_nx^n$$

If the function has an infinite number of terms, it is called a **power series.**

In Section 5.5 we verified the well-known binomial expansion

$$(1+x)^n = 1 + \binom{n}{1}x + \binom{n}{2}x^2 + \cdots + \binom{n}{r}x^r + \cdots + \binom{n}{n}x^n$$

Then $g(x) = (1+x)^n$ is the generating function for $a_r = C(n, r)$, the number of ways to select an r-subset from an n-set. Recall that we derived the expansion of $(1+x)^n$ by first considering the formal multiplication of $(a+x)^3$:

$$\begin{aligned}(a+x)(a+x)(a+x) &= aaa + aax + axa + axx + xaa \\ &\quad + xax + xxa + xxx\end{aligned}$$

When $a = 1$, we obtained

$$\begin{aligned}(1+x)(1+x)(1+x) &= 111 + 11x + 1x1 + 1xx + x11 \\ &\quad + x1x + xx1 + xxx \end{aligned} \tag{1}$$

Such a formal expansion lists all ways of multiplying a term in the first factor times a term in the second factor times a term in the third factor. The problem of determining the coefficient of x^r in $(1+x)^3$, and more generally in $(1+x)^n$, reduces to the problem of counting the number of different formal products with exactly r xs and $(n-r)$ 1s. So the coefficient of x^r in $(1+x)^3$ is $C(3, r)$, and in $(1+x)^n$ is $C(n, r)$.

It is very important that multiplication in a product of several polynomial factors be viewed as generating the collection of all formal products obtained by multiplying together a term from each polynomial factor. If the ith polynomial factor contains r_i different terms and there are n factors, then there will be $r_1 \times r_2 \times r_3 \times \cdots \times r_n$ different formal products. For example, there will be 2^n formal products in the expansion of $(1+x)^n$.

In the expansion of $(1+x+x^2)^4$, the set of all formal products will be sequences of the form:

$$\begin{Bmatrix} 1 \\ x \\ x^2 \end{Bmatrix} \cdot \begin{Bmatrix} 1 \\ x \\ x^2 \end{Bmatrix} \cdot \begin{Bmatrix} 1 \\ x \\ x^2 \end{Bmatrix} \cdot \begin{Bmatrix} 1 \\ x \\ x^2 \end{Bmatrix} \tag{2}$$

that is, a 1 or an x or an x^2 in each of the entries in the product, such as, $x1x^2x$.

In this chapter we are primarily concerned with multiplying polynomial factors in which the powers of x in each factor have coefficient 1, factors such as $(1+x+x^2+x^3)$ or $(1+x^2+x^4+x^6+\cdots)$. These factors are completely specified by the set of different exponents of x. Note that $1 = x^0$. Thus expansion (1) can be rewritten

$$\begin{aligned}(x^0+x^1)(x^0+x^1)(x^0+x^1) &= x^0x^0x^0 + x^0x^0x^1 + x^0x^1x^0 + x^0x^1x^1 + x^1x^0x^0 \\ &\quad + x^1x^0x^1 + x^1x^1x^0 + x^1x^1x^1\end{aligned}$$

And the formal products in expansion (2) can be written as

$$\begin{Bmatrix} x^0 \\ x^1 \\ x^2 \end{Bmatrix} \cdot \begin{Bmatrix} x^0 \\ x^1 \\ x^2 \end{Bmatrix} \cdot \begin{Bmatrix} x^0 \\ x^1 \\ x^2 \end{Bmatrix} \cdot \begin{Bmatrix} x^0 \\ x^1 \\ x^2 \end{Bmatrix} \quad \text{or} \quad x^{e_1}x^{e_2}x^{e_3}x^{e_4} \qquad 0 \le e_i \le 2 \tag{3}$$

The problem of determining the coefficient of x^r when we multiply several such polynomial factors together can be restated in terms of exponents. Consider the coefficient of x^5 in the expansion of $(1+x+x^2)^4$. It is the number of different formal products, such as $x^2x^0x^2x^1$, formed in expansion (3) whose sum of exponents is 5. Determining the coefficient of x^5 in $(1+x+x^2)^4$ can be modeled as an integer-solution-to-an-equation problem (Example 6 of Section 5.4). The number of formal products $x^{e_1}x^{e_2}x^{e_3}x^{e_4}$ $0 \le e_i \le 2$ equaling x^5 is the same as the number of integer solutions to

$$e_1 + e_2 + e_3 + e_4 = 5 \qquad 0 \le e_i \le 2$$

According to the equivalent forms of selection-with-repetition discussed in Section 5.4, the preceding integer-solution-to-an-equation problem is equivalent to the problem of selecting five objects from a collection of four types, with at most two objects of each type. It is also equivalent to the problem of distributing five identical objects into four distinct boxes with at most two objects in each box.

More generally, the coefficient of x^r in $(1+x+x^2)^4$, that is, the number of formal products $x^{e_1}x^{e_2}x^{e_3}x^{e_4}$ $0 \le e_i \le 2$ equaling x^r, will be the number of integer solutions to

$$e_1 + e_2 + e_3 + e_4 = r \qquad 0 \le e_i \le 2$$

This problem in turn equals the number of ways of selecting r objects from four types with at most 2 of each type (or of distributing r identical objects into four boxes with at most 2 objects in any box). Thus $(1+x+x^2)^4$ is the generating function for a_r, the number of ways to select r objects from four types with at most 2 of each type (or to perform the equivalent distribution).

At this stage, we are concerned only with how to build generating function models for counting problems. In the next section we will see how various algebraic manipulations of generating functions permit us to evaluate desired coefficients.

We have shown how the coefficients of $(1+x+x^2)^4$ can be interpreted as the solutions to a certain selection-with-repetition or distribution-of-identical-objects problem. This line of reasoning can be reversed: Given a certain selection-with-repetition or distribution problem, we can build a generating function whose coefficients are the answers to this problem. We now give some examples of how to build such generating functions. We use the intermediate model of an integer-solution-to-an-equation to aid in the construction of generating function models.

Example 1

Find the generating function for a_r, the number of ways to select r balls from a pile of three green, three white, three blue, and three gold balls.

This selection problem can be modeled as the number of integer solutions to

$$e_1+e_2+e_3+e_4=r \qquad 0 \le e_i \le 3$$

Here e_1 represents the number of green balls chosen, e_2 the number of white, e_3 blue, and e_4 gold. We want to construct a product of polynomial factors such that when multiplied out formally (as described above), we obtain all products of the form $x^{e_1}x^{e_2}x^{e_3}x^{e_4}$, with each exponent e_i between 0 and 3. Then we need four factors, and *each factor should consist of an "inventory" of the powers of x from which x^{e_i} is chosen*. That is, each factor should be $(x^0+x^1+x^2+x^3)$. The desired generating function is thus $(x^0+x^1+x^2+x^3)^4$ or $(1+x+x^2+x^3)^4$. ■

Example 2

Use a generating function to model the problem of counting all selections of six objects chosen from three types of objects with repetition of up to four objects of each type. Also model the problem with unlimited repetition.

This selection problem can be modeled as the number of integer solutions to

$$e_1+e_2+e_3=6 \qquad 0 \le e_i \le 4$$

This problem does not ask for the general solution of the ways to select r objects. However, in building a generating function, we automatically model all values of r, not just $r=6$. Wanting a solution to the problem for six objects means that we are interested only in the coefficient of x^6, that is, the ways $x^{e_1}x^{e_2}x^{e_3}$ can equal x^6. We build a generating function with a factor of $(1+x+x^2+x^3+x^4)$ for each x^{e_i}. The desired generating function is $(1+x+x^2+x^3+x^4)^3$, and we want the coefficient of x^6 in it.

Permitting unlimited repetition means that any number can be chosen of each type and so any exponent value is possible (although in this particular problem no exponent can exceed 6). From this point of view, the answer is the coefficient of x^6 in $(1+x+x^2+x^3+\cdots)^3$, where "$+\cdots$" means that the factor is an infinite series. ■

In the unlimited repetition case above, we could have used the generating function $(1+x+x^2+x^3+x^4+x^5+x^6)^3$, since we wanted only the coefficient of x^6 and thus greater powers of x could not be used. However, we shall see in the next section that it is easier to use infinite series in generating functions. In selection-with-repetition problems, when we ask for six objects from three types with unlimited repetition, we do not add the constraint of at most six of any type—it is implicit in the problem and there is no need to make it explicit. The same situation applies with generating functions.

Example 3

Find a generating function for a_r, the number of ways to distribute r identical objects into five distinct boxes with an even number of objects not exceeding 10 in the first two boxes and between three and five in the other boxes.

While the constraints may be a bit contrived in this example, it is still easy to model the problem as an integer-solution-to-an-equation problem with appropriate constraints. We want to count all integer solutions to

$$e_1+e_2+e_3+e_4+e_5=r \qquad e_1, e_2 \text{ even}, \qquad 0 \le e_1, e_2 \le 10,$$
$$3 \le e_3, e_4, e_5 \le 5$$

To generate all formal products of the form $x^{e_1}x^{e_2}x^{e_3}x^{e_4}x^{e_5}$, with the given constraints on the e_is, we need a product of five factors, each containing the inventory of the powers of x permitted for its x^{e_i}. For example, the inventory for x^{e_1} is $(1+x^2+x^4+x^6+x^8+x^{10})$. The required generating function is $g(x)=(1+x^2+x^4+x^6+x^8+x^{10})^2(x^3+x^4+x^5)^3$. ■

With a little practice, generating function models become simple to build. However, the concept behind generating functions is far from simple. In the previous chapter, we solved similar selection and distribution problems with explicit formulas involving binomial coefficients. Now we are modeling these problems as some coefficient, which represents the number of formal products in the multiplication of certain polynomial factors. All we write down is the polynomial factors.

A generating function cannot be designed to model a selection or distribution problem for just one amount, say, six objects. It must model the problem *for all possible numbers of objects*. A generating function model stores all the necessary information about these subproblems in one function.

In the next section, determining the value of a specified coefficient of such a function will be reduced to a series of rote algebraic operations. The fact that the function models a complex selection or distribution problem will be irrelevant. The combinatorial reasoning arises solely in the construction of generating functions.

6.1 EXERCISES

Summary of Exercises The first 21 exercises involve simple generating function modeling. The remaining problems involve more challenging modeling; Exercises 25–29 use multinomial generating functions.

1. For each of the following expressions, list the set of all formal products in which the exponents sum to 4.

(a) $(1+x+x^2)^2(1+x)^2$

(b) $(1+x+x^2+x^3+x^4)^3$

(c) $(1+x^2+x^4)^2(1+x+x^2)^2$

(d) $(1+x+x^2+x^3+\cdots)^3$

2. Build a generating function for a_r, the number of integer solutions to the equations:

(a) $e_1+e_2+e_3+e_4+e_5=r, \quad 0 \le e_i \le 4$

(b) $e_1+e_2+e_3=r, \quad 0 < e_i < 4$

(c) $e_1+e_2+e_3+e_4=r, \quad 2 \le e_i \le 8 \quad e_1$ even, $\quad e_2$ odd

(d) $e_1+e_2+e_3+e_4=r, \quad 0 \le e_i$

(e) $e_1+e_2+e_3+e_4=r, \quad 0 < e_i, \quad e_2, e_4$ odd, $\quad e_4 \le 3$

3. Build a generating function for a_r, the number of r selections from a pile of:

(a) Three red, four black, and four white balls.

(b) Five jelly beans, three licorice sticks, eight lollipops with at least one of each candy.

(c) Unlimited amounts of pennies, nickels, dimes, and quarters.

(d) Seven types of lightbulbs with an odd number of the first and second types.

4. Build a generating function for a_r, the number of distributions of r identical objects into:

(a) Five different boxes with at most four objects in each box.

(b) Four different boxes with between three and eight objects in each box.

(c) Seven different boxes with at least one object in each box.

(d) Three different boxes with at most five objects in the first box.

5. Use a generating function for modeling the number of 5-combinations of the letters M, A, T, H in which M and A can appear any number of times but T and H appear at most once. Which coefficient in this generating function do we want?

6. Use a generating function for modeling the number of different selections of r hot dogs when there are five types of hot dogs.

7. Use a generating function for modeling the number of distributions of 18 chocolate bunny rabbits into four Easter baskets with at least 3 rabbits in each basket. Which coefficient do we want?

8. **(a)** Use a generating function for modeling the number of different election outcomes in an election for class president if 27 students are voting among four candidates. Which coefficient do we want?

(b) Suppose each student who is a candidate votes for herself or himself. Now what is the generating function and the required coefficient?

(c) Suppose no candidate receives a majority of the vote. Repeat part (a).

9. Find a generating function for a_r, the number of r-combinations of an n-set with repetition.

10. Given one each of u types of candy, two each of v types of candy, and three each of w types of candy, find a generating function for the number of ways to select r candies.

11. Find a generating function for a_k, the number of k-combinations of n types of objects with an even number of the first type, an odd number of the second type, and any amount of the other types.

12. Find a generating function for a_r, the number of ways to distribute r identical objects into q distinct boxes with an odd number between r_1 and s_1 in the first box, an even number between r_2 and s_2 in the second box, and at most three in the other boxes.

13. Find a generating function for a_r, the number of ways n distinct dice can show a sum of r.

14. Find a generating function for a_r, the number of ways a roll of six distinct dice can show a sum of r if:

(a) The first three dice are odd and the second three even.

(b) The ith die does not show a value of i.

15. Build a generating function for a_r, the number of integer solutions to $e_1 + e_2 + e_3 + e_4 = r$, $-3 \leq e_j \leq 3$.

16. Find a generating function for the number of integers between 0 and 999,999 whose sum of digits is r.

17. Find a generating function for the number of selections of r sticks of chewing gum chosen from eight flavors if each flavor comes in packets of five sticks.

18. **(a)** Use a generating function for modeling the number of distributions of 20 identical balls into five distinct boxes if each box has between 2 and 7 balls.

(b) Factor out an x^2 from each polynomial factor in part (a). Interpret this revised generating function combinatorially.

19. Use a generating function for modeling the number of ways to select five integers from $1, 2, \ldots, n$, no two of which are consecutive. Which coefficient do we want for $n = 20$? For a general n?

20. Explain why $(1+x+x^2+\cdots+x^r)^4$ is not a proper generating function for a_r, the number of ways to select r objects from four types with repetition. What is the correct generating function?

21. Explain why $(1+x+x^2+x^3+x^4)^r$ is not a proper generating function for the number of ways to distribute r jelly beans among r children with no child getting more than four jelly beans.

22. Show that the generating function for the number of integer solutions to $e_1+e_2+e_3+e_4=r$, $0 \le e_1 \le e_2 \le e_3 \le e_4$, is

$$(1+x+x^2+\cdots)(1+x^2+x^4+\cdots)$$
$$(1+x^3+x^6+\cdots)(1+x^4+x^8+\cdots)$$

23. Find a generating function for the number of ways to make r cents change in pennies, nickels, and dimes.

24. A national singing contest has 5 distinct entrants from each state. Use a generating function for modeling the number of ways to pick 20 semifinalists if:

(a) There is at most 1 person from each state.

(b) There are at most 3 people from each state.

25. Find a generating function $g(x, y)$ whose coefficient of $x^r y^s$ is the number of ways to distribute r chocolate bars and s lollipops among five children such that no child gets more than three lollipops.

26. **(a)** Find a generating function $g(x, y, z)$ whose coefficient $x^r y^s z^t$ is the number of ways to distribute r red balls, s blue balls, and t green balls to n people with between three and six balls of each type to each person.

(b) Suppose also that each person gets at least as many red balls as blues.

(c) Suppose also that the first three people get equal numbers of reds and blues.

(d) Suppose also that no one gets the same number of green and red balls.

27. Find a generating function $g(x, y, z)$ whose coefficient of $x^r y^s z^t$ is the number of ways eight people can each pick two different fruits from a bowl of apples, oranges, and bananas for a total of r apples, s oranges, and t bananas.

28. Find a generating function $(x_1, x_2, \ldots, x_m)$ whose coefficient of $x_1^{r_1} x_2^{r_2} \ldots x_m^{r_m}$ is the number of ways n people can pick a total of r_1 chairs of type 1, r_2 chairs of type 2, $\ldots$ r_m chairs of type m if:

(a) Each person picks one chair.

(b) Each person picks either two chairs of one type or no chairs at all.

(c) Person i picks up to i chairs of exactly one type.

29. If $g(x_1, x_2, \ldots, x_p) = (x_1+x_2+\cdots+x_p)^n$, $p > n$, how many terms of $g(x_1, \ldots, x_p)$'s expansion have no exponent of any x_i greater than 1? What is the coefficient of one of these terms?

6.2 CALCULATING COEFFICIENTS OF GENERATING FUNCTIONS

We now develop algebraic techniques for calculating the coefficients of generating functions. All these methods seek to reduce a complex generating function to a simple binomial-type generating function or a product of binomial-type generating functions. For easy reference, we list in Table 6.1 all the polynomial identities and expansions to be used in this section.

The rule for multiplication of generating functions in Eq. (6) is simply the standard formula for polynomial multiplication. Identity (1) can be verified by polynomial "long division." We restate it, multiplying both sides of Eq. (1) by $(1-x)$, as $(1-x^{m+1}) = (1-x)(1+x+x^2+\cdots+x^m)$. We verify that the product of the right-hand side is $1-x^{m+1}$ by "long multiplication."

$$\begin{array}{l} 1+x+x^2+\cdots+x^m \\ \underline{1-x \qquad\qquad\qquad\quad} \\ 1+x+x^2+\cdots+x^m \\ \underline{\quad -x-x^2-x^3\cdots-x^m-x^{m+1}} \\ 1 \qquad\qquad\qquad\qquad\quad -x^{m+1} \end{array} \tag{*}$$

If m is made infinitely large, so that $1+x+x^2+\cdots+x^m$ becomes the infinite series $1+x+x^2+\cdots$, then the multiplication process (*) will yield a power series in which the coefficient of each x^k, $k>0$, is zero [the reader can confirm this in (*)]. We conclude that $(1-x)(1+x+x^2+\cdots)=1$. [Analytically, this equation is valid for $|x|<1$; the "remainder" term x^{m+1} in (*) goes to 0 as m becomes infinite.] Dividing both sides of this equation by $(1-x)$ yields identity (2).

Table 6.1 Polynomial Expansions

(1) $\dfrac{1-x^{m+1}}{1-x} = 1+x+x^2+\cdots+x^m$

(2) $\dfrac{1}{1-x} = 1+x+x^2+\cdots$

(3) $(1+x)^n = 1+\binom{n}{1}x+\binom{n}{2}x^2+\cdots+\binom{n}{r}x^r+\cdots+\binom{n}{n}x^n$

(4) $(1-x^m)^n = 1-\binom{n}{1}x^m+\binom{n}{2}x^{2m}+\cdots+(-1)^k\binom{n}{k}x^{km}+\cdots+(-1)^n\binom{n}{n}x^{nm}$

(5) $\dfrac{1}{(1-x)^n} = 1+\binom{1+n-1}{1}x+\binom{2+n-1}{2}x^2+\cdots+\binom{r+n-1}{r}x^r+\cdots$

(6) If $h(x)=f(x)g(x)$, where $f(x)=a_0+a_1x+a_2x^2+\cdots$ and $g(x)=b_0+b_1x+b_2x^2+\cdots$, then

$$h(x)=a_0b_0+(a_1b_0+a_0b_1)x+(a_2b_0+a_1b_1+a_0b_2)x^2+\cdots$$
$$+(a_rb_0+a_{r-1}b_1+a_{r-2}b_2+\cdots+a_0b_r)x^r+\cdots$$

Expansion (3), the binomial expansion, was explained at the start of Section 6.1. Expansion (4) is obtained from (3) by expanding $(1+y)^n$, when $y = -x^m$:

$$[1+(-x^m)]^n = 1 + \binom{n}{1}(-x^m) + \binom{n}{2}(-x^m)^2 + \cdots + \binom{n}{k}(-x^m)^k + \cdots + \binom{n}{n}(-x^m)^n$$

By identity (2), $(1-x)^{-n}$, or equivalently $\left(\frac{1}{1-x}\right)^n$, equals

$$(1+x+x^2+x^3+\cdots)^n \tag{7}$$

Let us determine the coefficient of x^r in Eq. (7) by counting the number of formal products whose sum of exponents is r. If e_i represents the exponent of the ith term in a formal product, then the number of formal products $x^{e_1}x^{e_2}x^{e_3}\ldots x^{e_n}$ whose exponents sum to r is the same as the number of integer solutions to the equation

$$e_1 + e_2 + e_3 + \cdots + e_n = r \quad e_i \geq 0$$

In Example 5 of Section 5.4, we showed that the number of nonnegative integer solutions to this equation is $C(r+n-1, r)$. Thus the coefficient of x^r in Eq. (7) is $C(r+n-1, r)$. This verifies expansion (5).

With formulas (1) to (6) we can determine the coefficients of a variety of generating functions: First, perform algebraic manipulations to reduce a given generating function to one of the forms $(1+x)^n$, $(1-x^m)^n$, or $(1-x)^{-n}$, or a product of two such expansions; then use expansions (3) to (5) and the product rule (6) to obtain any desired coefficient. We illustrate some common reduction methods in the following examples.

Example 1

Find the coefficient of x^{16} in $(x^2+x^3+x^4+\cdots)^5$. What is the coefficient of x^r?

To simplify the expression, we extract x^2 from each polynomial factor and then apply identity (2).

$$(x^2+x^3+x^4+\cdots)^5 = [x^2(1+x+x^2+\cdots)]^5$$
$$= x^{10}(1+x+x^2+\cdots)^5 = x^{10}\frac{1}{(1-x)^5}$$

Thus the coefficient of x^{16} in $(x^2+x^3+x^4+\cdots)^5$ is the coefficient of x^{16} in x^{10} $(1-x)^{-5}$. But the coefficient of x^{16} in this latter expression will be the coefficient of x^6 in $(1-x)^{-5}$ [i.e., the x^6 term in $(1-x)^{-5}$ is multiplied by x^{10} to become the x^{16} term in $x^{10}(1-x)^{-5}$]. From expansion (5), we see that the coefficient of x^6 in $(1-x)^{-5}$ is $C(6+5-1, 6)$.

More generally, the coefficient of x^r in $x^{10}(1-x)^{-5}$ equals the coefficient of x^{r-10} in $(1-x)^{-5}$, namely, $C((r-10)+5-1, (r-10))$. ∎

Observe that $(x^2+x^3+x^4+\cdots)^5$ is the generating function a_r, for the number of ways to select r objects with repetition from five types with at least 2 of each type.

In the last chapter, we solved such a problem by first picking two objects in each type—1 way—and then counting the ways to select the remaining $r - 10$ objects—$C((r-10)+5-1, (r-10))$ ways. In the generating function analysis in Example 1, we algebraically picked out an x^2 from each factor for a total of x^{10} and then found the coefficient of x^{r-10} in $(1+x+x^2+\cdots)^5$, the generating function for selection with unrestricted repetition of r from five types.

The standard algebraic technique of extracting the highest common power of x from each factor corresponds to the "trick" used to solve the associated selection problem. Such correspondences are a major reason for using generating functions: the algebraic techniques automatically do the combinatorial reasoning for us.

Example 2

Use generating functions to find the number of ways to collect \$15 from 20 distinct people if each of the first 19 people can give a dollar (or nothing) and the twentieth person can give either \$1 or \$5 (or nothing).

This collection problem is equivalent to finding the number of integer solutions to $x_1 + x_2 + \cdots + x_{19} + x_{20} = 20$ when $x_i = 0$ or $1, i = 1, 2, \ldots 19$, and $x_{20} = 0$ or 1 or 5. The generating function for this integer-solution-of-an-equation problem is $(1+x)^{19}(1+x+x^5)$. We want the coefficient of x^{15}. The first part of this generating function has the binomial expansion

$$(1+x)^{19} = 1 + \binom{19}{1}x + \binom{19}{2}x^2 + \cdots + \binom{19}{r}x^r + \cdots + \binom{19}{19}x^{19}$$

If we let $f(x)$ be this first polynomial and let $g(x) = 1 + x + x^5$, then we can use Eq. (6) to calculate the coefficient of x^{15} in $h(x) = f(x)g(x)$. Let a_r be the coefficient of x^r in $f(x)$ and b_r the coefficient of x^r in $g(x)$. We know that $a_r = \binom{19}{r}$ and that $b_0 = b_1 = b_5 = 1$ (other b_is are zero).

Then the coefficient of x^{15} in $h(x)$ is

$$a_{15}b_0 + a_{14}b_1 + a_{13}b_2 + \cdots + a_0 b_{15}$$

which reduces to

$$a_{15}b_0 + a_{14}b_1 + a_{10}b_5$$

since b_0, b_1, b_5 are the only nonzero coefficients in $g(x)$. Substituting the values of the various as and bs in Eq. (8), we have

$$\binom{19}{15} \times 1 + \binom{19}{14} \times 1 + \binom{19}{10} \times 1 = \binom{19}{15} + \binom{19}{14} + \binom{19}{10} \quad \blacksquare$$

Example 3

How many ways are there to distribute 25 identical balls into seven distinct boxes if the first box can have no more than 10 balls but any number can go into each of the other six boxes?

The generating function for the number of ways to distribute r balls into seven boxes with at most 10 balls in the first box is

$$(1+x+x^2+\cdots+x^{10})(1+x+x^2+\cdots)^6$$
$$=\left(\frac{1-x^{11}}{1-x}\right)\left(\frac{1}{1-x}\right)^6=(1-x^{11})\left(\frac{1}{1-x}\right)^7$$

using identities (1) and (2). Let $f(x)=1-x^{11}$ and $g(x)=(1-x)^{-7}$. Using expansion (5), we have

$$g(x)=(1-x)^{-7}=1+\binom{1+7-1}{1}x+\binom{2+7-1}{2}x^2$$
$$+\cdots+\binom{r+7-1}{r}x^r+\cdots$$

We want the coefficient of x^{25} (25 balls distributed) in $h(x)=f(x)g(x)$. As in Example 2, we need to consider only the terms in the product of the two polynomials $(1-x^{11})$ and $\left(\frac{1}{1-x}\right)^7$ that yield an x^{25} term. The only nonzero coefficients in $f(x)=(1-x^{11})$ are $a_0=1$ and $a_{11}=-1$. So the coefficient of x^{25} in $f(x)g(x)$ is:

$$a_0b_{25}+a_{11}b_{14}=1\times\binom{25+7-1}{25}+(-1)\times\binom{14+7-1}{14}$$

The combinatorial interpretation of the answer in Example 3 is that we count all the ways to distribute without restriction the 25 balls into the seven boxes, $C(25+7-1,25)$ ways, and then subtract the distributions that violate the first box constraint, that is, distributions with at least 11 balls in the first box, $C((25-11)+7-1,(25-11))$ (first put 11 balls in the first box and then distribute the remaining balls arbitrarily). Again, generating functions automatically performed this combinatorial reasoning. ■

The next example employs all the techniques used in the first three examples to solve a problem that cannot be solved by the combinatorial methods of the previous chapter.

Example 4

How many ways are there to select 25 toys from seven types of toys with between two and six of each type?

The generating function for a_r, the number of ways to select r toys from seven types with between 2 and 6 of each type, is

$$(x^2+x^3+x^4+x^5+x^6)^7$$

We want the coefficient of x^{25}. As in Example 1, we extract x^2 from each factor to get

$$[x^2(1+x+x^2+x^3+x^4)]^7 = x^{14}(1+x+x^2+x^3+x^4)^7$$

Now we reduce our problem to finding the coefficient of $x^{25-14} = x^{11}$ in $(1+x+x^2+x^3+x^4)^7$. Using identity (1), we can rewrite this generating function as

$$(1+x+x^2+x^3+x^4)^7 = \left(\frac{1-x^5}{1-x}\right)^7 = (1-x^5)^7\left(\frac{1}{1-x}\right)^7$$

Let $f(x) = (1-x^5)^7$ and $g(x) = (1-x)^{-7}$. By expansions (4) and (5), respectively, we have

$$f(x) = (1-x^5)^7 = 1 - \binom{7}{1}x^5 + \binom{7}{2}x^{10} - \binom{7}{3}x^{15} + \cdots$$

$$g(x) - \left(\frac{1}{1-x}\right)^7 = 1 - \binom{1+7-1}{1}x + \binom{2+7-1}{2}x^2 + \cdots$$
$$+ \binom{r+7-1}{r}x^r + \cdots$$

To find the coefficient of x^{11}, we need to consider only the terms in the product of the two polynomials $(1-x^5)^7$ and $\left(\frac{1}{1-x}\right)^7$ that yield x^{11}. The only nonzero coefficients in $f(x) = (1-x^5)^7$ with a subscript ≤ 11 (larger subscripts can be ignored) are a_0, a_5, and a_{10} [see the expansion of $f(x)$ above]. The products involving these three coefficients that yield x^{11} terms are:

$$a_0b_{11} \qquad + \qquad a_5b_6 \qquad + \qquad a_{10}b_1$$
$$= 1\times\binom{11+7-1}{11} + \left(-\binom{7}{1}\right)\times\binom{6+7-1}{6} + \binom{7}{2}\times\binom{1+7-1}{1} \quad \blacksquare$$

The following combinatorial interpretation can be given to the final answer in Example 4. The first term, $C(11+7-1, 11)$, counts the number of ways to select 11 toys from seven types of toys with no restriction, where 11 is the number of additional toys to select after we first pick 2 toys of each type. The next term, $-7C(6+7-1, 6)$, subtracts seven cases of a violation where we pick at least 5 additional toys of some type (we pick 5 of some type and then pick $11-5=6$ more toys from the 7 types). The final term, $C(7,2)C(1+7-1, 1)$ adds back all $C(7,2)$ cases where some *pair* of violations occurred, that is, with at least 5 chosen from a pair of types [we pick 5 of two types and then pick $11-(2\times 5)=1$ more toy from the 7 types]. The logic behind this combinatorial approach will be developed in Chapter 8.

We close this section by showing how generating functions can be used to verify binomial identities. We express the right side of the identity as a particular coefficient in some generating function $h(x)$ and the left side as the same coefficient in a product of generating functions $f(x)$ and $g(x)$, where $h(x) = f(x)g(x)$.

Example 5: Binomial Identity

Verify the binomial identity

$$\binom{n}{0}^2 + \binom{n}{1}^2 + \cdots + \binom{n}{n}^2 = \binom{2n}{n}$$

The right-hand side of the identity is the coefficient of x^n in $(1+x)^{2n}$. The left-hand side terms involves coefficients in $(1+x)^n$. The product of generating functions we want is $(1+x)^n(1+x)^n$. That is, let $f(x) = g(x) = (1+x)^n$ so that $f(x)g(x) = h(x) = (1+x)^{2n}$. Then $a_r = b_r = C(n, r)$. By the product rule (6), the coefficient of x^n in $f(x)g(x)$ is

$$a_0b_n + a_1b_{n-1} + \cdots + a_nb_0$$

$$= \binom{n}{0}\binom{n}{n} + \binom{n}{1}\binom{n}{n-1} + \cdots + \binom{n}{0}\binom{n}{n}$$

$$= \binom{n}{0}^2 + \binom{n}{1}^2 + \cdots + \binom{n}{n}^2 \quad \text{since} \binom{n}{r} = \binom{n}{n-r}$$

Equating coefficients of x^n in $(1+x)^n(1+x)^n = (1+x)^{2n}$, we obtain the required identity. ∎

The reader is encouraged to compare the generating function proof of the preceding combinatorial identity with the combinatorial proof of the same identity given in Example 3 of Section 5.5.

6.2 EXERCISES

Summary of Exercises The first 29 exercises are similar to the examples of coefficient calculation in this section. Exercises 31–36 involve binomial identities (several of these problems are quite tricky). Exercises 38–42 and 44 introduce the topic of probability generating functions (some background in probability is helpful).

1. Find the coefficient of x^8 in $(1+x+x^2+x^3+\cdots)^n$.

2. Find the coefficient of x^r in $(x^5+x^6+x^7+\cdots)^8$.

3. Find the coefficient of x^9 in $(1+x^2+x^4)(1+x)^m$.

4. Find the coefficient of x^{20} in $(x+x^2+x^3+x^4+x^5)(x^2+x^3+x^4+\cdots)^5$.

5. Find the coefficient of x^{18} in $(x^2+x^3+x^4+x^5+x^6+x^7)^4$.

6. Find the coefficient of x^{50} in $(x^{10}+x^{11}+\cdots+x^{25})(x+x^2+\cdots+x^{15})(x^{20}+\cdots+x^{45})$.

7. Find the coefficient of x^{35} in $(x^3+x^4+x^5+x^6+x^7)^7$.

8. Find the coefficient of x^{24} in $(x+x^2+x^3+x^4+x^5)^9$.

9. Find the coefficient of x^{19} in $(x+x^2+x^3+x^4+x^5+x^6+x^7)^4$.

10. Find the coefficient of x^{36} in $(x^2+x^3+x^4+x^5+x^6+x^7+x^8)^6$.

11. Find the coefficient of x^{12} in:

(a) $x^2(1-x)^{-10}$

(b) $\dfrac{x^2-3x}{(1-x)^4}$

(c) $\dfrac{(1-x^2)^5}{(1-x)^5}$

(d) $\dfrac{x+3}{1-2x+x^2}$

(e) $\dfrac{b^m x^m}{(1-bx)^{m+1}}$

12. Give a formula similar to Eq. (1) for:

(a) $1+x^4+x^8+\cdots+x^{24}$

(b) $x^{20}+x^{40}+\cdots+x^{180}$

13. Find the coefficient of x^9 in $(x^2+x^3+x^4+x^5)^5$.

14. Find the coefficient of x^{18} in $(1+x^3+x^6+x^9+\cdots)^7$.

15. Find the coefficient of x^{12} in:

(a) $(1-x)^8$

(b) $(1+x)^{-1}$

(c) $(1+x)^{-8}$

(d) $(1-4x)^{-5}$

(e) $(1+x^3)^{-4}$

16. Find the coefficient of x^{25} in $(1+x^3+x^8)^{10}$.

17. Use generating functions to find the number of ways to select 10 balls from a large pile of red, white, and blue balls if:

(a) The selection has at least two balls of each color.

(b) The selection has at most two red balls.

(c) The selection has an even number of blue balls.

18. Use generating functions to find the number of ways to distribute r jelly beans among eight children if:

(a) Each child gets at least one jelly bean.

(b) Each child gets an even number of beans.

19. How many ways are there to place an order for 12 chocolate sundaes if there are 5 types of sundaes, and at most 4 sundaes of one type are allowed?

20. How many ways are there to paint the 10 identical rooms in a hotel with five colors if at most three rooms can be painted green, at most three painted blue, at most three red, and no constraint on the other two colors, black and white?

21. How many ways are there to distribute 20 cents to n children and one parent if the parent receives either a nickel or a dime and:

(a) The children receive any amounts?

(b) Each child receives at most 1¢?

22. How many ways are there to get a sum of 25 when 10 distinct dice are rolled?

23. How many ways are there to select 300 chocolate candies from seven types if each type comes in boxes of 20 and if at least one but not more than five boxes of each type are chosen? (*Hint:* Solve in terms of boxes of chocolate.)

24. How many different committees of 40 senators can be formed if the two senators from the same state (50 states in all) are considered identical?

25. How many ways are there to split 6 copies of one book, 7 copies of a second book, and 11 copies of a third book between two teachers if each teacher gets 12 books and each teacher gets at least 2 copies of each book?

26. How many ways are there to divide five pears, five apples, five doughnuts, five lollipops, five chocolate cats, and five candy rocks into two (unordered) piles of 15 objects each?

27. How many ways are there to collect \$24 from 4 children and 6 adults if each person gives at least \$1, but each child can give at most \$4 and each adult at most \$7?

28. If a coin is flipped 25 times with eight tails occurring, what is the probability that no run of six (or more) consecutive heads occurs?

29. If 10 steaks and 15 lobsters are distributed among four people, how many ways are there to give each person at most 5 steaks and at most 5 lobsters?

30. Show that $(1-x-x^2-x^3-x^4-x^5-x^6)^{-1}$ is the generating function for the number of ways a sum of r can occur if a die is rolled any number of times.

31. Use generating functions to show that

$$\sum_{k=0}^{m} \binom{m}{k}\binom{n}{r-k} = \binom{m+n}{r}$$

32. Use the equation

$$\frac{(1-x^2)^n}{(1-x)^n} = (1+x)^n$$

to show that

$$\sum_{k=0}^{m/2} (-1)^k \binom{n}{k}\binom{n+m-2k-1}{n-1} = \binom{n}{m} \qquad m \le n \text{ and } m \text{ even}$$

33. Use binomial expansions to evaluate:

(a) $\displaystyle\sum_{k=0}^{m} \binom{m}{k}\binom{n}{r+k}$ **(b)** $\displaystyle\sum_{k=0}^{r} (-1)^k \binom{n}{k}\binom{n}{r-k}$ **(c)** $\displaystyle\sum_{k=0}^{n} 2^k \binom{n}{k}$

34. **(a)** Evaluate $\displaystyle\sum_{k=n_1}^{n_2} \binom{k+5}{k}$. **(b)** Evaluate $\displaystyle\sum_{k=0}^{m} \binom{n-k}{m-k}$.

35. **(a)** Show that $\displaystyle\sum_{k=0}^{\infty} \left(\frac{1}{2}\right)^k \binom{n+k-1}{k} = 2^n$. **(b)** Evaluate $\displaystyle\sum_{k=0}^{\infty} \left(\frac{1}{2}\right)^k k$.

36. Why cannot one set $x = -1$ in formula (5) to "prove" that

$$\left(\frac{1}{2}\right)^n = \sum_{k=0}^{\infty} (-1)^k \binom{n+k-1}{k}$$

37. If $g(x)$ is the generating function for a_r, then show that $g^{(k)}(0)/k! = a_r$.

38. A *probability generating function* $p_X(t)$ for a discrete random variable X has a polynomial expansion in which p_r, the coefficient of t^r, is equal to the probability that $X = r$.

(a) If X is the number of heads that occur when a fair coin is flipped n times, show that $p_X(t) = (\frac{1}{2})^n(1-t)^n$.

(b) If X is the number of heads that occur when a biased coin is flipped n times with probability p of heads (and $q = 1 - p$), show that $p_X(t) = (q + pt)^n$.

(c) If X is the number of times a fair coin is flipped until the fifth head occurs, find $p_X(t)$.

(d) Repeat part (c) until the mth head occurs and probability of a head is p.

39. **(a)** The *expected value* $E(X)$ of a discrete random variable X is defined to be $\sum p_r r$. Show that $E(X) = p'_X(1)$, that is, $(d/dt)p_X(t)$, with t set equal to 1.

(b) Find $E(X)$ for the random variables X in Exercise 38.

40. **(a)** The *second moment* $E_2(X)$ of a discrete random variable X is defined to be $\sum p_r r^2$. Show that $E_2(X) = p'_X(1) + P''_X(1)$.

(b) Find $E_2(X)$ for the random variables X in Exercises 38(b) and 38(c).

41. Suppose a fair coin is flipped until the mth head occurs and suppose that no more than s tails in a row occur. If X is the number of flips, find $P_X(t)$.

42. Experiments A' and A'' have probabilities p' and p'' of success in each trial and are performed n' and n'' times, respectively. Let X' and X'' be the number of successes in the respective experiments, and let X be the total number of successes on both experiments. Verify the following.

(a) $p_X(\mathrm{t}) = p'_X(t)p''_X(\mathrm{t})$

(b) $E(X) = E(X') + E(X'')$

(c) $E_2(X) = E_2(X') + E_2(X'') + E(X'X'')$

43. Suppose a red die is rolled once and then a green die is rolled as many times as the value on the red die. If a_r is the number of ways that the (variable length) sequence of rolls of the green die can sum to r, show that the generating function for a_r is $f(f(x))$, where $f(x) = (x + x^2 + x^3 + x^4 + x^5 + x^6)$.

44. Suppose X is the random variable of the number of minutes it takes to serve a person at a fast-food stand. Suppose Y is the random variable of the number of people who line up to be served at the stand in one minute. Let Z be the number of people who line up while a person is being served. Show that $p_Z(t) = p_X(p_Y(t))$.

6.3 PARTITIONS

In this section we discuss partitions and their generating functions. Unfortunately, there is no easy way to calculate the coefficients of most of these generating functions.

A **partition** of a group of r identical objects divides the group into a collection of (unordered) subsets of various sizes. Analogously, we define a partition of the integer r to be a collection of positive integers whose sum is r. Normally we write this collection as a sum and list the integers of the partition in increasing order. For example, the seven partitions of the integer 5 are

$$1+1+1+1+1 \qquad 1+1+1+2 \qquad 1+1+3$$
$$1+2+2 \qquad 1+4 \qquad 2+3 \qquad 5$$

Note that 5 is a "trivial" partition of itself.

Let us construct a generating function for a_r, the number of partitions of the integer r. A partition of an integer is described by specifying how many 1s, how many 2s, and so on, are in the sum. Let e_k denote the number of ks in a partition. Then

$$1e_1 + 2e_2 + 3e_3 + \cdots + ke_k + \cdots + re_r = r$$

Intuitively, we can think of picking r objects from an unlimited number of piles where the first pile contains single objects, the second pile contains objects stuck together in pairs, the third pile contains objects stuck together in triples, and so on. To model this integer-solution-to-an-equation problem with a generating function, we need polynomial factors whose formal multiplication yields products of the form

$$\begin{Bmatrix} x^0 \\ x^1 \\ x^2 \\ \vdots \end{Bmatrix} \begin{Bmatrix} (x^2)^0 \\ (x^2)^1 \\ (x^2)^2 \\ \vdots \end{Bmatrix} \begin{Bmatrix} (x^3)^0 \\ (x^3)^1 \\ (x^3)^2 \\ \vdots \end{Bmatrix} \cdots \begin{Bmatrix} (x^k)^0 \\ (x^k)^1 \\ (x^k)^2 \\ \vdots \end{Bmatrix} \cdots$$

Thus the generating function $g(x)$ must be

$$\begin{aligned} g(x) = {} & (1 + x + x^2 + x^3 + \cdots + x^n + \cdots) \\ & \cdot (1 + x^2 + x^4 + x^6 + \cdots + x^{2n} + \cdots) \\ & \cdot (1 + x^3 + x^6 + x^9 + \cdots + x^{3n} + \cdots) \\ & \quad \vdots \\ & \cdot (1 + x^k + x^{2k} + x^{3k} + \cdots + x^{kn} + \cdots) \\ & \quad \vdots \end{aligned}$$

If we set $y = x^2$, then the second factor in $g(x)$ becomes $1 + y + y^2 + \cdots = (1-y)^{-1}$. Thus

$$1 + x^2 + x^4 + x^6 + \cdots + x^{2n} + \cdots = (1 - x^2)^{-1}$$

Similarly, the kth factor can be written as $(1 - x^k)^{-1}$. For partitions up to $r = m$, we need the first m polynomial factors. But for arbitrary values of r, we need an infinite number of polynomial factors. Then

$$g(x) = \frac{1}{(1-x)(1-x^2)(1-x^3)\cdots(1-x^k)\cdots}$$

We now consider some more specialized partition problems.

Example 1

Find the generating function for a_r, the number of ways to express r as a sum of distinct integers.

We must constrain the standard partition problem not to allow any repetition of an integer. For example, there are three ways to write 5 as a sum of distinct integers $1+4$, $2+3$, and 5. The appropriate modification of the generating function for unrestricted partitions is

$$g(x) = (1+x)(1+x^2)(1+x^3)(1+x^4)\cdots(1+x^k)\cdots \quad \blacksquare$$

Example 2

Find a generating function for a_r, the number of ways that we can choose 2¢, 3¢, and 5¢ stamps adding to a net value of r¢.

The problem is equivalent to the number of integer solutions to

$$2e_2 + 3e_3 + 5e_5 = r \qquad 0 \le e_2, e_3, e_5$$

The appropriate generating function is

$$(1+x^2+x^4+x^6+\cdots)(1+x^3+x^6+x^9+\cdots)\cdot(1+x^5+x^{10}+x^{15}+\cdots) \quad \blacksquare$$

Example 3

Show with generating functions that every positive integer can be written as a unique sum of distinct powers of 2.

The generating function for a_r, the number of ways to write an integer r as a sum of distinct powers of 2, will be similar to the generating function for sums of distinct integers in Example 1, except that now only integers that are powers of 2 are used. The generating function is

$$g^*(x) = (1+x)(1+x^2)(1+x^4)(1+x^8)\cdots(1+x^{2^k})\cdots$$

To show that every integer can be written as a unique sum of distinct powers of 2, we must show that the coefficient of every power of x in $g^*(x)$ is 1. That is, show that

$$g^*(x) = 1 + x + x^2 + x^3 + \cdots = \frac{1}{1-x}$$

or equivalently

$$(1-x)g^*(x) = 1$$

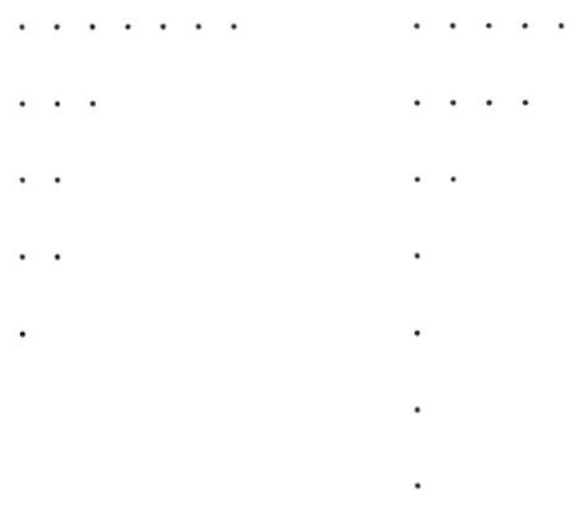

```
. . . . . . .      . . . . .
. . .              . . . .
. .                . .
. .                .
.                  .
                   .
                   .
```

Figure 6.1 **(*a*)** **(*b*)**

We prove this identity by repeatedly using the factorization $(1-x^k)(1+x^k) = 1-x^{2k}$.

$$
\begin{aligned}
(1-x)g^*(x) &= (1-x)(1+x)(1+x^2)\,(1+x^4)(1+x^8)\ \cdots \\
&= [\ \ (1-x^2)\ \ ](1+x^2)(1+x^4)(1+x^8)\ \cdots \\
&= [\ \ \ \ (1-x^4)\ \ \ \ ]\ \ (1+x^4)(1+x^8)\ \cdots \\
&\ \ \vdots \\
&= 1
\end{aligned}
$$

By successively making the replacement $(1-x^k)(1+x^k) = 1-x^{2k}$, we eventually eliminate all factors in $(1-x)g^*(x)$. Formally, the coefficient of any specific x^k in $(1-x)g^*(x)$ must be 0. So $(1-x)g^*(x) = 1$ and $g^*(x) = 1+x+x^2+\cdots$, as required. ■

A convenient tool for studying partitions is a diagram known as Ferrers diagram. A **Ferrers diagram** displays a partition of r dots in a set of rows listed in order of decreasing size. The partition $1+2+2+3+7$ of 15 is shown in the Ferrers diagram in Figure 6.1*a*. If we transpose the rows and columns of a Ferrers diagram of a partition of r, we get a Ferrers diagram of another partition of r. This diagram is called the **conjugate** of the original Ferrers diagram. For example, Figure 6.1*b* shows the conjugate of the Ferrers diagram in Figure 6.1*a*. This new Ferrers diagram represents the partition of 15, $1+1+1+1+2+4+5$. Clearly, transposing is unique: Two Ferrers diagrams have equal conjugates if and only if they are equal.

Example 4

Show that the number of partitions of an integer r as a sum of m positive integers is equal to the number of partions of r as a sum of positive integers, the largest of which is m.

If we draw a Ferrers diagram of a partition of r into m parts, then the Ferrers diagram will have m rows. The transposition of such a diagram will have m columns, that is, the largest row will have m dots. Thus there is a one-to-one correspondence between these two classes of partitions. ■

6.3 EXERCISES

Summary of Exercises Exercises 1–11 involve generating function models for partitions. The next six exercises use Ferrers diagrams to verify partition identities. The next five exercises are more difficult partition problems.

1. List all partitions of the integer: **(a)** 4, **(b)** 6.
2. Find a generating function for a_r, the number of partitions of r into:
   (a) Even integers. **(b)** Distinct odd integers.
3. Find a generating function for the number of ways to write the integer r as a sum of positive integers in which no integer appears more than three times.
4. Find a generating function for the number of integer solutions of $2x+3y+7z=r$ with:
   (a) $x, y, z \geq 0$ **(b)** $0 \leq z \leq 2 \leq y \leq 8 \leq x$
5. Find a generating function for the number of ways to make r cents' change in pennies, nickels, dimes, and quarters.
6. Find a generating function for the number of ways to distribute r identical objects into:
   (a) Three indistinguishable boxes.
   (b) n indistinguishable boxes ($n \leq r$).
7. **(a)** Show that the number of partitions of 10 into distinct parts (integers) is equal to the number of partitions of 10 into odd parts by listing all partitions of these two types.
   (b) Show algebraically that the generating function for partitions of r into distinct parts equals the generating function for partitions of r into odd parts, and hence the numbers of these two types of partitions are equal.
8. Show with generating functions that every positive integer has a unique decimal representation.
9. **(a)** Prove the result in Example 3 by induction.
   (b) Prove the result of Example 3 directly by recursively substituting $(1+x^k) = (1-x^{2k})/(1-x^k)$ in $g^*(x)$.
10. The equation in Example 3, $g^*(x)(1-x)=1$, can be rewritten $1-x=1/g^*(x)$. Use this latter equation to prove that among all partitions of an integer r, $r \geq 2$, into powers of 2, there are as many such partitions with an odd number of parts as with an even number of parts.
11. Let $R(r, k)$ denote the number of partitions of the integer r into k parts.
    (a) Show that $R(r, k) = R(r-1, k-1) + R(r-k, k)$.
    (b) Show that $\sum_{k=1}^{r} R(n-r, k) = R(n, r)$.

12. Use a Ferrers diagram to show that the number of partitions of an integer into parts of even size is equal to the number of partitions into parts such that each part occurs an even number of times.

13. Interpret the integer multiplication mn, "m times n," to be the sum of m ns. Prove that $mn = nm$.

14. Show that the number of partitions of the integer n into three parts equals the number of partitions of $2n$ into three parts of size $< n$.

15. Show that the number of partitions of n is equal to the number of partitions of $2n$ into n parts.

16. Show that any number of partitions of $2r + k$ into $r + k$ parts is the same for any k.

17. Show that the number of partitions of $r + k$ into k parts is equal to:

(a) The number of partitions of $r + \binom{k+1}{2}$ into k distinct parts.

(b) The number of partitions of r into parts of size $\leq k$.

18. Show that the number of *ordered* partitions of n is 2^{n-1}. For example, the ordered partitions of 4 are: 4, $1+3$, $2+2$, $3+1$, $1+1+2$, $1+2+1$, $2+1+1$, $1+1+1+1$. (*Hint:* Write n 1s in a row and determine all the ways to partition this sequence into clusters of 1s).

19. **(a)** Find a generating function for a_n, the number of partitions that add up to at most n.

(b) Find a generating function for a_n, the number of partitions of n into three parts in which no part is larger than the sum of the other two.

(c) Find a generating function for a_n, the number of different (incongruent) triangles with integral sides and perimeter n.

20. Show that $2(1-x)^{-3}[(1-x)^{-3} + (1+x)^{-3}]$ is the generating function for the number of ways to toss r identical dice and obtain an even sum.

21. **(a)** A partition of an integer r is *self-conjugate* if the Ferrers diagram of the partition is equal to its own transpose. Find a one-to-one correspondence between the self-conjugate partitions of r and the partitions of r into distinct odd parts.

(b) The largest square of dots in the upper left-hand corner of a Ferrers diagram is called the *Durfee square* of the Ferrers diagram. Find a generating function for the number of self-conjugate partitions of r whose Durfee square is size k (a $k \times k$ array of dots). (*Hint:* Use a $1-1$ correspondence between these and the partitions of $r - k^2$ into even parts of size at most $2k$.)

(c) Show that

$$(1+x)(1+x^3)(1+x^5)\dots$$
$$= 1 + \sum_{k=1}^{\infty} \frac{x^{k^2}}{(1-x^2)(1-x^4)(1-x^6)\dots(1-x^{2k})}$$

22. Let $\left\{ {n \atop k} \right\}$ denote the number of partitions of n objects into k nonempty subsets. Show that $\left\{ {n+1 \atop k} \right\} = k\left\{ {n \atop k} \right\} + \left\{ {n \atop k-1} \right\}$.

23. Write a computer program to determine the number of all partitions of an integer r:

(a) Into k parts.

(b) Into any number of parts.

(c) Into distinct parts.

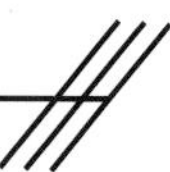

6.4 EXPONENTIAL GENERATING FUNCTIONS

In this section we discuss exponential generating functions. They are used to model and solve problems involving arrangements and distributions of distinct objects. The generating functions used in the previous sections to model selection with repetition problems are called **ordinary generating functions.** Exponential generating functions involve a more complicated modeling process than do ordinary generating functions. Consider the problem of finding the number of different words (arrangements) of four letters when the letters are chosen from an unlimited supply of as, bs and cs, and the word must contain at least two as. The possible sets of four letters to form the word are $\{a, a, a, a\}$, $\{a, a, a, b\}$, $\{a, a, a, c\}$, $\{a, a, b, b\}$, $\{a, a, b, c\}$, and $\{a, a, c, c\}$. The number of arrangements possible with each of these six sets is

$$\frac{4!}{4!0!0!} \quad \frac{4!}{3!1!0!} \quad \frac{4!}{3!0!1!} \quad \frac{4!}{2!2!0!} \quad \frac{4!}{2!1!1!} \quad \frac{4!}{2!0!2!}$$

respectively. So the total number of words will be the sum of these six terms.

The sets of letters used in such a word range over all sets of four letters chosen with repetition from as, bs, and cs with at least two as. Equivalently, the number of such sets equals the number of integer solutions to the equation

$$e_1 + e_2 + e_3 = 4 \qquad 2 \le e_1, \qquad 0 \le e_2, e_3 \tag{1}$$

At first sight, the four-letter words problem seems similar to previous problems that could be modeled by (ordinary) generating functions. However, in this case we do not want each integer solution of Eq. (1) to contribute 1 to the count of the number of possible words. Instead it must contribute $4!/(e_1!e_2!e_3!)$ words. In terms of generating functions, the coefficient of x^4 will count all formal products with associated coefficients

$$\frac{(e_1 + e_2 + e_3)!}{e_1!e_2!e_3!} x^{e_1} x^{e_2} x^{e_3} \qquad 2 \le e_1, \qquad 0 \le e_2, e_3$$

whose exponents sum to 4 (a much harder problem). Fortunately, exponential generating functions yield formal products of exactly this form.

An **exponential generating function** $g(x)$ for a_r, the number of arrangements with r objects, is a function with the power series expansion

$$g(x) = a_0 + a_1 x + a_2 \frac{x^2}{2!} + a_3 \frac{x^3}{3!} + \cdots + a_r \frac{x^r}{r!} + \cdots$$

We build exponential generating functions in the same way that we build ordinary generating functions: one polynomial factor for each type of object; each factor has a collection of powers of x that are an inventory of the choices for the number of objects of that type. However, now each power x^r is divided by $r!$.

As an example, let us consider the four-letter word problem with at least two as. We claim that the exponential generating function for the number of r-letter words formed from an unlimited number of as, bs, and cs containing at least two as is

$$g(x) = \left(\frac{x^2}{2!} + \frac{x^3}{3!} + \frac{x^4}{4!} + \cdots\right)\left(1 + x + \frac{x^2}{2!} + \frac{x^3}{3!} + \frac{x^4}{4!} + \cdots\right)^2 \qquad (2)$$

The coefficient of x^r in Eq. (2) will be the sum of all products $(x^{e_1}/e_1!)\,(x^{e_2}/e_2!)\,(x^{e_3}/e_3!)$, where $e_1 + e_2 + e_3 = r$, $2 \le e_1$, $0 \le e_2, e_3$. If we divide x^r by $r!$ and compensate by multiplying its coefficient by $r!$, then the x^r term in $g(x)$ becomes

$$\left(\sum_{e_1+e_2+e_3=r} \frac{r!}{e_1!e_2!e_3!}\right)\frac{x^r}{r!} \qquad 2 \le e_1, \qquad 0 \le e_2, e_3$$

This coefficient of $x^r/r!$ is just what we wanted. So the exponential generating function for the number of such r-letter words is indeed the expression (2). What makes exponential generating functions work is the "trick" of dividing x^r by $r!$ and multiplying the coefficient of x^r by $r!$.

Before we show how to calculate coefficients of an exponential generating function, let us give a few other examples of exponential generating function models.

Example 1

Find the exponential generating function for a_r, the number of r arrangements without repetition of n objects.

We know that the answer is $P(n, r)$. Since there is no repetition, the exponential generating function is $(1+x)^n$—our old binomial friend! The coefficient of x^r in $(1+x)^n$ is $\binom{n}{r}$. However, now we want the coefficient of $\frac{x^r}{r!}$ in $(1+x)^n$:

$$\binom{n}{r}x^r = \frac{n!}{(n-r)!r!}x^r = \frac{n!}{(n-r)!}\frac{x^r}{r!}$$

Thus $a_r = n!/(n-r)! = P(n, r)$, as expected. ∎

Example 2

Find the exponential generating function for a_r, the number of different arrangements of r objects chosen from four different types of objects with each type of object appearing at least two and no more than five times.

The number of ways to arrange the r objects with e_i objects of the ith type is $r!/e_1!e_2!e_3!e_4!$ To sum up all such terms with the given constraints, we want the coefficient of $x^r/r!$ in $(x^2/2!+x^3/3!+x^4/4!+x^5/5!)^4$. ■

Example 3

Find the exponential generating function for the number of ways to place r (distinct) people into three different rooms with at least one person in each room. Repeat with an even number of people in each room.

Recall that distributions of distinct objects are equivalent to arrangements with repetition; see Section 5.4. The required exponential generating functions are

$$\left(x+\frac{x^2}{2!}+\frac{x^3}{3!}+\frac{x^4}{4!}+\cdots\right)^3 \quad \text{and} \quad \left(1+\frac{x^2}{2!}+\frac{x^4}{4!}+\cdots\right)^3 \quad ■$$

There are few identities or expansions to use in conjunction with exponential generating functions. As a result, there are only a limited number of exponential generating functions whose coefficients can be easily evaluated. The fundamental expansion for exponential generating functions is

$$e^x = 1+x+\frac{x^2}{2!}+\frac{x^3}{3!}+\cdots+\frac{x^r}{r!}+\cdots \tag{3}$$

Replacing x by nx in Eq. (3), we obtain

$$e^{nx} = 1+nx+\frac{n^2x^2}{2!}+\frac{n^3x^3}{3!}+\cdots+\frac{n^rx^r}{r!}+\cdots \tag{4}$$

The power series in Eq. (3) is the Taylor series for e^x, which is derived in all calculus texts. This expansion and the companion expansion for e^{nx} are valid for all values of x. There is no way to factor out a common power of x in exponential generating functions, since the power of x must be matched with $r!$ in the denominator. The best that can be done is to subtract the missing (lowest) powers of x from e^x. For example, the exponential factor representing two or more of a certain type of objects can be written

$$\frac{x^2}{2!}+\frac{x^3}{3!}+\frac{x^4}{4!}+\cdots = e^x-1-x$$

Two useful expansions derived from Eq. (3) are

$$\frac{1}{2}(e^x+e^{-x}) = 1+\frac{x^2}{2!}+\frac{x^4}{4!}+\frac{x^6}{6!}+\cdots \tag{5}$$

$$\frac{1}{2}(e^x-e^{-x}) = x+\frac{x^3}{3!}+\frac{x^5}{5!}+\frac{x^7}{7!}+\cdots \tag{6}$$

Let us work some examples using Eqs. (3) to (6).

Example 4

Find the number of different r arrangements of objects chosen from unlimited supplies of n types of objects.

In Chapter 5 we would have solved this problem by arguing that there are n choices for the type of object in each of the r positions, for a total of n^r different arrangements. Now let us solve this problem with exponential generating functions. The exponential generating function for this problem is

$$\left(1+x+\frac{x^2}{2!}+\frac{x^3}{3!}+\cdots\right)^n = (e^x)^n = e^{nx}$$

By Eq. (4), the coefficient of $x^r/r!$ in this generating function is n^r. ■

Example 5

Find the number of ways to place 25 people into three rooms with at least one person in each room.

In Example 3, we found the exponential generating function for this problem

$$\left(x+\frac{x^2}{2!}+\frac{x^3}{3!}+\cdots\right)^3 = (e^x-1)^3$$

To find the coefficient of $x^r/r!$ in $(e^x-1)^3$, we first must expand this binomial expression in e^x

$$(e^x-1)^3 = e^{3x}-3e^{2x}+3e^x-1$$

From Eq. (4), we get

$$e^{3x}-3e^{2x}+3e^x-1=\sum_{r=0}^{\infty}3^r\frac{x^r}{r!}-3\sum_{r=0}^{\infty}2^r\frac{x^r}{r!}+3\sum_{r=0}^{\infty}\frac{x^r}{r!}-1$$

So the coefficient of $x^{25}/25!$ is $3^{25}-(3\times 2^{25})+3$. ■

Example 6

Find the number of r-digit quaternary sequences (whose digits are 0, 1, 2, and 3) with an even number of 0s and an odd number of 1s.

The exponential generating function for this problem is

$$\left(1+\frac{x^2}{2!}+\frac{x^4}{4!}+\frac{x^6}{6!}+\cdots\right)\left(x+\frac{x^3}{3!}+\frac{x^5}{5!}+\frac{x^7}{7!}+\cdots\right)\left(1+x+\frac{x^2}{2!}+\frac{x^3}{3!}+\cdots\right)^2$$

Using identities (5) and (6), we can write this expression as

$$\frac{1}{2}(e^x+e^{-x})\times\frac{1}{2}(e^x-e^{-x})e^xe^x=\frac{1}{4}(e^{2x}-e^0+e^0-e^{-2x})e^xe^x$$

$$=\frac{1}{4}(e^{2x}-e^{-2x})e^{2x}=\frac{1}{4}(e^{4x}-1)=\frac{1}{4}\left(\sum_{r=0}^{\infty}4^r\frac{x^r}{r!}-1\right)$$

Then for $r>0$, the coefficient of $x^r/r!$ is $\frac{1}{4}4^r=4^{r-1}$. The simple form of this answer suggests that there should be some combinatorial argument for obtaining this short answer directly. ■

6.4 EXERCISES

Summary of Exercises Most of the first 15 exercises are similar to the examples in this section. Exercises 21 and 22 are a continuation of the probability generating function exercises in Section 6.2.

1. Find the exponential generating function for the number of arrangements or r objects chosen from five different types with at most five of each type.

2. Find the exponential generating function for the number of ways to distribute r people into six different rooms with between two and four in each room.

3. Find an exponential generating function for a_r, the number of r-letter words with no vowel used more than once (consonants can be repeated).

4. **(a)** Find the exponential generating function for $s_{n,r}$, the number of ways to distribute r distinct objects into n distinct boxes with no empty box. Consider n a fixed constant.

(b) Determine $s_{n,r}$. The number $s_{n,r}/n!$ is called a *Stirling number of the second kind.*

5. Find the exponential generating function, and identify the appropriate coefficient, for the number of ways to deal a sequence of 13 cards (from a standard 52-card deck) if the suits are ignored and only the values of the cards are noted.

6. How many ways are there to distribute eight different toys among four children if the first child gets at least two toys?

7. How many r-digit ternary sequences are there with:

(a) An even number of 0s?

(b) An even number of 0s and even number of 1s?

(c) At least one 0 and at least one 1?

8. How many ways are there to make an r-arrangement of pennies, nickels, dimes, and quarters with at least one penny and an odd number of quarters? (Coins of the same denomination are identical.)

9. How many 10-letter words are there in which each of the letters e, n, r, s occur:

(a) At most once?

(b) At least once?

10. How many r-digit ternary sequences are there in which:

(a) No digit occurs exactly twice?

(b) 0 and 1 each appear a positive even number of times?

11. How many r-digit quaternary sequences are there in which the total number of 0s and 1s is even?

12. Find the exponential generating function for the number of ways to distribute r distinct objects into five different boxes when $b_1 < b_2 \leq 4$, where b_1, b_2 are the numbers of objects in boxes 1 and 2, respectively.

13. **(a)** Find the exponential generating functions for p_r, the probability that the first two boxes each have at least one object when r distinct objects are randomly distributed into n distinct boxes.

(b) Determine p_r.

14. Find an exponential generating function for the number of distributions of r distinct objects into n different boxes with exactly m nonempty boxes.

15. Find an exponential generating function with:

(a) $a_r = 1/(r+1)$ **(b)** $a_r = r!$

16. Show that if $g(x)$ is the exponential generating function for a_r, then $g^{(k)}(0) = a_k$, where $g^{(k)}(x)$ is the kth derivative of $g(x)$.

17. If

$$f(x) = \sum_{r=0}^{\infty} a_r \frac{x^r}{r!} \quad g(x) = \sum_{r=0}^{\infty} b_r \frac{x^r}{r!}$$

and

$$h(x) = f(x)g(x) = \sum_{r=0}^{\infty} c_r \frac{x^r}{r!}$$

then show that

$$c_r = \sum_{k=0}^{\infty} \binom{r}{k} a_k b_{r-k}$$

18. Show that $e^x e^y = e^{x+y}$ by formally multiplying the expansions of e^x and e^y together.

19. Find a combinatorial argument to show why the answer in Example 6 is 4^{r-1}.

20. Show that $e^x/(1-x)^n$ is the exponential generating function for the number of ways to choose some subset (possibly empty) of r distinct objects and distribute them into n different boxes with the order in each box counted.

21. A Poisson random variable X has $p_r = \text{Prob}(x = r) = \frac{\mu^r}{r!} e^{-\mu}$. Find the probability generating function for X (see Exercise 34 in Section 6.2).

22. Let $P(x) = \sum_{k=0}^{\infty} p_k x^k$ be the probability generating function for the discrete random variable X, that is, p_k is the probability that $X = k$.

(a) Show that the exponential generating function for m_k, the kth moment of X, $m_k = \sum_{j=0}^{\infty} j^k p_k$ is $P(e^x)$.

(b) The kth factorial moment of X, m_k^*, is defined to be equal to $\sum_{j=k}^{\infty} j!/(j-k)!$ Show that the exponential generating function for m_k^* is $P(x+1)$.

(c) If X is the number of heads when n coins are flipped, find m_1, m_2, m_2^*. (*Hint:* Use Exercise 16.)

(d) If X is Poisson (see Exercise 21), find m_1 and m_1^*.

6.5 A SUMMATION METHOD

In this section we show how to construct an ordinary generating function $h(x)$ whose coefficient of x^r is some specified function $p(r)$ of r, such as r^2 or $C(r, 3)$. Then we use $h(x)$ to calculate the sums $p(0) + p(1) + \cdots + p(n)$, for each positive n. The following four simple rules for constructing a new generating function from an old one will be used repeatedly in this section. Assume that $A(x) = \sum a_n x_n$, $B(x) = \sum b_n x_n$, and $C(x) = \sum c_n x_n$.

(i) If $b_n = da_n$, then $B(x) = dA(x)$, for any constant d.

(ii) If $c_n = a_n + b_n$, then $C(x) = A(x) + B(x)$.

(iii) If $c_n = \sum_{i=0}^{n} a_i b_{n-i}$, then $C(x) = A(x)B(x)$.

(iv) If $b_n = a_{n-k}$, except $b_i = 0$ for $i < k$, then $B(x) = x^k A(x)$.

Rule (iii) is simply expression (6) from Section 6.2. The other rules are immediate.

The other basic operation for our coefficient construction is to multiply each coefficient a_r in a generating function $g(x)$ by r. We claim that the new generating function $g^*(x)$ with $a_r^* = ra_r$ is obtained by differentiating $g(x)$ and then multiplying by x, that is, $g^*(x) = x\left(\frac{d}{dx} g(x)\right)$. If

$$g(x) = a_0 + a_1 x + a_2 x^2 + a_3 x^3 + \cdots + a_r x^r + \cdots \tag{1}$$

then differentiation of $g(x)$ yields

$$\frac{d}{dx} g(x) = a_1 + 2a_2 x + 3a_3 x^2 + \cdots + ra_r x^{r-1} + \cdots \tag{2}$$

and now multiplying by x [Rule (iv)] gives

$$g^*(x) = x\left[\frac{d}{dx} g(x)\right] = a_1 x + 2a_2 x^2 + 3a_3 x^3 + \cdots + ra_r x^r + \cdots \tag{3}$$

Note that the a_0 term in $g(x)$ disappears in Eq. (2) because $0a_0 = 0$.

Combining this operation with Rules (i) and (ii), we can repeatedly multiply the coefficient of x^r by r or a constant and can add such coefficients together. This permits us to form polynomials in r.

The natural question now is, To what coefficients do we apply these operations? The natural answer is, When in doubt start with the unit coefficients $a_r = 1$ of the generating function

$$\frac{1}{1-x} = 1 + x + x^2 + x^3 + \cdots + x^r + \cdots$$

Example 1

Build a generating function $h(x)$ with $a_r = 2r^2$.

Starting with $1/(1-x)$, we multiply its coefficients first by r using the generating function operations shown in Eqs. (2) and (3) to obtain

$$x\left(\frac{d}{dx}\frac{1}{1-x}\right) = x\left(\frac{1}{(1-x)^2}\right) = 1x + 2x^2 + 3x^3 + \cdots + rx^r + \cdots$$

Now we repeat these operations on $x/(1-x)^2$ to obtain

$$x\left(\frac{d}{dx}\frac{x}{(1-x)^2}\right) = \frac{x(1+x)}{(1-x)^3} = 1^2x + 2^2x^2 + 3^2x^3 + \cdots + r^2x^r + \cdots$$

Finally we multiply by 2 to obtain the required generating function

$$h(x) = \frac{2x(1+x)}{(1-x)^3} = (2\times 1^2)x + (2\times 2^2)x^2 + (2\times 3^2)x^3 + \cdots + (2\times r^2)x^r + \cdots \quad \blacksquare$$

Example 2

Build a generating function $h(x)$ with $a_r = (r+1)r(r-1)$.

We could multiply $(r+1)r(r-1)$ out getting $a_r = r^3 - r$, obtain generating functions for r^3 and r as an Example 1, and then subtract one generating function from the other. It is easier, however, to start with $3!(1-x)^{-4}$, whose coefficient a_r equals

$$a_r = 3!\binom{r+4-1}{r} = 3!\frac{(r+3)!}{r!3!} = \frac{(r+3)!}{r!} = (r+3)(r+2)(r+1)$$

Then the power series expansion of $3!(1-x)^{-4}$ is

$$\frac{3!}{(1-x)^4} = (3\times 2\times 1) + (4\times 3\times 2)x + (5\times 4\times 3)x^2 + \cdots + (r+3)(r+2)(r+1)x^r + \cdots \tag{4}$$

Compare Eq. (4) with the desired generating function

$$h(x) = (3\times 2\times 1)x^2 + (4\times 3\times 2)x^3 + (5\times 4\times 3)x^4 + \cdots + (r+1)r(r-1)x^r + \cdots$$

The generating function $h(x)$ that we seek is just the series in Eq. (4) multiplied by x^2. So $h(x) = 3!x^2(1-x)^{-4}$. ■

Generalizing the construction in Example 2, we see that $(n-1)!(1-x)^{-n}$ has a coefficient

$$a_r = (n-1)!C(r+n-1, r) = [r+(n-1)][r+(n-2)]\cdots(r+1)$$

Any coefficient involving a product of decreasing terms, such as $(r+1)r(r-1)$, can be built from $(n-1)!(1-x)^{-n}$ as in Example 2, where $n-1$ is the number of terms in the product.

Thus far the construction of an $h(x)$ with specified coefficients has been just an exercise in algebraic manipulations of generating functions. The following easily verified theorem gives some purpose to this exercise.

Theorem 1

If $h(x)$ is a generating function where a_r is the coefficient of x^r, then $h^*(x) = h(x)/(1-x)$ is a generating function of the sums of the a_rs. That is,

$$h^*(x) = a_0 + (a_0 + a_1)x + (a_0 + a_1 + a_2)x^2 + \cdots + \left(\sum_{i=0}^{r} a_i\right)x^r + \cdots$$

This theorem follows from Rule (iii) for the coefficients of the product $h^*(x) = f(x)h(x)$, where $f(x) = 1/(1-x)$. Now we return to the previous examples.

Example 1 (continued)

Evaluate the sum $2 \times 1^2 + 2 \times 2^2 + 2 \times 3^2 + \cdots + 2n^2$.

The generating function $h(x)$ for $a_r = 2r^2$ was found in Example 1 to be $2x(1+x)/(1-x)^3$. Then by Theorem 1, the desired sum $a_1 + a_2 + \cdots + a_n$ is the coefficient of x^n in

$$\begin{aligned} h^*(x) &= h(x)/(1-x) = 2x(1+x)/(1-x)^4 \\ &= 2x(1-x)^{-4} + 2x^2(1-x)^{-4} \end{aligned}$$

The coefficient of x^n in $2x(1-x)^{-4}$ is the coefficient of x^{n-1} in $2(1-x)^{-4}$, and the coefficient of x^n in $2x^2(1-x)^{-4}$ is the coefficient of x^{n-2} in $2(1-x)^{-4}$. Thus, the given sum equals

$$2\binom{(n-1)+4-1}{(n-1)} + 2\binom{(n-2)+4-1}{(n-2)} = 2\binom{n+2}{3} + 2\binom{n+1}{3} \quad \blacksquare$$

Example 2 (continued)

Evaluate the sum $3 \times 2 \times 1 + 4 \times 3 \times 2 + \cdots + (n+1)n(n-1)$.

The generating function $h(x)$ for $a_r = (r+1)r(r-1)$ was found in Example 2 to be $h(x) = 6x^2(1-x)^{-4}$. By Theorem 1 the desired sum is the coefficient of x^n in $h^*(x) = h(x)/(1-x) = 6x^2(1-x)^{-5}$. This coefficient is the x^{n-2} term in $6(1-x)^{-5}$, namely, $6C((n-2)+5-1, n-2) = 6C(n+2, 4)$. ■

Note that these two summation problems can also be solved with the summation method in Section 5.5 based on binomial identities. For comparison, see the analysis of the sum in Example 2 given in Example 4 of Section 5.5 [note that the general term of this sum in Section 5.5 is $(n-2)(n-1)n$ instead of $(n+1)n(n-1)$]. Both methods have their advantages.

6.5 EXERCISES

1. Find ordinary generating functions whose coefficient a_r equals:
   (a) r **(b)** 13 **(c)** $3r^2$ **(d)** $3r+7$ **(e)** $r(r-1)(r-2)(r-3)$.
2. Evaluate the sums (using generating functions):
   (a) $0+1+2+\cdots+n$
   (b) $13+13+\cdots+13$
   (c) $0+3+12+\cdots+3n^2$
   (d) $7+10+13+\cdots+(3n+7)$
   (e) $4\times3\times2\times1+5\times4\times3\times2+\cdots+n(n-1)(n-2)(n-3)$
3. Find a generating function with $a_r = r(r+2)$ (do not add together generating functions for r^2 and for $2r$).
4. **(a)** Show how r^2 and r^3 can be written as linear combinations of $P(r, 3)$, $P(r, 2)$, and $P(r, 1)$.
   (b) Use part (a) to find a generating function for $3r^3 - 5r^2 + 4r$.
5. Find a generating function for
   (a) $a_r = (r-1)^2$ **(b)** $a_r = 1/r$.
6. If $h(x)$ is the ordinary generating function for a_r, what is the coefficient of x^r in $h(x)(1-x)$ (give your answer in terms of the a_rs).
7. Verify Theorem 1 in this section.
8. If $h(x)$ is the ordinary generating function for a_r, find the generating function for $s_r = \sum_{k=r+1}^{\infty} a_k$, assuming all s_rs are finite and $a_r \to 0$ as $r \to \infty$.

6.6 SUMMARY AND REFERENCES

Generating functions are at once a simple-minded and a sophisticated mathematical model for counting problems; simple-minded because polynomial multiplication is a familiar, seemingly well-understood part of high school algebra and sophisticated because with standard algebraic manipulations on generating functions, one can solve complicated counting problems. These algebraic manipulations automatically perform the correct combinatorial reasoning for us! Note that generating functions are an elementary example of the algebraic approach that pervades the research frontiers of contemporary mathematics. Letting algebraic expressions, whether they model combinatorial, geometric, or functional information, do the work for us is what much of modern mathematics is all about.

In this chapter, generating functions were used to model selection and arrangement problems with constrained repetition. Partition problems were also modeled (but were not solved). Finally, we showed how to construct generating functions whose coefficient for x^r was a given function of r and used these generating functions to evaluate related sums. In the next three chapters, generating functions will be used to model and solve other combinatorial problems. Exercise 38 of Section 6.2 introduces one type of generating function used in probability theory (for more, see Feller [2]). Laplace and Fourier transforms in analysis are also generating functions [the Fourier transform of a function $f(t)$ can be viewed as a generating function for the Fourier coefficients of $f(t)$].

The first use of combinatorial generating functions was by DeMoivre around 1720. He used them to derive a formula for Fibonacci numbers (this derivation is given in Section 7.5). In 1748, Euler used generating functions in his work on partition problems. The theory of combinatorial generating functions, developed in the late eighteenth century, was primarily motivated by parallel work on probability generating functions (see Exercises 38 to 42 in Section 6.2 and Exercises 21 and 22 in Section 6.4). Laplace made many contributions to both theories and presented the first complete treatment of both in his 1812 classic *Théorie Analytique des Probabilités.*

For a full discussion of the use of generating functions in combinatorial mathematics, see MacMahon [3]. For a nice presentation of partition problems, also see Cameron [1].

1. P. Cameron, *Combinatorics: Topics, Techniques, Algorithms,* Cambridge University Press, Cambridge, 1994.

2. W. Feller, *An Introduction to Probability Theory and Its Applications,* vol. I, 2nd ed., John Wiley & Sons, New York, 1957.

3. P. MacMahon, *Combinatory Analysis,* vols. I and II (1915), reprinted in one volume, Chelsea Publishing, New York, 1960.

CHAPTER 7 RECURRENCE RELATIONS

7.1 RECURRENCE RELATION MODELS

In this chapter we show how a variety of counting problems can be modeled with recurrence relations. We then discuss methods of solving several common types of recurrence relations.

A **recurrence relation** is a recursive formula that counts the number of ways to do a procedure involving n objects in terms of the number of ways to do it with fewer objects. That is, if a_k is the number of ways to do the procedure with k objects, for $k = 0, 1, 2, \ldots,$ then a recurrence relation is an equation that expresses a_n as some function of preceding a_ks, $k < n$. The simplest recurrence relation is an equation such as $a_n = 2a_{n-1}$. The following equations display some of the forms of recurrence relations that we will build to model counting problems in the chapter:

$$\begin{aligned} a_n &= c_1 a_{n-1} + c_2 a_{n-2} + \cdots + c_r a_{n-r} \quad \text{where } c_i \text{ are constants} \\ a_n &= c a_{n-1} + f(n) \quad \text{where } f(n) \text{ is some function of } n \\ a_n &= a_0 a_{n-1} + a_1 a_{n-2} + \cdots + a_{n-1} a_0 \\ a_{n,m} &= a_{n-1,m} + a_{n-1,m-1} \end{aligned}$$

Just as mathematical induction is a proof technique that verifies a formula or assertion by inductively checking its validity for increasing values of n, so a recurrence relation is a counting technique that solves an enumeration problem by recursively computing the answer for successively larger values of n.

The observant reader will remember that mathematical induction also involves an initial step of verifying the formula or assertion for some starting (smallest) value of n. The same is true for a recurrence relation. We cannot recursively compute the next a_n unless some initial values are given. If the right-hand side of a recurrence relation involves the r preceding a_ks, then we need to be given the first r values, $a_0, a_1, \ldots, a_{r-1}$. For example, in the relation $a_n = a_{n-1} + a_{n-2}$, knowing only that $a_0 = 2$ is insufficient. If given also that $a_1 = 3$, then we use the relation to obtain $a_2 = a_1 + a_0 = 3 + 2 = 5$; $a_3 = a_2 + a_1 = 5 + 3 = 8$; $a_4 = 8 + 5 = 13$; and so on. The information about starting values needed to compute with a recurrence relation is called the **initial conditions.**

If we can devise a recurrence relation to model a counting problem we are studying and also determine the initial conditions, then it is usually possible to solve the problem for moderate sizes of n, such as $n = 20$, quickly by recursively computing successive values of a_n up to the desired n. For larger values of n, a programmable calculator or computer is needed.

For many common types of recurrence relations, there are explicit formulas for a_n. Sections 7.2–7.3 discuss some of these solutions. However, it is frequently easier to determine recursively the value of $a_0, a_1, \ldots$ up to the desired value, say, a_{12} than to compute a_{12} from a complicated general formula for a_n.

Example 1: Arrangements

Find a recurrence relation for the number of ways to arrange n distinct objects in a row. Find the number of arrangements of eight objects.

Let a_n denote the number of arrangements of n distinct objects. There are n choices for the first object in the row. This choice can be followed by any arrangement of the remaining $n-1$ objects; that is, by the a_{n-1} arrangements of the remaining $n-1$ objects. Thus $a_n = na_{n-1}$. Substituting recursively in this relation, we see that

$$a_n = na_{n-1} = n[(n-1)a_{n-2}] = \cdots = n(n-1)(n-2)\cdots \times 2 \times 1 = n!$$

In particular, $a_8 = 8!$ ∎

Of course, we already know that the number of arrangements of n objects is $n!$ It is often useful to try out a new technique on an old problem to see how it works before using it on new problems. Now for some new problems.

Example 2: Climbing Stairs

An elf has a staircase of n stairs to climb. Each step it takes can cover either one stair or two stairs. Find a recurrence relation for a_n, the number of different ways for the elf to ascend the n-stair staircase.

It is easy to check with a figure that $a_1 = 1$, $a_2 = 2$ (two one-steps or one two-steps) and $a_3 = 3$. For $n = 4$, Figure 7.1a depicts one way of climbing the stairs, taking successive steps of sizes 1, 2, 1. Other possibilities are two-stair steps either preceded or followed by two two-stair steps, or two two-stair steps, or four one-stair steps. In all, we count five ways to climb the four stairs; so $a_4 = 5$.

Is there some systematic way to enumerate the ways to climb four stairs that breaks the problem into parts involving the ways to climb three or fewer stairs? Clearly, once the first step is taken there are three or fewer stairs remaining to climb. Thus we see that after a first step of one stair, there are a_3 ways to continue the climb up the remaining three stairs. If the first step covers two stairs, then there are a_2 ways to continue up the remaining two stairs. So $a_4 = a_3 + a_2$. We confirm that the values for a_4, a_3, a_2 satisfy this relation: $5 = 3 + 2$. This argument applies to the first step when climbing any number of stairs, as is shown in Figures 7.1b and 7.1c. Thus $a_n = a_{n-1} + a_{n-2}$. ∎

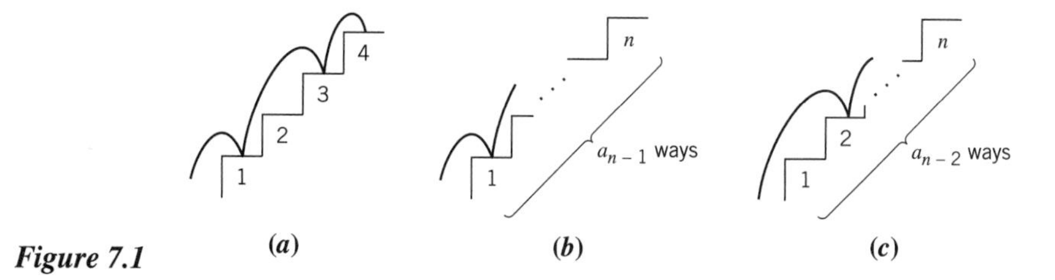

Figure 7.1 (*a*) (*b*) (*c*)

In Section 7.3 we obtain an explicit solution to this recurrence relation. The relation $a_n = a_{n-1} + a_{n-2}$ is called the **Fibonacci relation.** The numbers a_n generated by the Fibonacci relation with the initial conditions $a_0 = a_1 = 1$ are called the **Fibonacci numbers.** They begin $1, 1, 2, 3, 5, 8, 13, 21, 34, 55, 89$. Fibonacci numbers arise naturally in many areas of combinatorial mathematics. There is even a journal, *Fibonacci Quarterly*, devoted solely to research involving the Fibonacci relation and Fibonacci numbers. Fibonacci numbers have been applied to other fields of mathematics, such as numerical analysis. They occur in the natural world—for example, the arrangements of petals in some flowers. For more information about the occurrences of Fibonacci numbers in nature see [1].

Example 3: Dividing the Plane

Suppose we draw n straight lines on a piece of paper so that every pair of lines intersect (but no three lines intersect at a common point). Into how many regions do these n lines divide the plane?

Again we approach the problem initially by examining the situation for small values of n. With one line, the paper is divided into two regions. With two lines, we get four regions, that is, $a_2 = 4$. See Figure 7.2*a*. From Figure 7.2*b*, we see that $a_3 = 7$. The skeptical reader may ask, how do we know that three intersecting lines will always create seven regions? Let us go back one step then.

Clearly two intersecting lines will always yield four regions, as shown in Figure 7.2*a*. Now let us examine the effect of drawing the third line (labeled "3" in Figure 7.2*b*). It must cross each of the other two lines (at different points). Before, between, and after these two intersection points, the third line cuts through three of the regions formed by the first two lines (this action of the third line does not depend on how it is drawn, just that it intersects the other two lines). So in severing three regions, the third line must form three new regions, actually creating six new regions out of three old regions. Thus $a_3 = a_2 + 3 = 4 + 3 = 7$, independently of how the third line is drawn.

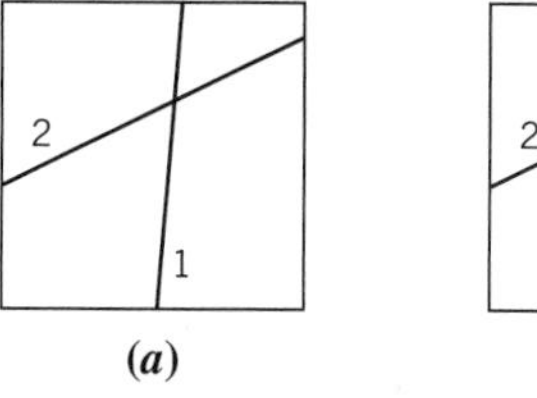

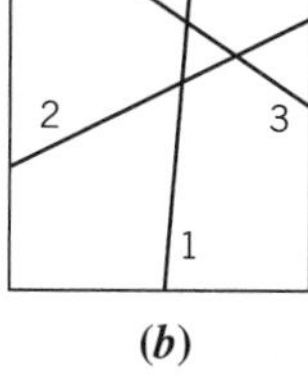

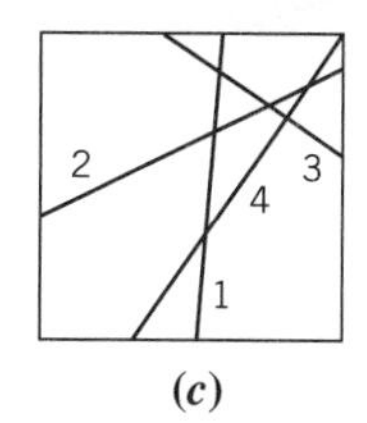

Figure 7.2 (*a*) (*b*) (*c*)

Similarly, the fourth line severs four regions before, between, and after its three intersection points with the first 3 lines (see Figure 7.2*c*), so that $a_4 = a_3 + 4 = 7 + 4 = 11$. In general, the *n*th line must sever *n* regions before, between, and after its $n - 1$ intersection points with the first $n - 1$ lines. So $a_n = a_{n-1} + n$. ■

Example 4: Tower of Hanoi

The Tower of Hanoi is a game consisting of *n* circular rings of varying size and three pegs on which the rings fit. Initially all the rings are placed on the first (leftmost) peg with the largest ring at the bottom covered by successively smaller rings. See Figure 7.3*a*. By transferring the rings among the pegs, one seeks to achieve a similarly tapered pile on the third (rightmost) peg. The complication is that each time a ring is transferred to a new peg, the transferred ring must be smaller than any of the rings already piled on this new peg; equivalently, at every stage in the game there must be a tapered pile (or no pile) on each peg.

Find a recurrence relation for a_n, the minimum number of moves required to play the Tower of Hanoi with *n* rings. How many moves are needed to play the six-ring game? Try playing this game with a dime, penny, nickel, and quarter (four rings) before reading our solution.

The key observation is that if, say, the six smallest rings are on peg A and we want to move them to peg C, we must first "play the five-ring Tower-of-Hanoi game" to get the five smallest rings from peg A to peg B, then move the sixth smallest ring from peg A to peg C, and then again "play the five-ring game" from peg B to peg C. See Figure 7.3. (Of course, to move the five smallest rings we must move the fifth smallest from A to B, which means playing four-ring games from A to C and from C to B, etc.) Thus to move the *n* rings from A to C, the $n - 1$ smallest rings must first be moved from A to B, then the largest (*n*th) ring moved from A to C, and then the $n - 1$ smallest rings moved from B to C.

If a_n is the number of moves needed to transfer a tapered pile of *n* rings from one peg to another peg, then the previous sentence yields the following recurrence relation: $a_n = a_{n-1} + 1 + a_{n-1} = 2a_{n-1} + 1$. The initial condition is $a_1 = 1$, and so $a_2 = 2a_1 + 1 = 3$; $a_3 = 2a_2 + 1 = 7$; $a_4 = 2a_3 + 1 = 15$; $a_5 = 2a_4 + 1 = 31$; and $a_6 = 2a_5 + 1 = 63$. So the six-ring game requires 63 moves. Note that the a_ns thus far fit the formula $a_n = 2^n - 1$. ■

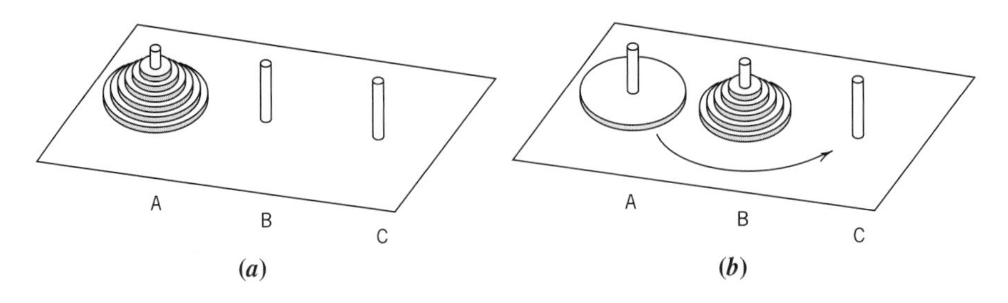

Figure 7.3 Tower of Hanoi

Example 5: Money Growing in a Savings Account

A bank pays 8 percent interest each year on money in savings accounts. Find recurrence relations for the amounts of money a gnome would have after n years if it follows the investment strategies of:

(a) Investing \$1000 and leaving it in the bank for n years.

(b) Investing \$100 at the end of each year.

If an account has x dollars at the start of a year, then at the end of the year (i.e., start of the next year) it will have x dollars plus the interest on the x dollars, provided no money was added or removed during the year. Then for part (a), the recurrence relation is $a_n = a_{n-1} + .08a_{n-1} = 1.08a_{n-1}$. The initial condition is $a_0 = 1000$. For part (b), the relation must reflect the \$100 added (which earns no interest since it comes at the end of the year). So $a_n = 1.08a_{n-1} + 100$, with $a_0 = 0$. ■

Example 6: Making Change

Find a recurrence relation for the number of different ways to hand out a piece of chewing gum (worth 1¢) or a candy bar (worth 10¢) or a doughnut (worth 20¢) on successive days until n¢ worth of food has been given away.

This problem can be treated similarly to the stair-climbing problem in Example 2. That is, if on the first day we hand out 1¢ worth of chewing gum, we are left with $(n-1)$¢ worth of food to give away on following days; if the first day we hand out 10¢ worth of candy, we have $(n-10)$¢ worth to dispense the next days; and if 20¢ then $(n-20)$¢ the next days. So $a_n = a_{n-1} + a_{n-10} + a_{n-20}$ with $a_0 = 1$ (one way to give nothing—by giving 0 pieces of each item), and implicitly, $a_k = 0$ for $k < 0$. ■

Example 7: A Forbidden Subsequence

Find a recurrence relation for a_n, the number of n-digit ternary sequences without any occurrence of the subsequence “012.”

Recall that a ternary sequence is a sequence composed of 0s, 1s and 2s. We start with the same analysis used in the elf stair-climbing problem. If the first digit in an n-digit ternary sequence is 1, then there are $a_{n-1}(n-1)$-digit ternary sequences without the pattern 012 that can follow that initial 1. Similarly if the first digit is 2.

However, there is a problem if the first digit is a 0. Among the $(n-1)$-digit ternary sequences without the pattern 012 that might follow the initial 0 are sequences that start with 12. While such $(n-1)$-digit sequences do not contain the pattern 012, the n-digit sequences of 0 followed by such $(n-1)$-digit sequences do start with the pattern 012. We correct this mistake by subtracting off all n-digit sequences starting with 012 (but with 012 not appearing later in the sequence). Such sequences are formed by 012 followed by any $(n-3)$-digit ternary sequence with no 012 pattern—there are a_{n-3} such sequences. Thus the desired recurrence relation is: $a_n = a_{n-1} + a_{n-1} + (a_{n-1} - a_{n-3}) = 3a_{n-1} - a_{n-3}$. ■

The preceding examples developed recurrence relation models either by breaking a problem into a first step followed by the same problem for a smaller set (Examples 1, 2, 6 and 7) or by observing the *change* in going from the case of $n-1$ to the case of n (Examples 3 and 5). Example 4 was a modified form of the former procedure in which the move of the nth ring (largest ring) was seen to be preceded and followed by an $(n-1)$-ring game.

There are two simple methods for solving some of the relations seen thus far. The first is recursive backward substitution: wherever a_{n-1} occurs in the relation for a_n, we replace a_{n-1} by the relation's formula for a_{n-1} (involving a_{n-2}) and then replace a_{n-2}, and so on. In Example 3, the relation $a_n = a_{n-1} + n$ becomes $a_n = (a_{n-2} + n - 1) + n = \cdots = 1 + 2 + 3 + \cdots + n - 1 + n$. We used this method in Example 1 to obtain $a_n = n(n-1)(n-2) \times \cdots \times 2 \times 1$.

The second method is to guess the solution to the relation and then to verify it by mathematical induction. In Example 4, we noted that $a_n = 2^n - 1$ for the first six values of the recurrence relation $a_n = 2a_{n-1} + 1, a_1 = 1$. We prove $a_n = 2^n - 1$ by induction as follows. It is seen to be true for $n = 1$. Assuming $a_{n-1} = 2^{n-1} - 1$, we then have $a_n = 2a_{n-1} + 1 = 2(2^{n-1} - 1) + 1 = 2^n - 2 + 1 = 2^n - 1$.

Example 8: Chromatic Polynomial

Use the following recurrence relation to determine the *chromatic polynomial* $P_k(G)$ of the graph G in Figure 7.4*a*. Recall that $P_k(G)$ gives the number of different k-colorings of G (see Example 7 in Section 2.3). For any two nonadjacent vertices x, y in G, G^+_{xy} is the graph obtained from G by adding the edge (x, y), and G^c_{xy} is the graph obtained from G by coalescing vertices x and y into a single vertex [and deleting edge (x,y)] that is adjacent to all vertices formerly adjacent to x or y. G^+_{xy} corresponds to colorings of G in which x and y receive different colors, and G^c_{xy} corresponds to colorings of G in which x and y have the same color. Then

$$P_k(G) = P_k(G^+_{xy}) + P_k(G^c_{xy})$$

To apply this recurrence relation to the graph in Figure 7.4*a*, we need some "initial condition" graphs whose chromatic polynomial is known. The required building blocks are the collection of complete graphs on n vertices K_n (all vertices mutually adjacent). In Example 7 of Section 2.3 we showed that $P_k(K_n) = P(k,n) = k!/(k-n)!$

Consider the nonadjacent pair b, d in G. Then $P_k(G) = P_k(G^+_{bd}) + P_k(G^c_{bd})$. Figure 7.4*b* shows G^+_{bd} and Figure 7.4*c* shows G^c_{bd}. To determine $P_k(G^+_{bd})$ we use the

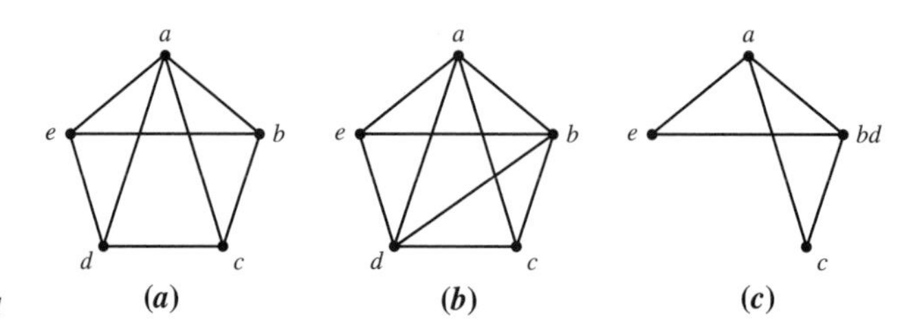

Figure 7.4

nonadjacent pair c, e : $P_k(G_{bd}^+) = P_k((G_{bd}^+)_{ce}^+) + P_k(G_{bd}^+)_{ce}^c) = P_k(K_5) + P_k(K_4) = k!/(k-5)! + k!/(k-4)!$ Similarly, $P_k(G_{bd}^c) = P_k((G_{bd}^c)_{ce}^+) + P_k((G_{bd}^c)_{ce}^c) = P_k(K_4) + P_k(K_3) = k!/(k-4)! + k!/(k-3)!$ In total, $P_k(G) = k!/(k-5)! + 2[k!/(k-4)!] + k!/(k-3)!$ ■

We now consider more complex recurrence relations involving two variable relations and simultaneous relations.

Example 9: Selection Without Repetition

Let $a_{n,k}$ denote the number of ways to select a subset of k objects from a set of n distinct objects. Find a recurrence relation for $a_{n,k}$.

Observe that $a_{n,k}$ is simply $\binom{n}{k}$. We break the problem into two subcases based on whether or not the first object is used. There are $a_{n-1,k}$ k-subsets that do not use the first object, and there are $a_{n-1,k-1}$ k-subsets that do use the first object. So we have $a_{n,k} = a_{n-1,k} + a_{n-1,k-1}$. This is the Pascal's Triangle Identity [identity (3) from Section 5.5]. The initial conditions are $a_{n,0} = a_{n,n} = 1$ for all $n \geq 0$ (and $a_{n,k} = 0, k > n$). ■

Example 10: Distributions

Find a recurrence relation for the ways to distribute n identical balls into k distinct boxes with between two and four balls in each box. Repeat the problem with balls of three colors.

In the spirit of the previous example, we consider how many balls go into the first box. If there are two in the first box, then there are $a_{n-2,k-1}$ ways to put the remaining $n-2$ identical balls in the remaining $k-1$ boxes. Continuing this line of reasoning, we see that $a_{n,k} = a_{n-2,k-1} + a_{n-3,k-1} + a_{n-4,k-1}$. The initial conditions are $a_{2,1} = a_{3,1} = a_{4,1} = 1$ and $a_{n,1} = 0$.

If now three colors are allowed, there are $C(2+3-1, 2) = 6$ ways to pick a subset of two balls for the first box from three types of colors with repetition. (Review the start of Section 5.3 for a discussion of selection with repetition.) Similarly, there are $C(3+3-1, 3) = 10$ ways to pick three balls from 3 types and $C(4+3-1, 4) = 15$ ways to pick four balls from 3 types. Then $a_{n,k} = 6a_{n-2,k-1} + 10a_{n-3,k-1} + 15a_{n-4,k-1}$ with initial conditions $a_{2,1} = 6$, $a_{3,1} = 10$, $a_{4,1} = 15$, and $a_{n,1} = 0$.

The problem with all balls identical can be solved by generating functions (see Example 3 of Section 6.1), but recurrence relations are the only practical approach with the extra constraint of different types of balls. ■

Example 11: Placing Parentheses

Find a recurrence relation for a_n, the number of ways to place parentheses to multiply the n numbers $k_1 \times k_2 \times k_3 \times k_4 \times \cdots \times k_n$ on a calculator.

To clarify the problem, observe that there is one way to multiply $(k_1 \times k_2)$; so $a_2 = 1$. There are two ways to multiply $k_1 \times k_2 \times k_3$, namely, $[(k_1 \times k_2) \times k_3]$ and $[k_1 \times (k_2 \times k_3)]$; so $a_3 = 2$. It is not clear what a_0 and a_1 should be, but to make the eventual recurrence relation have a simple form we let $a_0 = 0$ and $a_1 = 1$. To find a recurrence relation for a_n, we look at the last multiplication (the outermost parenthesis) in the product of the n numbers. This last multiplication involves the products of two multiplication subproblems:

$$(k_1 \times k_2 \times \cdots \times k_i) \times (k_{i+1} \times k_{i+2} \times \cdots \times k_n)$$

where i can range from 1 to $n-1$. The numbers of ways to parenthesize the two respective subproblems are a_i and a_{n-i}, and so there are $a_i a_{n-i}$ ways to parenthesize both subproblems. Summing over all i, we obtain the recurrence relation (for $n \geq 2$). ∎

Example 12: Systems of Recurrence Relations

Find recurrence relations for:

(a) The number of n-digit binary sequences with an even number of 0s, and

(b) The number of n-digit ternary sequences with an even number of 0s and an even number of 1s.

(a) We use the first-step analysis of the stair-climbing model to find a recurrence relation for a_n, the number of n-digit binary sequences with an even number of 0s. If an n-digit binary sequence starts with a 1, then we require an even number of 0s in the remaining $(n-1)$-digit sequence—a_{n-1} such sequences. If the n-digit sequence starts with a 0, we require an *odd* number of 0s in the remaining $n-1$ digits—$2^{n-1} - a_{n-1}$ such sequences, since all $2^{n-1}(n-1)$-digit sequences minus the even-0s $(n-1)$-digit sequences yields the odd-0s $(n-1)$-digit sequences. In sum, $a_n = a_{n-1} + (2^{n-1} - a_{n-1}) = 2^{n-1}$. Our recurrence relation reduces into a formula for a_n. In words, every $(n-1)$-digit sequence yields an even-0s n-digit sequence by adding a 1 in front if the $(n-1)$-digit sequence has an even number of 0s or adding a 0 in front if the sequence has an odd number of 0s.

(b) We will need simultaneous recurrence relations for a_n, the number of n-digit ternary sequences with even 0s and even 1s; b_n, the number of n-digit ternary sequences with even 0s and odd 1s; and c_n, the number of n-digit sequences with odd 0s and even 1s. Observe that $3^n - a_n - b_n - c_n$ is the number of n-digit ternary sequences with odd 0s and odd 1s. An n-digit ternary sequence with even 0s and even 1s is obtained either by having a 1 for the first digit followed by an $(n-1)$-digit sequence with even 0s and odd 1s, or a 0 followed by an $(n-1)$-digit sequence with odd 0s and even 1s, or a 2 followed by an $(n-1)$-digit sequence with even 0s and even 1s. Thus $a_n = b_{n-1} + c_{n-1} + a_{n-1}$. Similar analyses yield $b_n = a_{n-1} + (3^{n-1} - a_{n-1} - b_{n-1} - c_{n-1}) + b_{n-1} = 3^{n-1} - c_{n-1}$ and $c_n = a_{n-1} + (3^{n-1} - a_{n-1} - b_{n-1} - c_{n-1}) + c_{n-1} = 3^{n-1} - b_{n-1}$. The initial conditions are $a_1 = b_1 = c_1 = 1$. To recursively compute values for a_n, we must simultaneously compute b_n and c_n. ∎

We close this section with a few words about difference equations. The **first (backward) difference** Δa_n of the sequence $(a_0, a_1, a_2, \ldots)$ is defined to be $\Delta a_n = a_n - a_{n-1}$. The second difference is $\Delta^2 a_n = \Delta a_n - \Delta a_{n-1} = a_n - 2a_{n-1} + a_{n-2}$, and so on. A difference equation is an equation involving a_n and its differences, such as $2\Delta^2 a_n - 3\Delta a_n + a_n = 0$. Observe that

$$a_{n-1} = a_n - (a_n - a_{n-1}) = a_n - \Delta a_n$$
$$a_{n-2} = a_{n-1} - \Delta a_{n-1} = (a_n - \Delta a_n) - \Delta(a_n - \Delta a_n) = a_n - 2\Delta a_n + \Delta^2 a_n (^*)$$

Similar equations can express a_{n-k} in terms of $a_n, \Delta a_n, \ldots, \Delta^k a_n$. Thus any recurrence relation can be rewritten as a difference equation, by expressing the a_{n-k}s on the right-hand side of (*) in terms of a_n and its differences. Conversely, by writing Δa_n as $a_n - a_{n-1}$, and so on, any difference equation can be written as a recurrence relation.

Difference equations are commonly used to approximate differential equations when solving differential equations on a computer. Difference equations have wide use in their own right as models for dynamical systems for which differential equations (which require continuous functions) are inappropriate. They are used in economics in models for predicting the gross national product in successive years. They are used in ecology to model the numbers of various species in successive years. As noted above, any difference equation model can also be formulated as a recurrence relation; however, the behavior of a dynamical system is easier to analyze and explain in terms of differences. Refer to Sandefur [4] for further information about difference equations and their applications.

Example 13: Two-Animal Population Model

Assume that if undisturbed by foxes, the number of rabbits increases each year by an amount αr_n, where r_n is the number of rabbits, but when foxes are present, each rabbit has probability βf_n of being eaten by a fox (f_n is the number of foxes). Foxes alone decrease by an amount γf_n each year, but when rabbits are present, each fox has probability δr_n of feeding and raising up a new young fox (death of foxes is included in the γf_n term). Give a pair of simultaneous difference equations describing the number of rabbits and foxes in successive years.

The information given about yearly changes in the two populations yields the difference equations:

$$\Delta r_n = \alpha r_n - \beta r_n f_n$$
$$\Delta f_n = \gamma f_n + \delta r_n f_n \quad \blacksquare$$

7.1 EXERCISES

Summary of Exercises The first 39 exercises call for recurrence relation modeling similar to that in the examples, with multiple indices and equations required in Exercises 28–40 and difference equations in Exercises 41–43. The remaining exercises are more advanced problems.

1. Find a recurrence relation for the number of ways to distribute n distinct objects into five boxes. What is the initial condition?

2. **(a)** Find a recurrence relation for the number or ways the elf in Example 2 can climb n stairs if each step covers either one or two or three stairs?

(b) How many ways are there for the elf to climb five stairs?

3. Find a recurrence relation for the number of ways to arrange cars in a row with n spaces if we can use Cadillacs or Continentals or Fords. A Cadillac or Continental requires two spaces, whereas a Ford requires just one space.

4. **(a)** Find a recurrence relation for the number of ways to go n miles by foot walking at 3 miles per hour or jogging at 6 miles per hour or running at 10 miles per hour; at the end of each hour a choice is made of how to go the next hour.

(b) How many ways are there to go 12 miles?

5. Find a recurrence relation for the number of ways to distribute a total of n cents on successive days using 1947 pennies, 1958 pennies, 1971 pennies, 1951 nickels, 1967 nickels, 1959 dimes, and 1975 quarters.

6. **(a)** Find a recurrence relation for the number of n-digit binary sequences with no pair of consecutive 1s.

(b) Repeat for n-digit ternary sequences.

(c) Repeat for n-digit ternary sequences with no consecutive 1s or consecutive 2s.

7. Find a recurrence relation for the number of pairs of rabbits after n months if: (1) initially there is one pair of rabbits who were just born; and (2) every month each pair of rabbits that are over one month old have a pair of offspring (a male and a female).

8. Show that the binomial sum

$$s_n = \binom{n+1}{0} + \binom{n}{1} + \binom{n-1}{2} + \cdots$$

satisfies the Fibonacci relation.

9. Find a recurrence relation for the number of ways to arrange n dominoes to fill a 2-by-n checkerboard.

10. Find a recurrence relation for a_n for the number of bees in the nth previous generation of a male bee, if a male bee is born asexually from a single female and a female bee has the normal male and female parents. The ancestral chart at the right shows that $a_1 = 1, a_2 = 2, a_3 = 3$.

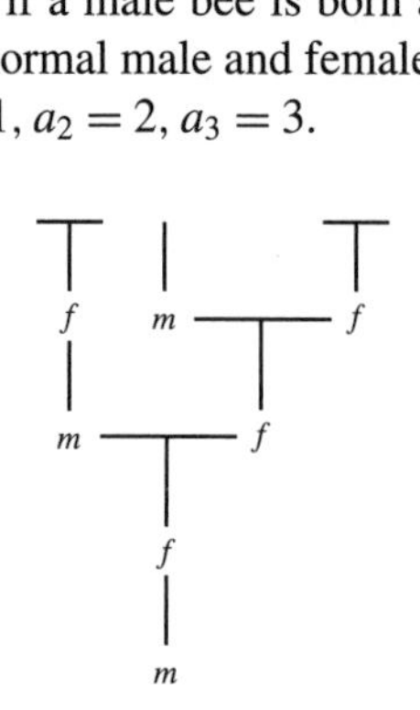

11. Find a recurrence relation for a_n for the number of ways for an image to be reflected n times by internal faces of two adjacent panes of glass. The diagram below shows that $a_0 = 1$, $a_1 = 2$, and $a_2 = 3$.

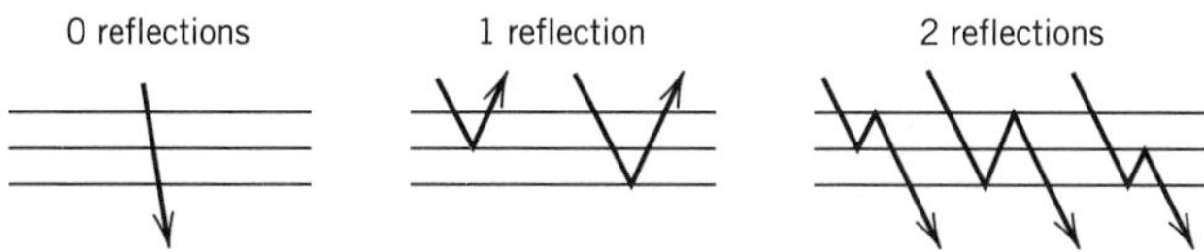

12. Find a recurrence relation for the number of regions created by n mutually intersecting circles on a piece of paper (no three circles have a common intersection point).

13. **(a)** Find a recurrence relation for the number of regions created by n lines on a piece of paper if k of the lines are parallel and the other $n - k$ lines intersect all other lines (no three lines intersect at one point).

(b) If $n = 9$ and $k = 3$, find the number of regions.

14. Show that each of the following rules for playing the Tower of Hanoi works.

(a) On odd-numbered moves, move the smallest ring clockwise one peg (think of the pegs being at the corners of a triangle); and on even-numbered moves, make the only legal move not using the smallest ring.

(b) Number the rings from 1 to n in order of increasing size. Never move the same ring twice in a row. Always put even-numbered rings on top of odd-numbered rings (or on an empty peg) and put odd-numbered rings on top of even-numbered rings (or on an empty peg).

15. Find a recurrence relation for the amount of money in a savings account after n years if the interest rate is 6 percent and \$50 is added at the start of each year.

16. **(a)** Find a recurrence relation for the amount of money outstanding on a \$30,000 mortgage after n years if the interest rate is 8 percent and the yearly payment (paid at the end of each year after interest is computed) is \$3000.

(b) Use a calculator or computer to determine how many years it will take to pay off the mortgage.

17. **(a)** Find a recurrence relation for the number of sequences of 1s, 3s and 5s whose terms sum to n.

(b) Repeat part (a) with the added condition that no 5 can be followed by a 1.

18. **(a)** Find a recurrence relation for the number of ways to arrange three types of flags on a flagpole n feet high: red flags (1 foot high), gold flags (1 foot high), and green flags (2 feet high).

(b) Repeat part (a) with the added condition that there may not be three 1-foot flags (red or gold) in a row.

19. Find a recurrence relation for a_n, the number of ways to give away \$1 or \$2 or \$3 for n days with the constraint that there is an even number of days when \$1 is given away.

20. Find a recurrence relation to count the number of n-digit binary sequences with at least one instance of consecutive 0s.

21. Find a recurrence relation for the number of n-digit quaternary (0, 1, 2, 3) sequences with at least one 1 and the first 1 occurring before the first 0 (possibly no 0s).

22. Find a recurrence relation for the number of n-digit ternary (0, 1, 2) sequence in which no 1 appears anywhere to the right of any 2.

23. Find a recurrence relation for the number of n-digit ternary sequences that have the pattern "012" occurring for the first time at the end of the sequence.

24. A switching game has n switches, all initially in the OFF position. In order to be able to flip the ith switch, the $(i-1)$st switch must be ON and all earlier switches OFF. The first switch can always be flipped. Find a recurrence relation for the total number of times the n switches must be flipped to get the nth switch ON and all others OFF.

25. **(a)** Let T be a binary tree with all leaves at level n. Find a recurrence relation for the number of leaves in T. Repeat for T an m-ary tree.

(b) Let T be a binary tree with all leaves at level n. Find a recurrence relation for the total number of vertices in T. Repeat for T an m-ary tree.

26. Find a recurrence relation for the number of ways to pair off $2n$ people for tennis matches.

27. Determine the chromatic polynomial for the following graphs:

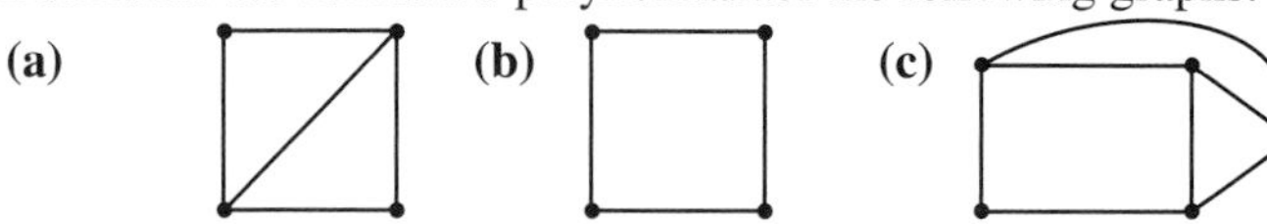

28. Find a recurrence relation for the number of ways to pick k objects with repetition from n types.

29. Find a recurrence relation for the number of ways to select n objects from k types with at most three of any one type.

30. Find a recurrence relation for $a_{n,k}$, the number of ways to order n doughnuts from k different types of doughnuts if two or four or six doughnuts must be chosen of each type.

31. Find a recurrence relation for the number of partitions of the integer n into k parts.

32. Find a recurrence relation for the number $a_{n,m,k}$ of distributions of n identical objects into k distinct boxes with at most four objects in a box and with exactly m boxes having four objects.

33. Find a system of recurrence relations for computing the number of n-digit binary sequences with an even number of 0s and an even number of 1s.

34. Find a system of recurrence relations for computing the number of n-digit quaternary sequences with:

(a) An even number of 0s.

(b) An even total number of 0s and 1s.

(c) An even number of 0s and an even number of 1s.

35. Find a system of recurrence relations for computing the number of n-digit binary sequences with exactly one pair of consecutive 0s.

36. Find a system of recurrence relations for the number of n-digit binary sequences with k adjacent pairs of 1s and no adjacent pairs of 0s.

37. Find a system of recurrence relations for computing the number of ways to hand out a penny or a nickel or a dime on successive days until n cents are given such that the same amount of money is not handed out on two consecutive days.

38. Find a recurrence relation for the number of ways to pair off $2n$ points on a circle with nonintersecting chords. (*Hint:* The recurrence involves products of a_ks as in Example 11.)

39. Find a recurrence relation for the number of ways to divide an n-gon into triangles with noncrossing diagonals.

40. Find a recurrence relation for the number of binary trees with n labeled leaves.

41. Find Δa_n and $\Delta^2 a_n$ if a_n equals:

(a) $3n+2$ **(b)** n^2 **(c)** n^3

42. Let f_n be the amount of food that can be bought with n dollars. Let p_n be the "perceived" value of the \$$n$ of food. Suppose the increase in perceived value with one dollar more of food equals the relative, or percentage, increase in the actual amount of food. Find a difference equation relating Δp_n and Δf_n.

43. **(a)** Find a recurrence relation for the number of permutations of the first n integers such that each integer differs by one (except for the first integer) from some integer to the left of it in the permutation. What is the initial condition?

(b) Solve the relation in part (a) by guessing and verifying the guess by induction.

(c) Give a direct combinatorial answer to this problem.

44. **(a)** Find a recurrence relation for $f(n, k)$, the number of k subsets of the integers 1 through n with no pair of consecutive integers.

(b) Show that $\Sigma_{k=0}^{n/2} f(n, k) = F_{n+1}$, the $(n+1)$st Fibonacci number (n even), where $F_0 = F_1 = 1$.

45. Find a system of recurrence relations for computing a_n, the number of (unordered) collections of (identical) pennies, (identical) nickels, (identical) dimes, and (identical) quarters whose value is n cents.

46. Find a recurrence relation for the number of ways a coin can be flipped $2n$ times with n heads and n tails and:

(a) The number of heads at any time never being less than the number of tails.

(b) The number of heads equaling the number of tails only after all $2n$ flips.

47. Find a recurrence relation for the number of incongruent integral-sided triangles whose perimeter is n (the relation is different for n odd and n even).

48. Find a system of recurrence relations for computing the number of spanning trees in the "ladder" graph with $2n$ vertices.

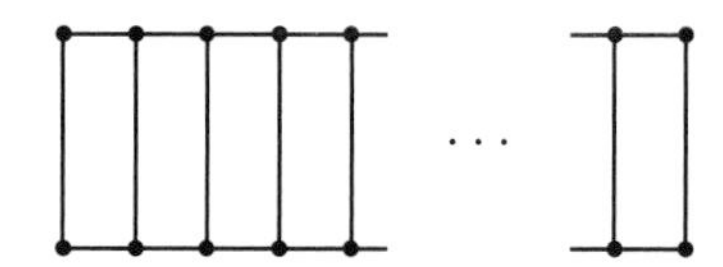

49. Verify the following identities for Fibonacci numbers (F_i is the ith Fibonacci number) by induction or combinatorial argument. Here $F_0 = F_1 = 1$.

(a) $\sum_{i=0}^{n} F_i = F_{n+2} - 1$ **(b)** $\sum_{i=0}^{n} F_i^2 = F_n F_{n+1}$

(c) $\sum_{k=0}^{n} F_{2k} = F_{2n+1}$ **(d)** $F_n F_{n+2} = F_{n+1}^2 + (-1)^n$

(e) $F_1 - F_2 + F_3 - \cdots - F_{2n} = -F_{2n-1}$

50. **(a)** Show that $F_{n+m} = F_m F_n + F_{m-1} F_{n-1}$.

(b) From part (a) conclude that F_{n-1} divides F_{kn-1}.

51. A set of cards is numbered from 1 to n. If the top card is k, we invert the order of the top k cards in the pile and again look at the new top card and invert again, and so on. Show that the maximum number of rounds until 1 is at the top is F_{n+1}. (*Hint:* Induct on the number k of the top card.)

7.2 DIVIDE-AND-CONQUER RELATIONS

In this section we present a special class of recurrence relations that arise frequently in the analysis of recursive computer algorithms. These are algorithms that use a "divide-and-conquer" approach to recursively split a problem into two subproblems of half the size. The dictionary search in Example 3 of Section 3.1 is such an algorithm. A binary tree is explicitly or implicitly associated with most "divide-and-conquer" algorithms. Conversely, many counting problems involving trees (the subject of Chapter 3) are most easily solved with divide-and-conquer recurrence relations.

The total number of steps a_n required by a divide-and-conquer algorithm to process an n-element problem frequently satisfies a recurrence relation of the form

$$a_n = ca_{n/2} + f(n) \tag{1}$$

The following table indicates the form of the solution of Eq. (1) for some common values of c and $f(n)$ ($\lceil r \rceil$ denotes the smallest integer m with $m \geq r$):

c	$f(n)$	a_n
$c=1$	d	$d\lceil\log_2 n\rceil + A$
$c=2$	d	$An-d$
$c>2$	dn	$An^{\log_2 c} + \left(\dfrac{2d}{2-c}\right)n$
$c=2$	dn	$dn(\lceil\log_2 n\rceil + A)$

The constant A is to be chosen to fit the initial condition.

If a problem is recursively split into k parts instead of two parts, then one should replace 2 by k everywhere in the foregoing table, except the solution for $c=k$ and $f(n)=d$ becomes $An-d/(k-1)$. For example, the recurrence relation

$$a_n = ca_{n/k} + dn \quad c \neq k$$

has the solution

$$a_n = An^{\log_k c} + \left(\frac{kd}{k-c}\right)n \tag{2}$$

The solutions for a_n given in the foregoing table are easily verified by substitution. Consider the case: $a_n = ca_{n/2} + dn$, $c \neq 2$. Substituting the table's solution of $a_n = An^{\log_2 c} + [2d/(2-c)]n$ into $ca_{n/2} + dn$, we have

$$\begin{aligned} ca_{n/2} + dn &= c\left[A\left(\frac{n}{2}\right)^{\log_2 c} + \left(\frac{2d}{2-c}\right)\frac{n}{2}\right] + dn \\ &= \frac{cAn^{\log_2 c}}{2^{\log_2 c}} + \frac{cdn}{2-c} + \frac{(2-c)dn}{2-c} \\ &= \frac{cAn^{\log_2 c}}{c} + \frac{cdn + (2-c)dn}{2-c} = An^{\log_2 c} + \left(\frac{2d}{2-c}\right)n = a_n \end{aligned}$$

The following examples illustrate such "divide-and-conquer" recurrences and their solution.

Example 1: Rounds in a Tournament

In a tennis tournament, each entrant plays a match in the first round. Next, all winners from the first round play a second-round match. Winners continue to move on to the next round, until finally only one player is left—the tournament winner. Assuming that tournaments always involve $n = 2^k$ players, for some k, find and solve a recurrence relation for the number of rounds in a tournament of n players.

In terms of binary trees, a_n is the height of a balanced binary tree with $n = 2^k$ leaves. Since half the players are eliminated in each round, the number of rounds increases by 1 when the number of players doubles. The recurrence relation for a_n, the number of rounds, is thus

$$a_n = a_{n/2} + 1$$

From the foregoing table, the solution of this recurrence relation is $a_n = \log_2 n + A$. To determine A, we observe that $0 = a_1 = \log_2 1 + A = 0 + A$, and so $A = 0$. ■

Example 2: Finding the Largest and Smallest Numbers in a Set

Build a recurrence relation model to count the number of comparisons that must be made in the following algorithm for finding the largest number l and the smallest number s in a set S of n distinct integers. Then solve this recurrence relation.

Initially suppose that n is an even number. Assume that we have already found l_1 and s_1, the largest and smallest numbers, respectively, in the first half of S (the first $n/2$ numbers) and have found l_2 and s_2, the largest and smallest numbers in the second half of S. Then make two comparisons, one between l_1 and l_2 and the other between s_1 and s_2 to find the largest number l and smallest number s in S.

The associated recurrence relation for the number of comparisons in this procedure is $a_n = 2a_{n/2} + 2$, for $n \geq 4$ and even. If n is odd and we split S almost equally, the relation would be $a_n = a_{(n+1)/2} + a_{(n-1)/2} + 2$. Observe that $a_1 = 0$, since the one number is both largest and smallest. And $a_2 = 1$, since we can determine the larger and smaller number in a two-element set with one comparison. With these two relations along with a_1 and a_2, we can recursively determine the number of comparisons needed for any n.

Next we solve the recurrence relation $a_n = 2a_{n/2} + 2, n \geq 4$, with $a_2 = 1$. The foregoing table tells us that the solution will be of the form $a_n = An - 2$. We confirm this by substituting $a_n = An - 2$ into both sides of the relation $a_n = 2a_{n/2} + 2$:

$$\begin{aligned} An - 2 = a_n = 2a_{n/2} + 2 &= 2(A\frac{n}{2} - 2) + 2 \\ &= An - 4 + 2 = An - 2 \end{aligned}$$

We can use the initial condition $a_2 = 1$ to determine A:

$$1 = a_2 = A_1(2) - 2 \quad \text{or} \quad A_1 = \frac{3}{2}$$

So $a_n = \frac{3}{2}n - 2$ is the number of comparisons needed to find the largest and smallest number according to the procedure given previously. (It is possible to prove that one cannot do better than $\frac{3}{2}n - 2$ comparisons.) ■

Example 3: Efficient Multidigit Multiplication

Normally one must do n^2 digit-times-digit multiplications to multiply two n-digit numbers. Use a divide-and-conquer approach to develop a faster algorithm.

Let us initially assume that n is a power of 2. Let the two n-digit numbers be g and h. We split each of these numbers into two $n/2$-digit parts:

$$g = g_1 10^{n/2} + g_2 \quad h = h_1 10^{n/2} + h_2$$

Then

$$g \times h = (g_1 \times h_1)10^n + (g_1 \times h_2 + g_2 \times h_1)10^{n/2} + g_2 \times h_2 \tag{2}$$

Observe that

$$g_1 \times h_2 + g_2 \times h_1 = (g_1 + g_2) \times (h_1 + h_2) - g_1 \times h_1 - g_2 \times h_2$$

and so we need to make only three $n/2$-digit multiplications, $g_1 \times h_1$, $g_2 \times h_2$, and $(g_1 + g_2) \times (h_1 + h_2)$ to determine $g \times h$ in Eq. (2) (actually $(g_1 + g_2)$ or $(h_1 + h_2)$ may be $(n/2 + 1)$-digit numbers, but this slight variation does not affect the general magnitude of our solution). If a_n represents the number of digit-times-digit multiplications needed to multiply two n-digit numbers by the foregoing procedure, then the procedure yields the recurrence $a_n = 3a_{n/2}$.

By the table at the start of this section (where $d = 0$), a_n is proportional to $n^{\log_2 3} = n^{1.6}$, a substantial improvement over n^2. ■

In some settings, we are not so interested in an exact formula for a_n as we are in the general rate of growth for a_n. The following theorem from Cormen, Leiserson, and Rivest [2] gives bounds on such growth:

Theorem

Let $a_n = ca_{n/k} + f(n)$ be a recurrence relation with positive constant c and the positive function $f(n)$.

(a) If for large n, $f(n)$ grows proportional to $n^{\log_k c}$ [that is, there are positive constants p and p' such that $pn^{\log_k r} \le f(n) \le p'n^{\log_k c}$], then a_n grows proportional to $n^{\log_k c} \log_2 c$.

(b) If for large n, $f(n) \le pn^q$, where p is a positive constant and $q < \log_k c$, then a_n grows at most at a rate proportional to $n^{\log_k c}$.

7.2 EXERCISES

1. Solve the following recurrence relations assuming that n is a power of 2 (leaving a constant A to be determined):

(a) $a_n = 2a_{n/2} + 5$

(b) $a_n = 2a_{n/4} + n$

(c) $a_n = a_{n/2} + 2n - 1$

(d) $a_n = 3a_{n/3} + 4$

(e) $a_n = 16a_{n/2} + 5n$

(f) $a_n = 4a_{n/2} + 3n$

2. Find and solve a recurrence relation for the number of matches played in a tournament with n players, where n is a power of 2.

3. In a large corporation with n salespeople, every 10 salespeople report to a local manager, every 10 local managers report to a district manager, and so forth until finally 10 vice-presidents report to the firm's president. If the firm has n salespeople, where n is a power of 10, find and solve recurrence relations for:

(a) The number of different managerial levels in the firm.

(b) The number of managers (up through president) in the firm.

4. In a tennis tournament, each player wins k hundreds of dollars, where k is the number of people in the subtournament won by the player (the subsection of the tournament including the player, the player's victims, and their victims, and so forth; a player who loses in the first round gets \$100). If the tournament has n contestants, where n is a power of 2, find and solve a recurrence relation for the total prize money in the tournament.

5. Consider the following method for rearranging the n distinct numbers $x_1, x_2, \ldots, x_n$ in order of increasing size (n is a power of 2). Pair the integers off $\{x_1, x_2\}$, $\{x_3, x_4\}$, and so forth. Compare each pair and put the smaller number first. Next pair off the pairs into sets of four numbers and merge the ordered pairs to get ordered four-tuples. Continue this process until the whole set is ordered. Find and solve a recurrence relation for the total number of comparisons required to rearrange n distinct numbers. (*Hint:* First find the number of comparisons needed to merge two ordered k-tuples into an ordered $2k$-tuple).

6. In a standard elimination tournament, a player wins \$100k when he/she wins a match in the kth round (e.g., first round win earns \$100, second round win \$200). Develop and solve a recurrence relation for a_n, the total amount of money given away in a tournament with n entrants, where n is assumed to be a power of 2.

7. Verify by substitution the form of solution given in the text to the recurrence relations

(a) $a_n = a_{n/2} + d$

(b) $a_n = 2a_{n/2} + d$

(c) $a_n = 2a_{n/2} + dn$

8. **(a)** Use a divide-and-conquer approach to devise a procedure to find the largest and next-to-largest numbers in a set of n distinct integers.

(b) Give a recurrence relation for the number of comparisons performed by your procedure.

(c) Solve the recurrence relation obtained in part (b).

9. **(a)** Use a divide-and-conquer approach to devise a procedure to find the largest number in a set of n distinct integers.

(b) Give a recurrence relation for the number of comparisons performed by your procedure.

(c) Solve the recurrence relation obtained in part (b).

7.3 SOLUTION OF LINEAR RECURRENCE RELATIONS

In this section we show how to solve recurrence relations of the form

$$a_n = c_1 a_{n-1} + c_2 a_{n-2} + \cdots + c_r a_{n-r} \tag{1}$$

where the c_is are given constants. There is a simple technique for solving such relations. Readers who have studied linear differential equations with constant coefficients will see a great similarity between their solution and the form of solutions we discuss here. The general solution to Eq. (1) will involve a sum of individual solutions of the form $a_n = \alpha^n$. To determine what α is, we simply substitute α^k for a_k in Eq. (1), yielding

$$\alpha^n = c_1\alpha^{n-1} + c_2\alpha^{n-2} + \cdots + c_r\alpha^{n-r} \tag{2}$$

We can reduce the power of α in all terms in Eq. (2) by dividing both sides by α^{n-r}:

$$\alpha^r = c_1\alpha^{r-1} + c_2\alpha^{r-2} + \cdots + c_r \tag{3}$$

or equivalently

$$\alpha^r - c_1\alpha^{r-1} - c_2\alpha^{r-2} - \cdots - c_r = 0 \tag{4}$$

Equation (4) is called the **characteristic equation** of the recurrence relation (1). It has r roots, some of which may be complex (but we shall initially assume that there are no multiple roots).

If $\alpha_1, \alpha_2, \ldots, \alpha_r$ are the r roots of Eq. (4), then, for any i, $0 \le i \le r$, $a_n = \alpha_i^n$ is a solution to the recurrence relation (1). It is easy to check that any linear combination of such solutions is also a solution (Exercise 8). That is,

$$a_n = A_1\alpha_1^n + A_2\alpha_2^n + \cdots + A_r\alpha_r^n \tag{5}$$

is a solution to Eq. (1), for any choice of constants A_i, $1 \le i \le r$.

Recall that for a recurrence relation involving $a_{n-1}, a_{n-2}, \ldots, a_{n-r}$ on the right side, we need to be given the initial conditions of the first r values $a_0, a_1, a_2, \ldots, a_{r-1}$. Let us denote such a set of initial values by $a_0', a_1', a_2', \ldots, a_{r-1}'$. Then the A_is must be chosen to satisfy the r constraints:

$$A_1\alpha_1^k + A_2\alpha_2^k + \cdots + A_r\alpha_r^k = a_k' \quad 0 \le k \le r-1 \tag{6}$$

The r linear equations in Eq. (6) can be solved by Gaussian elimination to determine the r constants A_i (remember that at this stage the α_is are known). With the A_is determined, we will have the desired solution for a_n, a solution that satisfies Eq. (4), and hence the recurrence relation (1), and satisfies the initial conditions $a_0 = a_0'$, $a_1 = a_1', \ldots, a_{r-1} = a_{r-1}'$. If the solution of the characteristic equation (4) has a root α_* of multiplicity m, then $\alpha_*^n, n\alpha_*^n, n^2\alpha_*^n, \ldots, n^{(m-1)}\alpha_*^n$ can be shown to be the m associated individual solutions to be used in Eqs. (5) and (6).

Example 1: Doubling Rabbit Population

Every year Dr. Finch's rabbit population doubles. He started with six rabbits. How many rabbits does he have after eight years? After n years?

If a_n is the number of rabbits, then a_n satisfies the relation $a_n = 2a_{n-1}$. We are also given $a_0 = 6$. Substituting $a_n = \alpha^n$, we obtain $\alpha^n = 2\alpha^{n-1}$ or, dividing by α^{n-1}, $\alpha = 2$. So $a_n = 2^n$ is the one individual solution, and $a_n = A2^n$ is the general solution. The initial condition is $6 = a_0 = A2^0$, or $A = 6$. The desired solution is then $a_n = 6 \times 2^n$. After 8 years, we have $a_8 = 6 \times 2^8 = 6 \times 256 = 1536$ rabbits. ■

Example 2: Second-Order Linear Recurrence Relation

Solve the recurrence relation $a_n = 2a_{n-1} + 3a_{n-2}$ with $a_0 = a_1 = 1$.

Setting $a_n = \alpha^n$, we get the characteristic equation:

$$\alpha^n = 2\alpha^{n-1} + 3\alpha^{n-2}$$

which yields

$$\alpha^2 = 2\alpha + 3$$

This may be written $\alpha^2 - 2\alpha - 3 = 0$ or $(\alpha - 3)(\alpha + 1) = 0$. That is, the roots are $+3$ and -1. So the basic solutions of the recurrence relation are $a_n = 3^n$ and $a_n = (-1)^n$, and the general solution is

$$a_n = A_1 3^n + A_2(-1)^n$$

Now we determine A_1 and A_2 by using the initial conditions:

$$1 = a_0 = A_1 3^0 + A_2(-1)^0 = A_1 + A_2$$
$$1 = a_1 = A_1 3^1 + A_2(-1)^1 = 3A_1 - A_2$$

We solve these two simultaneous equations to obtain $A_1 = \frac{1}{2}$, $A_2 = \frac{1}{2}$ (add the two equations together to eliminate A_2 yielding $2 = 4A_1$ or $A_1 = \frac{1}{2}$, and then determine A_2). The required solution to the recurrence relation with the given initial conditions is

$$a_n = \frac{1}{2} \times 3^n + \frac{1}{2} \times (-1)^n. \ ■$$

Example 3: Solution of Fibonacci Relation

Find a formula for the number of ways for the elf in Example 2 of Section 7.1 to climb the n stairs.

The recurrence relation obtained in Example 2 of Section 7.1 was the Fibonacci relation $a_n = a_{n-1} + a_{n-2}$, with the initial conditions $a_1 = 1$, $a_2 = 2$. The associated characteristic equation is obtained by setting $a_n = \alpha^n$:

$$\alpha^n = \alpha^{n-1} + \alpha^{n-2}$$

which reduces to

$$\alpha^2 = \alpha + 1 \quad \text{or} \quad \alpha^2 - \alpha - 1 = 0$$

Using the quadratic formula, we get

$$\alpha = \frac{1}{2(1)}[-(-1) \pm \sqrt{(-1)^2 - 4(1)(-1)}] = \frac{1}{2}(1 \pm \sqrt{5})$$

That is, we have roots $\frac{1}{2} + \frac{1}{2}\sqrt{5}$ and $\frac{1}{2} - \frac{1}{2}\sqrt{5}$, and the general solution of the problem is

$$a_n = A_1 \left(\frac{1}{2} + \frac{1}{2}\sqrt{5}\right)^n - A_2 \left(\frac{1}{2} - \frac{1}{2}\sqrt{5}\right)^n$$

It is left as an exericse to show that $A_1 = \frac{1}{\sqrt{5}}\left(\frac{1+\sqrt{5}}{2}\right)$ and $A_2 = \frac{1}{\sqrt{5}}\left(\frac{1-\sqrt{5}}{2}\right)$. We note the surprising fact that to generate the Fibonacci sequence of integers 1, 1, 2, 3, 5, 8, 13, ..., we need powers of $\left(\frac{1}{2} + \frac{1}{2}\sqrt{5}\right)$ and $\left(\frac{1}{2} - \frac{1}{2}\sqrt{5}\right)$. ∎

Optional: The following example illustrates a solution of a recurrence relation that has complex and multiple roots.

Example 4: Complex and Multiple Roots

Find a formula for a_n satisfying the relation $a_n = -2a_{n-2} - a_{n-4}$ with $a_0 = 0, a_1 = 1, a_2 = 2$, and $a_3 = 3$.

Substituting $a_n = \alpha^n$, we obtain $\alpha^n = -2\alpha^{n-2} - \alpha^{n-4}$, which yields the characteristic equation $\alpha^4 + 2\alpha^2 + 1 = (\alpha^2 + 1)^2 = 0$. The roots of this equation are $\alpha = +i$ and $\alpha = -i$ (where $i = \sqrt{-1}$) and each root has multiplicity 2. Recall that when α is a double root, the associated recurrence relation solutions are $a_n = \alpha^n$ and $a_n = n\alpha^n$. So the general solution is

$$a_n = A_1 i^n + A_2 n i^n + A_3(-i)^n + A_4 n(-1)^n$$

The initial conditions yield the equations

$$\begin{aligned}
0 = a_0 &= A_1 i^0 + A_2 0 i^0 + A_3(-i)^0 + A_4 0(-i)^0 \\
&= A_1 + 0 + A_3 + 0 \\
1 = a_1 &= A_1 i^1 + A_2 1 i^1 + A_3(-i)^1 + A_4 1(-i)^1 \\
&= i(A_1 + A_2 - A_3 - A_4) \\
2 = a_2 &= A_1 i^2 + A_2 2 i^2 + A_3(-i)^2 + A_4 2(-i)^2 \\
&= -A_1 - 2A_2 - A_3 - 2A_4 \\
3 = a_3 &= A_1 i^3 + A_2 3 i^3 + A_3(-i)^3 + A_4 3(-i)^3 \\
&= i(-A_1 - 3A_2 + A_3 + 3A_4)
\end{aligned}$$

Solving these four simultaneous equations in four unknown A_is, we obtain

$$A_1 = -\frac{3}{2}i \quad A_2 = -\frac{1}{2} + i \quad A_3 = \frac{3}{2}i \quad A_4 = -\frac{1}{2} - i$$

Then the solution of the recurrence relation is

$$a_n = -\frac{3}{2}i^{n+1} + \left(-\frac{1}{2} + i\right) n i^n + \frac{3}{2}i(-i)^n + \left(-\frac{1}{2} - i\right) n(-i)^n \quad ∎$$

We remind the reader that for specific values of n, such as $n = 12$, it is easier to determine a_{12} in the two preceding examples by recursively calculating a_3, a_4, a_5 up to a_{12} from the recurrence relation than to solve the initial-condition equations.

7.3 EXERCISES

1. If $500 is invested in a savings account earning 8 percent a year, give a formula for the amount of money in the account after n years.
2. Find and solve a recurrence relation for the number of n-digit ternary sequences with no consecutive digits being equal.
3. Solve the following recurrence relations:
   (a) $a_n = 3a_{n-1} + 4a_{n-2}$, $a_0 = a_1 = 1$
   (b) $a_n = a_{n-2}$, $a_0 = a_1 = 1$
   (c) $a_n = 2a_{n-1} - a_{n-2}$, $a_0 = a_1 = 2$
   (d) $a_n = 3a_{n-1} - 3a_{n-2} + a_{n-3}$, $a_0 = a_1 = 1$, $a_2 = 2$
4. Determine the constants A_1 and A_2 in Example 3. First show that the initial conditions $a_1 = 1$, $a_2 = 2$ are equivalent to the initial conditions $a_0 = 1$, $a_1 = 1$.
5. Find and solve a recurrence relation for the number of ways to arrange flags on an n-foot flagpole using three types of flags: red flags 2 feet high, yellow flags 1 foot high, and blue flags 1 foot high.
6. Find and solve a recurrence relation for the number of ways to make a pile of n chips using red, white, and blue chips and such that no two red chips are together.
7. Find and solve a recurrence relation for p_n, the value of a stock market indicator that obeys the rule that the change this year (from the previous year) equals twice last year's change. Suppose $p_0 = 1$, $p_1 = 4$.
8. Show that any linear combination of solutions to Eq. (1) is itself a solution to Eq. (1).
9. Show that if the characteristic equation (4) has a root α_* of multiplicity 3, then $n^j \alpha_*^n$, for $j = 0, 1, 2$, are solutions of Eq. (1).
10. Show that if F_n is the nth Fibonacci number in the Fibonacci sequence starting $F_0 = F_1 = 1$, then

$$\lim_{n\to\infty} \frac{F_{n+1}}{F_n} = \frac{1+\sqrt{5}}{2}$$

11. If the recurrence relation $a_n = c_1 a_{n-1} + c_2 a_{n-2}$ has a general solution $a_n = A_1 3^n + A_2 6^n$, find c_1 and c_2.

7.4 SOLUTION OF INHOMOGENEOUS RECURRENCE RELATIONS

A recurrence relation is called **homogeneous** if all the terms of the relation involve some a_k. In the preceding section, we presented a method for solving any homogeneous recurrence relation of the form $a_n = c_1 a_{n-1} + c_2 a_{n-2} + \cdots + c_r a_{n-r}$. When an additional term involving a constant or function of n appears in the recurrence relation, such as

$$a_n = c a_{n-1} + f(n) \tag{1}$$

where c is a constant and $f(n)$ is a function of n, then the recurrence relation is said to be **inhomogeneous.**

In this section, we discuss methods for solving inhomogeneous recurrence relations of the form of Eq. (1). The key idea in solving these relations is that a general solution for an inhomogeneous relation is made up of a general solution to the associated homogeneous relation [obtained by deleting the $f(n)$ term] plus *any* one particular solution to the inhomogeneous relation.

For Eq. (1), the homogeneous relation is $a_n = c a_{n-1}$, whose general solution is $a_n = Ac^n$. Suppose a_n^* is some particular solution of Eq. (1); that is, $a_n^* = c a_{n-1}^* + f(n)$. Then we see that $a_n = Ac^n + a_n^*$ satisfies Eq. (1):

$$\begin{aligned} a_n &= Ac^n + a_n^* \\ &= Ac^n + [c a_{n-1}^* + f(n)] \\ &= c(Ac^{n-1} + a_{n-1}^*) + f(n) = c a_{n-1} + f(n) \end{aligned}$$

The constant A in the general solution is chosen to satisfy the initial condition, as in the previous section (A cannot be determined until a_n^* is found).

There is one special case of Eq. (1) that we can restate as an enumeration problem treated in previous chapters. If $c = 1$, then Eq. (1) becomes

$$a_n = a_{n-1} + f(n) \tag{2}$$

We can iterate Eq. (2) to get

$$\begin{aligned} a_1 &= a_0 + f(1) \\ a_2 &= a_1 + f(2) = [a_0 + f(1)] + f(2) \\ a_3 &= a_2 + f(3) = [a_0 + f(1) + f(2)] + f(3) \\ &\vdots \\ a_n &= a_0 + f(1) + f(2) + f(3) + \cdots + f(n) = a_0 + \sum_{k=1}^{n} f(k) \end{aligned}$$

So a_n is just the sum of the $f(k)$s plus a_0. In Sections 5.5 and 6.5 we presented methods for summing functions of n. Either method can be used to solve Eq. (2).

Example 1: Summation Recurrence

Solve the recurrence relation $a_n = a_{n-1} + n$ with initial condition $a_1 = 2$, obtained in Example 3 of Section 7.1, for the number of regions created by n mutually intersecting lines.

The initial condition of $a_1 = 2$ can be replaced by the initial condition $a_0 = 1$ (no lines means that the plane is one big region). By the foregoing discussion, we see that $a_n = 1 + (1 + 2 + 3 + \cdots + n)$. The expression to be summed can be written as

$$\binom{1}{1} + \binom{2}{1} + \binom{3}{1} + \cdots + \binom{n}{1} \tag{3}$$

By identity (7) of Section 5.5, this sum equals $\binom{n+1}{2} = \frac{1}{2}n(n+1)$. Then $a_n = 1 + \frac{1}{2}n(n+1)$. ■

When $c \neq 1$ in Eq. (1), there are known solutions to Eq. (1) to use for specific functions $f(n)$, similar to the situation for "divide-and conquer" recurrences presented in Section 7.2. We present a table for the simplest $f(n)$s. These solutions can be derived by generating function methods introduced in the next section.

$f(n)$	*Particular solution* $p(n)$
d, a constant	B
dn	$B_1n + B_0$
dn^2	$B_2n^2 + B_1n + B_0$
ed^n	Bd^n

The Bs are constants to be determined. If $f(n)$ were a sum of several different terms, we would separately solve the relation for each separate $f(n)$ term and then add these solutions together to get a particular solution for the composite $f(n)$.

There is one case in which the particular solution for $f(n) = ed^n$ will not work. This involves the relation $a_n = da_{n-1} + ed^n$—note here that the homogeneous solution $a_n = Ad^n$ has the same form as $f(n)$. Then one must try $a_n^* = Bnd^n$ as the particular solution.

Example 2: Solving Tower of Hanoi Puzzle

Solve the recurrence relation $a_n = 2a_{n-1} + 1$ with $a_1 = 1$ obtained in Example 4 of Section 7.1 for the number of moves required to play the n-ring Tower of Hanoi puzzle. The general solution to the homogeneous equation $a_n = 2a_{n-1}$ is $a_n = A2^n$.

We find a particular solution to the inhomogeneous relation by setting $a_n^* = B$ [this is the form of a particular solution given in the foregoing table when $f(n)$ is a constant]. Substituting in the relation, we have

$$B = a_n^* = 2a_{n-1}^* + 1 = 2B + 1 \quad \text{or} \quad B = -1$$

So $a_n^* = -1$ is the particular solution, and the general inhomogeneous solution is $a_n = A2^n + a_n^* = A2^n - 1$. We now can determine A from the initial condition: $1 = a_1 = A2^1 - 1$, or $1 = 2A - 1$. Hence $A = 1$, and the desired solution is $a_n = 2^n - 1$. ■

Example 3: Compound Inhomogeneous Term

Solve the recurrence relation $a_n = 3a_{n-1} - 4n + 3 \times 2^n$ to find its general solution. Also find the solution when $a_1 = 8$.

The general solution to the homogeneous equation $a_n = 3a_{n-1}$ is $a_n = A3^n$. We solve for a particular solution of the relation separately for each inhomogeneous term. For $a_n = 3a_{n-1} - 4n$, we try the form $a_n^* = B_1 n + B_0$, obtaining

$$B_1 n + B_0 = a_n^* = 3a_{n-1}^* - 4n = 3[B_1(n-1) + B_0] - 4n. \quad (4)$$

We now equate the constant terms and the coefficients of n on each side of Eq. (4):

$$\text{Constant terms:} \quad B_0 = -3B_1 + 3B_0 \quad (5)$$

$$n \text{ terms:} \quad B_1 n = 3B_1 n - 4n \quad \text{or} \quad B_1 = 3B_1 - 4 \quad (6)$$

Solving for B_1 in Eq. (6), we obtain $B_1 = 2$. Substituting $B_1 = 2$ in Eq. (5), we obtain $B_0 = 3$. So $a_n^* = 2n + 3$ is a particular solution of $a_n = 3a_{n-1} - 4n$.

Next for $a_n = 3a_{n-1} + 3 \times 2^n$, we try $a_n^+ = B2^n$, obtaining

$$B2^n = a_n^+ = 3a_{n-1}^+ + 3 \times 2^n = 3(B2^{n-1}) + 3 \times 2^n \quad (7)$$

Dividing both sides of (7) by 2^{n-1}, we get $2B = 3B + 6$, or $B = -6$. So $a_n^+ = -6 \times 2^n$ is a particular solution of $a_n = 3a_{n-1} + 3 \times 2^n$. Combining our particular solutions with the general homogeneous solution, we obtain the general inhomogeneous solution

$$a_n = A3^n + 2n + 3 - 6 \times 2^n$$

When $a_1 = 8$, we can determine A:

$$8 = a_1 = A3^1 + 2(1) + 3 - 6 \times 2^1$$
$$= 3A - 7$$

Thus $A = 5$, and the solution is $5 \times 3^n + 2n + 3 - 6 \times 2^n$. ■

7.4 EXERCISES

1. Solve the following recurrence relations:
   (a) $a_n = a_{n-1} + 3(n-1)$, $a_0 = 1$
   (b) $a_n = a_{n-1} + n(n-1)$, $a_0 = 3$
   (c) $a_n = a_{n-1} + 3n^2$, $a_0 = 10$
2. Find and solve a recurrence relation for the number of infinite regions formed by n infinite lines drawn in the plane so that each pair of lines intersects at a different point.
3. Find and solve a recurrence relation for the number of different square subboards of any size that can be drawn on an $n \times n$ chessboard.

4. Find and solve a recurrence relation for the number of different regions formed when n mutually intersecting planes are drawn in three-dimensional space such that no four planes intersect at a common point and no two planes have parallel intersection lines in a third plane. [*Hint:* Reduce to a two-dimensional problem (Example 1).]

5. Find and solve a recurrence relation for the number of regions into which a convex n-gon is divided by all its diagonals, assuming no three diagonals intersect at a common point. (*Hint:* Sum the inhomogeneous term using a special case of an identity from Section 5.5.)

6. If the average of two successive years' production $\frac{1}{2}(a_n + a_{n-1})$ is $2n + 5$ and $a_0 = 3$, find a_n.

7. Solve the recurrence relation $a_n = 1.08a_{n-1} + 100$, $a_0 = 0$, from part (b) of Example 5 in Section 7.1.

8. Suppose a savings account earns 5 percent a year. Initially there is \$1000 in the account, and in year k, \$$10k$ are withdrawn. How much money is in the account at the end of n years if:

(a) Annual withdrawal is at year's end?

(b) Withdrawal is at start of year?

9. Solve the following recurrence relations:

(a) $a_n = 3a_{n-1} - 2$, $a_0 = 0$

(b) $a_n = 2a_{n-1} + (-1)^n$, $a_0 = 2$

(c) $a_n = 2a_{n-1} + n$, $a_0 = 1$

(d) $a_n = 2a_{n-1} + 2n^2$, $a_0 = 3$

10. Solve the recurrence relation $a_n = 3a_{n-1} + n^2 - 3$, with $a_0 = 1$.

11. Solve the recurrence relation $a_n = 3a_{n-1} - 2a_{n-2} + 3$, $a_0 = a_1 = 1$.

12. Find and solve a recurrence relation for the number of n-digit ternary sequences in which no 1 appears to the right of any 2.

13. Find and solve a recurrence relation for the earnings of a company when the *rate of increase* of earnings increases by $\$10 \times 2^k$ in the kth year from the previous year, where $a_0 = 20$ and $a_1 = 1020$.

14. Show that the general solution to any inhomogeneous linear recurrence relation is the general solution to the associated homogeneous relation plus one particular inhomogenous solution.

15. Show that the form of the particular solution of Eq. (1) given in the table in this section is correct for:

(a) $f(n) = d$

(b) $f(n) = dn$

(c) $f(n) = dn^2$

(d) $f(n) = d^n$

16. Show that if $f(n)$ in Eq. (1) is the sum of several different terms, a particular solution for this $f(n)$ may be obtained by summing particular solutions for the individual terms.

17. Find a general solution to $a_n - 5a_{n-1} + 6a_{n-2} = 2 + 3n$.

18. If the recurrence relation $a_n - c_1 a_{n-1} + c_2 a_{n-2} = c_3 n + c_4$ has a general solution $a_n = A_1 2^n + A_2 5^n + 3n - 5$, find c_1, c_2, c_3, c_4.

19. Solve the following recurrence relations when $a_0 = 1$:

(a) $a_n^2 = 2a_{n-1}^2 + 1$ (*Hint:* Let $b_n = a_n^2$.)

(b) $a_n = -na_{n-1} + n!$ [*Hint:* Define an appropriate b_n as in part (a).]

7.5 SOLUTIONS WITH GENERATING FUNCTIONS

Most recurrence relations for a_n can be converted into an equation involving the generating function $g(x) = a_0 + a_1 x + \cdots + a_n x^n + \cdots$. This associated functional equation for $g(x)$ can often be solved algebraically and the resulting expression for $g(x)$ expanded in a power series to obtain a_n as the coefficient of x^n. Some of the algebraic manipulations of the functional equations may be new to the reader.

We will treat $g(x)$ as if it were a standard single variable, such as y, and treat other functions of x as constants. For example, the functional equation $g(x) = x^2 g(x) - 2x$ can be solved by rewriting the equation as $g(x)(1 - x^2) = -2x$ and hence $g(x) = -2x(1 - x^2)^{-1}$. Similarly, the functional equation

$$(1 - x^2)[g(x)]^2 - 4xg(x) + 4x^2 = 0 \tag{1}$$

can be solved by the quadratic formula that we normally apply to equations such as $ay^2 + by + c = 0$. Now $a = (1 - x^2)$, $b = -4x$, and $c = 4x^2$. Intuitively, for each particular value of x, $g(x)$ is the solution of Eq. (1). Thus, by the quadratic formula, the solution to Eq. (1) is

$$g(x) = \frac{1}{2(1 - x^2)}[4x \pm \sqrt{16x^2 - 16x^2(1 - x^2)}] = \frac{1}{2(1 - x^2)}(4x \pm 4x^2)$$

So $g(x) = 2(x + x^2)/(1 - x^2)$ or $2(x - x^2)/(1 - x^2)$. If there are two (or more) possible solutions, only one will normally make sense as a generating function for a_n (e.g., have a power series expansion with the correct value for the initial condition a_0).

Now let us show by example how the various recurrence relations obtained in Section 7.1 can be converted into functional equations for an associated generating function.

Example 1: Summation Recurrence

Find a functional equation for $g(x) = a_0 + a_1 x + \cdots + a_n x^n + \cdots$ where a_n satisfies the recurrence relation $a_n = a_{n-1} + n$, when $n \geq 1$, obtained in Example 3 of Section 7.1. The initial condition was $a_0 = 1$. Solve the functional equation and expand $g(x)$ to find a_n.

Using this recurrence relation for every term in $g(x)$ except a_0, we have $a_n x^n = a_{n-1}x^n + nx^n$, $n \geq 1$. Summing the terms, we can write

$$g(x) - a_0 = \sum_{n=1}^{\infty} a_n x^n = \sum_{n=1}^{\infty}(a_{n-1}x^n + nx^n) \tag{2}$$

$$= x\sum_{n=1}^{\infty} a_{n-1}x^{n-1} + \sum_{n=1}^{\infty} nx^n \tag{3}$$

$$= x\sum_{m=0}^{\infty} a_m x^m + \sum_{n=0}^{\infty} \binom{n}{1} x^n \tag{4}$$

$$= xg(x) + \frac{x}{(1-x)^2} \tag{5}$$

Line (3) is obtained from line (2) by breaking up the sum of the two x^n terms into sums of each term, and by rewriting $a_{n-1}x^n$ as $xa_{n-1}x^{n-1}$ (in order to make the power of x correspond with the subscript of a_{n-1}). Line (4) is obtained from line (3) by reindexing the first sum with $m = n - 1$, and by adding the "phantom" (zero) term $0x^0$ to the second sum and rewriting n as $C(n, 1)$. The first series is the generating function $g(x)$ multiplied by x. The second series has a generating function obtained by the construction presented in Section 6.5. Equating line (5) with $g(x) - a_0$ [the left side of line (2)] and setting $a_0 = 1$, we have the required functional equation for $g(x)$:

$$g(x) - 1 = xg(x) + \frac{x}{(1-x)^2} \tag{6}$$

Solving for $g(x)$, we rewrite Eq. (6) as

$$g(x) - xg(x) = 1 + \frac{x}{(1-x^2)} \quad \text{or} \quad g(x)(1-x) = 1 + \frac{x}{(1-x)^2}$$

Thus

$$g(x) = \frac{1}{(1-x)} + \frac{x}{(1-x)^3}$$

The coefficient of x^n in $(1-x)^{-1}$ is just 1 and in $x(1-x)^{-3}$ is $C((n-1)+3-1, n-1) = C(n+1, n-1) = C(n+1, 2)$. Then

$$g(x) = \frac{1}{(1-x)} + \frac{x}{(1-x)^3} = \sum_{n=0}^{\infty} x^n + \sum_{n=0}^{\infty} \binom{n+1}{2} x^n = \sum_{n=0}^{\infty} \left[1 + \binom{n+1}{2}\right] x^n$$

and so $a_n = 1 + C(n+1, 2)$—the same answer as obtained for this recurrence relation in Example 1 of Section 7.4. ■

Example 2: Fibonacci Relation

Use generating functions to solve the recurrence relation $a_n = a_{n-1} + a_{n-2}$, with $a_1 = 1$, $a_2 = 2$ obtained in Example 2 of Section 7.1.

The initial conditions, $a_1 = 1, a_2 = 2$, are equivalent to $a_0 = 1, a_1 = 1$. Then using the same power series summation approach as in the previous example, we obtain

$$\begin{aligned} g(x) - a_0 - a_1 x = \sum_{n=2}^{\infty} a_n x^n &= \sum_{n=2}^{\infty} (a_{n-1} x^n + a_{n-2} x^n) \\ &= x \sum_{n=2}^{\infty} a_{n-1} x^{n-1} + x^2 \sum_{n=2}^{\infty} a_{n-2} x^{n-2} \\ &= x \sum_{m=1}^{\infty} a_m x^m + x^2 \sum_{k=0}^{\infty} a_k x^k \\ &= x[g(x) - a_0] + x^2 g(x) \end{aligned}$$

Setting $a_0 = 1$ and $a_1 = 1$, we have the functional equation

$$g(x) - 1 - x = x[g(x) - 1] + x^2 g(x) \quad \text{or} \quad g(x) - xg(x) - x^2 g(x) - 1 = 0$$

and so

$$g(x)(1 - x - x^2) = 1 \quad \text{or} \quad g(x) = 1/(1 - x - x^2)$$

Observe that this denominator is closely related to the characteristic equation of this recurrence relation, given in Example 3 of Section 7.3. *In general, $g(x)$ will have a denominator $1 + c_1 x + c_2 x^2 + \cdots + c_r x^r$ if and only if $x^r + c^{r-1} x^{r-1} + c_2 x^{r-2} + \cdots + c_r$ is the characteristic equation of the associated recurrence relation, and so $1 - \alpha x$ is a factor of the denominator of g(x) if and only if α is a root of the characteristic equation.* The numerator will depend on the initial conditions.

By the quadratic formula, we can factor

$$1 - x - x^2 = [1 - \frac{1}{2}(1 + \sqrt{5})x][1 - \frac{1}{2}(1 - \sqrt{5})x]$$

For simplicity, let us define $\alpha_1 = \frac{1}{2}(1 + \sqrt{5})$ and $\alpha_2 = \frac{1}{2}(1 - \sqrt{5})$. Then we have

$$g(x) = \frac{1}{(1 - \alpha_1 x)(1 - \alpha_2 x)} = \frac{\alpha_1/\sqrt{5}}{1 - \alpha_1 x} - \frac{\alpha_2/\sqrt{5}}{1 - \alpha_2 x} \tag{7}$$

The decomposition of $g(x)$ into two fractions in Eq. (7) is obtained by the method of partial fractions, the reverse process of combining two fractions into a common fraction. Setting $y = \alpha_1 x$ in the first fraction on the right side in

Eq. (7), we have

$$\frac{\alpha_1/\sqrt{5}}{1 - \alpha_1 x} = \frac{\alpha_1}{\sqrt{5}} \left(\frac{1}{1 - y} \right) = \frac{\alpha_1}{\sqrt{5}} \sum_{n=0}^{\infty} y^n = \frac{\alpha_1}{\sqrt{5}} \sum_{n=0}^{\infty} \alpha_1^n x^n$$

The same type of expansion exists for $y = \alpha_2 x$. Then a_n, the coefficient of x^n in the power series expansion of $g(x)$, is

$$\begin{aligned} a_n &= \frac{1}{\sqrt{5}}\alpha_1^{n+1} - \frac{1}{\sqrt{5}}\alpha_2^{n+1} \\ &= \frac{1}{\sqrt{5}}\left(\frac{1+\sqrt{5}}{2}\right)^{n+1} - \frac{1}{\sqrt{5}}\left(\frac{1-\sqrt{5}}{2}\right)^{n+1} \end{aligned}$$ ■

Observe that it is much easier to determine a_{10} by recursively computing $a_2, a_3, \ldots$ up to a_{10} using the Fibonacci relation than by setting $n = 10$ in the foregoing formula for a_n.

Example 3: Selection Without Repetition

Let $g_n(x)$ be a family of generating functions $g_n(x) = a_{n,0} + a_{n,1}x + \cdots + a_{n,k}x^k + \cdots + a_{n,n}x^n$ satisfying the relation $a_{n,k} = a_{n-1,k} + a_{n-1,k-1}$, with $a_{n,0} = a_{n,n} = 1$ (and $a_{n,k} = 0$, $k > n$) obtained in Example 9 of Section 7.1 for $a_{n,k}$, the number of k-subsets of an n-set. Find a functional relation among the $g_n(x)$s and solve it to obtain a formula for $a_{n,k}$. Using the power series summation method, we obtain

$$\begin{aligned} g_n(x) - 1 = \sum_{k=1}^{n} a_{n,k}x^k &= \sum_{k=1}^{n}(a_{n-1,k}x^k + a_{n-1,k-1}x^k) \\ &= \sum_{k=1}^{n} a_{n-1,k}x^k + x\sum_{h=0}^{n-1} a_{n-1,h}x^h \\ &= g_{n-1}(x) - 1 + xg_{n-1}(x) \end{aligned}$$

Thus

$$g_n(x) = g_{n-1}(x) + xg_{n-1}(x) = (1+x)g_{n-1}(x)$$

The resulting recurrence $g_n(x) = (1+x)g_{n-1}(x)$ is solved just like the recurrence $a_n = ca_{n-1}$. The solution is

$$g_n(x) = (1+x)^n g_0(x) = (1+x)^n$$

since $g_0(x) = a_{0,0} = 1$. Now by the binomial theorem, we have $a_{n,k} = C(n, k)$. ■

Optional: The rest of this section involves more complicated computations. The reader may skip this material. First we consider a nonlinear recurrence relation and solve it using generating functions.

Example 4: Placing Parentheses

Solve the recurrence relation $a_n = a_1a_{n-1} + a_2a_{n-2} + \cdots + a_ia_{n-i} + \cdots + a_{n-1}a_1$ obtained in Example 11 of Section 7.1 for the number of ways to place parentheses when multiplying n numbers.

Observe that if $g(x) = a_0 + a_1x + \cdots + a_nx^n + \cdots$, then the right-hand side of this equation is simply the coefficient of x^n, for $n \geq 2$, in the product $g(x)g(x) =$

$(0+a_1x+\cdots+a_nx^n+\cdots)^2$. Using the power series summation method, we have (recall that $a_0=0$ and $a_1=1$)

$$g(x)-1x=\sum_{n=2}^{\infty}a_nx^n=\sum_{n=2}^{\infty}(a_1a_{n-1}+a_2a_{n-2}+\cdots+a_{n-1}a_1)x^n=[g(x)]^2$$

Solving this quadratic equation in $g(x)$ as described at the start of this section, we obtain $g(x)=\frac{1}{2}(1\pm\sqrt{1-4x})$. To make $a_0=0$ [i.e., $g(0)=0$], we want the solution $g(x)=\frac{1}{2}-\frac{1}{2}\sqrt{1-4x}$.

This $g(x)$ requires a new type of generating function expansion called the *generalized binomial theorem*. The power series expansion $(1+y)^q=\binom{q}{0}+\binom{q}{1}x+\binom{q}{2}x^2+\cdots+\binom{q}{n}x^n+\cdots$, where q is any real number, has a coefficient $\binom{q}{n}$ of x^n defined as

$$\binom{q}{n}=\frac{q(q-1)(q-2)\times\cdots\times[q-(n-1)]}{n!} \tag{8}$$

[The formula for this generalized coefficient $\binom{q}{n}$ arises from the Taylor series for $(1+y)^q$; see any standard calculus text.]

Using Eq. (8), the coefficient of x^n in $\sqrt{1-4x}$ is [we think of $\sqrt{1-4x}$ as $(1+y)^{1/2}$, where $y=-4x$]:

$$\begin{aligned}\binom{\frac{1}{2}}{n}(-4)^n &= \frac{\frac{1}{2}(-\frac{1}{2})(-\frac{3}{2})\times\cdots\times[\frac{1}{2}(2n-3)]}{n!}(-4)^n\\ &= \frac{-1\times1\times3\times5\times\cdots\times(2n-3)}{n!}2^n\\ &= -\frac{2}{n}\binom{2n-2}{n-1}\end{aligned} \tag{9}$$

The last step in Eq. (9) is obtained by multiplying certain numbers in the numerator by appropriately selected powers of 2 (details are left to Exercise 6). Multiplying the final expression in Eq. (9) by $-\frac{1}{2}$, we obtain the coefficient of x^n in $-\frac{1}{2}\sqrt{1-4x}$,

$$a_n=\frac{1}{n}\binom{2n-2}{n-1}\quad n\geq 1\ \blacksquare$$

The expression $\frac{1}{n}C(2n-2,n-1)$ arises in various combinatorial settings and is called the nth *Catalan number*. We note, as an aside, that while where parentheses are placed makes no real difference when multiplying numbers, if we were working with a complex product of matrices then the placement of parentheses has an important impact on the amount of computation required. Next let us consider generating functions for simultaneous recurrence relations.

Example 5: Simultaneous Recurrence Relations

Use generating functions to solve the set of simultaneous recurrence relations obtained in Example 11 of Section 7.1.

$$a_n=a_{n-1}+b_{n-1}+c_{n-1},\qquad b_n=3^{n-1}-c_{n-1},\qquad c_n=3^{n-1}-b_{n-1},$$
$$a_1=b_1=c_1=1$$

Let $A(x)$, $B(x)$, and $C(x)$ be the generating functions for a_n, b_n, and c_n, respectively. We use the power series summation method to obtain

$$\begin{aligned}
A(x) - a_0 = \sum_{n=1}^{\infty} a_n x^n &= \sum_{n=1}^{\infty} a_{n-1} x^n + \sum_{n=1}^{\infty} b_{n-1} x^n + \sum_{n=1}^{\infty} c_{n-1} x^n \\
&= x \sum_{m=0}^{\infty} a_m x^{m+} x \sum_{m=0}^{\infty} b_m x^m + x \sum_{m=0}^{\infty} c_m x^m \\
&= xA(x) + xB(x) + xC(x) \\
B(x) - b_0 = \sum_{n=1}^{\infty} b_n x^n &= \sum_{n=1}^{\infty} 3^{n-1} x^n - \sum_{n=1}^{\infty} c_{n-1} x^n \\
&= x(1-3x)^{-1} - xC(x) \\
C(x) - c_0 = \sum_{n=1}^{\infty} c_n x^n &= \sum_{n=1}^{\infty} 3^{n-1} x^n - \sum_{n=1}^{\infty} b_{n-1} x^n \\
&= x(1-3x)^{-1} - xB(x)
\end{aligned}$$

It is always desirable to state initial conditions in terms of a_0, b_0, c_0. Solving our three recurrence relations for a_0, b_0, c_0 given $a_1 = b_1 = c_1 = 1$, we get $1 = b_1 = 3^0 - c_0$ or $c_0 = 0$. Similarly we find that $b_0 = 0$ and $a_0 = 1$. Then our functional equations are

$$A(x) - 1 = xA(x) + xB(x) + xC(x)$$

or

$$A(x) = \frac{1}{1-x}[xB(x) + xC(x) + 1] \tag{10}$$

and

$$B(x) = \frac{x}{1-3x} - xC(x) \tag{11}$$

$$C(x) = \frac{x}{1-3x} - xB(x) \tag{12}$$

We can solve Eqs. (11) and (12) for $B(x)$ and $C(x)$ simultaneously. Multiplying Eq. (12) by x and using this expression for $xC(x)$ in Eq. (11), we have

$$\begin{aligned}
B(x) &= \frac{x}{1-3x} - xC(x) = \frac{x}{1-3x} - x\left[\frac{x}{1-3x} - xB(x)\right] \\
\Rightarrow B(x)(1-x^2) &= \frac{x-x^2}{1-3x} \\
\Rightarrow B(x) &= \frac{(1-x)x}{(1-x^2)(1-3x)} = \frac{x}{(1+x)(1-3x)} = \frac{\frac{1}{4}}{1-3x} - \frac{\frac{1}{4}}{1+x}
\end{aligned}$$

The last step is again a partial fraction decomposition. The coefficient of x^n in $\frac{1}{4}(1-3x)^{-1}$ is $\frac{1}{4}3^n$ and in $-\frac{1}{4}(1+x)^{-1}$ is $-\frac{1}{4}(-1)^n$. So b_n, the coefficient of x^n in $B(x)$, is $\frac{1}{4}[3^n - (-1)^n]$. Equations (11) and (12) are symmetric with respect to $B(x)$ and $C(x)$, and so $C(x) = B(x)$ and $c_n = b_n = \frac{1}{4}[3^n - (-1)^n]$.

Next we solve for $A(x)$ in Eq. (10):

$$A(x) = \frac{1}{1-x}[xB(x) + xC(x) + 1] = \frac{2x}{1-x}B(x) + \frac{1}{1-x}$$

$$[\text{since} B(x) = C(x)]$$

$$= \frac{2x}{1-x}\left(\frac{\frac{1}{4}}{1-3x} - \frac{\frac{1}{4}}{1+x}\right) + \frac{1}{1-x}$$

$$= \frac{\frac{1}{2}x}{(1-x)(1-3x)} - \frac{\frac{1}{2}x}{1-x^2} + \frac{1}{1-x}$$

$$= \left(\frac{\frac{1}{4}}{1-3x} - \frac{\frac{1}{4}}{1-x}\right) - \frac{\frac{1}{2}x}{1-x^2} + \frac{1}{1-x}$$

The coefficient of x^n in $\frac{1}{4}(1-3x)^{-1}$ is $\frac{1}{4}3^n$, in $-\frac{1}{4}(1-x)^{-1}$ is $-\frac{1}{4}$, in $-\frac{x}{2}(1-x^2)^{-1}$ is $-\frac{1}{2}$, n odd, or 0, n even, and in $(1-x)^{-1}$ is 1. Collecting these terms, we get $a_n = \frac{1}{4}(3^n + 3)$, n even, and $= \frac{1}{4}(3^n + 1)$, n odd. ■

7.5 EXERCISES

1. Find functional equations for the generating functions whose coefficients satisfy the relations:
   (a) $a_n = a_{n-1} + 2,\ a_0 = 1$
   (b) $a_n = 3a_{n-1} - 2a_{n-2} + 2,\ a_0 = a_1 = 1$
   (c) $a_n = a_{n-1} + n(n-1),\ a_0 = 1$
   (d) $a_n = 2a_{n-1} + 2^n,\ a_0 = 1$
2. Solve the recurrence relations in Exercise 1 using generating functions.
3. Find functional equations for the generating functions whose coefficients satisfy the relations:
   (a) $a_n = \sum_{i=0}^{n-1} a_i a_{n-1-i}\ (n \geq 1),\ a_0 = 1$
   (b) $a_n = \sum_{i=2}^{n-2} a_i a_{n-i}\ (n \geq 3),\ a_0 = a_1 = a_2 = 1$
   (c) $a_n = \sum_{i=1}^{n-1} 2^i a_{n-i}\ (n \geq 2),\ a_0 = a_1 = 1$
4. Find a functional equation and solve it for the sequence of generating functions $F_n(x) = \sum_{k=0}^{n} a_{n,k}x^k$ whose coefficients satisfy (assume $F_0 = 1$ and $a_{n,0} = 1$):
   (a) $a_{n,k} = a_{n,k-1} - 2a_{n-1,k-1}$
   (b) $a_{n,k} = 2a_{n-1,k} - 3a_{n,k-1}$

5. Verify the form of particular solutions to inhomogeneous recurrence relations in the table in Section 7.4.

6. Verify in Eq. (9) that

$$\frac{-1 \times 1 \times 3 \times 5 \times \cdots \times (2n-3)}{n!} = -\frac{2}{n}\binom{2n-2}{n-1}$$

7. Find a recurrence relation and solve it with generating functions for the number of ways to divide an n-gon into triangles with noncrossing diagonals. (*Hint:* Use reasoning similar to Example 4.)

8. Find a recurrence relation and associated generating function for the number of n-digit ternary sequences that have the pattern "012" occurring for the first time at the end of the sequence.

9. Find a recurrence relation and associated generating function for the number of different binary trees with n leaves.

10. Find a recurrence relation for $a_{n,k}$, the number of k-subsets of an n set with repetition. Find an equation for $F_n(x) = \sum_{k=0}^{\infty} a_{n,k}x^k$, and solve for $F_n(x)$ and $a_{n,k}$.

11. Let $a_{n,k}$ be the probability that k successes occur in an experiment with n trials if each trial has probability p of success. Find a recurrence relation for $a_{n,k}$. Use this relation to find and solve an equation for $F_n(x) = \sum_{k=0}^{n} a_{n,k}x^k$.

12. **(a)** Find a recurrence relation for $a_{n,k}$, the number of k-permutations of n elements.

(b) Show that $F_n(x) = \sum_{k=0}^{n} a_{n,k}x^k$ satisfies the differential equation [$F'_{n-1}(x)$ denotes the derivative of $F_{n-1}(x)$]

$$F_n(x) = (1+x)F_{n-1}(x) + x^2F'_{n-1}(x)$$

(c) Find a functional equation for $G_n(x) = \sum_{k=0}^{n} a_{n,k}x^k/k!$

13. Find and solve a system of recurrence relations allowing one to determine the number of n-digit quaternary sequences with an odd number of 1s and an odd number of 2s.

14. Find and solve simultaneous recurrence relations for determining the number of n-digit ternary sequences whose sum of digits is a multiple of 3.

15. **(a)** Find a recurrence relation for a_n, the number of ways to partition n distinct objects among n indistinguishable boxes (some boxes may be empty).

(b) Let $g(x) = \sum_{n=0}^{\infty} a_nx^n/n!$ where $a_0 = 1$. Show that $g(x)$ satisfies the differential equation $g'(x) = g(x)e^x$. Solve this equation for $g(x)$.

16. **(a)** Define $s_{n,r}$ as numbers such that $\sum_{r=0}^{n} s_{n,r}x^r = x(x-1)(x-2)\cdots(x-r+1)$. Find a recurrence relation for $s_{n,r}$.

(b) Find a differential equation for $F_n(x) = \sum_{r=0}^{n} s_{n,r}x^r/r!$

7.6 SUMMARY AND REFERENCES

In this chapter we saw that recurrence relations are one of the simplest ways to solve counting problems. Without fully understanding the combinatorial process, as was required in Chapter 5, we now need only express a given problem for n objects in terms of the problem posed for fewer numbers of objects. Once a recurrence relation has been found, then starting with a_1 (the solution for one object), we can successively compute the solutions for 2, 3, ... up to any (moderate) value of n. Or we can try one of the techniques in the later sections of this chapter to solve the recurrence relation explicitly.

The first recurrence relation in mathematical writings was the Fibonacci relation. In his work *Liber abaci*, published in 1220, Leonardo di Pisa, known also as Fibonacci, posed a counting problem about the growth of a rabbit population (Exercise 7 in Section 7.1). The number a_n of rabbits after n months was shown to satisfy the Fibonacci relation $a_n = a_{n-1} + a_{n-2}$. As mentioned in Section 6.6, DeMoivre gave the first solution of this relation 500 years later in 1730 using the generation function derivation given in Example 2 of Section 7.5. The Fibonacci relation and numbers have proven to be amazingly ubiquitous. For example, it has been shown that the ratios of Fibonacci numbers provide an optimal way (in a certain sense) to divide up an interval when searching for the minimum of a function in this interval (see Kiefer [3]). The appearance of Fibonacci numbers in rings of leaves around flowers is discussed in Adler [1].

The methods for solving recurrence relations appeared originally in the development of the theory of difference equations, cousins of differential equations. For a good presentation of the methods and applications of difference equations, see Sandefur [4].

1. I. Adler, "The consequence of constant pressure in phyllotaxis," *J. Theoretical Biology* 65 (1977), 29–77.
2. T. Cormen, C. Leiserson, and R. Rivest, *An Introduction to Algorithms,* MIT Press, Cambridge, MA 1990.
3. J. Kiefer, "Sequential minimax search for a maximum," *Proceedings of American Math. Society* 4 (1953), 502–506.
4. J. Sandefur, *Discrete Dynamical Modeling,* Oxford University Press, New York, 1993.

CHAPTER 8
INCLUSION–EXCLUSION

8.1 COUNTING WITH VENN DIAGRAMS

In this chapter we develop a set-theoretic formula for counting problems involving several interacting properties in which either all properties must hold or none must hold. An example is counting all five-card hands with at least one card in each suit, or equivalently, all five-card hands with no void in any suit. In the process of solving such problems, we have to count the subsets of outcomes in which various combinations of the properties hold. We use Venn diagrams to depict these different combinations. See Appendix A.1 for a review of the essentials of sets and Venn diagrams.

Let us begin with a one-property Venn diagram, and then progress to two- and three-property problems. In Figure 8.1 we show a set A within a universe $\mathcal{U}$. The complementary set $\overline{A}$ consists of all elements of $\mathcal{U}$ not in A. Let $N(S)$ denote the number of elements in set S. We define $N = N(\mathcal{U})$. Then $N(A) = N - N(\overline{A})$, or $N(\overline{A}) = N - N(A)$. This is similar to the situation in probability where the probability of an event E is one minus the probability of the complementary event $\overline{E}$.

Suppose, for example, that there is a "universe" of 100 students in a math course and there are 30 students who are not math majors in the course, that is, $N(\overline{A}) = 30$, where A is the set of math majors. Then the number $N(A)$ of math majors in the course is $N(A) = N - N(\overline{A}) = 100 - 30 = 70$.

Next consider a problem with two sets. Let the universe $\mathcal{U}$ be all students in a school, let F be the set of students taking French, and let L be the set of students taking Latin. See Figure 8.2. We want formulas for the number of students taking French or Latin $N(F \cup L)$ and the number taking neither language $N(\overline{F} \cap \overline{L})$ in terms of N, $N(F)$, $N(L)$, and $N(F \cap L)$. Note that $N(F \cup L)$ is not simply $N(F) + N(L)$, because $N(F) + N(L)$ counts each student taking both languages two times. Thus, we must know how many students take both languages, the number $N(F \cap L)$. Subtracting $N(F \cap L)$ from $N(F) + N(L)$ corrects the double counting of students taking two languages. That is,

$$N(F \cup L) = N(F) + N(L) - N(F \cap L) \tag{1}$$

By de Morgan's Law (Eq. BA3 of Appendix A.1), $\overline{F} \cap \overline{L} = \overline{F \cup L}$, and so

$$N(\overline{F} \cap \overline{L}) = N(\overline{F \cup L}) = N - N(F \cup L)$$

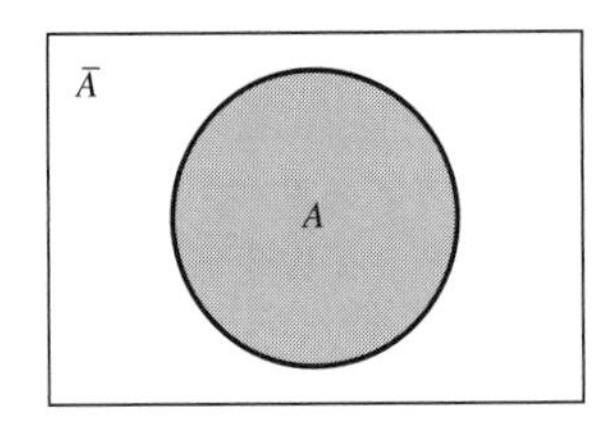

Figure 8.1

Combining this equation with Eq. (1), we have

$$N(\overline{F} \cap \overline{L}) = N - N(F \cup L) = N - N(F) - N(L) + N(F \cap L) \tag{2}$$

Example 1: Students Taking Neither Language

If a school has 100 students with 50 students taking French, 40 students taking Latin, and 20 students taking both languages, how many students take no language?

In this problem, $N = 100$, $N(F) = 50$, $N(L) = 40$, and $N(F \cap L) = 20$. We need to determine $N(\overline{F} \cap \overline{L})$. By Eq. (2), we have

$$N(\overline{F} \cap \overline{L}) = N - N(F) - N(L) + N(F \cap L) = 100 - 50 - 40 + 20 = 30 \quad \blacksquare$$

The next example applies this set-theoretic formula to a counting problem that has no obvious set-theoretic structure.

Example 2: Restricted Arrangements

How many arrangements of the digits $0, 1, 2, \ldots, 9$ are there in which the first digit is greater than 1 and the last digit is less than 8?

Before using formula (2) to solve this problem, let us consider a more direct approach and see why it fails. For the first digit, there are 8 choices. Then for the last digit, there are ... the number of choices depends on whether or not an 8 or 9 was chosen for the first digit. If an 8 or 9 were chosen for the first digit, there will be 8 choices for the last digit, while if neither 8 nor 9 were chosen for the first digit, there will be 7 choices for the last digit. Because of this difficulty, we shall solve the problem with formula (2).

Let $\mathcal{U}$ be all arrangements of $0, 1, 2, \ldots, 9$. Let F be the set of all arrangements with a 0 or 1 in the first digit, and let L be the set of all arrangements with an 8 or 9 in the last digit. Then the number of arrangements with first digit greater than 1 and the last digit less than 8 is $N(\overline{F} \cap \overline{L})$.

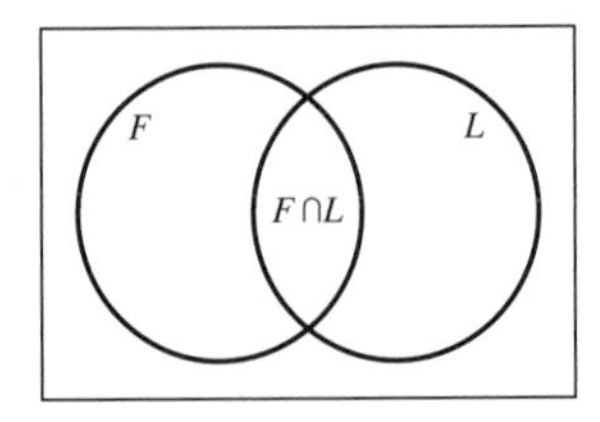

Figure 8.2

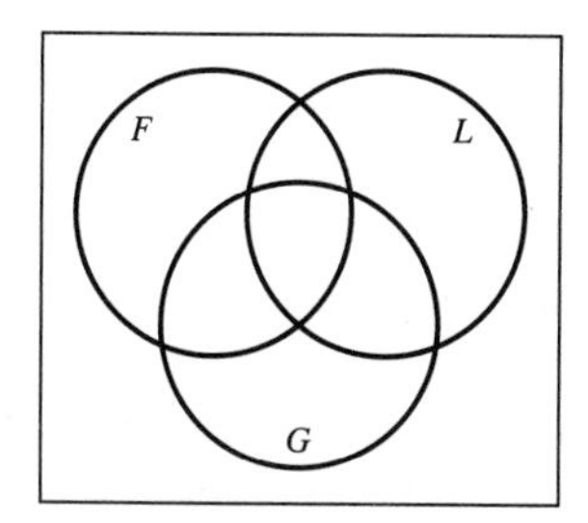

Figure 8.3

We have $N = 10!$, $N(F) = 2 \times 9!$ (two choices for the first digit followed by any arrangement for the remaining 9 digits), and $N(L) = 2 \times 9!$ Similarly, $N(F \cap L) = 2 \times 2 \times 8!$. Then by Eq. (2),

$$N(\overline{F} \cap \overline{L}) = 10! - (2 \times 9!) - (2 \times 9!) + (2 \times 2 \times 8!) \quad \blacksquare$$

Consider next a problem with three sets. We extend Figure 8.2 with the additional set G of students taking German, as shown in Figure 8.3. We want a formula for $N(\overline{F} \cap \overline{L} \cap \overline{G})$. A first guess might be

$$N(\overline{F} \cap \overline{L} \cap \overline{G}) \stackrel{?}{=} N - N(F) - N(L) - N(G)$$

As in Figure 8.2, this formula double counts (that is, subtracts twice) the students in two of the sets, F, L, and G. We can correct this first formula by adding the number of students taking two languages. Thus we propose the formula

$$\begin{aligned} N(\overline{F} \cap \overline{L} \cap \overline{G}) \stackrel{?}{=}\ & N - N(F) - N(L) - N(G) \\ & + N(F \cap L) + N(L \cap G) + N(F \cap G) \end{aligned} \qquad (3)$$

Figure 8.4 shows how many times a student will be added and subtracted in parts of this formula. A student taking no language is counted once (by the term N)—such students are exactly the ones we want to count. The challenge is to make sure that all other students are counted a net of 0 times. The students taking one language are counted once by N and subtracted once by the term $-[N(F) + N(L) + N(G)]$, for a net count of 0. The students taking two languages are counted once by N, subtracted

Number of Languages Taken by Student	N	$-[N(F) + N(L) + N(G)]$	$+[N(F \cap L) + N(L \cap G) + N(F \cap G)]$	$-N(F \cap L \cap G)$
0	+1	0	0	0
1	+1	−1	0	0
2	+1	−2	+1	0
3	+1	−3	+3	−1

Figure 8.4

twice by $-[N(F)+N(L)+N(G)]$ (since they are in exactly two of the three sets), and then added once by the term $+[N(F\cap L)+N(L\cap G)+N(F\cap G)]$ (since they are in exactly one of the three pairwise intersections), for a net count of 0. Finally, we consider the students taking all three languages. They are counted once by N, then subtracted three times by the sum of the three sets (since they are in all three sets), then added three times by the pairwise intersections (since they are in all three of these subsets). This yields a net count of $1-3+3=1$. Then we must correct formula (3) by subtracting $N(F\cap L\cap G)$ to make the net count of students with all three languages 0:

$$N(\overline{F}\cap\overline{L}\cap\overline{G}) = N-[N(F)+N(L)+N(G)] \\ +[N(F\cap L)+N(L\cap G)+N(F\cap G)]-N(F\cap L\cap G) \quad (4)$$

For general sets A_1, A_2, A_3, we rewrite Eq. (4) as

$$N(\overline{A}_1\cap\overline{A}_2\cap\overline{A}_3)=N-\sum_i N(A_i)+\sum_{ij} N(A_i\cap A_j)-N(A_1\cap A_2\cap A_3) \quad (5)$$

where the sums are understood to run over all possible i and all i,j pairs, respectively.

Example 3: Students Taking None of Three Languages

If a school has 100 students with 40 taking French, 40 taking Latin, and 40 taking German, 20 students are taking any given pair of languages, and 10 students are taking all three languages, then how many students are taking no language?

Here $N=100$, $N(F)=N(L)=N(G)=40$, $N(F\cap L)=N(L\cap G)=N(F\cap G)=20$, and $N(F\cap L\cap G)=10$. Then by Eq. (4), the number of students taking no language is $N(\overline{F}\cap\overline{L}\cap\overline{G})=100-(40+40+40)+(20+20+20)-10=30$. ∎

Next we apply Eq. (5) to two counting problems that cannot be solved by methods developed in the three previous chapters.

Example 4: Relatively Prime Numbers

How many positive integers ≤ 70 are relatively prime to 70?

Let $\mathcal{U}$ be the set of integers between 1 and 70. The phrase "relatively prime to 70" means "have no common divisors with 70." The prime divisors of 70 are 2, 5, and 7. Then we want to count the number of integers ≤ 70 that do not have 2 or 5 or 7 as divisors. Let A_1 be the set of integers in $\mathcal{U}$ that are divisible by 2, or equivalently, integers in $\mathcal{U}$ that are multiples of 2; A_2 be integers divisible by 5; and A_3 be integers divisible by 7. Then the number of positive integers ≤ 70 that are relatively prime to 70 equals $N(\overline{A}_1\cap\overline{A}_2\cap\overline{A}_3)$. We find

$$N=70 \quad N(A_1)=70/2=35 \quad N(A_2)=70/5=14 \quad N(A_3)=70/7=10$$

The integers divisible by 2 and 5 are simply the integers divisible by 10. Thus $N(A_1\cap A_2)=70/10=7$. By similar reasoning, $N(A_2\cap A_3)=70/(5\times 7)=2$,

$N(A_1 \cap A_3) = 70/(2 \times 7) = 5$, and $N(A_1 \cap A_2 \cap A_3) = 70/(2 \times 5 \times 7) = 1$. So by Eq. (5):

$$N(\overline{A}_1 \cap \overline{A}_2 \cap \overline{A}_3) = 70 - (35 + 14 + 10) + (7 + 2 + 5) - 1 = 24 \quad \blacksquare$$

Example 5: Ternary Sequences With No Voids

How many n-digit ternary (0, 1, 2) sequences are there with at least one 0, at least one 1, and at least one 2?

Let $\mathcal{U}$ be the set of all n-digit ternary sequences. Formula (5) counts outcomes in which none of a set of properties holds. So we must formulate this problem in terms of outcomes for which none of a set of properties holds. The solution is to define A_i to be the number of n-digit ternary sequences with *no* is. (Note that instead of numbering the sets A_1, A_2, A_3, we are using A_0, A_1, A_2.) Then the number of sequences with at least one of each digit will be $N(\overline{A}_0 \cap \overline{A}_1 \cap \overline{A}_2)$.

The number of n-digit ternary sequences is $N = 3^n$. The number of n-digit ternary sequences with no 0s is simply the number of n-digit sequences of 1s and 2s. Thus, $N(A_0) = 2^n$. Similarly, $N(A_1) = N(A_2) = 2^n$. The only n-digit sequence with no 0s and no 1s is the sequence of all 2s. Then $N(A_0 \cap A_1) = 1$; also $N(A_1 \cap A_2) = N(A_0 \cap A_2) = 1$. Finally, there is no ternary sequence with no 0s and no 1s and no 2s. Then by Eq. (5):

$$\begin{aligned} N(\overline{A}_0 \cap \overline{A}_1 \cap \overline{A}_2) &= 3^n - (2^n + 2^n + 2^n) + (1 + 1 + 1) - 0 \\ &= 3^n - 3 \times 2^n + 3 \quad \blacksquare \end{aligned}$$

Observe that whereas polynomial algebra was used in Chapter 6 to model counting problems and recurrence relations were used in Chapter 7, now we are using a set-theoretic model. This approach does not eliminate combinatorial enumeration as the other models did. We still must solve the subproblems of finding $N(A_i)$, $N(A_i \cap A_j)$, and so forth, but these are much easier problems.

Sometimes non-standard input data for a Venn diagram are given. The following example shows how to determine the sizes of all possible subsets in a Venn diagram in such circumstances.

Example 6: Non-Standard Constraints

Suppose there are 100 students in a school and there are 40 students taking each language, French, Latin, and German. Twenty students are taking only French, 20 only Latin, and 15 only German. In addition, 10 students are taking French and Latin. How many students are taking all three languages? No language?

We draw the Venn diagram for this problem and number each region as shown in Figure 8.5. Let N_i denote the number of students in region i, for $i = 1, 2, \ldots, 8$. Students taking only French are the subset $F \cap \overline{L} \cap \overline{G}$, region 1; so $N_1 = 20$. Similarly, the other information given us says $N_5 = 20$ and $N_7 = 15$. Students taking both French and Latin are the subset $F \cap L = (F \cap L \cap \overline{G}) \cup (F \cap L \cap G)$, regions 2 and 3. So

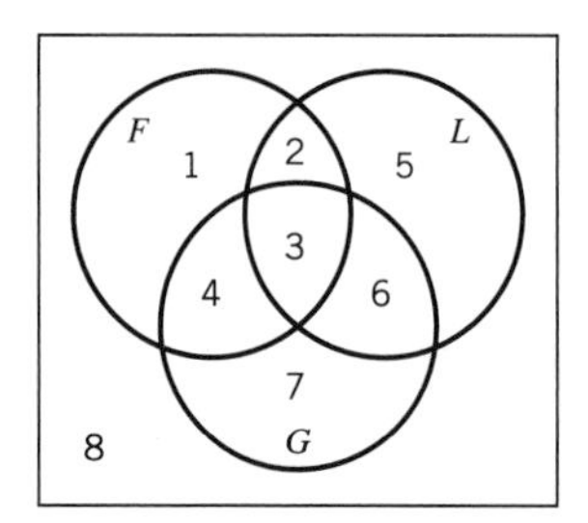

Figure 8.5

$N_2 + N_3 = 10$. The set of students taking French is F, which consists of regions 1, 2, 3, and 4. So

$$40 = N(F) = N_1 + (N_2 + N_3) + N_4 = 20 + (10) + N_4, \qquad \text{or} \qquad N_4 = 10$$

Similarly, set L consists of regions 2, 3, 5, and 6, and so $40 = N(L) = (N_2 + N_3) + N_5 + N_6 = (10) + 20 + N_6$, implying $N_6 = 10$.

Since G consists of regions 3, 4, 6, and 7, and since we were given $N_7 = 15$ and have just found that $N_4 = N_6 = 10$, then

$$40 = N(G) = N_3 + N_4 + N_6 + N_7 = N_3 + 10 + 10 + 15 \qquad \text{or} \qquad N_3 = 5$$

But region 3 is the subset $F \cap L \cap G$ of students taking all three languages. Thus, there are five trilingual students.

The general line of attack in these problems is to break the Venn diagram into the eight regions shown in Figure 8.5. For each subset whose size is given, write that number as a sum of N_i's of regions i in that subset. By combining these equations (and sometimes solving them simultaneously), one can eventually determine the number of elements in each region. Then the size of any subset is readily found as the sum of the sizes of the regions in that subset. For example, in the preceding paragraph we determined all N_is except N_2 and N_8. But $N_2 + N_3 = 10$ and N_3 was found to be 5; thus $N_2 = 5$. $N_8 = N(\overline{F \cap L \cap G})$ is the number of students taking no language. Since all regions total to N, then

$$\begin{aligned} N_8 = N - \sum_{i=1}^{7} N_i &= 100 - (20 + 5 + 5 + 10 + 20 + 10 + 15) \\ &= 100 - 85 = 15 \quad \blacksquare \end{aligned}$$

Note that because there are 8 underlying variables (the sizes of the 8 basic regions), we need to be given 8 pieces of information to be able to solve such a problem.

8.1 EXERCISES

Summary of Exercises The first 28 exercises are similar to the examples in this section. The last five exercises involve more complicated Venn diagram arguments; see the last paragraph in Example 6 for the strategy for solving these problems.

1. How many 5-letter "words" using the 26-letter alphabet (letters can be repeated) either begin or end with a vowel?

2. How many 9-digit Social Security numbers are there with repeated digits?

3. How many n-digit ternary sequences are there in which at least one pair of consecutive digits are the same?

4. What is the probability that at least two heads (not necessarily consecutive) will appear when a coin is flipped 8 times?

5. What is the probability that a 5-card hand has at least one pair (possibly two pairs, three of a kind, full house, or four of a kind)?

6. If n people of different heights are lined up in a queue, what is the probability that at least one person is just behind a taller person?

7. How many 4-digit numbers (including leading 0s) are there with exactly one 8 and no digit appearing exactly two times?

8. Suppose 60 percent of all families own a dishwasher, 30 percent own a trash compacter, and 20 percent own both. What percentage of all families own neither of these two appliances?

9. Among 600 families, 100 families have no children, 200 have only boys, and 200 have only girls. How many families have boy(s) and girl(s)?

10. How many arrangements of the digits $0, 1, \ldots, 9$ are there that do not end with an 8 and do not begin with a 3?

11. Suppose a bookcase has 200 books, 70 in French and 100 about mathematics. How many non-French books not about mathematics are there if:

(a) There are 30 French mathematics books?

(b) There are 60 French nonmathematics books?

12. How many arrangements of the 26 different letters are there that:

(a) Contain either the sequence "the" or the sequence "aid"?

(b) Contain neither the sequence "the" nor the sequence "math"?

13. A school has 200 students with 80 students taking each of the three subjects: trigonometry, probability, and basket-weaving. There are 30 students taking any given pair of these subjects, and 15 students taking all three subjects.

(a) How many students are taking none of these three subjects?

(b) How many students are taking only probability?

14. Suppose 60 percent of all college professors like tennis, 65 percent like bridge, and 50 percent like chess; 45 percent like any given pair of recreations.

(a) Should you be suspicious if told 20 percent like all three recreations?

(b) What is the smallest percentage who could like all three recreations?

15. How many numbers between 1 and 30 are relatively prime to 30?

16. How many numbers between 1 and 280 are relatively prime to 280?

17. How many ways are there to assign 20 different people to three different rooms with at least one person in each room?

18. How many n-digit numbers are there with at least one of the digits 1 or 2 or 3 absent? [*Hint:* This is a union problem; find $N(A_1 \cup A_2 \cup A_3)$.]

19. How many arrangements are there of MURMUR with no pair of consecutive letters the same?

20. If three couples are seated around a circular table, what is the probability that no wife and husband are beside one another?

21. How many ways are there to form a committee of 10 mathematical scientists from a group of 15 mathematicians, 12 statisticians, and 10 operations researchers with at least one person of each different profession on the committee?

22. Find the number of ways to arrange the six numbers 1, 2, 3, 4, 5, 6 such that either in the arrangement 1 is immediately followed by 2, or 3 is immediately followed by 4, or 5 is immediately followed by 6.

23. How many ways are there to deal a 6-card hand that contains at least one Jack, at least one 8, and at least one 2?

24. How many ways are there to arrange the letters in the word MISSISSIPPI so that either all the Is are consecutive *or* all the Ss are consecutive *or* all the Ps are consecutive?

25. How many arrangements are there of TAMELY with either T before A, or A before M, or M before E? By "before," we mean anywhere before, not just immediately before. [*Hint:* This is a union problem; find $N(A_1 \cup A_2 \cup A_3)$.]

26. How many arrangements are there of MATHEMATICS with both Ts before both As, *or* both As before both Ms, *or* both Ms before the E? Note that "before" is used as in Example 25.

27. The Bernsteins, Hendersons, and Smiths each have 5 children. If the 15 children of these three families camp out in five different tents, where each tent holds 3 children, and the 15 children are randomly assigned to the five tents, what is the probability that every family has at least two of its children in the same tent?

28. Suppose 45 percent of all newspaper readers like wine, 60 percent like orange juice, and 55 percent like tea. Suppose 35 percent like any given pair of these beverages and 25 percent like all three beverages.

(a) What percentage of the readers likes only wine?

(b) What percentage of the readers likes exactly two of these three beverages?

29. Suppose that among 40 toy robots, 28 have a broken wheel or are rusted but not both, 6 are not defective, and the number with a broken with a broken wheel equals the number with rust. How many robots are rusted?

30. Suppose a school with n students offers two languages, PASCAL and BASIC. If 30 students take no language, 70 students do not take just PASCAL (i.e., either they do not take PASCAL or they take both languages), 80 students do not take just BASIC, and 20 students take both languages, determine n.

31. Suppose a school with 120 students offers Yoga and Karate. If the number of students taking Yoga alone is twice the number taking Karate (possibly, Karate and Yoga), if 25 more students study neither skill than study both skills, and if 75 students take at least one skill, then how many students study Yoga?

32. In a class of 30 children, 20 take Latin, 14 take Greek, and 10 take Hebrew. If no child takes all three languages and 8 children take no language, how many children take Greek and Hebrew? [*Hint:* Use formula (5) to determine the value of the expression $N(L \cap G) + N(L \cap H) + N(G \cap H)$.]

33. Suppose among 150 people on a picnic, 90 bring salads or sandwiches, 80 bring sandwiches or cheese, 100 bring salads or cheese, 50 bring cheese and either salad or sandwiches (possibly both), 60 bring at least two foods, and 20 bring all three foods. How many people bring just one of the three foods? How many people bring sandwiches?

8.2 INCLUSION–EXCLUSION FORMULA

In this section we generalize the formula for counting $N(\overline{A}_1 \cap \overline{A}_2 \cap \overline{A}_3)$ to n sets $A_1, A_2, \ldots, A_n$. To simplify notation, we will omit the intersection symbol "$\cap$" in expressions and write intersected sets as a product. For example, $A_1 \cap A_2 \cap A_3$ would be written $A_1A_2A_3$. Using this new notation, the number of elements in none of the sets $A_1, A_2, \ldots, A_n$ will be written $N(\overline{A}_1\overline{A}_2 \cdots \overline{A}_n)$. The following formula is known as the Inclusion–Exclusion Formula because of the way it successively includes (adds) and excludes (subtracts) the various k-tuple intersections of sets.

Theorem 1 **Inclusion–Exclusion Formula**

Let $A_1, A_2, \cdots A_n$, be n sets in a universe $\mathcal{U}$ of N elements. Let S_k denote the sum of the sizes of all k-tuple intersections of the A_is. Then

$$N(\overline{A}_1\overline{A}_2 \cdots \overline{A}_n) = N - S_1 + S_2 - S_3 + \cdots + (-1)^k S_k + \cdots + (-1)^n S_n \qquad (1)$$

Proof

To clarify the definition of the S_ks, $S_1 = \sum_i N(A_i)$, $S_2 = \sum_{ij} N(A_iA_j)$, S_k is the sum of the $N(A_{j_1}A_{j_2} \cdots A_{j_k})$s for all sets of k A_js, and finally $S_n = N(A_1A_2 \cdots A_n)$. We prove this formula by the same method used for $N(\overline{F}\,\overline{L}\,\overline{G})$ in the previous section: we shall show that the net effect of Eq. (1) is to count any element in none of the sets A_j once and to count elements in one or more A_js a net of 0 times.

If an element is in none of the A_js, that is, is in $\overline{A}_1\overline{A}_2\cdots\overline{A}_n$, then it is counted once in the right-hand side of Eq. (1) by the term N and is not counted in any of the S_ks. So the count is 1 for each element in $\overline{A}_1\overline{A}_2\cdots\overline{A}_n$, as required. An element in exactly one A_j is counted once by N, is subtracted once by S_1 (since it is in one of the A_js), and is counted in none of the other S_ks—for a count of $1-1=0$, as required. Now more generally let us show that an element x that is in exactly m of the A_js has a net count of 0 in Eq. (1). Element x is counted once by N, is counted m times by S_1 (since x is in m A_is), is counted $C(m, 2)$ times by S_2 [since x is in the intersection A_iA_j for the $C(m, 2)$ pairwise intersections involving two of the m sets containing x],..., and, in general, is counted $C(m, k)$ times by S_k, $k \le m$. It is not counted in S_h when $h > m$. So the net count of x in Eq. (1) is

$$1-\binom{m}{1}+\binom{m}{2}-\binom{m}{3}+\cdots+(-1)^k\binom{m}{k}+\cdots+(-1)^m\binom{m}{m} \tag{2}$$

This alternating sum of binomial coefficients can be evaluated from the binomial expansion

$$\begin{aligned}(1+x)^m = 1+\binom{m}{1}x+\binom{m}{2}x^2+\cdots+\binom{m}{k}x^k \\ +\cdots+\binom{m}{m}x^m\end{aligned} \tag{3}$$

If we set $x=-1$ in Eq. (3), the right side of Eq. (3) is now the expression in Eq. (2), whereas the left side of Eq. (3) becomes $[1+(-1)]^m = 0^m = 0$. Thus, Eq. (2) equals 0, as required. ◆

Corollary

Let $A_1, A_2 \cdots A_n$ be sets in the universe $\mathcal{U}$. Then

$$\begin{aligned}N(A_1 \cup A_2 \cup \cdots \cup A_n) = S_1 - S_2 + S_3 \\ -\cdots+(-1)^{k-1}S_k+\cdots+(-1)^{n-1}S_n\end{aligned} \tag{4}$$

Proof

We write formula (1) as

$$N(\overline{A}_1\overline{A}_2\cdots\overline{A}_n) = N - [S_1 - S_2 + S_3 - \cdots + (-1)^{n-1}S_n] \tag{5}$$

Next we observe that the number of elements in none of the sets equals the total number of elements minus the number of elements in one or more sets, that is,

$$N(\overline{A}_1\overline{A}_2\cdots\overline{A}_n) = N - N(A_1 \cup A_2 \cup \cdots \cup A_n) \tag{6}$$

Comparing Eqs. (5) and (6), we see that the expression in brackets on the right side of Eq. (5) is $N(A_1 \cup A_2 \cup \cdots \cup A_n)$. ◆

Before giving examples of the inclusion–exclusion formula, we want to emphasize an important logical point about applying this formula. To use this formula in a counting problem, one must select a universe $\mathcal{U}$ and a collection of sets A_i in that universe such that the outcomes to be counted are the subset of elements in $\mathcal{U}$ that are in *none* of the A_is. That is, the A_is represent properties *not* satisfied by the outcomes being counted.

Example 1: Card Hands With No Suit Voids

How many ways are there to select a 5-card hand from a regular 52-card deck such that the hand contains at least one card in each suit?

The universe $\mathcal{U}$ should be the set of all 5-card hands. We need to define the sets A_i such that hands with at least one card in each suit are in none of the A_is. With a moment's thought, we see that at least one card in a suit is equivalent to no void in the suit. Thus, we let A_1 be the set of 5-card hands with a void in spades; A_2 a void in hearts; A_3 a void in diamonds; and A_4 a void in clubs. Now the question asks for $N(\overline{A}_1\overline{A}_2\overline{A}_3\overline{A}_4)$, and we can use the inclusion–exclusion formula.

We must next calculate N, S_1, S_2, S_3, and S_4. As noted in Chapter 5, a 5-card hand is simply a subset of 5 cards, and so $N = C(52, 5)$. The size of A_1, the set of hands with a void in spades, is simply the number of 5-card hands chosen from the $52 - 13 = 39$ nonspade cards. So $N(A_1) = C(39, 5)$. Likewise, $N(A_i) = (39, 5)$, $i = 2, 3, 4$, and so $S_1 = 4 \times C(39, 5)$. The hands in A_1A_2 are hands chosen from the 26 non-(spades or hearts) cards, and so $N(A_1A_2) = C(26, 5)$. There are $C(4, 2) = 6$ different intersections of two out of the four sets, and so $S_2 = 6 \times C(26, 5)$. There are $C(4, 3) = 4$ different triple intersections of the sets and each has $C(13, 5)$ hands. So $S_3 = 4 \times C(13, 5)$. Finally, a hand cannot be void in all four suits, and so $S_4 = 0$. Then by Eq. (1),

$$N(\overline{A}_1\overline{A}_2\overline{A}_3\overline{A}_4) = \binom{52}{5} - 4\binom{39}{5} + 6\binom{26}{5} - 4\binom{13}{5} + 0 \quad \blacksquare$$

We note that, in general, S_k is a sum of $C(n, k)$ different k-tuple intersections of the n A_is.

Example 2: Distributions With an Empty Box

How many ways are there to distribute r distinct objects into five (distinct) boxes with at least one empty box?

We do not need to determine $N(\overline{A}_1\overline{A}_2\overline{A}_3\overline{A}_4\overline{A}_5)$ in this problem, because it does not concern outcomes where some property does not hold for all boxes. Rather, this is a union problem, using the corollary's formula.

Let $\mathcal{U}$ be all distributions of r distinct objects into five boxes. Let A_i be the set of distributions with a void in box i. Then the required number of distributions with at least one void is $N(A_1 \cup A_2 \cup \cdots A_5)$. We have $N = 5^r$, $N(A_i) = 4^r$ (distributions with each object going into one of the other 4 boxes), $N(A_iA_j) = 3^r$, and so forth. As just noted, there are $C(5, k)$ subsets in S_k, $k = 1, 2, \ldots, 5$. Thus by Eq. (4),

$$\begin{aligned} N(A_1 \cup A_2 \cup \cdots A_5) &= S_1 - S_2 + S_3 - S_4 + S_5 \\ &= \binom{5}{1}4^r - \binom{5}{2}3^r + \binom{5}{3}2^r - \binom{5}{4}1^r + 0 \quad \blacksquare \end{aligned}$$

It is easy to mistake union problems, which use phrases such as "with at least one empty box," with standard inclusion–exclusion problems, which use phrases such as "at least one object in every box." The former ask for at least one of a set of properties to hold, whereas the latter ask for every property to hold. Moreover, the latter problems must be solved by using the complementary properties, such as "box i is empty," and determining all ways for none of these complementary properties to hold. The reader has to reason carefully through a counting problem and determine whether to frame the solution as counting outcomes where none of a set of properties hold or outcomes where one or more properties hold.

Another common source of confusion is the subscripts of the S_is and the subscripts of the A_is. In example 1, students sometimes define A_1 to be all hands with a void in one suit, A_2 to be all hands with a void in two suits, and so on. This is wrong. The subscript of the As is an ordinal (ordering) number. A_1 denotes the *first* set (in Example 1, the set of all hands with a void in spades), A_2 the *second* set, A_3 the *third* set, ... The subscript of the Ss is a cardinal (magnitude) number. S_1 is the sum of the sizes of all single sets, S_2 the sum of the sizes of all pairwise intersections of sets, S_3 the sum of the sizes of all 3-way intersections of sets, ...

Example 3: Upper Bounds on Integer Solutions

How many different integer solutions are there to the equation

$$x_1 + x_2 + x_3 + x_4 + x_5 + x_6 = 20 \quad 0 \le x_i \le 8$$

This type of problem was solved with generating functions in Section 6.2. Now we solve it with an inclusion–exclusion approach. Let $\mathcal{U}$ be all integer solutions with $x_i \ge 0$, and let A_i be the set of integer solutions with $x_i > 8$, or equivalently $x_i \ge 9$. Then the number of solutions with $0 \le x_i \le 8$ will be $N(\overline{A}_1\overline{A}_2 \cdots \overline{A}_6)$.

Recalling the formulas for integer solutions of equations from Section 5.5 (Example 5) we have

$$N = \binom{20+6-1}{20} = \binom{25}{20}$$

To count $N(A_1)$, the outcomes with $x_1 \ge 9$, we consider the x_is to be amounts of objects chosen when a total of 20 objects are chosen from 6 types. The outcomes with at least 9 of the first type can be generated by first picking 9 of the first type and then picking the remaining $20 - 9$ objects without restriction from the 6 types. This reasoning applies to all $N(A_i)$:

$$N(A_i) = \binom{(20-9)+6-1}{(20-9)} = \binom{16}{11}$$

By a similar reasoning, now first choosing 9 from two types and then choosing the remaining $20 - 9 - 9$ objects without restriction, we have

$$N(A_iA_j) = \binom{(20-9-9)+6-1}{(20-9-9)} = \binom{7}{2}$$

For a solution to be in three or more A_is, the sum of the respective x_is would exceed 20—impossible. So $S_j = 0$ for $j \geq 3$, and

$$N(\overline{A}_1\overline{A}_2 \cdots \overline{A}_6) = N - S_1 + S_2 = \binom{25}{20} - \binom{6}{1}\binom{16}{11} + \binom{6}{2}\binom{7}{2} \blacksquare$$

Recall that to use generating functions to solve the preceding problem, we would seek the coefficient of x^{20} in

$$(1 + x + \cdots + x^8)^6 = [(1 - x^9)/(1 - x)]^6 = (1 - x^9)^6(1 - x)^{-6}$$

The coefficient of x^{20} in this product is $a_0b_{20} + a_9b_{11} + a_{18}b_2$, where a_k is the coefficient of x^k in $(1 - x)^{-6}$ and b_k is the coefficient of x^k in $(1 - x^9)^6$. This coefficient of x^{20} turns out to be exactly the foregoing expression for $N(\overline{A}_1\overline{A}_2 \cdots \overline{A}_6)$. The factor $(1 - x^9)^6 = 1 - C(6, 1)x^9 + C(6, 2)x^{18} \ldots$ does the inclusion–exclusion task of subtracting cases where one x_i is at least 9 and adding back cases where two x_is are at least 9. By using generating functions to solve this problem, we did not need to know anything about the inclusion–exclusion complexities of this problem. Generating functions automatically performed the required inclusion–exclusion calculations!

Example 4: Retrieving Hats

What is the probability that if n people randomly reach into a dark closet to retrieve their hats, no person will pick his own hat?

The probability will be the fraction of outcomes in which no person gets her own hat. Our universe $\mathcal{U}$ will be all ways for the n people to successively select a different hat. So $N = n!$ If A_i is the set of outcomes in which person i gets her own hat, then $N(\overline{A}_1\overline{A}_2 \cdots \overline{A}_n)$ counts the required outcomes where no one gets her own hat.

Then $N(A_i) = (n - 1)!$, since given that person i gets her hat, the number of possible outcomes is all ways for the other $n - 1$ people to select hats. Similarly $N(A_iA_j) = (n - 2)!$, and generally $N(A_{j_1}A_{j_2} \cdots A_{j_k}) = (n - k)!$ for k-way intersections. Since S_k is a sum of $C(n, k)$ k-way terms, we obtain by Eq. (1)

$$\begin{aligned} N(\overline{A}_1\overline{A}_2 \cdots \overline{A}_n) &= N - S_1 + S_2 + \cdots + (-1)^k S_k + \cdots + (-1)^n S_n \\ &= n! - \binom{n}{1}(n-1)! + \binom{n}{2}(n-2)! + \cdots \\ &\quad + (-1)^k \binom{n}{k}(n-k)! + \cdots + (-1)^n \binom{n}{n} 0! \\ &= \sum_{k=0}^{n} (-1)^k \binom{n}{k}(n-k)! \end{aligned}$$

Recalling that $\binom{n}{k} = \dfrac{n!}{k!(n-k)!}$, we see that $\binom{n}{k}(n-k)! = \dfrac{n!}{k!}\cdot$ So

$$N(\overline{A}_1\overline{A}_2 \cdots \overline{A}_n) = \sum_{k=0}^{n} \frac{(-1)^k n!}{k!} = n! \sum_{k=0}^{n} \frac{(-1)^k}{k!}$$

and now the probability that no person gets her own hat is

$$\frac{N(\overline{A}_1\overline{A}_2\cdots\overline{A}_n)}{N} = n!\sum_{k=0}^{n}\frac{(-1)^k}{k!}\Big/ n! = 1 - 1 + \frac{1}{2!} - \frac{1}{3!} + \cdots + \frac{(-1)^n}{n!} \qquad (7)$$

This alternating series is the first $n+1$ terms of $\sum_{k=0}^{\infty}(-1)^k/k!$, which is the power series for e^x when $x = -1$. The series converges very fast. The difference between e^{-1} and Eq. (7) is always less than $1/n!$ For example, $e^{-1} = 0.367879\ldots$ and for $n = 8$, the series in Eq. (7) equals 0.367888 (even for $n = 5$, it is 0.366). Thus, for all but very small n, the desired probability is essentially e^{-1}. The answer is independent of n. ■

The problem treated in Example 4 is equivalent to asking for all permutations of the sequence $1, 2, \ldots, n$ such that no number is left fixed, that is, no number i is still in the ith position. Such rearrangements of a sequence are called **derangements.** The symbol D_n is used to denote the number of derangements of n integers. From Example 4, we have

$$D_n = n!\sum_{k=0}^{n}\frac{(-1)^k}{k!} \approx n!e^{-1}$$

For exact values of D_n, it is usually easier to use the following simple recurrence relation:

$$D_n = nD_{n-1} + (-1)^n \qquad n \geq 2$$

See Exercise 36 for details on deriving this recurrence.

Example 5: Graph Coloring

How many ways are there to color the four vertices in the graph shown in Figure 8.6 with n colors (such that vertices with a common edge must be different colors)?

We label the edges e_1, e_2, e_3, e_4, e_5, as shown in Figure 8.6. The universe should be the set of all ways to color the vertices. So $N = n^4$ (n choices for each vertex). The situation we must avoid is coloring adjacent vertices with the same color. For each edge e_i, we define the set A_i to be all colorings of the graph in which the vertices at either end of edge e_i have the same color. Then $N(\overline{A}_1\overline{A}_2\overline{A}_3\overline{A}_4\overline{A}_5)$ will be the desired number of proper colorings with n colors.

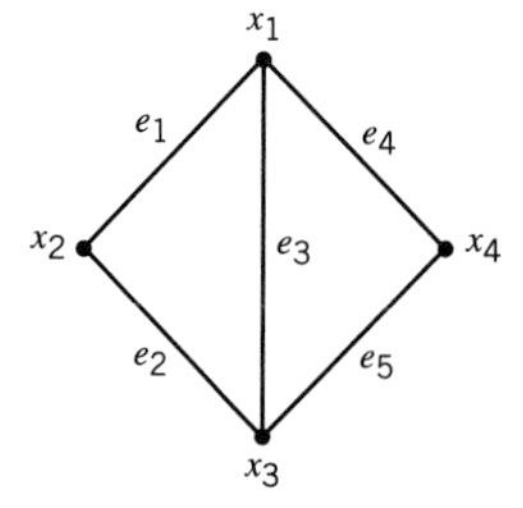

Figure 8.6

Then $N(A_i) = n^3$, since the two vertices connected by e_i get one color and the two other vertices each get any color, and $S_1 = 5 \times n^3$. Similarly, $N(A_i A_j) = n^2$ and $S_2 = C(5, 2) \times n^2$. A little care must be used with three-way intersections. Specifically, edges e_1, e_2, e_3 interconnect three, not four, vertices, and so $N(A_1 A_2 A_3) = n^2$ (one color for vertices x_1, x_2, x_3 and one color for vertex x_4). Likewise, $N(A_3 A_4 A_5) = n^2$. Any other set of three edges interconnect to all four vertices, and so the associated $N(A_i A_j A_k)$ equals n. Then $S_3 = 2 \times n^2 + [C(5, 3) - 2] \times n$. Four or five edges always interconnect all four vertices, and so $S_4 = C(5, 4) \times n$ and $S_5 = n$. Then the answer to our problem is

$$
\begin{aligned}
N(\overline{A}_1\overline{A}_2\overline{A}_3\overline{A}_4\overline{A}_5) &= n^4 - \binom{5}{1} n^3 + \binom{5}{2} n^2 \\
&\quad - \left[2n^2 + \left(\binom{5}{3} - 2\right) \times n\right] + \binom{5}{4} n - n \\
&= n^4 - 5n^3 + 8n^2 - 4n \quad \blacksquare
\end{aligned}
$$

Note that the preceding expression is the chromatic polynomial of this graph (see Example 7 in Section 2.3).

We conclude this section with two generalizations of the inclusion–exclusion formula. *Many readers may want to skip this material.*

Theorem 2

If $A_1, A_2, \ldots, A_n$ are n sets in a universe $\mathcal{U}$ of N elements, then the number N_m of elements in exactly m sets and the number N_m^* of elements in at least m sets are given by:

$$
\begin{aligned}
N_m &= S_m - \binom{m+1}{m} S_{m+1} + \binom{m+2}{m} S_{m+2} + \cdots + (-1)^{k-m} \binom{k}{m} S_k \\
&\quad + \cdots + (-1)^{n-m} \binom{n}{m} S_n \qquad (8)
\end{aligned}
$$

$$
\begin{aligned}
N_m^* &= S_m - \binom{m}{m-1} S_{m+1} + \binom{m+1}{m-1} S_{m+2} + \cdots + (-1)^{k-m} \binom{k-1}{m-1} S_k \\
&\quad + \cdots + (-1)^{n-m} \binom{n-1}{m-1} S_n \qquad (9)
\end{aligned}
$$

Proof

The formula for N_m can be proved in a fashion similar to the proof of the inclusion–exclusion formula. All elements in exactly m sets will be counted once in S_m, and not counted in any other term in (8), as required. We must show that elements in more than m sets are counted a net of 0 times in Eq. (8). In this formula the count is slightly trickier to sum. Readers who dislike such technicalities should skip this proof.

The count for an element in r sets, $m \le r \le n$, is $C(r, k)$ in S_k, and so the net count of this element in Eq. (8) is

$$\binom{r}{m} - \binom{m+1}{m}\binom{r}{m+1} + \cdots + (-1)^{k-m}\binom{k}{m}\binom{r}{k} \qquad (10)$$
$$+ \cdots + (-1)^{r-m}\binom{r}{m}\binom{r}{r}$$

Remember that the element is not counted in S_k for $k > r$. Recall from Example 1 of Section 5.5 that the number of ways to pick k objects from r and then pick m special objects from those k is equal to the number of ways to pick the m special objects from the r first and then pick $k - m$ more from the remaining $r - m$. So

$$\binom{k}{m}\binom{r}{k} = \binom{r}{m}\binom{r-m}{k-m}$$

Using this substitution, Eq. (10) now becomes

$$\binom{r}{m} - \binom{r}{m}\binom{r-m}{1} + \cdots + (-1)^{k-m}\binom{r}{m}\binom{r-m}{r-m}$$
$$+ \cdots + (-1)^{r-m}\binom{r}{m}\binom{r-m}{r-m}$$
$$= \binom{r}{m}\left[1 - \binom{r-m}{1} + \cdots + (-1)^{k-m}\binom{r-m}{k-m} + \cdots + (-1)^{r-m}\binom{r-m}{k-m}\right]$$

As in the proof of Theorem 1, the expression in brackets here is just the expansion of $(1+x)^{r-m}$ with $x = -1$ So Eq. (10) sums to 0, as required.

The formula for N_m^* can be verified with induction by showing that formula (9) for N_m^* satisfies $N_m^* = N_m + N_{m+1}^*$: "In at least m sets" means "in exactly m sets" or "in at least $m + 1$ sets." ◆

Example 6

Find the number of 4-digit ternary sequences with exactly two 1s. Also find the number with at least two 1s.

Let $\mathcal{U}$ be all 4-digit ternary sequences. If A_i is the set of 4-digit ternary sequences with a 1 in position *i,* then N_2 and N_2^* are the numbers of 4-digit ternary sequences with exactly two 1s and at least two 1s, respectively. Then $N = 3^4$ and $S_1 = C(4, 1)3^3$, $S_2 = C(4, 2)3^2$, $S_3 = C(4, 3)3^1$, and $S_4 = C(4, 4)3^0 = 1$. Now by formulas (8) and (9), the desired numbers are

$$N_2 = S_2 - \binom{4}{3}S_3 + \binom{4}{2}S_4 = \binom{4}{2}3^2 - \binom{3}{2}\left[\binom{4}{3}3^1\right] + \binom{4}{2}1 = 24$$
$$N_2^* = S_2 - \binom{2}{1}S_3 + \binom{3}{1}S_4 = \binom{4}{2}3^2 - \binom{2}{1}\left[\binom{4}{3}3^1\right] + \binom{3}{1}1 = 33 \quad \blacksquare$$

The observant reader may have noted that N_2 can be computed directly by a simple combinatorial argument.

8.2 EXERCISES

Summary of Exercises These exercises require a combination of inclusion–exclusion modeling and Chapter 5 type of enumeration skills (to solve the subproblems, some of which are tricky). Exercises 37–47 use Theorem 2.

1. How many n-digit decimal sequences (using digits $0, 1, 2, \ldots, 9$) are there in which digits 1, 2, 3 all appear?

2. How many ways are there to roll 10 distinct dice so that all 6 faces appear?

3. What is the probability that a 10-card hand has at least one 4 of a kind?

4. How many positive integers ≤ 420 are relatively prime to 420?

5. What is the probability that a 13-card bridge hand has:

(a) At least one card in each suit?

(b) At least one void in a suit?

(c) At least one of each type of face card (face cards are Aces, Kings, Queens, and Jacks)?

6. Given five pairs of gloves, how many ways are there for five people to each choose two gloves with no one getting a matching pair?

7. How many arrangements are there of $a, a, a, b, b, b, c, c, c,$ without three consecutive letters the same?

8. Given $2n$ letters, two of each of n types, how many arrangements are there with no pair of consecutive letters the same?

9. How many permutations of the 26 letters are there that contain none of the sequences MATH, RUNS, FROM, or JOE?

10. How many integer solutions of $x_1 + x_2 + x_3 + x_4 = 30$ are there with:

(a) $0 \leq x_i \leq 10$?

(b) $-10 \leq x_i \leq 20$?

(c) $0 \leq x_i, x_1 \leq 5, x_2 \leq 10, x_3 \leq 15, x_4 \leq 21$?

11. How many ways are there to distribute 25 identical balls into six distinct boxes with at most 6 balls in any of the first three boxes?

12. How many 4-digit numbers (including leading 0s) are there with no digit appearing exactly two times?

13. How many ways are there for a child to take 12 pieces of candy with four types of candy if the child does not take exactly two pieces of any type of candy?

14. Santa Claus has five toy airplanes of each of n plane models. How many ways are there to put one airplane in each of $r(r \geq n)$ identical stockings such that all models of planes are used?

15. A wizard has five friends. During a long wizards' conference, it met any given friend at dinner 10 times, any given pair of friends 5 times, any given threesome of friends 3 times, any given foursome 2 times, and all five friends together once. If in addition it ate alone 6 times, determine how many days the wizards' conference lasted.

16. How many secret codes can be made by assigning each letter of the alphabet a (unique) different letter? Give an approximate answer using Euler's number e.

17. How many ways are there to distribute 10 books to 10 children (one to a child) and then collect the books and redistribute them with each child getting a new one?

18. How many arrangements of $1, 2, \ldots, n$ are there in which only the odd integers must be deranged (even integers may be in their own positions)?

19. How many ways are there to assign each of five professors in a math department to two courses in the fall semester (i.e., 10 different math courses in all) and then assign each professor two courses in the spring semester such that no professor teaches the same two courses both semesters?

20. Consider the following game with a pile of n cards numbered 1 through n. Successively pick a different (random) number between 1 and n and remove all cards in the pile down to, and including, the card with this number until the pile is empty. If the chosen card number has already been removed, pick another number. What is the approximate probability that at some stage the number chosen is the card at the top of the pile? (*Hint:* Approach as an arrangement with repetition.)

21. There are 15 students, three (distinct) students each from 5 different high schools. There are 5 admissions officers, one from each of 5 colleges. Each of the officers successively picks 3 of the students to go to their college. How many ways are there to do this so that no officer picks 3 students from the same high school?

22. The rooms in the circular house plan shown below are to be painted using eight colors such that rooms with a common doorway must be different colors. In how many ways can this be done?

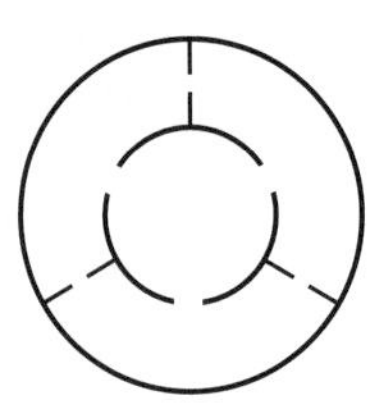

23. How many ways are there to color the vertices with n colors in the following graphs such that adjacent vertices get different colors?

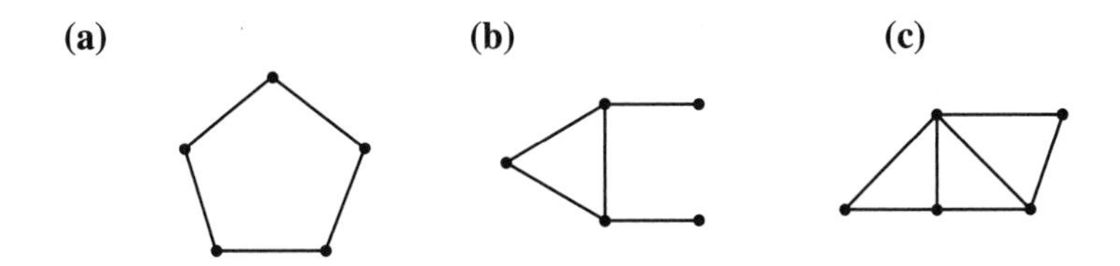

24. The four walls and ceiling of a room are to be painted with five colors available. How many ways can this be done if bordering sides of the room must have different colors?

25. **(a)** How many arrangements of the integers 1 through n are there in which i is never immediately followed by $i+1$, for $i = 1, 2, \ldots, n-1$?

(b) Show that your answer equals $D_n + D_{n-1}$.

26. Repeat Exercise 25(a) now considering n to be followed by 1.

27. If two identical dice are rolled n successive times, how many sequences of outcomes contain all doubles (a pair of 1s, of 2s, etc.)?

28. How many ways are there to seat n couples around a circular table such that no couple sits next to each other?

29. If there are n families with five members each, how many ways are there to seat all $5n$ people at a circular table so that each person sits beside another member of his/her family? (*Hint:* This is not a standard inclusion–exclusion problem but uses a type of inclusion–exclusion argument.)

30. How many arrangements of a, a, a, b, b, b, c, c, c have no adjacent letters the same? (*Hint:* This is *tricky*—not a normal inclusion–exclusion problem.)

31. Suppose that a person with seven friends invites a different subset of three friends to dinner every night for one week (seven days). How many ways can this be done so that all friends are included at least once?

32. There are 6 tennis players and each week for a month (4 weeks), a *different* pair of the 6 play a tennis match. How many ways are there to form the sequence of 4 matches so that every player plays at least once?

33. A company produces 8 different designs for sweaters. Each sweater is made from 3 different pieces of cloth (top piece, middle piece, bottom piece). There are 6 different colors available for each piece, and the 3 pieces in a sweater must each be a different color. How many collections (subsets) of 8 designs are possible if each color appears in at least one of the designs?

34. Find the number of ways to give each of six different people seated in a circle one of m different types of entrees if adjacent people must get different entrees.

35. How many ways are there to distribute r distinct objects into n indistinguishable boxes with no box empty?

36. **(a)** Show that D_n satisfies the recurrence $D_n = (n-1)(D_{n-1} + D_{n-2})$.

(b) Rewriting the recurrence in part (a) as $D_n - nD_{n-1} = -[D_{n-1} - (n-1)D_{n-2}]$, iterate backwards to obtain the recurrence $D_n = nD_{n-1} + (-1)^n$.

(c) Use part (b) to make a list of D_n values up to $n = 10$.

(d) Use part (a) to show that $D(x) = e^{-x}(1-x)^{-1}$ is the exponential generating function for D_n.

37. How many ways are there to distribute r distinct objects into n distinct boxes with exactly three empty boxes? With at least three empty boxes?

38. How many ways are there to deal a six-card hand with at most one void in a suit?

39. How many ways are there to arrange the letters in INTELLIGENT with at least two consecutive pairs of identical letters?

40. If n balls labeled 1, 2, . . . , n are successively removed from an urn, a *rencontre* is said to occur if the ith ball removed is numbered i. If the n balls are removed in random order, what is the probability that exactly k rencontres occur? Show that this probability is about $e^{-1}/k!$.

41. Show that the number of ways to place r different balls in n different cells with m cells having exactly k balls is

$$\frac{(-1)^n n! r!}{m!} \sum_{j=m}^{n} (-1)^j \frac{(n-j)^{r-jk}}{(j-m)!(n-j)!(r-jk)!(k!)^j}$$

42. **(a)** If $g(x)$ is the (ordinary) generating function for N_m (see Theorem 2), show that $g(x) = \sum_{k=0}^{n} S_k(x-1)^k$. This $g(x)$ is called the *hit polynomial.*

(b) Show that $2[g(1) + g(-1)]$ is the number of elements in an even number of A_is.

(c) Use part (b) to determine the number of n-digit ternary sequences with an even number of 0s. Simplify your answer with a binomial expansion summation.

43. Show that

$$S_m = \sum_{k=m}^{n} \binom{k}{m} N_m.$$

44. Use a combinatorial argument (with inclusion–exclusion) to prove:

(a) $\sum_{k=0}^{m} (-1)^k \binom{m}{k}\binom{n-k}{r} = 0, \quad n \geq r \geq m$

(b) $\sum_{k=0}^{m} (-1)^k \binom{n}{k}\binom{n-k}{m-k} = 0$

(c) $\sum_{k=m}^{n} (-1)^{k-m} \binom{n}{k} = \binom{n-1}{m-1}$

(d) $\sum_{k=0}^{n} (-1)^k \binom{n}{k}\binom{n-k+r-1}{r} = \binom{r-1}{n-1}$

45. Use Theorem 2 to show that

$$\binom{n}{m} = \sum_{k=m}^{n} (-1)^{k-m} \binom{k}{m}\binom{n}{k} 2^{n-k}$$

46. Prove formula (9) in Theorem 2.

47. In Example 4, let the random variable X = number of people who get their own hat. Find $E(X)$.

8.3 RESTRICTED POSITIONS AND ROOK POLYNOMIALS

In this section we consider the special problem of counting arrangements of n objects when particular objects can appear only in certain positions. We will solve one such problem involving five objects in careful detail. In the process we will indicate how to generalize our analysis to any restricted-positions problem.

Consider the problem of finding all arrangements of a, b, c, d, e with the restrictions indicated in Figure 8.7. That is, a may *not* be put in position 1 or 5; b may not be put in 2 or 3; c not in 3 or 4; and e not in 5. There is no restriction on d. A permissible arrangement can be represented by picking five unmarked squares in Figure 8.7, with one square in each row and each column. For example, the permissible arrangement *badec* corresponds to picking squares $(a, 2)$, $(b, 1)$, $(c, 5)$, $(d, 3)$, $(e, 4)$.

When viewed in terms of the 5×5 array of squares, the arrangement problem can be thought of as a matching problem, matching letters with positions. In Section 4.4 we also considered matching problems. There we wanted to determine whether any complete matching existed. Here we want to count how many complete matchings there are.

We use the inclusion–exclusion formula, expression (1) of the previous section, to count the number of permissible arrangements for Figure 8.7. Let $\mathcal{U}$ be the set of all arrangements of the five letters without restrictions. So $N = 5!$. Let A_i be the set of arrangements with a forbidden letter in position i (note that we could equally well define the properties in terms of the ith letter being in a forbidden position). The number of permissible arrangements will then be $N(\overline{A}_1\overline{A}_2\overline{A}_3\overline{A}_4\overline{A}_5)$. In terms of Figure 8.7, A_i is the set of all collections of five squares, each in a different row and column such that the square in column i is a darkened square. We obtain $N(A_i)$ by counting the ways to put a forbidden letter in position i times the $4!$ ways to arrange the remaining four letters in the other four positions (we do not worry about forbidden positions

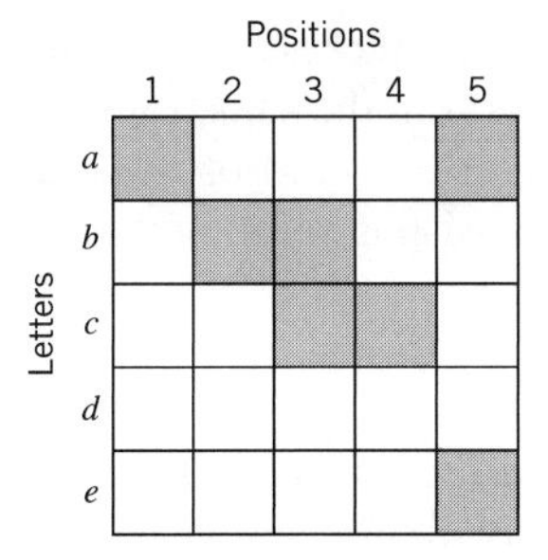

Figure 8.7

for these letters). Then $N(A_1) = 1 \times 4!$, $N(A_2) = 1 \times 4!$, $N(A_3) = 2 \times 4!$, $N(A_4) = 1 \times 4!$, and $N(A_5) = 2 \times 4!$. Collecting terms, we obtain

$$\begin{aligned} S_1 &= \sum_{i=1}^{5} N(A_i) = 1 \times 4! + 1 \times 4! + 2 \times 4! + 1 \times 4! + 2 \times 4! \\ &= (1+1+2+1+2)4! = 7 \times 4! \end{aligned}$$

Observe that $(1+1+2+1+2) = 7$ is just the number of the darkened squares in Figure 8.7. Since each choice of a darkened square (i.e., some letter in some forbidden position) leads to 4! possibilities, then

$$S_1 = (\text{number of darkened squares}) \times 4!$$

for any restricted-positions problem with a 5×5 family of darkened squares similar to Figure 8.7.

Next, $N(A_iA_j)$ will be the number of ways to put (different) forbidden letters in positions i and j times the 3! ways to arrange the remaining three letters. Or equivalently, the ways to pick two darkened squares, one in column i and one in column j (and in different rows), times 3!. The reader can check that

$$\begin{array}{lll} N(A_1A_2) = 1 \times 3! & N(A_1A_3) = 2 \times 3! & N(A_1A_4) = 1 \times 3! \\ N(A_1A_5) = 1 \times 3! & N(A_2A_3) = 1 \times 3! & N(A_2A_4) = 1 \times 3! \\ N(A_2A_5) = 2 \times 3! & N(A_3A_4) = 1 \times 3! & N(A_3A_5) = 4 \times 3! \\ N(A_4A_5) = 2 \times 3! & & \end{array}$$

Collecting terms, we obtain

$$S_2 = \sum_{ij} N(A_iA_j) = (1+2+1+1+1+1+2+1+4+2)3! = 16 \times 3!$$

The number 16 counts the ways to select two darkened squares, each in a different row and column. Generalizing, we will have

$$S_k = \begin{pmatrix} \text{number of ways to pick } k \text{ darkened squares} \\ \text{each in a different row and column} \end{pmatrix} \times (5-k)! \qquad (1)$$

Since letter d's row in Figure 8.7 has no darkened squares, there is no way to pick five darkened squares, each in a different row (and column). Thus $S_5 = 0$. On the other hand, tedious case-by-case counting apparently awaits us for S_3 and S_4. Instead, let us try to develop a theory for determining the number of ways to pick k darkened squares, each in a different row and column.

This darkened squares selection problem can be restated in terms of a recreational mathematics question about a chess-like game. A chess piece called a **rook** can capture any opponent's piece on the chessboard in the same row or column as the rook (provided there are no intervening pieces). Instead of using a normal 8×8 chessboard, we "play chess" on the "board" consisting solely of the darkened squares in Figure 8.7.

Counting the number of ways to place k mutually noncapturing rooks on this board of darkened squares is equivalent to our original subproblem of counting the number of ways to pick k darkened squares in Figure 8.7, each in a different row and column. The phrase "k mutually noncapturing rooks" is simpler to say and more pictorial.

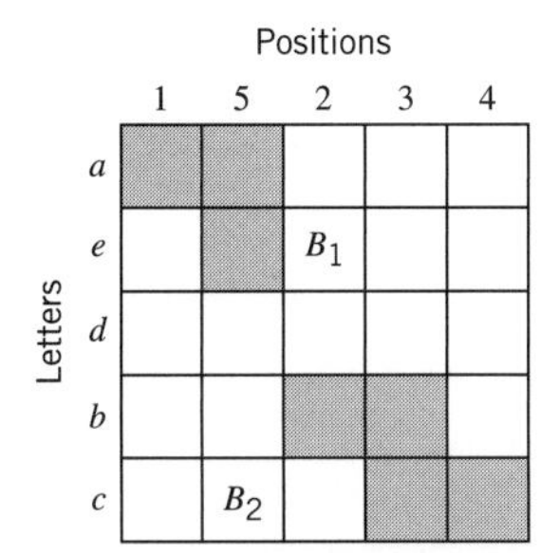

Figure 8.8

A common technique in combinatorial analysis is to break a big messy problem into smaller manageable subproblems. We will now develop two breaking-up operations to help us count noncapturing rooks on a given board B.

The first operation applies to a board B that can be decomposed into **disjoint** subboards B_1 and B_2, that is, subboards involving different sets of rows and columns. Often a board has to be properly rearranged before the disjoint nature of the two subboards can be seen.

When the rows and columns of Figure 8.7 are rearranged as shown in Figure 8.8, it is obvious that the three darkened squares in rows a and e and columns 1 and 5 are disjoint from the four darkened squares in rows b and c and columns 2, 3, and 4. Let B be the board of darkened squares in Figure 8.8, let B_1 be the three darkened squares in rows a and e, and let B_2 be the four darkened squares in rows b and c.

Define $r_k(B)$ to be the number of ways to place k noncapturing rooks on board B, $r_k(B_1)$ the number of ways to place k noncapturing rooks on subboard B_1, and $r_k(B_2)$ the number of ways to place k noncapturing rooks on subboard B_2. There are three ways to place one rook on subboard B_1 in Figure 8.8, since B_1 has three squares, and similarly four ways to place one rook on subboard B_2. Then $r_1(B_1) = 3$ and $r_1(B_2) = 4$. A little thought shows that there is only one way to place two rooks on subboard B_1 and three ways to place two rooks on subboard B_2, so that $r_2(B_1) = 1$ and $r_2(B_2) = 3$. Note that $r_k(B_1) = r_k(B_2) = 0$ for $k \geq 3$, since each subboard has only two rows. It will be convenient to define $r_0 = 1$ for all boards.

Observe next that since B_1 and B_2 are disjoint, placing, say, two noncapturing rooks on the whole board B can be broken into three cases: placing two noncapturing rooks on B_1 (and none on B_2), placing one rook on each subboard, or placing two noncapturing rooks on B_2. Thus we see that

$$r_2(B) = r_2(B_1) + r_1(B_1)r_1(B_2) + r_2(B_2)$$

or, using that fact that $r_0(B_2) = r_0(B_1) = 1$,

$$\begin{aligned} r_2(B) &= r_2(B_1)r_0(B_2) + r_1(B_1)r_1(B_2) + r_0(B_1)r_2(B_2) \\ &= 1 \times 1 + 3 \times 4 + 1 \times 3 = 16 \end{aligned} \tag{2}$$

Recall that 16 is the number obtained earlier when summing all $N(A_iA_j)$ to count all ways to pick two darkened squares each in a different row and column.

The reasoning leading to Eq. (2) applies to $r_k(B)$ for any k and for any board B that decomposes into two disjoint subboards B_1 and B_2.

Lemma

If B is a board of darkened squares that decomposes into the two disjoint subboards B_1 and B_2, then

$$r_k(B) = r_k(B_1)r_0(B_2) + r_{k-1}(B_1)r_1(B_2) + \cdots + r_0(B_1)r_k(B_2) \tag{3}$$

The observant reader may notice that Eq. (3) is very similar to formula (6) in Section 6.2 for the coefficient of a product of two generating functions. That is, if $f(x) = \sum a_r x^r$ and $g(x) = \sum b_r x^r$, then the coefficient of x^k in $h(x) = f(x)g(x)$ is $a_k b_0 + a_{k-1}b_1 + \cdots + a_0 b_k$. We will now exploit this similarity.

We define the **rook polynomial** *R(x, B)* of the board B of darkened squares to be

$$R(x, B) = r_0(B) + r_1(B)x + r_2(B)x^2 + \cdots$$

Remember that $r_0(B) = 1$ for all B. Note that the rook polynomial depends only on the darkened squares, not on the size of the original assignment diagram. Then for B_1 and B_2 as defined above, we found that

$$R(x, B_1) = 1 + 3x + 1x^2 \quad \text{and} \quad R(x, B_2) = 1 + 4x + 3x^2$$

Moreover, by the correspondence between Eq. (3) and the formula for the product of two generating functions, we see that $r_k(B)$, the coefficient of x^k in the rook polynomial *R(x, B)* of the full board, is simply the coefficient of x^k in the product $R(x, B_1)R(x, B_2)$. That is,

$$\begin{aligned} R(x, B) &= R(x, B_1)R(x, B_2) = (1 + 3x + 1x^2)(1 + 4x + 3x^2) \\ &= 1 + [(3 \times 1) + (1 \times 4)]x + [(1 \times 1) + (3 \times 4) + (1 \times 3)]x^2 \\ &\quad + [(1 \times 4) + (3 \times 3)]x^3 + (1 \times 3)x^4 \\ &= 1 + 7x + 16x^2 + 13x^3 + 3x^4 \end{aligned}$$

This product relation is true for any such B, B_1, and B_2.

Theorem 1

If B is a board of darkened squares that decomposes into the two disjoint subboards B_1 and B_2 then

$$R(x, B) = R(x, B_1)R(x, B_2)$$

Without meaning to belittle the role of generating functions in Theorem 1, we should observe that the generating functions were used here solely because polynomial multiplication corresponds to the subboard composition rule given in Eq. (3). That is, plugging the $r_k(B_i)$s into two polynomials and multiplying them together is a more familiar way to organize the computation required by Eq. (3). Shortly we will decompose a board B into nondisjoint subboards, and rook polynomials will then play a truly essential role.

Now that $R(x, B)$ has been determined for the board of darkened squares in Figure 8.8 and hence in Figure 8.7, we can solve our original problem about the permissible arrangements of a, b, c, d, e. Expression (1) for S_k can now be rewritten

$$S_k = r_k(B)(5-k)!$$

By the inclusion–exclusion formula, the number of permissible arrangements is

$$\begin{aligned} N(\overline{A}_1\overline{A}_2\overline{A}_3\overline{A}_4\overline{A}_5) &= N - S_1 + S_2 - S_3 + S_4 - S_5 \\ &= 5! - r_1(B)4! + r_2(B)3! - r_3(B)2! + r_4(B)1! - r_5(B)0! \\ &= 5! - 7 \times 4! + 16 \times 3! - 13 \times 2! + 3 \times 1! - 0 \times 0! \\ &= 120 - 168 + 96 - 26 + 3 - 0 = 25 \end{aligned}$$

The values for the $r_k(B)$s came from the rook polynomial $R(x, B)$ for Figure 8.8 computed above. This rook-polynomial-based variation on the inclusion–exclusion formula is valid for any arrangement problem with restricted positions.

Theorem 2

The number of ways to arrange n distinct objects when there are restricted positions is equal to

$$\begin{aligned} n! - r_1(B)(n-1)! + r_2(B)(n-2)! + \cdots + (-1)^k r_k(B)(n-k)! \\ + \cdots + (-1)^n r_n(B)0! \end{aligned} \tag{3}$$

where the $r_k(B)$s are the coefficients of the rook polynomial $R(x, B)$ for the board B of forbidden positions.

Let us summarize the little theory we have developed.

1. Given a problem of counting arrangements or matchings with restricted positions, display the constraints in an array with darkened squares for forbidden positions, as in Figure 8.7.
2. Try to rearrange the array so that the board B of darkened squares can be decomposed into disjoint subboards B_1 and B_2.
3. By inspection, determine the $r_k(B_i)$s, the number of ways to place k noncapturing rooks on subboard B_i.
4. Use the $r_k(B_i)$s to form the rook polynomials $R(x, B_1)$ and $R(x, B_2)$, and multiply $R(x, B_1)R(x, B_2)$ to obtain $R(x, B)$.
5. Insert the coefficient $r_k(B)$ of $R(x, B)$ in formula (3).

Example 1: Sending Birthday Cards

How many ways are there to send six different birthday cards, denoted $C_1, C_2, C_3, C_4, C_5, C_6$, to three aunts and three uncles, denoted $A_1, A_2, A_3, U_1, U_2, U_3$, if aunt A_1 would not like cards C_2 and C_4; if A_2 would not like C_1 or C_5; if A_3 likes all cards; if U_1 would not like C_1 or C_5; if U_2 would not like C_4; and if U_3 would not like C_6?

The forbidden positions information is displayed in Figure 8.9. We rearrange the board by putting together rows with a darkened square in the same column, and

	A_1	A_2	A_3	U_1	U_2	U_3
C_1						
C_2						
C_3						
C_4						
C_5						
C_6						

Figure 8.9

putting together columns with a darkened square in the same row. For example, rows C_1 and C_5 both have darkened squares in columns A_2 and U_1, and so we put rows C_1 and C_5 beside one another and columns A_2 and U_1 beside one another; similarly for rows C_2 and C_4 and columns A_1 and U_2. We get the rearrangement shown in Figure 8.10. Thus the original board B of darkened squares decomposes into the two disjoint subboards, B_1 in rows C_1 and C_5, and B_2 in rows C_2, C_4, and C_6. Actually B_2 itself decomposes into two disjoint subboards B_2' and B_2'', where B_2'' is the single square (C_6, U_3). By inspection we see that

$$R(x, B_1) = 1 + 4x + 2x^2$$
$$R(x, B_2) = R(x, B_2')R(x, B_2'') = (1 + 3x + x^2)(1 + x)$$

So

$$\begin{aligned} R(x, B) &= R(x, B_1)R(x, B_2) \\ &= (1 + 4x + 2x^2)(1 + 3x + x^2)(1 + x) \\ &= 1 + 8x + 22x^2 + 25x^3 + 12x^4 + 2x^5 \end{aligned}$$

Then the answer to the card-mailing problem is

$$\begin{aligned} &\sum_{k=0}^{6}(-1)^k r_k(B)(6-k)! \\ &\quad = 6! - 8 \times 5! + 22 \times 4! - 25 \times 3! + 12 \times 2! - 2 \times 1! + 0 \times 0! \\ &\quad = 720 - 960 + 528 - 150 + 24 - 2 + 0 = 160 \quad \blacksquare \end{aligned}$$

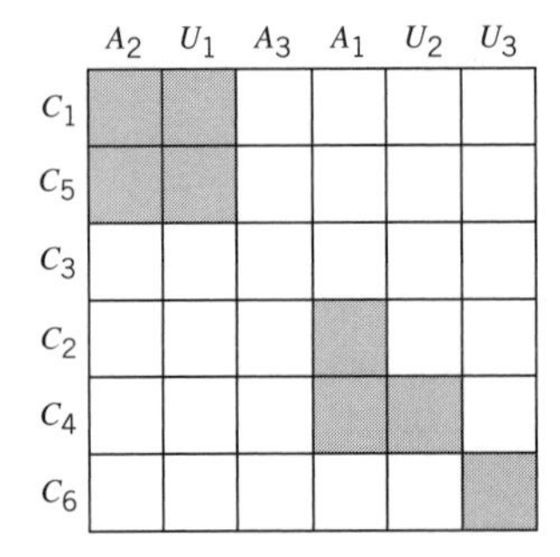

Figure 8.10

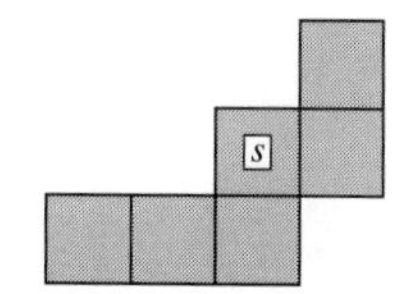

Figure 8.11

Now let us consider the problem of determining the coefficients of $R(x, B)$ when the board B does not decompose into two disjoint subboards. Consider the board B shown in Figure 8.11. Let us break the problem of determining $r_k(B)$ into two cases, depending on whether or not a certain square s is one of the squares chosen for the k noncapturing rooks. Let B_s be the board obtained from B by deleting square s, and let B_s^* be the board obtained from B by deleting square s plus all squares in the same row or column as s. If s is the square indicated in Figure 8.11, then B_s and B_s^* are as shown in Figure 8.12. If square s is not used, we must place k noncapturing rooks on B_s. If square s is used, then we must place $k-1$ noncapturing rooks on B_s^*. Hence we conclude that

$$r_k(B) = r_k(B_s) + r_{k-1}(B_s^*) \tag{4}$$

Using the generating function methods introduced in Section 7.5 for turning a recurrence relation into a generating function, we obtain from Eq. (4)

$$\begin{aligned} R(x, B) &= \sum_k r_k(B)x^k = \sum_k r_k(B_s)x^k + \sum_k r_{k-1}(B_s^*)x^k \\ &= \sum_k r_k(B_s)x^k + x\sum_h r_h(B_s^*)x^h \\ &= R(x, B_s) + xR(x, B_s^*) \end{aligned}$$

Observe that B_s and B_s^* both break into disjoint subboards whose rook polynomials are easily determined by inspection:

$$\begin{aligned} R(x, B_s) &= (1+3x)(1+2x) = 1+5x+6x^2 \\ R(x, B_s^*) &= (1+2x)(1+x) = 1+3x+2x^2 \\ R(x, B) = R(x, B_s) + xR(x, B_s^*) &= (1+5x+6x^2) + x(1+3x+2x^2) \\ &= 1+6x+9x^2+2x^3 \end{aligned}$$

These results apply to any board B and any square s in B.

Theorem 3

Let B be any board of darkened squares. Let s be one of the squares of B, and let B_s and B_s^* be as defined above. Then

$$R(x, B) = R(x, B_s) + xR(B_s^*)$$

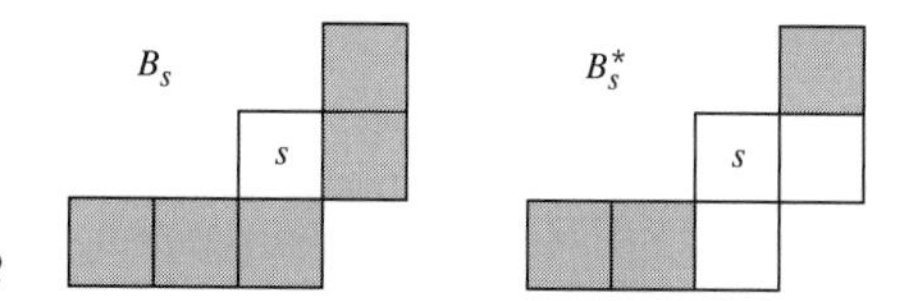

Figure 8.12

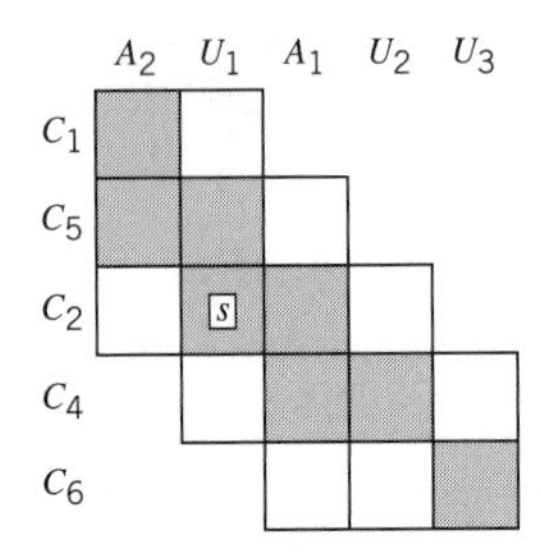

Figure 8.13

Theorem 3 provides a way to simplify any board's rook polynomial. If the boards B_s and B_s^* do not break up into disjoint subboards, we can reapply Theorem 3 to B_s and B_s^*. It is important that the square s be chosen to split up B as much as possible.

Example 2: Nondecomposable Constraints

In Example 1, suppose that the tastes of uncle U_1 change and now he would not like card C_2 but would like C_1. The new board of forbidden positions is shown in Figure 8.13. The square in the bottom right corner, call it t, is still disjoint from the other squares, call them board B_1. Board B_1 cannot be decomposed into disjoint subboards, and so we must use Theorem 3. The square s that breaks up B_1 most evenly is (C_2, U_1), indicated in Figure 8.13.

Boards B_s and B_s^* shown in Figure 8.14 both decompose into simple disjoint subboards:

$$R(x, B_s) = (1+3x+x^2)(1+3x+x^2) = 1+6x+11x^2+6x^3+x^4$$
$$R(x, B_s^*) = (1+2x)(1+2x) = 1+4x+4x^2$$

Then

$$\begin{aligned} R(x, B_1) = R(x, B_s) + xR(x, B_s^*) &= (1+6x+11x^2+6x^3+x^4) \\ &\quad +x(1+4x+4x^2) \\ &= 1+7x+15x^2+10x^3+x^4 \end{aligned}$$

and

$$\begin{aligned} R(x, B) &= R(x, B_1)R(x, t) \\ &= (1+7x+15x^2+10x^3+x^4)(1+x) \\ &= 1+8x+22x^2+25x^3+11x^4+x^5 \end{aligned}$$

Now by Theorem 2, the number of ways to send birthday cards is

$$6! - 8 \times 5! + 22 \times 4! - 25 \times 3! + 11 \times 2! - 1 \times 1! + 0 \times 0! = 159$$

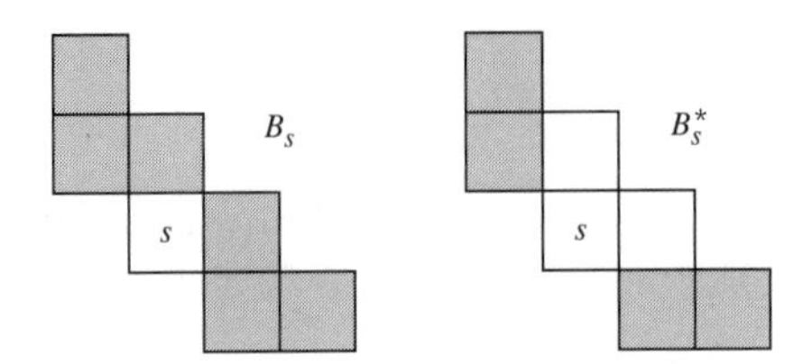

Figure 8.14

Note that uncle U_1's change in tastes changed the final rook polynomial only slightly: $r_4(B)$ changed from 12 to 11, $r_5(B)$ changed from 2 to 1, and the final answer changed from 160 to 159. ■

8.3 EXERCISES

Summary of Exercises The first nine exercises are similar to the examples in this section; the next seven exercises develop theory about rook polynomials and combinatorial theory based on rook polynomials.

1. Describe the associated chessboard of darkened squares for finding all derangements of 1, 2, 3, 4, 5.
2. Find the rook polynomial for the following boards:

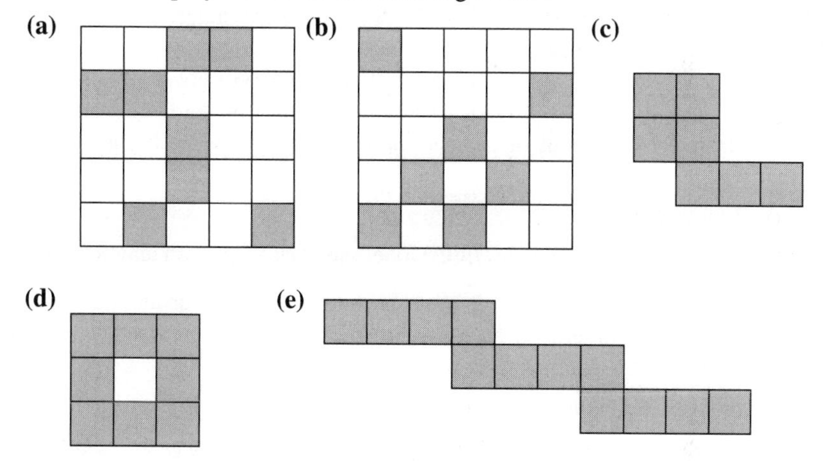

3. Find the number of matchings of 5 men with 5 women given the constraints in the figure below on the left, where the rows represent the men and the columns represent the women.

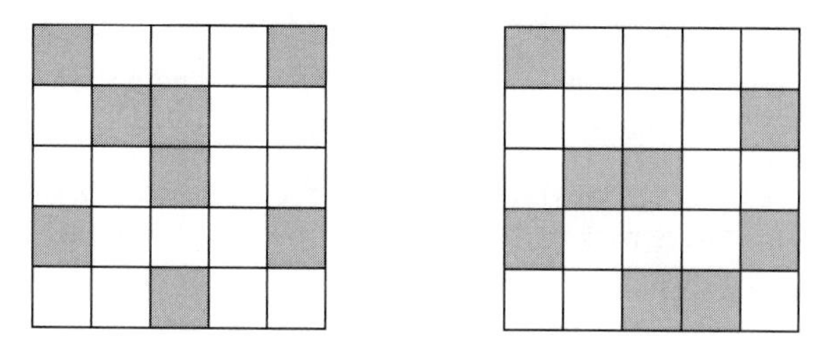

4. Find the number of matchings of 5 men with 5 women given the constraints in the figure above on the right, where the rows represent the men and the columns represent the women.

5. Seven dwarfs $D_1, D_2, D_3, D_4, D_5, D_6, D_7$ each must be assigned to one of seven jobs in a mine, $J_1, J_2, J_3, J_4, J_5, J_6, J_7$. D_1 cannot do jobs J_1 or J_3; D_2 cannot do J_1 or J_5; D_4 cannot do J_3 or J_6; D_5 cannot do J_2 or J_7; D_7 cannot do J_4. D_3 and D_6 can do all jobs. How many ways are there to assign the dwarfs to different jobs?

6. A pair of two distinct dice are rolled six times. Suppose none of the ordered pairs of values (1, 5), (2, 6), (3, 4), (5, 5), (5, 3), (6, 1), (6, 2) occur. What is the probability that all six values on the first die and all six values on the second die occur once in the six rolls of the two dice?

7. A computer dating service wants to match four women each with one of five men. If woman 1 is incompatible with men 3 and 5; woman 2 is incompatible with men 1 and 2; woman 3 is incompatible with man 4; and woman 4 is incompatible with men 2 and 4, how many matches of the four women are there?

8. Suppose five officials O_1, O_2, O_3, O_4, O_5 are to be assigned five different city cars, an Escort, a Lexus, a Nissan, a Taurus, and a Volvo. O_1 will not drive an Escort or a Nissan; O_2 will not drive a Taurus; O_3 will not drive a Lexus or a Volvo; O_4 will not drive a Lexus; and O_5 will not drive an Escort or a Nissan. If a feasible assignment of cars is chosen randomly, what is the probability that:

(a) O_1 gets the Volvo?

(b) O_2 or O_5 get the Volvo? (*Hint:* Model this constraint with an altered board.)

9. Calculate the number of words that can be formed by rearranging the letters EERRIE so that no letters appears at one of its original positions, for example, no E as the first, second or sixth letter.

10. Find the rook polynomial for a full $n \times n$ board.

11. An *ascent* in a permutation is a consecutive pair of the form $i, i+1$. The ascents in the following permutation are underlined: $\underline{12}5\underline{34}$.

(a) Design a chessboard to represent a permutation of 1, 2, 3, 4, 5 so that a check in entry (i, j) means i is followed immediately by j in the permutation. Darken entries so as to exclude all ascents.

(b) Find the rook polynomial for the darkened squares in part (a).

(c) Find the number of permutations of 1, 2, 3, 4, 5 with k ascents. (*Hint:* See Theorem 2 in Section 8.2.)

12. Let $R_{n,m}(x)$ be the rook polynomial for an $n \times m$ chessboard (n rows, m columns, all squares may have rooks).

(a) Show that $R_{n,m}(x) = R_{n-1,m}(x) + mxR_{n-1,m-1}(x)$

(b) Show that $\frac{d}{dx}R_{n,m}(x) = nmR_{n-1,m-1}(x)$. [*Hint:* Use identity (5) in Section 5.5.]

13. Find two different chessboards (not row or column rearrangements of one another) that have the same rook polynomial.

14. Consider all permutations of 1, 2, . . ., n in which i appears in neither position i nor $i+1$ (n not in n or 1). Such a permutation is called a *menage.* Let $M_n(x)$ be the rook polynomial for the forbidden squares in a menage. Let $M_n^*(x)$ be the rook polynomial when n may appear in position 1, and let $M_n^0(x)$ be the rook polynomial when both 1 and n may appear in position 1.

(a) Show that $M_n^*(x) = xM_n^*(x) + M_n^0(x)$, $M_n^0(x) = xM_{n-1}^0(x) + M_n^*(x)$, and $M_n(x) = M_n^*(x) + xM_{n-1}^*(x)$.

(b) Using the initial conditions $M_1^*(x) = 1 + x$, $M_1^0(x) = 1$, show by induction that

$$M_n^*(x) = \sum_{k=0}^{n} \binom{2n-k}{k} x^k \quad \text{and} \quad M_n^0(x) = \sum_{k=0}^{n-1} \binom{2n-k-1}{k} x^k$$

(c) Find $M_n(x)$.

15. Given an $n \times m$ chessboard $C_{n,m}$ (see Exercise 12) and a board C of darkened squares in $C_{n,m}$, the *complement* C' of C in $C_{n,m}$ is the board of nondarkened squares.

(a) Show that

$$r_k(C') = \sum_{j=0}^{k} (-1)^j \binom{n-j}{k-j} \binom{m-j}{k-j} (k-j)! r_k(C)$$

(b) Show that $R(x, C') = x^n R(1/x, C)$.

16. Use Theorem 2 in Section 8.2 to derive a formula for counting arrangements when exactly k elements appear in forbidden positions.

8.4 SUMMARY AND REFERENCE

Frequently the combinatorial complexities in the counting problems in Chapter 5 arose from simultaneous constraints such as "with at least one card in each suit." In this chapter, we formed a simple set-theoretic model for such problems and solved once and for all the combinatorial logic of this model. The resulting formula, the inclusion–exclusion formula, was then applied to various counting problems. This formula requires the proper set-theoretic restatement of a problem and the solution of some fairly straightforward subproblems, but in return the formula eliminates all the worry about logical decomposition (and worry about counting some outcomes twice). After having no help in their problem-solving in Chapter 5, readers should find it easy to appreciate fully the power of a formula that does much of the reasoning for them. This is what mathematics is all about!

The last section on rook polynomials provides a nice mini-theory about organizing the inclusion–exclusion computations in arrangements with restricted positions. The inclusion–exclusion formula was obtained by J. Sylvester about 100 years ago, although in the early eighteenth century the number of derangements had been calculated by Montmort and a (noncounting) set union and intersection version of the formula was published by De Moivre. Rook polynomials were not invented until the mid-twentieth century (see Riordan [1]).

1. J. Riordan, *An Introduction to Combinatorial Analysis,* John Wiley & Sons, New York, 1958.

PART THREE
ADDITIONAL TOPICS

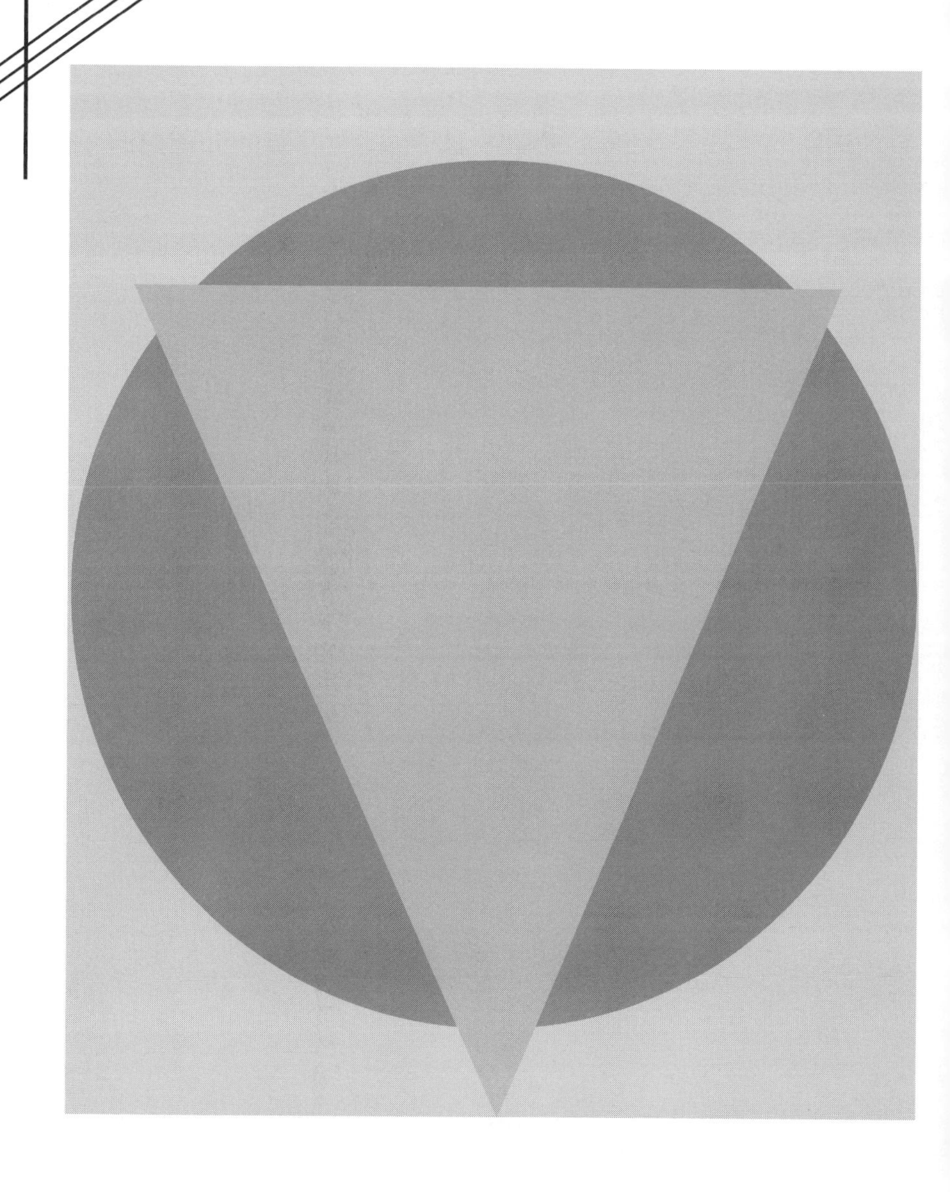

CHAPTER 9

POLYA'S ENUMERATION FORMULA

9.1 EQUIVALENCE AND SYMMETRY GROUPS

In this chapter we examine a special class of counting problems. Consider the ways of coloring the corners of a square black, white, or red: There are three color choices at each of the four corners, giving $3 \times 3 \times 3 \times 3 = 81$ different colorings. Suppose, however, that the square is not in a fixed position but is unoriented, like a square molecule in a liquid. Now how many different corner colorings are there? The unoriented figure being colored could be a n-gon or a cube, and the edges or faces could be colored instead of the corners. The floating square problem is equivalent to finding the number of different (unoriented) necklaces of four beads colored black, white, or red. Polya's motivation in developing his formula for counting distinct colorings of unoriented figures came from a problem in chemistry, the enumeration of isomers.

The difficulty in these problems comes from the geometric symmetries of the figure being colored. We develop a special formula, based on this set of symmetries, to count all distinct colorings of a figure. With a little more work, we also obtain a generating function that gives a **pattern inventory** of the distinct colorings. For example, the pattern inventory of black–white colorings of the corners of a cube with all geometric symmetries allowed is

$$b^8 + b^7w + 3b^6w^2 + 3b^5w^3 + 7b^4w^4 + 3b^3w^5 + 3b^2w^6 + bw^7 + w^8$$

where the coefficient of b^iw^j is the number of nonequivalent colorings with i black corners and j white corners.

We develop our formula and associated theory around the sample problem of coloring the corners of an unoriented square (floating in three dimensions) with black and white. It is impossible to draw an unoriented object; any picture shows it in a fixed position. Thus, we start our analysis with the $2^4 = 16$ black–white colorings of the fixed square. See Figure 9.1. We can partition these 16 colorings into subsets of colorings that are equivalent when the square is floating. There are six such subsets (see the groupings of colored squares shown in Figure 9.1), and so there are six different 2-colorings of the floating square. The set of fixed colorings would be too large if a harder sample problem were used, such as 2-colorings of a cube or 3-colorings of the square.

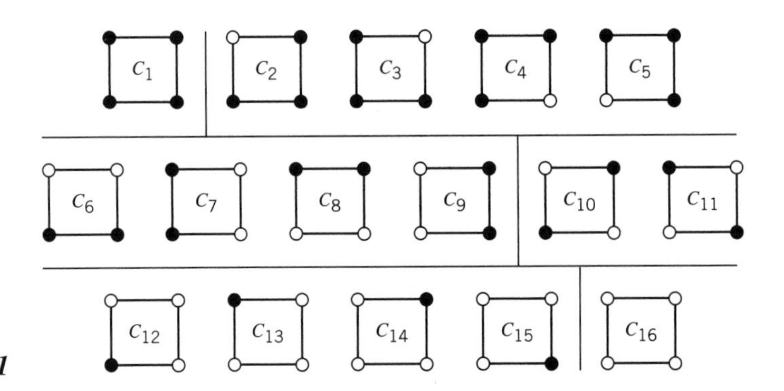

Figure 9.1

We seek a theory and formula to explain why there are six such distinct 2-colorings of the square. Note that the six subsets of equivalent colorings vary in size.

To define the partition of a set into subsets of equivalent elements, we first define the general concept of the equivalence of two elements a and b. We write this equivalence as $a \sim b$. The fundamental properties of an **equivalence relation** are:

(i) Transitivity: $a \sim b, b \sim c \Rightarrow a \sim c$.

(ii) Reflexivity: $a \sim a$.

(iii) Symmetry: $a \sim b \Rightarrow b \sim a$.

All other properties of equivalence can be derived from these three. Any binary relation with these three properties is called an equivalence relation. Such a relation defines a partition into subsets of mutually equivalent elements called **equivalence classes.**

Example 1: Equivalence Relations

(a) For a set of people, being the same weight is an equivalence relation; all people of a given weight form an equivalence class.

(b) For a set of numbers, differing by an even number is an equivalence relation; the even numbers form one equivalence class and the odd numbers the other class.

(c) For a set of figures, having the same number of corners is an equivalence relation; for each n, an equivalence class consists of all n-corner figures. ∎

Next we turn our attention to the motions that map the square onto itself (see Figure 9.2). These motions are what make the foregoing colorings equivalent to another one. Before we develop a theory about symmetries and their relation to coloring equivalence, let us take a closer geometric look at the symmetries of a square and of some other figures.

Example 2: Symmetries of Even n-gons

Figure 9.2 displays the set of motions that map the square onto itself, the symmetries of the square. How was this set obtained? More generally, what is the set of symmetries of an n-gon, for n even?

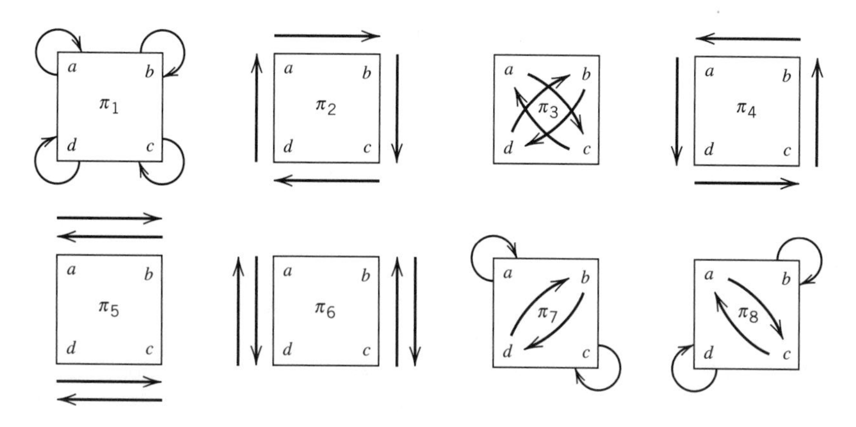

Figure 9.2

The symmetries of the square divide into two classes: rotations—circular motions in the plane; and reflections (flips)—motions using the third dimension. The rotations, about the center of the square, are easy to find. Each rotation is an integral multiple of the smallest (nonzero) rotation. The rotations are $\pi_2 = 90°$ rotation, $\pi_3 = 180°$ rotation, $\pi_4 = 270°$ rotation, and $\pi_1 = 360°$ (or $0°$) rotation.

Reflections are a little harder to visualize since they are motions in three dimensions. The reflections are $\pi_5 =$ reflection about the vertical axis, $\pi_6 =$ reflection about the horizontal axis, $\pi_7 =$ reflection about opposite corners a and c, and $\pi_8 =$ reflection about opposite corners b and d.

In a regular n-gon, the smallest rotation is $(360/n)°$. Any multiple of this $(360/n)°$ rotation is again a rotation, and so there are n rotations in all. There are two types of reflections for a regular even n-gon: flipping about the middles of two opposite sides and flipping about two opposite corners. Since there are $n/2$ pairs of opposite sides and $n/2$ pairs of opposite corners, a regular even n-gon will have $n/2 + n/2 = n$ reflections. Summing rotations and reflections, we find that a regular even n-gon has $2n$ symmetries.

We leave it as an exercise to show that these $2n$ symmetries just described are distinct. To show that there are at most $2n$ symmetries of an even n-gon, consider a particular corner, call it x, and the edge e incident to x on the clockwise side. The position of x and the relative position of e after the action of a symmetry totally determine the symmetry (readers should convince themselves of this). A symmetry could map x to any of the n corners of the n-gon—n choices—and e could be mapped to either side of x—2 choices. In total, there are $2n$ possible different symmetries of the even n-gon. ■

Example 3: Symmetries of Odd n-gons

Describe the symmetries of a pentagon, and more generally, of an n-gon for odd n.

As noted in Example 2, any regular n-gon has n rotational symmetries. A pentagon will have five rotational symmetries of $0°$, $72°$, $144°$, $216°$, and $288°$. However, the

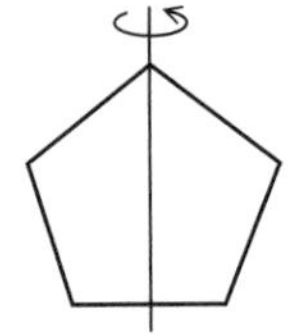

Figure 9.3 **Symmetric reflection of a pentagon**

reflections discussed in Example 2 about opposite sides or opposite corners do not exist in the pentagon. Instead, we reflect about an axis of symmetry running from one corner to the middle of an opposite side (see Figure 9.3). There are five such reflections, for a total of 10 symmetries.

In a regular odd n-gon, there are n such flips, along with n rotations. Summing rotations and reflections, we find that a regular odd n-gon also has $2n$ symmetries. We leave it as an exercise for the reader to show that these $2n$ symmetries are all distinct. The same argument used in Example 2 shows that there are only these $2n$ symmetries. ■

Example 4: Symmetries of a Tetrahedron

Describe the symmetries of a tetrahedron.

A tetrahedron consists of four equilateral triangles that meet at six edges and four corners (see Figure 9.4). Besides the motion leaving all corners fixed, call it the 0° motion, we can revolve 120° or 240° about a corner and the center of the opposite face (see Figure 9.4*a*), or we can revolve 180° about the middle of opposite edges (see Figure 9.4*b*). Since there are four pairs of a corner and opposite face and three pairs of opposite edges, we have a total of 1 (0° motion) $+ 4 \times 2 + 3 = 12$ symmetries. It is an exercise to check that these 12 symmetries are distinct and that no other symmetries exist. ■

The symmetries of a square are naturally characterized by the way they permute the corners of the square. Thus, the 180° rotation π_3 (see Figure 9.2) can be described as the corner permutation: $a \to c, b \to d, c \to a, d \to b$; in tabular form, we write $\binom{a\ b\ c\ d}{c\ d\ a\ b}$. The 90° rotation π_2 can be described as: $a \to b, b \to c, c \to d, d \to a$, or $a \to b \to c \to d \to a$.

A permutation of the form $x_1 \to x_2 \to x_3 \cdots \to x_n \to x_1$ is called a cyclic permutation or **cycle.** Thus, π_2 is a cycle of length 4. Cycles are usually written in the form $(x_1x_2x_3 \ldots x_nx_1)$. So $\pi_2 = (abcd)$. Any permutation can be expressed

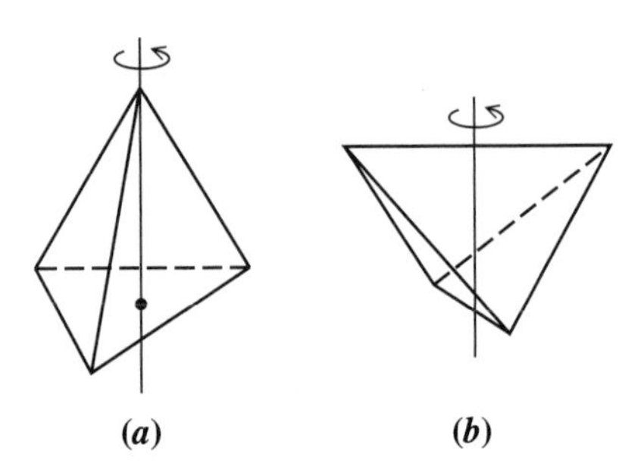

Figure 9.4 **Symmetric revolutions of a tetrahedron**

as a product of disjoint cycles (proof of this claim is an exercise). For example, $\pi_3 = (ac)(bd)$, $\pi_4 = (adcb)$, and $\pi_7 = (a)(bd)(c)$. The reader should find cycle decompositions for the other π_is. The depiction of a motion as in Figure 9.2, with arrows indicating the mapping at each corner, will make it easier to trace out cycles in later calculations.

In permuting the corners, the symmetries create permutations of the colorings of the corners. For example, if C_i is the ith square in Figure 9.1, then π_3 is the following permutation of colorings:

$$\pi_3 = \begin{pmatrix} C_1 & C_2 & C_3 & C_4 & C_5 & C_6 & C_7 & C_8 & C_9 & C_{10} & C_{11} & C_{12} & C_{13} & C_{14} & C_{15} & C_{16} \\ C_1 & C_4 & C_5 & C_2 & C_3 & C_8 & C_9 & C_6 & C_7 & C_{10} & C_{11} & C_{14} & C_{15} & C_{12} & C_{13} & C_{16} \end{pmatrix} \quad (1)$$

The point is that while a symmetry π_i is easily visualized by how it moves the corners of the square, what we are really interested in is the way π_i takes one coloring into another (making them equivalent). Thus, we formally define our coloring equivalence as follows:

$$\text{colorings C and } C' \text{ are } \textit{equivalent, } C \sim C', \\ \text{if there exists a symmetry } \pi_i \text{ such that } \pi_i(C) = C' \quad (2)$$

The properties of the set G of symmetries that interest us are the ones that make the relation $C \sim C'$ in Eq. (2) an equivalence relation. These properties of G are (here $\pi_i \cdot \pi_j$ means *applying motion π_i followed by motion π_j*):

1. *Closure:* If $\pi_i, \pi_j \in G$, then $\pi_i \cdot \pi_j \in G$; for example, in Figure 9.2, $\pi_2 \cdot \pi_5 = \pi_7$.
2. *Identity:* G contains an identity motion π_{I} such that $\pi_{\mathrm{I}} \cdot \pi_i = \pi_i$ and $\pi_i \cdot \pi_{\mathrm{I}} = \pi_i$; in Figure 9.2, π_{I} is π_1.
3. *Inverses:* For each $\pi_i \in G$, there exists an inverse in G, denoted π_i^{-1}, such that $\pi_i^{-1} \cdot \pi_i = \pi_{\mathrm{I}}$ and $\pi_i \cdot \pi_i^{-1} = \pi_{\mathrm{I}}$; for example, in Figure 9.2, $\pi_2^{-1} = \pi_4$.

Observe that closure makes our coloring relation $\sim$ satisfy transitivity [property (i) of an equivalence relation]. For suppose $C \sim C'$ and $C' \sim C''$. Since $C \sim C''$, there must exist $\pi_i \in G$ such that $\pi_i(C) = C'$. Similarly, there is a $\pi_j \in G$ such that $\pi_j(C') = C''$. Then by closure, there exists $\pi_k = \pi_i \cdot \pi_j \in G$ with $\pi_k(C) = (\pi_i \cdot \pi_j)(C) = C''$. Thus $C \sim C''$. Similarly, properties (2) and (3) of the symmetries imply that our coloring relation satisfies properties (ii) and (iii) of an equivalence relation, respectively (see Exercise 16).

A collection G of mathematical objects with a binary operation is called a **group** if it satisfies properties 1, 2, and 3 along with the associativity property—$(\pi_i \cdot \pi_j) \cdot \pi_k = \pi_i \cdot (\pi_j \cdot \pi_k)$. Thus we have the following theorem.

Theorem

Let G be a group of permutations of the set S (corners of a square) and T be any collection of colorings of S (2-colorings of the corners). Then G induces a partition of T into equivalence classes with the relation $C \sim C' \Leftrightarrow$ some $\pi \in G$ takes C to C'.

Note that S could be any set of objects and T could be any possible collection of colorings. The following simple lemma about groups lies at the heart of the counting formula developed in the next section.

Lemma

For any two permutations π_i, π_j in a group G, there exists a unique permutation $\pi_k = \pi_i^{-1} \cdot \pi_j$ in G such that $\pi_i \cdot \pi_k = \pi_j$.

Proof

First we show that $\pi_i \cdot \pi_k = \pi_j$. Since $\pi_k = \pi_i^{-1} \cdot \pi_j$, then

$$\begin{aligned}\pi_i \cdot \pi_k = \pi_i \cdot \left(\pi_i^{-1} \cdot \pi_j\right) &= \left(\pi_i \cdot \pi_i^{-1}\right) \cdot \pi_j \quad \text{(by associativity)} \\ &= \pi_{\mathrm{I}} \cdot \pi_j = \pi_j\end{aligned}$$

as claimed. Next suppose there also exists a permutation π_k' such that $\pi_i \cdot \pi_k' = \pi_j$. Then $\pi_i \cdot \pi_k = \pi_i \cdot \pi_k'$. Multiplying the equation by π_i^{-1}, we have

$$\begin{aligned}\pi_i^{-1} \cdot (\pi_i \cdot \pi_k) = \pi_i^{-1} \cdot (\pi_i \cdot \pi_k') &\Rightarrow (\pi_i^{-1} \cdot \pi_i) \cdot \pi_k = \left(\pi_i^{-1} \cdot \pi_i\right) \cdot \pi_k' \\ &\Rightarrow \pi_{\mathrm{I}} \cdot \pi_k = \pi_{\mathrm{I}} \cdot \pi_k' \Rightarrow \pi_k = \pi_k' \quad \blacklozenge\end{aligned}$$

9.1 EXERCISES

Summary of Exercises The first 13 exercises continue the examples of equivalence and symmetries given in this section. The remaining exercises develop basic aspects of group theory associated with symmetries. Prior experience with modern algebra is needed for most of these latter problems.

1. Which of the following relations are equivalence relations? State your reasons.
   - **(a)** $\leq$ (less than or equal to), for a set of numbers.
   - **(b)** $=$ (equal to), for a set of numbers.
   - **(c)** "Difference is odd," for a set of numbers.
   - **(d)** "Being blood relations," for a group of people.
   - **(e)** "Having a common friend," for a group of people.
2. Which of the following collections with given operations are groups? For those collections that are groups, which elements are the identities?
   - **(a)** The nonnegative integers 0, 1, 2, ... with addition.
   - **(b)** The integers 0, 1, 2, ..., $n-1$ with addition modulo n.
   - **(c)** All polynomials (with integer coefficients) with polynomial addition.
   - **(d)** All nonzero fractions with regular multiplication.
   - **(e)** All invertible 2×2 real-valued matrices with matrix multiplication.

3. Find all symmetries of the following figures (indicate with arrows where the corners move in each symmetry, as in Figure 9.2):

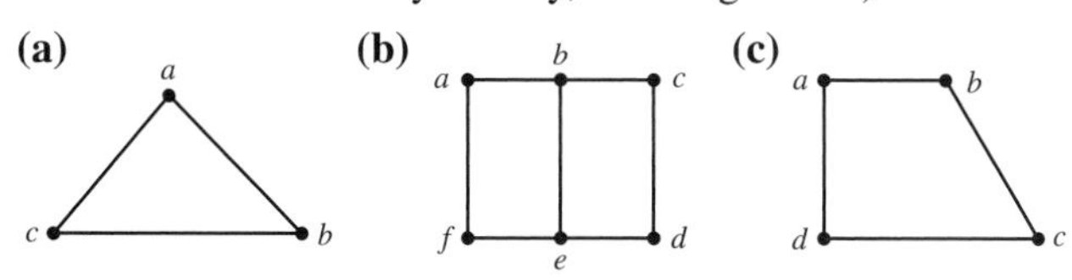

4. Write the following symmetries or permutations as a product of cyclic permutations:

(a) π_2 **(b)** π_3 **(c)** π_6 **(d)** π_I
(e) $\pi_4 \cdot \pi_7$ **(f)** $\pi_7 \cdot \pi_4$ **(g)** $\begin{pmatrix} 1 & 2 & 3 & 4 & 5 & 6 & 7 \\ 3 & 5 & 4 & 6 & 7 & 1 & 2 \end{pmatrix}$

5. For the symmetries of the square listed in Figure 9.2, give the associated permutation of 2-colorings, as in Eq. (1), for:

(a) π_1 **(b)** π_2 **(c)** π_5 **(d)** π_7

6. **(a)** List all 2-colorings of the three corners of a triangle.

(b) For the following symmetries of a triangle, give the associated permutation of 2-colorings, as in Eq. (1), for:

(i) $\pi = 120°$ rotation **(ii)** $\pi =$ flip about vertical axis.

7. Show that the 8 symmetries of a square listed in Figure 9.2 are all distinct.

8. Show that the $2n$ symmetries of an n-gon (even or odd) mentioned in Examples 2 and 3 are all distinct.

9. Show that the 12 symmetries of a tetrahedron listed in Example 4 are all distinct. Show that there are exactly 12 symmetries of a tetrahedron.

10. Find the symmetry of the square equal to the following products (remember that $\pi_i \cdot \pi_j$ means applying motion π_i followed by motion π_j):

(a) $\pi_2 \cdot \pi_4$ **(b)** $\pi_2 \cdot \pi_5$ **(c)** $\pi_7 \cdot (\pi_2 \cdot \pi_8)$ **(d)** $(\pi_7 \cdot \pi_6) \cdot \pi_3$

11. **(a)** Write out the 6×6 multiplication table for the product of all pairs of symmetries of a triangle.

(b) Repeat part (a) for integers 1, 2, 3, 4 with multiplication modulo 5.

(c) Repeat part (a) for the following group of permutations of 1, 2, 3, 4:

$$\begin{pmatrix} 1 & 2 & 3 & 4 \\ 1 & 2 & 3 & 4 \end{pmatrix} \begin{pmatrix} 1 & 2 & 3 & 4 \\ 2 & 1 & 4 & 3 \end{pmatrix} \begin{pmatrix} 1 & 2 & 3 & 4 \\ 3 & 4 & 1 & 2 \end{pmatrix} \begin{pmatrix} 1 & 2 & 3 & 4 \\ 4 & 3 & 2 & 1 \end{pmatrix}$$

12. Find two symmetries of the square π_i, π_j such that $\pi_i \cdot \pi_j \neq \pi_j \cdot \pi_i$ (this means the group of symmetries of a square is noncommutative).

13. Many organic compounds consist of a basic structure formed by carbon atoms (carbon atoms are the corners of a floating figure), plus submolecular groups called radicals that are attached to each carbon atom (the carbons are like corners and the radicals like colors). Suppose such an organic molecule has six carbon atoms, with four radicals of type A and two radicals of type B. Suppose this molecule has three *isomers,* that is, three different ways that the two types of radicals can be distributed. Which of the following two hypothetical carbon structures would have three isomers with four As and two Bs?

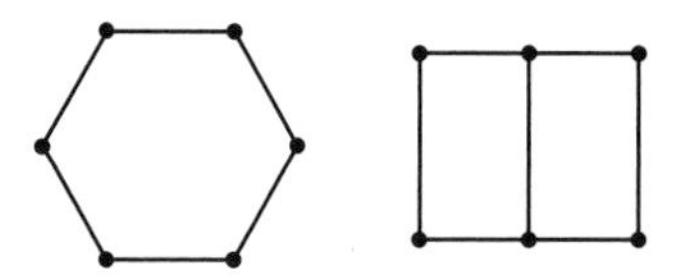

14. Prove that an equivalence relation partitions a set into disjoint subsets of mutually equivalent elements.

15. Give a procedure for decomposing any permutation into a product of cycles.

16. Show that properties (2) and (3) of a group G of permutations imply properties (ii) and (iii), respectively, of the associated equivalence relation defined in the theorem.

17. Let S be a set and G a group of permutations of S. For any two subsets S_1, S_2 of S, define $S_1 \sim S_2$ to mean that for some $\pi \in G$, $S_1 = \pi(S_2) (= \{\pi(s) \mid s \in S_2\})$. Show that $\sim$ is an equivalence relation.

18. Prove that the set of permutations of the 2-colorings of a square [see Eq. (1)] forms a group.

19. Prove that for any prime p, the integers $1, 2, \ldots, p-1$ with multiplication modulo p form a group.

20. A subset of elements in a group G is said to *generate* the group if all elements in G can be obtained as (repeated) products of elements in the subset.

(a) Which of the following subsets generate the group of symmetries of a square?

(i) π_1, π_2, π_3 **(ii)** π_2, π_5 **(iii)** π_3, π_6 **(iv)** π_6, π_7.

(b) Show that the group of symmetries of a regular n-gon can be generated by a subset of two elements.

21. If a subset G' of elements in a group G is itself a group, then G' is called a *subgroup* of G.

(a) Show that $G' = \{\pi_1, \pi_2, \pi_3, \pi_4\}$ and $G'' = \{\pi_1, \pi_7\}$ are subgroups of the group G of symmetries of a square.

(b) Find another 4-element subgroup of G containing π_3.

(c) Find all subgroups of the group G of symmetries of a square.

22. **(a)** How many different binary relations on n elements are possible?

(b) How many symmetric binary relations are possible?

23. Show that $\pi_i \sim \pi_j$ if there exists $\pi \in G$ such that $\pi_i = \pi^{-1}\pi_j\pi$ is an equivalence relation.

24. A *transposition* is a cycle of size 2, that is, a permutation that interchanges the positions of just two elements (and leaves all other elements fixed). Show by induction that any permutation of n elements can be written as a composition of transpositions.

25. For a given group G of n elements, define the function f_π on G as follows: for each $\pi' \in G$, $f_\pi(\pi') = \pi \cdot \pi'$.

(a) Show that f_π is a one-to-one mapping for any $\pi \in G$.

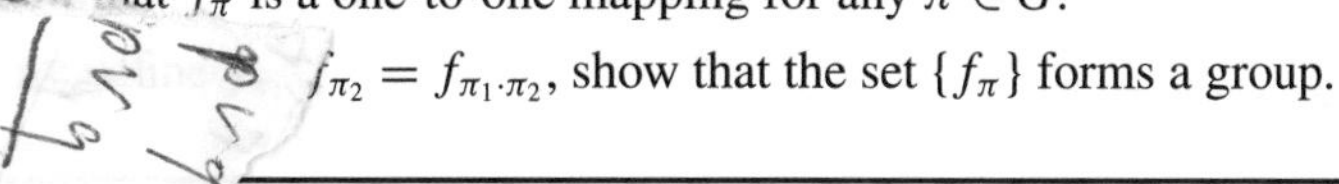

$f_{\pi_2} = f_{\pi_1 \cdot \pi_2}$, show that the set $\{f_\pi\}$ forms a group.

9.2 BURNSIDE'S THEOREM

We now develop a theory for counting the number of different (nonequivalent) 2-colorings of the square. More generally, in a set T of colorings of the corners (or edges or faces) of some figure, we seek the number N of equivalence classes of T induced by a group G of symmetries of this figure.

Suppose there is a group of s symmetries acting on the c colorings in T. Let E_C be the equivalence class consisting of C and all colorings C' equivalent to C, that is, all C' such that for some $\pi \in G$, $\pi(C) = C'$. If each of the s πs takes C to a different coloring $\pi(C)$, then E_C would have s colorings. Note that the set of $\pi(C)$s includes C since $\pi_I(C) = C (\pi_I$ is the identity symmetry). If every equivalence class is like this with s colorings, then

$$sN = c: \quad \text{(number of symmetries)} \times \text{(number of equivalence classes)} = \text{(total number of colorings)}$$

Solving for N, we have $N = c/s$.

Consider, for example, the $c = n!$ oriented seatings of n people around a round table. There are $s = n$ cyclic rotations of the seatings, and each equivalence class consists of n seatings. Thus, the number of equivalence classes (cyclicly nonequivalent seatings) is $N = n!/n = (n-1)!$

On the other hand, suppose we have a small round table with three positions for chairs (each 120° apart), and white and black chairs are available. There are $2^3 = 8$ ways to place a white or black chair in each position. See Figure 9.5. There are three cyclic rotations of the table possible, 0°, 120°, and 240°. We have $c = 8$ "colorings" and $s = 3$ symmetries, but the number of equivalence classes cannot be $N = \frac{8}{3}$, a fraction!

It is true that the three arrangements of one black and two white chairs (or vice versa) form an equivalence class, since 0°, 120°, and 240° rotations move the one black chair to different positions. However, an arrangement of three black chairs (or three white chairs) forms an equivalence class by itself. See Figure 9.5. Any rotation maps this arrangement of 3 black chairs into itself, that is, leaves it fixed.

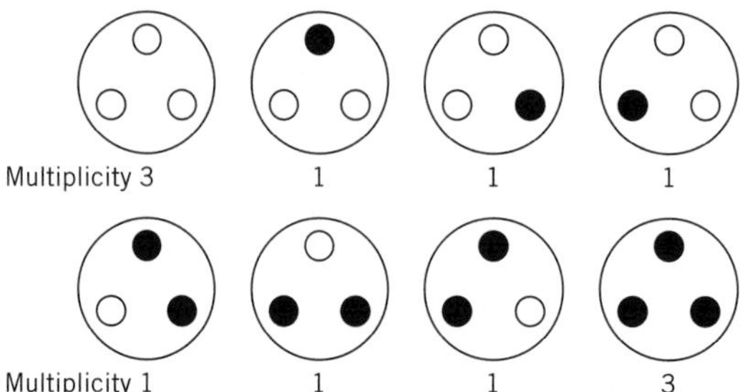

Figure 9.5

We need to correct the numerator in the formula $N = c/s$ by adding the multiplicities of an arrangement, when one or more symmetries map the arrangement to itself instead of to other arrangements. In this way when multiplicities are counted, every equivalence class will have s members. Since two symmetries, along with the 0° symmetry, leave the all-black-chair arrangement fixed and similarly for the all-white, then counting arrangements in Figure 9.5 with multiplicities we have the correct answer

$$N = (3+1+1+1+1+1+1+3)/3 = 12/3 = 4$$

The "multiplicity" correction is even more complicated for 2-colorings of our square. Here the size of an equivalence class of colorings may be 1 or 2 or 4, but never 8 (= the number of symmetries of the square). The first problem is that several πs, besides the identity symmetry π_1, may leave a coloring C_i fixed—that is, $\pi(C_i) = C_i$. The other problem is that if C_k is another coloring in C_i's equivalence class, there may be several πs all taking C_i to C_k. For example, the coloring C_{10} (see Figure 9.1) is fixed by symmetries $\pi_1, \pi_3, \pi_7, \pi_8$, and is mapped to C_{11} by symmetries $\pi_2, \pi_4, \pi_5, \pi_6$.

By the lemma in Section 9.1, the symmetries $\pi_2, \pi_4, \pi_5, \pi_6$ taking C_{10} to C_{11} can be written in the form $\pi = \pi_2 \cdot \pi'$, where π' is a symmetry that leaves C_{11} fixed. For example,

$$\pi_5 = \pi_2 \cdot \pi_8 \colon \begin{pmatrix} a\,b\,c\,d \\ b\,a\,d\,c \end{pmatrix} = \begin{pmatrix} a\,b\,c\,d \\ b\,c\,d\,a \end{pmatrix} \cdot \begin{pmatrix} a\,b\,c\,d \\ c\,b\,a\,d \end{pmatrix}$$

Similarly $\pi_2 = \pi_2 \cdot \pi_1, \pi_4 = \pi_2 \cdot \pi_3, \pi_6 = \pi_2 \cdot \pi_7$. Conversely, given any π^* that leaves C_{11} fixed, $\pi_2 \cdot \pi^*$ takes C_{10} to C_{11} and so $\pi_2 \cdot \pi^*$ must be one of $\pi_2, \pi_4, \pi_5, \pi_6$. Thus there is a $1-1$ correspondence between the πs that take C_{10} to C_{11} and the πs that leave C_{11} fixed. Therefore, *to count the colorings in an equivalence class E with appropriate multiplicities* (i.e., coloring C_{11} has multiplicity 4 since four different πs take C_{10} to C_{11}), *it suffices for each coloring in E to sum the number of* π*s that leave the coloring fixed.*

In the case of the equivalence class consisting of C_{10} and C_{11}, each of C_{10} and C_{11} have multiplicity 4, so that the size of their equivalence class including multiplicities is $4+4=8$ (= s, the number of symmetries), as required.

In general, when multiplicities are counted, each equivalence class E will have s elements. If $\phi(x)$ denotes the number of πs that leave the coloring x fixed, then $\sum_{x \in E} \phi(x) = s$.

Formal proof that $\sum_{x \in E} \phi(x) = s$: Let $x_1, x_2, \ldots, x_m$ be the colorings in equivalence class E and let $\pi_1, \pi_2, \ldots, \pi_s$ be the group of symmetries. These πs can be divided into m groups, $R_1, R_2, \ldots, R_m$, where R_i is the set of πs that map x_1 to x_i. As shown above for $x_1 = C_{10}$ and $x_i = C_{11}$, the number of πs mapping x_1 to x_i equals $\phi(x_i)$, the number of πs that leave x_i fixed. Thus, $\sum_{x \in E} \phi(x)$ sums the number of πs in R_1, in $R_2, \ldots,$ and in R_m. But the sum of the Rs is just the total number of symmetries, s.

Summing over all equivalence classes, we c
proved by Burnside about 100 years ago.

Theorem **(Burnside, 1897)**

Let G be a group of permutations of the set S (corners of a square). Let T be any collection of colorings of S (2-colorings of the corners) that is closed under G. Then the number N of equivalence classes is

$$N = \frac{1}{|G|} \sum_{x \in T} \phi(x)$$

or

$$N = \frac{1}{|G|} \sum_{\pi \in G} \Psi(\pi) \qquad (*)$$

where $|G|$ is the number of permutations and $\Psi(\pi)$ is the number of colorings in T left fixed by π.

By "closed under G," we mean that for all $\pi \in G$ and $x \in T$, $\pi(x) \in T$. This closure property is automatic when T is the set of all corner 2-colorings of the square. In Section 9.4 we need to apply formula (*) to special subsets of S, such as the set of colorings with two corners black and two corners white, that are closed under G.

The two sums in the theorem both count all instances of some coloring being left fixed by some π, the first sums over the different colorings, the second sums over different πs. Formula (*) will turn out to be more useful in later computations.

We informally summarize the spirit behind formula (*) as follows. The total number c of all colorings of our square is equal to $\Psi(\pi_1)$, since the identity symmetry π_1 leaves all colorings fixed. If each of the 8 symmetries mapped a coloring C into 8 different colorings, then we would have 8 colorings in each equivalence class and hence a total $c/8 = \Psi(\pi_1)/|G|$ equivalence classes. However, the colorings in the collection $\{\pi(C) : \pi \in G\}$ forming an equivalence class of colorings of an unoriented square are never distinct for an unoriented square. The terms $\Psi(\pi_i)$ in (*) add the "multiplicities" of repeated colorings so that each equivalence class has 8 colorings in the sum in (*).

Example 1: 2-Colored Batons

A baton is painted with equal-sized cylindrical bands. Each band can be painted black or white. If the baton is unoriented as when spun in the air, how many different 2-colorings of the baton are possible if the baton has 2 bands? 3 bands? 4 bands?

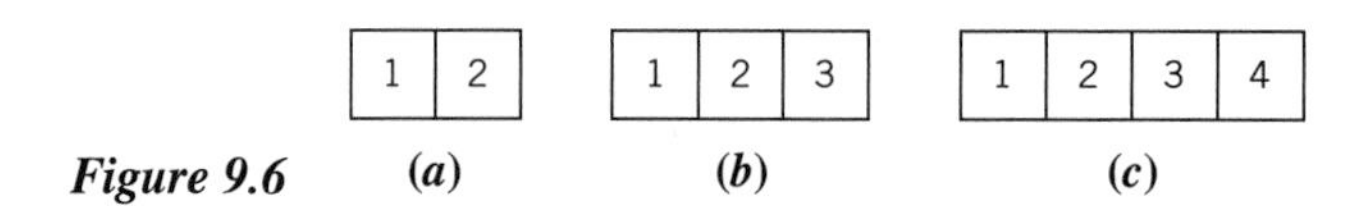

Figure 9.6 (*a*) (*b*) (*c*)

The batons with 2 bands, 3 bands, and 4 bands are pictured in Figure 9.6. Irrespective of the number of bands, there are two symmetries of a baton: π_1 is a $0°$ revolution of the baton—π_1 is the identity symmetry—and π_2 is a $180°$ revolution of the baton.

(a) For the 2-band baton, the set of 2-colorings left fixed by π_1 is all 2-colorings of the baton. There are $2^2 = 4$ 2-colorings, and so $\Psi(\pi_1) = 4$. The set of 2-colorings left fixed by π_2 consists of the all-black and all-white coloring, and so $\Psi(\pi_2) = 2$. By Burnside's Theorem, the number of different colorings is $\frac{1}{2}[\Psi(\pi_1) + \Psi(\pi_2)] = \frac{1}{2}(4+2) = 3$.

(b) For the 3-band baton, all 2^3 2-colorings are left fixed by π_1, and so $\Psi(\pi_1) = 2^3 = 8$. The set of 2-colorings left fixed by π_2 can have any color in the middle band (band 2) and a common color in the two end bands, and so $\Psi(\pi_2) = 2 \times 2 = 4$. The number of different colorings is $\frac{1}{2}[\Psi(\pi_1) + \Psi(\pi_2)] = \frac{1}{2}(8+4) = 6$.

(c) For the 4-band baton, all 2^4 2-colorings are left fixed by π_1, and so $\Psi(\pi_1) = 2^4 = 16$. The set of 2-colorings left fixed by π_2 have a common color for the end bands and a common color for the inner bands, so $\Psi(\pi_2) = 2 \times 2 = 4$. The number of different colorings is $\frac{1}{2}[\Psi(\pi_1) + \Psi(\pi_2)] = \frac{1}{2}(16+4) = 10$. ∎

Example 2: 3-Colored Batons

How many different 3-colorings of the bands of an n-band baton are there if the baton is unoriented (as in Example 1)?

As in Example 1, the symmetries of the baton are a $0°$ revolution and a $180°$ revolution. We apply formula (*) for the group of the $0°$ and $180°$ revolution acting on an n-band baton with 3 colors. There are 3^n colorings of the fixed baton and so $\Psi(0°) = 3^n$. The number of colorings left fixed by a $180°$ spin depends on whether n is even or odd.

If n is even, each of the $n/2$ bands on one half of the baton can be any color—$3^{n/2}$ choices—and then for the coloring to be fixed by a $180°$ spin, each of the symmetrically opposite bands must be the corresponding color. So $\Psi(180°) = 3^{n/2}$ and we have from formula (*): $N = \frac{1}{2}(3^n + 3^{n/2})$.

To enumerate batons left fixed by a $180°$ spin when n is odd, we can use any color for the "odd" band in the middle of the baton—3 choices. Each of the $(n-1)/2$ bands on one side of the middle band can be any color—$3^{(n-1)/2}$ choices—and again the other $(n-1)/2$ bands must be colored symmetrically. So $\Psi((180°) = 3 \times 3^{(n-1)/2} = 3^{(n+1)/2}$ and $N = \frac{1}{2}(3^n + 3^{(n+1)/2})$. ∎

Example 3: 3-Colored Necklaces

Suppose a necklace can be made from beads of three colors, black, white, and red. How many different necklaces with n beads are there?

When the n beads are positioned symmetrically about the circle, the beads occupy the positions of the corners of a regular n-gon. Thus, our question asks for the number of corner colorings of an n-gon using three colors. The answer depends on what is meant by "different." If the beads are not allowed to move about the necklace, that is, the n-gon is fixed, the answer is $3 \times 3 \times \cdots \times 3 = 3^n$ (three color choices at each of n corners). A more realistic interpretation of our problem would allow the beads to move freely about the circle, that is, the n-gon rotates freely (but in this case flips will not be allowed). We employ formula (*) to count the number N of equivalence classes of these 3-colorings induced by the rotational symmetries of an n-gon. We shall now do the calculations for $n = 3$. A more general technique for larger n is developed in the next section.

There are $3^3 = 27$ 3-colorings of a 3-bead necklace, and three rotations of $0°$, $120°$, $240°$. The $0°$ rotation leaves all colorings fixed, and so $\Psi(0°) = 27$. The $120°$ rotation cannot fix colorings in which some color occurs at only one corner. It follows that the $120°$ rotation fixes just the monochromatic colorings. Thus, $\Psi(120°) = 3$. The $240°$ rotation is a reverse $120°$ rotation, and so $\Psi(240°) = 3$. By formula (*), we have

$$N = \frac{1}{3}(27 + 3 + 3) = 11 \quad \blacksquare$$

9.2 EXERCISES

Summary of Exercises The exercises continue the application of Burnside's Theorem to count colorings introduced in Example 1, 2, and 3.

1. How many different n-bead necklaces are there using beads of red, white, blue, and green (assume necklaces can rotate but cannot flip over)?

(a) $n = 3$ **(b)** $n = 4$

2. How many different ways are there to place a diamond, sapphire, or ruby at each of the four vertices of this pin?

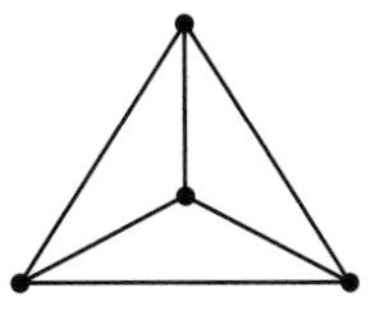

3. Fifteen balls are put in a triangular array as shown. How many different arrays can be made using balls of three colors if the array is free to rotate?

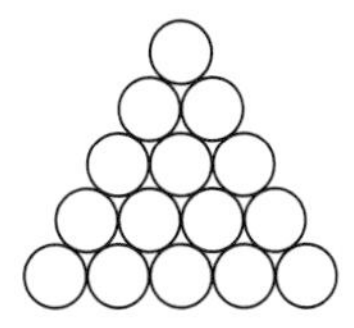

4. How many different ways are there to 2-color the 64 squares of an 8×8 chessboard that rotates freely?

5. A merry-go-round can be built with three different styles of horses. How many five-horse merry-go-rounds are there?

6. A domino is a thin rectangular piece of wood with two adjacent squares on one side (the other side is black). Each square is either blank or has 1, 2, 3, 4, 5, or 6 dots.

   (a) How many different dominoes are there?

   (b) Check your answer by modeling a domino as an (unordered) subset of 2 numbers chosen with repetition for 0, 1, 2, 3, 4, 5, 6.

7. Two n-digit decimal sequences consisting of digits 0, 1, 6, 8, 9 are vertically equivalent if reading one upside down produces the other, that is, 0068 and 8900. How many different (vertically inequivalent) n-digit (0, 1, 6, 8, 9) sequences are there?

8. How many different ways are there to color the five faces of an unoriented pyramid (with a square base) using red, white, blue, and yellow?

9. **(a)** Find the group of all possible permutations of three objects.

   (b) Find the number of ways to distribute 12 identical balls in three indistinguishable boxes. [*Hint:* First let boxes be distinct, and then use part (a).]

10. How many ways are there to 3-color the n bands of a baton if adjacent bands must have different colors?

11. How many ways are there to 3-color the corners of a square with rotations and reflections allowed if adjacent corners must have different colors?

12. **(a)** How many ways are there to distribute 12 red jelly beans to four children, a pair of identical female twins and a pair of identical male twins?

   (b) Repeat part (a) with the requirement that each child must have at least one jelly bean.

13. **(a)** Show that for any given i, the subset G_i of symmetries of the square that leave coloring C_i (of Figure 9.1) fixed is a group (a subgroup of all symmetries).

   (b) Find G_4.

   (c) Find G_7.

9.3 THE CYCLE INDEX

Without further theory, we would find it very difficult to apply Burnside's Theorem to counting different colorings of an unoriented figure. Recall that Burnside's Theorem in Section 9.2 says that the number N of equivalence classes in a set T of colorings of a figure with respect to a group of symmetries G is

$$N = \frac{1}{|G|} \sum_{\pi \in G} \Psi(\pi)$$

where $\Psi(\pi)$ is the number of colorings in T left fixed by π. If the set T were all 3-colorings of the corners of a 10-gon or a cube, it would seem close to impossible to determine $\Psi(\pi)$s, the number of colorings left fixed by various symmetries π of the figure. However, we shall show that $\Psi(\pi)$ can be determined easily from the structure of π. We develop the theory for this simplified calculation of $\Psi(\pi)$ in terms of 2-colorings of a square.

Let us apply Burnside's Theorem to the 2-colorings of the square. Initially we determine the number of 2-colorings left fixed by motion π_i by inspection (using Figures 9.1 and 9.2). As we count $\Psi(\pi_i)$ for each π_i, we look for a pattern that would enable us to predict mathematically which colorings must be left fixed by each π_i. It is helpful to make a table of the π_is and the colorings that they leave fixed; see columns (i) and (ii) in Figure 9.7 [columns (iii) and (iv) are developed later].

The 0° rotation π_1 leaves each corner fixed and hence it leaves all colorings fixed. So $\Psi(\pi_1) = 16$. The 90° rotation π_2 cyclicly permutes corners a, b, c, d. Being left fixed by the 90° rotation means that each corner in a coloring has the same color after the 90° rotation as it did before. Since π_2 takes a to b, then a coloring left fixed by π_2 must have the same color at a as at b. Similarly, such a coloring must have the same color at b as at c, the same color at c as at d, and the same at d as at a. Taken together, these conditions imply that only the colorings of all white or all black corners, C_1 and C_{16}, are left fixed. Thus $\Psi(\pi_2) = 2$. In general, *a coloring C will be left fixed by π if and only if for each corner v, the color of C at v is the same as the color at $\pi(v)$ so that the symmetry leaves the color at $\pi(v)$ unchanged.*

Next we consider the 180° rotation π_3. Looking at the depiction of π_3 in Figure 9.2, we see that π_3 causes corners a and c to interchange and corners b and d to

(i) Motion π_i	*(ii)* Colorings Left Fixed by π_i	*(iii)* Cycle Structure Representation
π_1	16—all colorings	x_1^4
π_2	2—C_1, C_{16}	x_4
π_3	2—$C_1, C_{10}, C_{11}, C_{16}$	x_2^2
π_4	2—C_1, C_{16}	x_4
π_5	2—C_1, C_6, C_8, C_{16}	x_2^2
π_6	2—C_1, C_7, C_9C_{16}	x_2^2
π_7	8—C_1, C_2, C_4, C_{10} $C_{11}, C_{12}, C_{14}, C_{16}$	$x_1^2x_2$
π_8	8—C_1, C_3, C_5, C_{10} $C_{11}, C_{13}, C_{15}, C_{16}$	$x_1^2x_2$

Figure 9.7

interchange. It follows that a coloring left fixed by π_3 must have the same color at corners a and c and the same color at b and d (no further conditions are needed). With two color choices for a, c and with two color choices for b, d, we can construct $2 \times 2 = 4$ colorings that will be left fixed, namely, $C_1, C_{10}, C_{11}, C_{16}$. Hence $\Psi(\pi_3) = 4$.

Symmetry π_4 is a rotation of 270° or −90°, and so the symmetry is similar to π_2. Hence $\Psi(\pi_4) = \Psi(\pi_2) = 2$. The horizontal rotation π_5 interchanges corners a, b and interchanges corners c, d. Like the 180° rotation π_3, symmetry π_5 will leave a coloring fixed if corners a and b have the same color—2 choices, white or black—and corners c and d have the same color—2 choices. Then like π_3, we have $\Psi(\pi_5) = 2 \times 2 = 4$ colorings fixed, namely, C_1, C_6, C_8, C_{16}.

A pattern is becoming clear. The reader should now be able quickly to predict that π_6 will also leave $2 \times 2 = 4$ 2-colorings of the square fixed, for again this symmetry interchanges two pairs of corners. Formally, an interchange is a cyclic permutation on two elements. All our enumeration of fixed colorings has been based on the fact that if π_i *cyclicly permutes a subset of corners (that is, the corners form a cycle of* π_i*), then those corners must be the same color in any coloring left fixed by* π_i.

As mentioned in Section 9.1, any π_i can be represented as a product of disjoint cycles. For example, $\pi_3 = (ac)(bd)$ and $\pi_4 = (adcb)$. For each symmetry, we need to get such a cyclic representation and count the number of ways to assign a color to each cycle of corners. For future use, let us also classify the cycles by their length. It will prove convenient to encode a symmetry's cycle information in the form of a product containing one x_1 for each cycle of 1 corner in π_i, one x_2 for each cycle of size 2, and so forth. This expression is called the **cycle structure representation** of a symmetry.

The cycle structure representation of π_2 and π_4 is x_4, since each consists of one 4-cycle: $\pi_2 = (abcd)$ and $\pi_4 = (adcb)$. The cycle structure representation for π_3, π_5, and π_6 is x_2x_2 or x_2^2, since each consists of two 2-cycles. Column (iii) in Figure 9.7 gives the cycle structure representation of each symmetry.

What about π_1? Previously, it sufficed to say that π_1 leaves all colorings fixed. Now it is time to point out that a corner left fixed by a permutation is classified as a 1-cycle. Thus π_1 consists of four 1-cycles. Its cycle structure representation is then π_1^4. We "predict" that π_1 leaves $2^4 = 16$ colorings fixed, that is, π_1 leaves all 2-colorings fixed.

For any π_i, *the number of colorings left fixed will be given by setting each* x_j *equal to 2 (or, in general, the number of colors available) in the cycle structure representation of* π_j, *that is,*

$$\Psi(\pi) = 2^{\text{number of cycles in } \pi}$$

Let us apply this theory to π_7 and π_8. For each of these reflections, the cycle structure representation is seen to be $x_1^2x_2$ (follow the arrows in Figure 9.2) and thus for each we can find $2^2 \times 2 = 8$ colorings that are left fixed. Note that for 180° flips around opposite corners and midpoints of opposite edges, all corners will be in cycles of size 2, unless a corner is left fixed by the flip.

To obtain the number of different 2-colorings of the floating square with Burnside's Theorem, we sum the numbers in column (ii) of Figure 9.7 and divide by 8:

$$\frac{1}{|G|}\sum_{\pi \in G} \Psi(\pi) = \frac{1}{8}(16+2+4+2+4+4+8+8) = \frac{1}{8}(48) = 6$$

There is a slightly simpler way to get this result. First, algebraically sum the cycle structure representations of each symmetry, collecting like terms together, and then divide by 8. From column (iii) of Figure 9.7, we obtain $\frac{1}{8}(x_1^4 + 2x_4 + 3x_2^2 + 2x_1^2 x_2)$. This expression is called the **cycle index** $P_G(x_1, x_2, \ldots, x_k)$ for a group G of symmetries. By setting each $x_i = 2$ in this cycle index, that is, $P_G(2, 2, \ldots, 2)$, we get the same answer. (Before, the steps were reversed: We first set $x_i = 2$ in each cycle structure representation and then added.)

Suppose that instead of 2 colors, we had 3 colors. Then the same reasoning applies, but now there are 3 choices for the color of the corners in each cycle. If a symmetry has k cycles, then it will leave 3^k 3-colorings of the square fixed, and the number of different 3-colorings will be $P_G(3, 3, \ldots, 3)$. More generally, for any m, $P_G(m, m, \ldots, m)$ will be the number of nonequivalent m-colorings of an unoriented square. The argument used to derive this coloring counting formula with the cycle index of a square is valid for colorings of any set with associated symmetries.

Theorem

Let S be a nonempty set of elements and G be a group of symmetries of S that acts to induce an equivalence relation on the set of m-colorings of S. Then the number of nonequivalent m-colorings of S is given by $P_G(m, m, \ldots, m)$.

Example 1: Coloring Necklaces

Use this theorem to re-solve the problem of Example 3 in Section 9.2 of counting n-bead necklaces with black, white, and red beads.

Recall that beads on the necklace can rotate freely around the circle but reflections are not permitted and that the number of different 3-colored strings of n beads is equal to the number of 3-colorings of a cyclicly unoriented n-gon. For $n = 3$, the rotations are of 0°, 120°, and 240° with cycle structure representations of x_1^3, x_3, and x_3, respectively. Thus, $P_G = \frac{1}{3}(x_1^3 + 2x_3)$. The number of 3-colored strings of three beads is $P_G(3, 3, 3) = \frac{1}{3}(3^3 + 2 \times 3) = 11$. More generally, the number of m-colored necklaces of three beads is $P_G(m, m, m) = \frac{1}{3}(m^3 + 2m)$.

Let us try a more complicated case: $n = 8$ (see Figure 9.8). The rotations are of 0°, 45°, 90°, 135°, 180°, 225°, 270°, and 315°. The 0° rotation consists of eight 1-cycles. The 45° rotation is the cyclic permutation $(abcdefgh)$. The 90° rotation has the cycle decomposition $(aceg)$ $(bdfh)$. The 135° rotation is the cyclic permutation $(adgbehcf)$. The 180° rotation has the cyclic decomposition (ae) (bf) (cg) (dh). The cycle structure representations are thus: 0° rotation, x_1^8; 45° rotation, x_8; 90° rotation,

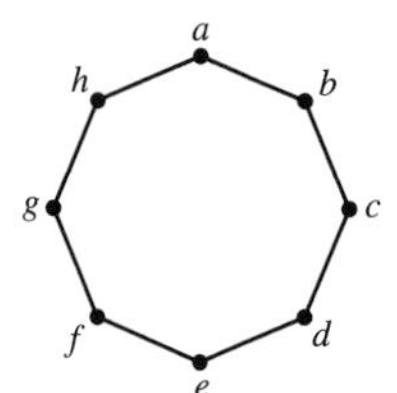

Figure 9.8

x_4^2; 135° rotation, x_8; and 180° rotation, x_2^4. The 225°, 270°, and 315° rotations are reverse rotations of 135°, 90°, 45°, respectively, and have the corresponding cycle structure representations. Collecting terms, we obtain

$$P_G = \frac{1}{8}\left(x_1^8 + 4x_8 + 2x_4^2 + x_2^4\right)$$

The number of different m-colored necklaces of eight beads is

$$\frac{1}{8}(m^8 + 4m + 2m^2 + m^4)$$

For $m = 3$, we have

$$\frac{1}{8}(3^8 + 4 \times 3 + 2 \times 3^2 + 3^4) = \frac{1}{8}(6561 + 12 + 18 + 81) = 834 \;\blacksquare$$

Example 2: Coloring Corners of a Tetrahedron

Use the theorem to determine the number of 3-colorings of the 4 corners of a floating tetrahedron. In Example 4 in Section 9.1 we listed the 12 symmetries of the tetrahedron (see Figure 9.4 in Section 9.1): the 0° revolution, the eight revolutions of 120° and 240° about a corner and the middle of the opposite face, and the three revolutions of 180° about the middle of opposite edges.

The 0° revolution has the cycle structure representation x_1^4. The 120° revolution about corner a and the middle of face bcd has the cyclic decomposition $(a)(bcd)$ and its cycle structure representation is x_1x_3. By symmetry, the other 120° and 240° revolutions have this same cycle structure representation. The 180° revolution about the middle of edges ab and cd has the cyclic decomposition $(ab)(cd)$ and its cycle structure representation is x_2^2. By symmetry, the other 180° revolutions have the same cycle structure representation. Thus we have

$$P_G = \frac{1}{12}\left(x_1^4 + 8x_1x_3 + 3x_2^2\right)$$

The number of different corner 3-colorings is

$$\begin{aligned} P_G(3, 3, 3, 3) &= \frac{1}{12}(3^4 + 8 \times 3 \times 3 + 3 \times 3^2) \\ &= \frac{1}{12}(81 + 72 + 27) = 15 \;\blacksquare \end{aligned}$$

9.3 EXERCISES

Summary of Exercises The first 11 exercises count distinct colorings of unoriented figures. The remaining problems involve associated theory. Note that "floating" means that all possible rotations and reflection are allowed.

1. How many ways are there to color the corners of a floating square using four different colors?

2. How many ways are there to 4-color the corners of a pentagon that is:
   (a) Distinct with respect to rotations only?
   (b) Distinct with respect to rotations and reflections?

3. How many ways are there to 3-color the corners of a hexagon that is:
   (a) Distinct with respect to rotations only?
   (b) Distinct with respect to rotations and reflections?
   (c) Find two 3-colorings of the hexagon that are different in part (a) but equivalent in part (b).

4. How many different n-bead necklaces (cyclicly distinct) can be made from three colors of beads when:
   (a) $n = 7$ **(b)** $n = 9$ **(c)** $n = 10$ **(d)** $n = 11$

5. Find the number of different m-colorings of the vertices of the following floating figures.

(a)

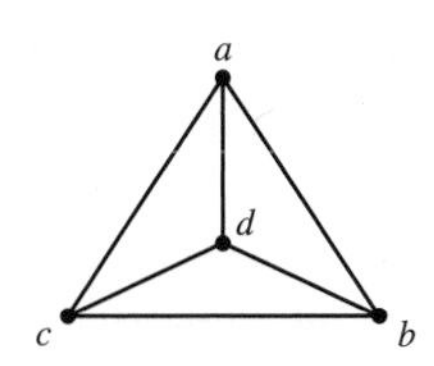

(b)

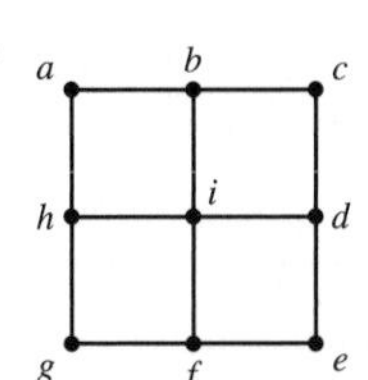

(c)

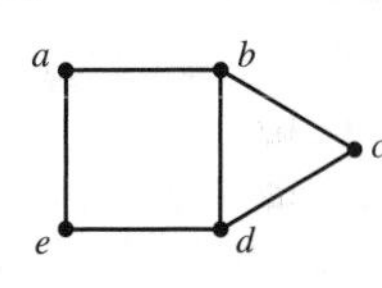

(d)

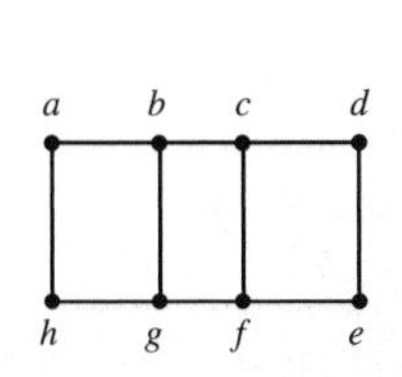

(e)

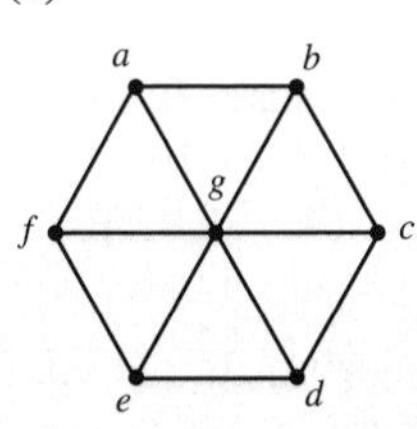

6. **(a)** Find the cycle index for the group of symmetries of a square in terms of permutations of edges, not corners.
   (b) How many ways are there to 3-color the edges of a floating square?
   (c) How many ways are there to 3-color both edges and corners of a floating square?

(d) Why does not: (number of floating 3-colorings of corners) (number of floating 3-colorings of edges) = (number of floating 3-colorings of edges and corners)? Explain.

7. How many ways are there to 3-color the *edges* of the floating figures in Exercise 5? (*Hint:* The cycle index now is for symmetries of the edges.)

8. **(a)** Find the number of different n-bead 3-colored necklaces (cyclicly distinct) in which each color appears at least once when (i) $n = 3$, (ii) $n = 4$, (iii) $n = 7$.

(b) Repeat part (a) when necklaces may reflect as well as rotate.

9. Find the number of different 2-sided dominoes (two squares of 1 to 6 dots or a blank on each side of the domino).

10. **(a)** Let G be the group of all 4! permutations of 1, 2, 3, 4. Find P_G.

(b) Use part (a) to find the number of ways to paint four identical marbles each one of three colors (check your answer by modeling this problem as a selection-with-repetition problem).

(c) Use part (a) to find the number of ways to put 12 balls chosen from three colors into four indistinguishable boxes with 3 balls in each box.

11. **(a)** Find the number of 2×4 chessboards distinct under rotation whose squares are colored red or black.

(b) Suppose that two chessboards are also considered equivalent (aside from rotational symmetry) if one can be obtained from the other by complementing red and black colors. How many different 2×4 chessboards are there?

12. Show that if $x_1^{k_1} x_2^{k_2} \cdots x_m^{k_m}$ is the cycle index for a symmetry π of an n-gon (expressed in terms of a permutation of the corners), then $1k_1 + 2k_2 + \cdots + mk_m = n$.

13. In solving for the number of corner 2-colorings of some unoriented figure, suppose we are given the cycle index P_{G*} of the group G^* of induced permutations of the 2-colorings [as in Eq. (1) in Section 9.1], instead of the usual cycle index P_G of the group of symmetries of the figure. What integer values should be substituted now for each x_i in P_{G*} to get the number of 2-colorings, or will no substitution work? Explain.

14. Let S be some set of n objects and G a group of permutations of S. For subsets S_1, S_2 of S, define the equivalence relation $S_1 \sim S_2$ if for some $\pi \in G$, $S_1 = \pi(S_2)$ $(= \{\pi(s) \mid s \in S_2\})$. Show that the number of equivalence classes equals $P_G(2, 2, \ldots)$. (There are 2^n subsets of S in all.)

15. Find the number of m-colorings of the corners of a p-gon, where p is a prime $(p > 2)$, if:

(a) Only rotations are allowed.

(b) Rotations and reflections are allowed.

9.4 POLYA'S FORMULA

We are now ready to address our ultimate goal of a formula for the pattern inventory. Recall that the pattern inventory is a generating function that tells how many colorings of an unoriented figure there are using different possible collections of colors. For black-white colorings of the unoriented square, the pattern inventory is $1b^4 + 1b^3w + 2b^2w^2 + 1bw^3 + 1w^4$. For example, the term $1b^3w$ tells us that there is one nonequivalent coloring with 3 black (b) corners and 1 white (w) corner. The coefficients in the pattern inventory can be viewed as the results of several Burnside Theorem–type counting problems. Recall that Burnside's Theorem says that the number N of equivalence classes in T, a collection of colorings of S (where S is the corners of a square or the faces of a tetrahedron, etc.), caused by a group of symmetries of S is

$$N = \frac{1}{|G|} \sum_{\pi \in G} \Psi(\pi) \qquad (*)$$

In the case of 2-colorings of the floating square, we divide the set of all colorings in Figure 9.1 into sets based on the numbers of black and white corners:

$$\begin{aligned}
T_0 &= \{C_1\} \\
T_1 &= \{C_2, C_3, C_4, C_5\} \\
T_2 &= \{C_6, C_7, C_8, C_9, C_{10}, C_{11}\} \\
T_3 &= \{C_{12}, C_{13}, C_{14}, C_{15}\} \\
T_4 &= \{C_{16}\}
\end{aligned}$$

The coefficient of b^3w can be obtained from Eq. (*) if we let the group G of symmetries of the square act on just the set T_1. Recall that formula (*) from Burnside's Theorem applies to any collection of colorings that is closed under G. The term "closed under G" means that for any $\pi \in G$, and any coloring $C \in T$, $\pi(C)$ equals another coloring in T. T_1 is closed under G, since any symmetry acting on a coloring with 3 black corners and 1 white corner yields another coloring with 3 black corners and 1 white corner. The same is true for the other T_ks. Thus, *the coefficient of* $b^{4-k}w^k$ *in the pattern inventory is the result of Eq.* (*) *when* T_k *is the set on which* G *acts.*

Let us try to solve these five subproblems simultaneously. In Figure 9.9 we have duplicated the table in Figure 9.7 and added a new column (iv). In the first row of column (iv), we write a polynomial whose coefficients give the numbers of 2-colorings in each T_k left fixed by π_1, then in the second row of the table we write a polynomial for the numbers of 2-colorings in each T_k left fixed by π_2, then by π_3, and so forth. Then we total up the b^4 term in each row (the number of 2-colorings with 4 blacks) and divide by 8 to get the coefficient of b^4 in the pattern inventory, total up the b^3w term in each row and divide by 8 to get the coefficient of b^3w, and so forth.

Since the action of π_1 leaves all C_s fixed, the first row's coefficients are $1, 4, 6, 4, 1$. We write: $b^4 + 4b^3w + 6b^2w^2 + 4bw^3 + w^4$; this is an *inventory of fixed colorings.* For π_1, the inventory of fixed colorings is an inventory of all colorings. Observe that this inventory is simply $(b+w)^4 = (b+w)(b+w)(b+w)(b+w)$, one $(b+w)$ for

(i) Motion π_i	*(ii)* Colorings Left Fixed by π_i	*(iii)* Cycle Structure Representation	*(iv)* Inventory of Colorings Left Fixed by π_i					
π_1	16—all colorings	x_1^4	$(b+w)^4$	$= 1b^4$	$+4b^3w$	$+6b^2w^2$	$+4bw^3$	$+1w^4$
π_2	2—C_1, C_{16}	x_4	(b^4+w^4)	$= 1b^4$				$+1w^4$
π_3	2—$C_1, C_{10}, C_{11}, C_{16}$	x_2^2	$(b^2+w^2)^2$	$= 1b^4$		$+2b^2w^2$		$+1w^4$
π_4	2—C_1, C_{16}	x_4	(b^4+w^4)	$= 1b^4$				$+1w^4$
π_5	2—C_1, C_6, C_8, C_{16}	x_2^2	$(b^2+w^2)^2$	$= 1b^4$		$+2b^2w^2$		$+1w^4$
π_6	2—C_1, C_7, C_9, C_{16}	x_2^2	$(b^2+w^2)^2$	$= 1b^4$		$+2b^2w^2$		$+1w^4$
π_7	8—C_1, C_2, C_4, C_{10} $C_{11}, C_{12}, C_{14}, C_{16}$	$x_1^2x_2$	$(b+w)^2(b^2+w^2)$	$= 1b^4$	$+2b^3w$	$+2b^2w^2$	$+2bw^3$	$+1w^4$
π_8	8—C_1, C_3, C_5, C_{10} $C_{11}, C_{13}, C_{15}, C_{16}$	$x_1^2x_2$	$(b+w)^2(b^2+w^2)$	$= 1b^4$	$+2b^3w$	$+2b^2w^2$	$+2bw^3$	$+1w^4$

Figure 9.9

each corner. For π_2, the inventory is $b^4 + w^4$. For π_3, we find by observation that the inventory is $b^4 + 2b^2w^2 + w^4$. This expression factors into $(b^2 + w^2)^2$.

Just as we did before when counting the total number of colorings fixed by the action of some π, let us look for a pattern in the inventories of fixed colorings. Again the key to the pattern is the fact that in a coloring left fixed by π, all corners in a cycle of π must have the same color. Since π_2 has one cycle involving all four corners, the possibilities are thus all corners black or all corners white; hence the inventory is $b^4 + w^4$. The motion π_3 has two 2-cycles (ac) and (bd). Each 2-cycle uses two blacks or two whites in a fixed coloring. Hence the inventory of a cycle of size two is $b^2 + w^2$. The possibilities with two such cycles have the inventory $(b^2 + w^2)(b^2 + w^2)$.

The inventory of fixed colorings for π_i will be a product of factors $(b^j + w^j)$, one factor for each j-cycle of the π_i. So we need to know the number of cycles in π_i of each size. But this is exactly the information encoded in the cycle structure representation. Indeed, setting $x_j = (b^j + w^j)$ in the representation yields precisely the inventory of fixed colorings for π_i. By this method we compute the rest of the inventories of fixed colorings. See Figure 9.9. For π_7 especially, the inventory should be checked against the list of colorings in column (ii). The pattern inventory is obtained by adding together the inventories of fixed colorings, collecting like-power terms, and dividing by 8.

As before in Section 9.3, we get a more compact formula and save some computation by first adding together the cycle structure representations and dividing by 8, and then setting each $x_j = (b^j + w^j)$ and doing the polynomial algebra all at once. Again the first step in this approach yields the cycle index $P_G(x_1, x_2, \ldots, x_k)$. Thus by setting $x_j = (b^j + w^j)$ in P_G, we obtain the pattern inventory.

If three colors, black, white, and red, were permitted, each cycle of size j would have an inventory of $(b^j + w^j + r^j)$ in a fixed coloring. So we would set $x_j = (b^j + w^j + r^j)$ in P_G. The preceding argument applies for any number of colors and any figure. In greater generality we have the following theorem.

Theorem **(Polya's Enumeration Formula)**

Let S be a set of elements and G be a group of permutations of S that acts to induce an equivalence relation on the colorings of S. The inventory of nonequivalent colorings of S using two colors is given by the generating function $P_G((b+w), (b^2+w^2),$

$(b^3 + w^3), \ldots, (b^k + w^k))$. The inventory using colors $c_1, c_2, \ldots, c_m$ is

$$P_G\left(\sum_{j=1}^{m} c_j, \sum_{j=1}^{m} c_j^2, \ldots, \sum_{j=1}^{m} c_j^k\right)$$

For a moment, let us return to the problem of counting the *total* number of nonequivalent 2-colorings. This number is simply the sum of the coefficients in the pattern inventory. To sum coefficients, we set the indeterminants, b and w (and hence their powers), equal to 1, or, equivalently, set $x_j = 2$ in P_G. If m colors were allowed, we would set $x_j = m$ in P_G, obtaining the same formula as in the theorem in Section 9.3.

As in many other generating function problems, the actual expansion of the generating function for a pattern inventory can be quite tedious. We are expanding expressions of the form $(c_1^i + c_2^i + \cdots + C_m^i)^r$. When m and r get large, it is time to turn to computer algebra software.

Example 1: Pattern Inventory for 3-Bead Necklaces

Determine the pattern inventory for 3-bead necklaces distinct under rotations using black and white beads. Repeat using black, white, and red beads.

From Example 1 of Section 9.3, we know $P_G = \frac{1}{3}(x_1^3 + 2x_3)$. Substituting $x_j = (b^j + w^j)$, we get

$$\begin{aligned}\frac{1}{3}[(b+w)^3 + 2(b^3+w^3)] &= \frac{1}{3}[(b^3 + 3b^2w + 3bw^2 + w^3) + (2b^3 + 2w^3)] \\ &= \frac{1}{3}(3b^3 + 3b^2w + 3bw^2 + 3w^3) \\ &= b^3 + b^2w + bw^2 + w^3\end{aligned}$$

This result could be obtained empirically. There is only one way to color all beads black or all white. If one bead is white and the others black, then by rotation the white bead can occur anywhere; thus there is only one necklace with one white and two blacks. The same is true by symmetry for a necklace with one black and two whites.

Now consider 3-bead necklaces using three colors. We substitute $x_j = (b^j + w^j + r^j)$, in P_G to obtain $\frac{1}{3}[(b+w+r)^3 + 2(b^3+w^3+r^3)]$. Instead of expanding the polynomials in this expression, we use indirect means. Perhaps again each inventory term has coefficient 1. There is a general test for whether all inventory coefficients are 1: Compare N^*, the number of terms in the pattern inventory, with N, the total number of patterns [i.e., $P_G(m, m, \ldots, m)$]. Since N equals the sum of the coefficients in the inventory, then $N^* = N$ if and only if each term has coefficient 1. The number N^* of terms in the pattern inventory when n elements (corner, beads, etc.) are colored with m colors is $C(n+m-1, n)$ (see Exercise 19). For the case at hand, $m = 3, n = 3$, and so $N^* = \binom{3+3-1}{3} = 10$. From Example 1 of Section 9.3, we know that $N = 11$. Since $N^* \neq N$, we know that all terms do not have coefficient 1.

On the other hand, the only way for $N^* = N + 1$ is that nine terms in the inventory must have coefficient 1 and one term coefficient 2. But as argued above, there is only one necklace with all beads the same color and only one necklace with one bead of color A and the other two beads of color B. The only other possibility for a 3-bead necklace using three colors is to have one bead of each color. Thus, there must be two necklaces with one bead of each color (the necklaces are any cyclic order of the three colors and the reverse cyclic order), and the pattern inventory for black, white and red 3-bead necklaces is

$$b^3 + w^3 + r^3 + b^2w + b^2r + w^2b + w^2r + r^2b + r^2w + 2bwr \quad \blacksquare$$

Example 2: Pattern Inventory for 7-Bead Necklaces

Find the number of 7-bead necklaces distinct under rotations using three black and four white beads.

We need to determine the coefficient of b^3w^4 in the pattern inventory. Each rotation, except the 0° rotation, is a cyclic permutation when the number of beads is a prime [see Exercise 13(a)], so $P_G = \frac{1}{7}(x_1^7 + 6x_7)$. The pattern inventory is $\frac{1}{7}[(b+w)^7 + 6(b^7 + w^7)]$.

Since the factor $6(b^7 + w^7)$ in the pattern inventory contributes nothing to the b^3w^4 term, we can neglect it. Thus the number of 3-black, 4-white necklaces is simply

$$\frac{1}{7}[(\text{coefficient of } b^3w^4 \text{ in } (b+w)^7] = \frac{1}{7}\binom{7}{3} \quad \blacksquare$$

Example 3: Pattern Inventory for Edge 2-Colorings of a Tetrahedron

Find the pattern inventory of black–white edge colorings of a tetrahedron.

Although we calculated the cycle index for corner symmetries of the tetrahedron in Example 2 of Section 9.3, we need a different cycle index for edge symmetries. Since the set of objects to be colored is the six edges, we need to consider the symmetries of the tetrahedron as permutations of the edges.

The 0° revolution clearly leaves all edges fixed and thus has cycle structure representation x_1^6. The 120° (or 240°) revolution about a corner and the middle of the opposite face cyclicly permutes the edges incident to that corner and cyclicly permutes the edges bounding the opposite face (see Figure 9.4 in Section 9.1). Thus, the 120° revolution has cycle structure representation x_3^2.

The 180° revolution about opposite edges leaves those two edges fixed (see Figure 9.4). Since two applications of a 180° revolution return the tetrahedron to its original position, the other 4 edges not left fixed must be in 2-cycles. Thus, the 180° revolution has cycle structure representation $x_1^2x_2^2$. Then $P_G = \frac{1}{12}(x_1^6 + 8x_3^2 + 3x_1^2x_2^2)$.

Substituting $x_j = (b^j + w^j)$, we get

$$\begin{aligned}
&\frac{1}{12}[(b+w)^6 + 8(b^3+w^3)^2 + (b+w)^2(b^2+w^2)^2] \\
&\quad = \frac{1}{12}[(b^6 + 6b^5w + 15b^4w^2 + 20b^3w^3 + 15b^2w^4 + 6bw^5 + w^6) + (8b^6 + 16b^3w^3 \\
&\qquad + 8w^6) + (3b^6 + 6b^5w + 9b^4w^2 + 12b^3w^3 + 9b^2w^4 + 6bw^5 + 3w^6)] \\
&\quad = \frac{1}{12}(12b^6 + 12b^5w + 24b^4w^2 + 48b^3w^3 + 24b^2w^4 + 12bw^5 + 12w^6) \\
&\quad = b^6 + b^5w + 2b^4w^2 + 4b^3w^3 + 2b^2w^4 + bw^5 + w^6 \quad \blacksquare
\end{aligned}$$

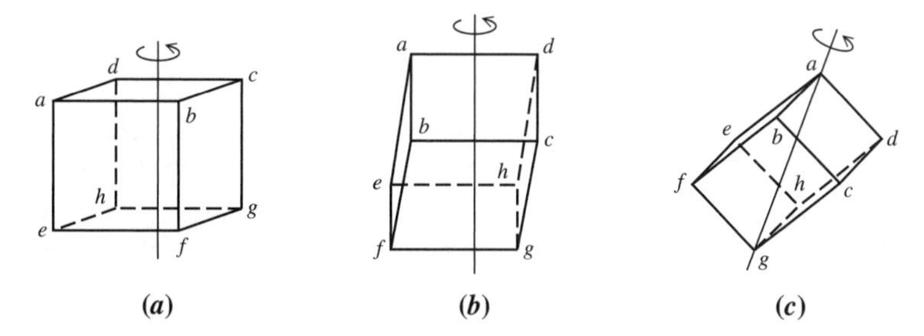

Figure 9.10 **Revolutions of the cube (a) Revolution about opposits faces (b) Revolution about opposite edges (c) Revolution about opposite corners**

Example 4: Pattern Inventory for Corner 2-Colorings of a Cube

Find the pattern inventory for corner 2-colorings of a floating cube.

The symmetries of the cube involve revolutions about opposite faces, about opposite edges, and about opposite corners. First, of course, there is the identity symmetry, with cycle structure representation x_1^8.

(a) *Opposite faces:* As a concrete example, we revolve about the center point of the pair of opposite faces *abcd* and *efgh*; see Figure 9.10*a*. A 90° revolution yields the permutation (*abcd*) (*efgh*) with cycle structure representation x_4^2. A 270° revolution has the same structure. A 180° revolution yields the permutation (*ac*)(*bd*)(*eg*)(*fh*) with cycle structure representation x_2^4. There are three pairs of opposite faces and so the total contribution to the cycle index of opposite-face revolutions is $6x_4^2 + 3x_2^4$.

(b) *Opposite edges:* As a concrete example, we revolve about the middle of opposite edges *ab* and *gh*; see Figure 9.10*b*. A 180° revolution yields the permutation (*ab*)(*bc*)(*eh*)(*fg*) with cycle structure representation x_2^4. There are six pairs of opposite edges, and so the total contribution of opposite edge revolutions is $6x_2^4$.

(c) Opposite corners: As a concerete example, we revolve about the opposite corners *a* and *g*; see Figure 9.10*c*. A 120° revolution yields the permutation (*a*)(*bde*)(*chf*)(*g*) with cycle structure representation $x_1^2x_3^2$. One way to see what this permutation does is by noting that the three corners *b*, *d*, *e* adjacent to *a* must be cyclically permuted in any motion that leaves *a* fixed (and similarly for the corners adjacent to *g*). A 240° revolution has the same structure. There are four pairs of opposite corners and so the contribution of opposite-corner revolutions is $8x_1^2x_3^2$.

We leave it to the reader (see Exercise 17) to verify that we have enumerated all symmetries of the cube and that these symmetries are all distinct.

Collecting terms, we find $P_G = \frac{1}{24}(x_1^8 + 6x_4^2 + 9x_2^4 + 8x_1^2x_3^2)$. The pattern inventory for corner colorings of the cube using black and white is thus

$$\frac{1}{24}[(b+w)^8 + 6(b^4+w^4)^2 + 9(b^2+w^2)^4 + 8(b+w)^2(b^3+w^3)^2]$$

As in previous examples, the coefficients of the terms b^8, b^7w, bw^7, and w^8 are readily seen to be 1. The b^6w^2, b^5w^3, and b^4w^4 terms in the four factors of the generating function are $(\cdots + 28b^6w^2 + 56b^5w^3 + 70b^4w^4 + \cdots)$, $6(\cdots + 2b^4w^4 + \cdots)$, $9(\cdots + 4b^6w^2 + 6b^4w^4 + \cdots)$, and $8(\cdots + b^6w^2 + 2b^5w^3 + 4b^4w^4 + \cdots)$, respectively. Summing and dividing by 24 we have $\frac{1}{24}(\cdots + 72b^6w^2 + 72b^5w^3 + 168b^4w^4 + \cdots) = 3b^6w^2 + 3b^5w^3 + 7b^4w^4$. (It is easy to detect errors in these calculations, since most errors will result in a noninteger coefficient.) By symmetry, we fill out the pattern inventory to obtain

$$b^8 + b^7w + 3b^6w^2 + 3b^5w^3 + 7b^4w^4 + 3b^3w^5 + 3b^2w^6 + bw^7 + w^8 \quad \blacksquare$$

9.4 EXERCISES

Summary of Exercises The first 16 exercises use Polya's enumeration formula, and the remaining problems involve associated theory. Note that "floating" means that all possible rotations and reflections are allowed.

1. Find the pattern inventory for black and white corner colorings of a floating pentagon.
2. Find an expression for the pattern inventory for black–white, n-bead necklaces (rotations only) and find the number of necklaces with 3 white beads and the rest black:

   (a) $n = 6$ **(b)** $n = 9$ **(c)** $n = 10$ **(d)** $n = 11$

3. Find the pattern inventory for black, white, and red corner colorings of a floating square.
4. Find an expression for the pattern inventory for the 2-colorings (rotations only) of the 16 squares in a 4×4 chessboard.
5. Find an expression for the pattern inventory for black–white corner colorings of the following floating figures:

   (a)
   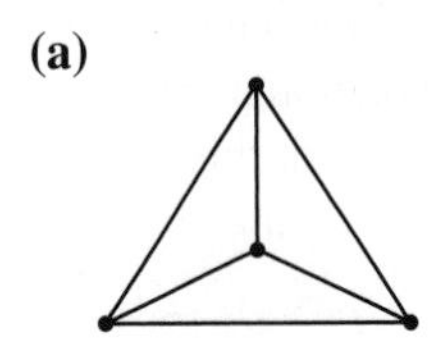
   (b)

   (c)
   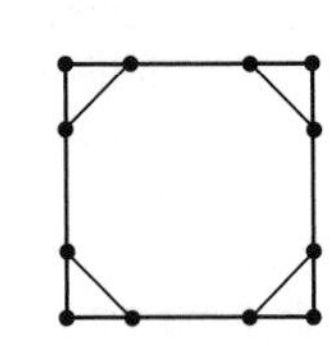

6. **(a)** Find the pattern inventory for corner 2-colorings of a floating pyramid (with a square base).

   (b) Repeat part (a) for edge colorings.

   (c) Repeat part (a) for face colorings.

7. Find an expression for the pattern inventory for edge 2-colorings of the floating figures in parts (a) and (b) in Exercise 5.

8. Find an expression for the pattern inventory for edge 2-colorings of a floating cube and find the number of edge 2-colorings with 3 white and 9 black edges.

9. Find an expression for the pattern inventory for face 2-colorings of:

(a) A floating tetrahedron. **(b)** A floating cube.

10. **(a)** Find the pattern inventory for the corner 3-colorings of a floating pentagon with adjacent corners different colors.

(b) Repeat part (a) for a floating tetrahedron.

(c) Repeat part (a) for a floating cube.

11. Find an expression for the pattern inventory for black–white colorings of four indistinguishable balls (see Exercise 10 in Section 9.3).

12. Give an empirical argument (without use of the cycle index) to show that there are $\lfloor \frac{n}{2} \rfloor$ different n-bead necklaces with 2 white beads and $n-2$ black beads ($\lfloor r \rfloor$ is the largest integer $\leq r$).

13. Given that p is a prime and p-bead necklaces are made of black and white beads:

(a) Show that each rotation except $0°$ is a cyclic permutation of the corners.

(b) What is the number of such necklaces with exactly k white beads?

14. **(a)** Suppose that instead of coloring the corners of a floating square, we attach 0 or 1 or 2 identical jelly beans at a corner. Find a generating function, and expand it, for the number of different squares with a total of k beans at its corners. (*Hint:* In the inventory of patterns left fixed, the exponents within a factor will vary.)

(b) Repeat part (a) with each corner getting a red jelly bean or a white jelly bean or a red and a white jelly bean. Now find a generating function for the number of squares with j whites and k reds.

15. Suppose n batons with small holes drilled through their midpoints are strung along a piece of wire and each end of each baton is painted one of three possible colors (red, white, blue). The wire is fixed but the batons revolve about their centers. How many indistinguishable configurations are there for:

(a) $n = 2$ **(b)** $n = 3$

16. How many distinct (nonisomorphic) unlabeled graphs are there with four vertices? (*Hint:* Each possible edge has two possible colors: "edge present" and "edge absent.")

17. Show that the 24 symmetries of a cube listed in Example 4 are all distinct and include all symmetries of a cube.

18. Characterize the geometric figures that when floating have only one coloring with one black corner and the others white.

19. Show that there are $C(n+m-1, n)$ different terms in the pattern inventory for m-colorings of corners of an n-gon.

20. Let $a_{k,n}$ denote the minimum number of coefficients in the pattern inventory of all k-colorings of the corners of an n-gon needed so that using symmetry all other coefficients in the pattern inventory are known. Find a generating function $g_k(x)$ for $a_{k,n}$.

9.5 SUMMARY AND REFERENCES

Polya's enumeration formula is important, practically because it solves important problems, mathematically because it is an elegant marriage of group theory and generating functions, and pedagogically because it is the only truly powerful combinatorial formula we will see in this book (where a difficult problem can be solved by "plugging" the right numbers into a specialized formula). Of equal importance is the manner in which Polya's formula has been developed: Sections 9.3 and 9.4 could be read as a case study in the experimental derivation of a mathematical theory. (This approach follows the pedagogical style Polya used when he taught this material.) Students interested in a more rigorous development of this theory and its extensions are referred to [2], which includes an English translation of Polya's original 1937 paper, "Kombinatorische Anzahlbestimmungen für Gruppen, Graphen und chemische Verbindungen."

In a more formal development, one defines a coloring as a function f from a set S (of corners) to a set R (of colors). The function f_6, corresponding to C_6 (in Figure 9.1), would be written in tabular form as $\binom{a\,b\,c\,d}{w\,w\,b\,b}$. We define the composition $\pi \cdot f$ of a motion and a color function to be a new color function f'; for example, $\pi_2 \cdot f_6$ maps $a \to b \to w, b \to c \to b, c \to d \to b, d \to a \to w$, or $\pi_2 \cdot f_6 = \binom{a\,b\,c\,d}{w\,b\,w\,b} = f_7$. The coloring permutation induced by a symmetry π maps each f to $\pi \cdot f$. Although tedious, a formal calculation of the coloring permutation is thus possible (before we did it by inspection). We define coloring equivalence by: $f \sim f'$ if and only if there exists π such that $\pi \cdot f = f'$. Our theory can readily be restated in terms of this new definition.

Polya's enumeration formula has an important application to another field of combinatorial mathematics. It is used to enumerate families of graphs (see Exercise 16 in Section 9.4). This application was pioneered by F. Harary; see Harary and Palmer [1].

1. F. Harary and E. Palmer, *Graphical Enumeration,* Academic Press, New York, 1973.

2. G. Polya and R. C. Reade, *Combinatorial Enumeration of Groups, Graphs, and Chemical Compounds,* Springer-Verlag, New York, 1987.

CHAPTER 10
GAMES WITH GRAPHS

10.1 PROGRESSIVELY FINITE GAMES

In this chapter we develop the theory of progressively finite games and apply it to Nim-type games. While clearly not an important use of graphs, our graph-theoretic analysis of these games yields some interesting results. This chapter is meant as a sort of mathematical "dessert" following the more serious topics in preceding chapters. It is essential to study how graph models can be used to solve a variety of real-world problems, but there is a more personal reward in learning how graphs can permit one to win at certain games.

A game in which two players take turns making a move until one player wins (no ties are allowed) is called **progressively finite** if: (1) there are a finite number of different positions in the game; and (2) the play of the game must end after a finite number of moves.

Our objective in this section is winning: how to determine winning strategies for progressively finite games. We can model a progressively finite game by a directed graph with a vertex for each position that can occur in the play of the game and a directed edge for each possible move from one position to another.

Observe that the graph of a progressively finite game cannot contain any directed circuits, since players could move around and around a circuit of positions forever (violating the constraint on finite play). Thus, no positions can ever be repeated in the play of a game. Rather, the game moves inexorably toward some final position that is a win for one of the players. Games such as checkers and chess that permit ties and repetition of positions are not progressively finite. Most progressively finite games are "takeaway"-type games of the sort illustrated by the games in Examples 1 and 2.

Example 1: Restricted Takeaway Game

A set of 16 objects is placed on a table. Two players take turns removing 1, 2, 3, or 4 objects. The winner is the player who removes the last object. The graph of this game is shown in Figure 10.1 (all edges are directed from left to right). ∎

Example 2: Inverted Takeaway Game

Starting with an empty pile, two players add 1 penny or 2 pennies or 1 nickel to the pile until the value of the pile is the square of a positive integer ≥ 2 or until the value

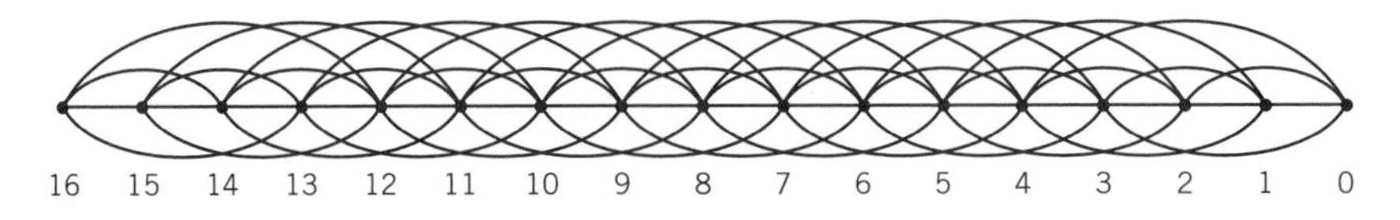

Figure 10.1

exceeds 40. The player whose addition brings the value of the pile to one of these critical amounts takes all the money in the pile. The graph of this game is shown in Figure 10.2 (all edges are directed from left to right; arrows point to the game-winning positions). This game is an inverted form of a "takeaway" game in which 1, 2, or 5 objects are withdrawn from a pile until the pile is reduced to certain critical sizes. ∎

The reader is encouraged to try playing these games with a friend. A winning strategy for the first player (who makes the first move) in the game in Example 1 can be found with a little thought. The game in Example 2 is harder. The details of this second game only serve to confuse systematic attempts to find a winning strategy. It is easier to develop a general theory of winning strategies in progressively finite games and then apply the theory to Example 2.

A **winning position** in a progressively finite game is a position at which play stops and the player who moved to this position is declared the winner.

In the graph of a progressively finite game, vertices with 0 out-degree must represent winning positions, for if no edges leave a vertex, the game must stop at that vertex. We call such vertices **winning vertices.** The graph of the game in Example 1 has just one winning vertex, numbered 0, while the graph of the game in Example 2 has six winning vertices, numbered 4, 9, 16, 25, 36, and "over 40." Every progressively finite game must have at least one winning position or else play would go on without end.

A **winning strategy** for a player is a rule that tells the player which move to make at each stage in a game so as to ensure that the player will eventually win, that is, finally move to a winning position. Obviously only one player can have a winning strategy.

Any vertex adjacent to a winning vertex is a "losing" vertex in the sense that if one player moves to a "losing" vertex, the other player can then move to a winning vertex (and win the game). The vertices numbered 1, 2, 3, and 4 are losing vertices in the game in Example 1. The vertices numbered 2, 3, 7, 8, 11, 14, 15, 20, 23, 24, 31, 34, 35, 37, 38, 39, and 40 are losing vertices in the game in Example 2.

Stepping back one more move from the end of the game, we see that if all edges from a vertex x go to losing vertices, then x is a "prewinning" vertex. Whenever player A moves to such a prewinning vertex, then player B's next move must be to a losing vertex and now player A can move to a winning vertex. Vertex 5 in the game

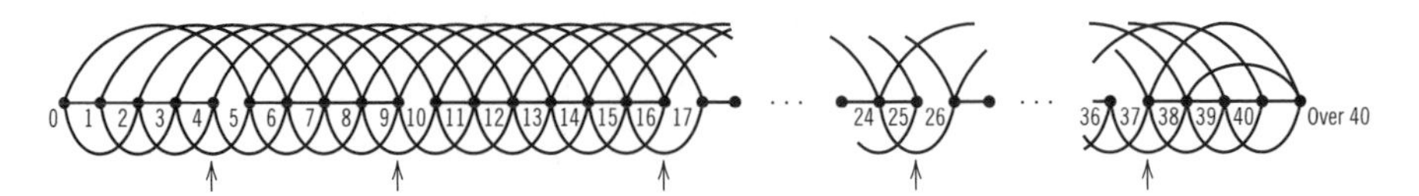

Figure 10.2

in Example 1 is a prewinning vertex, since all edges from 5 go to losing vertices (1, 2, 3, and 4). Vertices 6 and 33 are the prewinning vertices in the game in Example 2; all other vertices adjacent to losing vertices are also adjacent to nonlosing vertices.

The general theory of winning strategies in progressively finite games is based on a recursive extension of the preceding reasoning. We seek good vertices, such as winning and prewinning vertices, that win or lead only to bad vertices. When our opponent is forced to move to a bad vertex, such as losing vertices, then we will always be able to move to another good vertex. A winning strategy for the first player will tell the player how to find good vertices to move to from any bad vertex. This pattern of play, with the first player moving to successive good vertices and the second player forced to move to bad vertices, will continue until finally the first player reaches a good vertex that is a winning vertex. If the game starts at a good vertex, and so the first player's initial move must be to a bad vertex, then roles are reversed and it is the second player who now has a winning strategy using the good vertices.

Example 1 (continued)

The winning strategy for the first player in this game is to move to a vertex whose number is a multiple of 5. These are the good vertices. Thus, the move from the starting vertex 16 is to vertex 15 (i.e., the first player removes one object). From 15, the second player must move to one of the vertices 11, 12, 13, or 14. From any of these bad vertices the first player now moves to vertex 10. Whatever the second player's next move is, the first player will always be able to move to the prewinning vertex 5, and one round later the first player will win. Note that if the game started with only 15 objects, then the roles would be reversed and the second player would be able to use this winning strategy. ■

We formalize the concept of good vertices with the following definition. A **kernel** in a directed graph is a set of vertices such that:

1. There is no edge joining any two vertices in the kernel; and
2. There is an edge from every nonkernel vertex to some kernel vertex.

The kernel of the graph in Example 1 is the set of vertices numbered 0, 5, 10, and 15.

Theorem 1

If the graph of a progressively finite game has a kernel K, then a winning strategy for the first player is to move to a kernel vertex on every turn. However, if the starting vertex is in the kernel, then the second player can use this winning strategy.

Proof

First we show that all winning vertices must be in K. The reason is that vertices not in a kernel must have an edge directed to a vertex in the kernel. But winning vertices have 0 out-degree. Thus, all winning vertices must be in any kernel.

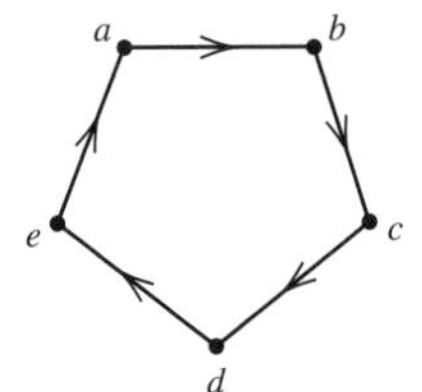

Figure 10.3

Next we show that moving to a kernel vertex on every turn is a winning strategy for the first player. By property (1) of a kernel, when the first player moves to a kernel vertex, the second player must then move to a nonkernel vertex. Then by property (2), the first player can always move from the nonkernel vertex to a kernel vertex. The play proceeds in this fashion with the first player always moving to a kernel vertex and the second player always moving to a nonkernel vertex. Since the game is progressively finite, the play must eventually end. Since all winning vertices are in the kernel, the first player must win.

If the starting vertex is in the kernel, then the roles of the first and second players are reversed and the second player can always move to a kernel vertex for an eventual win. ◆

The preceding proof does not show explicitly how successively moving to kernel vertices leads to a winning vertex. The proof is existential. It shows only that by moving to kernel vertices the first player must eventually arrive at a winning vertex.

A more immediate problem is to show that the graph of every progressively finite game has a kernel and then to find this kernel. Not all graphs have kernels. For example, the graph in Figure 10.3 has no kernel. To show that this graph has no kernel, we argue thus. If there were a kernel, we could assume by the symmetry in this graph that vertex a is in the kernel. Then vertex e, which has an edge to a, cannot be in the kernel. Vertex d's only outward edge goes to nonkernel vertex e, and so d must be in the kernel. Similarly, c cannot be in the kernel and then b must be in the kernel. But now the edge $(\vec{a, b})$ joins two kernel vertices—a contradiction.

Fortunately, graphs of progressively finite games do have kernels. To demonstrate the existence of kernels in such graphs, we need to organize the vertices in a progressively finite graph into levels based on the "distance" of the vertices from a winning vertex. We recursively define the level $l(x)$ of vertex x in a directed graph and the sets L_k of vertices at level $\leq k$ as follows. Let $s(x) = \{y \mid x \text{ has an edge to } y\}$ be the set of **successors** of x. Then

$$l(x) = 0 \Leftrightarrow s(x) \text{ is empty} \quad \text{and} \quad L_0 = \{x \mid l(x) = 0\}$$
$$l(x) = 1 \Leftrightarrow x \notin L_0 \text{ and } s(x) \subseteq L_0 \quad \text{and} \quad L_1 = L_0 \cup \{x \mid l(x) = 1\}$$

and in general,

$$l(x) = k \Leftrightarrow x \notin L_{k-1} \text{ and } s(x) \subseteq L_{k-1} \quad \text{and} \quad L_k = L_{k-1} \cup \{x \mid l(x) = k\}$$

Observe that $L_k - L_{k-1}$ is the set of vertices at level k.

It can be shown that $l(x)$ is the length of the longest path in the directed graph that starts at x (see Exercise 12). Since all paths in a progressively finite graph have finite length, the longest path from a vertex x has finite length. Further, the longest path from x must end at a vertex of out-degree 0; otherwise the path could be extended. Then starting from vertices of out-degree 0, the assignment of level numbers by this recursive definition will eventually reach all vertices.

It follows from this definition of level that every vertex at level 0 is a winning vertex and that every vertex at level 1 is a losing vertex. A vertex at level 2 is also a losing vertex if it is adjacent to a vertex at level 0; it is a prewinning vertex only if all its successors are at level 1. In general, every vertex at level $k (k > 0)$ must be adjacent to a vertex at level $k - 1$ and possibly other vertices at lower levels, but cannot be adjacent to any other vertex at level k (or greater). Now we can prove the fundamental theorem for progressively finite games.

Theorem 2

Every progressively finite game has a unique winning strategy. That is, the graph of every progressively finite game has a unique kernel.

Proof

The proof is by induction on the levels or, more precisely, on the sets L_k. Let K_k be the set of kernel vertices in L_k. First consider the set L_0. L_0 consists of vertices with 0 out-degree. These are the winning vertices. As noted in the proof of Theorem 1, all winning vertices must be in the kernel. Thus $K_0 = L_0$.

Next let us inductively assume for $n \geq 1$ that K_{n-1} is the unique, well-defined set of kernel vertices in L_{n-1}. We show that we can find a unique set of level-n vertices that must be added to K_{n-1} to form the kernel K_n for L_n. By the way that level numbers were defined, $l(x) = n$ means that $s(x) \subseteq L_{n-1}$. If a level-n vertex x is adjacent to no kernel vertex of K_{n-1}, this x must be in K_n (since by the definition of a kernel, any vertex with no successors in the kernel must itself be in the kernel). On the other hand, if x is adjacent to a kernel vertex of K_{n-1}, x cannot be in the kernel. Hence $K_n = K_{n-1} \cup \{x \mid l(x) = n \text{ and } s(x) \cap k_{n-1} = \emptyset\}$ is the unique, well-defined set of kernel vertices in L_n. It follows by mathematical induction that the graph has a unique kernel. By Theorem 1, this kernel is the unique winning strategy. ◆

The proof of Theorem 2 tells us how to build a kernel. First put the winning vertices in the kernel. Then recursively add the vertices at increasing levels not adjacent to the current set of kernel vertices. We implement this procedure for finding kernels using a labeling rule called a Grundy function.

Definition of Grundy Function $g(x)$

For each vertex x in a directed graph, $g(x)$ is the smallest nonnegative integer not assigned to any of x's successors.

We shall prove shortly that vertices with Grundy number 0 are exactly the set of kernel vertices. In the next section, Grundy numbers will be seen to play a fundamental role in more complex games.

In the graph of a progressively finite game, Grundy values can be easily determined using a level-by-level approach. The vertices on level 0 have no successors and so their Grundy number will be 0 (the smallest nonnegative integer). Next we determine $g(x)$ for vertices x at level 1, then vertices at level 2, and so forth. In this way, all of a vertex x's successors are assigned Grundy numbers before it is time to determine the Grundy number of x, since x's successors are at lower levels. When we come to x, we can check the Grundy function values of $s(x)$, x's successors, and set $g(x)$ equal to the smallest nonnegative integer not assigned to any vertex in $s(x)$. Actually we do not need to proceed in a totally level-by-level fashion. We can use any method that does not try to assign a vertex its Grundy number until all the vertex's successors have Grundy numbers.

No Grundy function can be defined for the graph in Figure 10.3. Each vertex x in Figure 10.3 has a successor whose Grundy number must be defined before $g(x)$ can be determined. Even if we try to invent simultaneously Grundy numbers for all vertices in Figure 10.3 at once, no Grundy function exists (details are left to the reader).

The Grundy numbers for the vertices in Figures 10.1 and 10.2 are given in the following tables. Note that the vertices in Figure 10.1 that are in the kernel—namely, 0, 5, 10, 15—all have Grundy number 0. This is no accident.

Table of Grundy Numbers for Figure 10.1

Vertex x	16	15	14	13	12	11	10	9	8	7	6	5	4	3	2	1	0
$g(x)$	1	0	4	3	2	1	0	4	3	2	1	0	4	3	2	1	0

Table of Grundy Numbers for Figure 10.2

Vertex x	0	1	2	3	4	5	6	7	8	9	10	11	12	13	14	15	16
$g(x)$	0	3	1	2	0	1	0	2	1	0	3	1	0	3	1	2	0
	17	18	19	20	21	22	23	24	25	26	27	28	29	30			
	1	0	2	1	0	3	2	1	0	2	0	1	2	0			
	31	32	33	34	35	36	37	38	39	40	over 40						
	1	2	0	1	2	0	1	3	2	1	0						

Theorem 3

The graph of a progressively finite game has a unique Grundy function. Further, the vertices with Grundy number 0 are the vertices in the kernel.

Proof

The recursive level-by-level construction of a Grundy function for the graph of a progressively finite game gives each vertex a unique Grundy number, as described above. By the definition of a Grundy function, a vertex x with Grundy number 0 cannot have a successor with Grundy number 0 (or else x would have to have a different number). Similarly any vertex y with Grundy number $g(y) = k > 0$ must have an edge to some vertex with Grundy number 0 (or else y's number would be 0). Thus, the set of vertices with Grundy number 0 satisfies the two defining properties of a kernel. ◆

Example 2 (continued)

By Theorem 3 and the foregoing table of Grundy numbers for Figure 10.2, we see that the kernel is the set 0, 4, 6, 9, 12, 16, 18, 21, 25, 27, 30, 33, 36, "over 40." Since the starting vertex is in the kernel, the second player has the winning strategy in this game. A play of the game might proceed as follows (let player A be the first player and B be the second player): first A moves to 1, then B moves to kernel vertex 6, then A must move to 11 (a move to 7 or 8 lets B win at 9), then B moves to kernel vertex 12, then A must move to 17, then B moves to kernel vertex 18, then A moves to 20, and then B moves to the winning vertex 25 (and collects the 25 cents). ■

10.1 EXERCISES

Summary of Exercises The first 10 exercises involve finding kernels and Grundy functions in various graphs of progressively finite games. Exercises 11–17 involve proofs of properties of kernels, Grundy functions, and level numbers. Exercises 18 and 19 present two more complicated progressively finite games.

1. Find a kernel in the following graphs or show why none can exist:

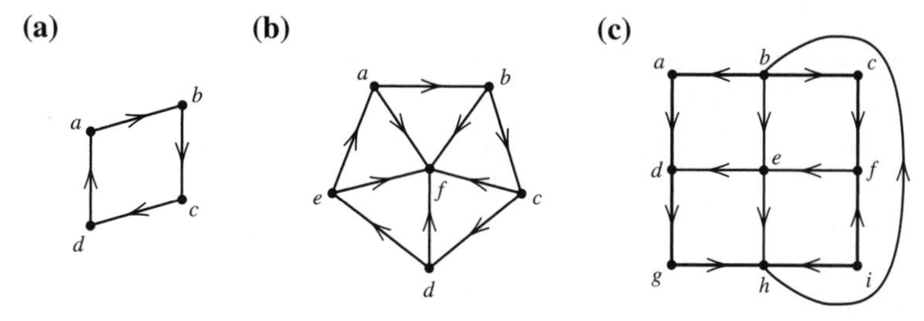

2. Suppose that there are 25 sticks and two players take turns removing up to 5 sticks. The winner is the player who removes the last stick. Which player has a winning strategy? Describe this strategy.
3. Repeat Example 2 with 2, 3, or 7 cents added each time. Find the set of positions in the kernel.
4. Show that in Example 2 the second player can always win by the second move.
5. Suppose in Example 2 that the first player A knows he/she will lose and wants to minimize his/her loss. What is the smallest winning amount A can force B

(the second player) to accept (such that if B tried to make the game go longer, then A could get into the kernel and win)?

6. **(a)** Suppose we have a pile of 7 red sticks and 10 blue sticks. A player can remove any number of red sticks or any number of blue sticks or an equal number of red and blue sticks. he winner is the player who removes the last stick. Find the set of positions in the kernel.

(b) Repeat part (a) with a limit of removing at most 5 sticks of one color (or both colors) on a move.

7. Repeat Example 1 but now the player to remove the last object loses. Describe the winning strategy for this game.

8. Find the Grundy function for graphs or games in:

(a) Exercise 1(a) **(b)** Exercise 1(b) **(c)** Exercise 3

9. Show that there is no Grundy function for the graph in Figure 10.10.

10. Find a directed graph possessing a kernel but no Grundy function.

11. If $W(S)$ is the set of vertices without an edge directed to any vertex in the set S, show that a set S is a kernel if and only if $S = W(S)$.

12. **(a)** Show that if $l(x) = k$ for a vertex x in the progressively finite graph G, then k is the length of the longest path starting at x in G.

(b) Show that if $g(x) = k$, there is a path of length k starting at x in G.

13. Show that for any vertex x in a progressively finite graph, $g(x) \leq l(x)$.

14. Show that if a directed graph G and every subgraph of G (obtained by deleting various vertices) have kernels, then G has a Grundy function.

15. Show that both the level numbers and the Grundy function in a progressively finite graph G constitute proper colorings of G.

16. Show that no matter how the edges are directed in a bipartite graph, it will always have a Grundy function.

17. Show that the graph of a progressively finite game can have only a finite number of vertices. [*Hint:* Show that if there are an infinite number of vertices, then there must be an infinite path (infinite play).]

18. Consider the following graph game. Player A tries to make a path with a set of vertices from the left to the right side of this graph. Player B tries to make a path from top to bottom. The players take turns picking vertices until one player gets the desired path.

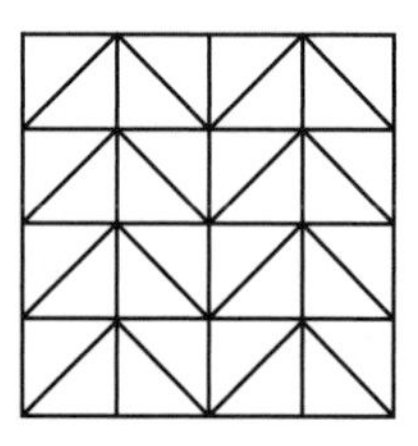

(a) Show that this is a progressively finite game.

(b) Find a winning strategy for Player A.

19. The game of *kayles* has a row of n equally spaced stones. Two players alternate turns of removing one or two consecutive stones (with no intervening spaces). The player to remove the last stone wins.

(a) Draw the graph and find its Grundy function for a four-stone game.

(b) Repeat part (a) for a six-stone game.

10.2 NIM-TYPE GAMES

In this section we extend the theory of progressively finite games to takeaway games involving several piles of objects. The simplest game of this form is called Nim, a game in which two players take turns removing any number they wish from one of the piles. The winner is the player who removes the last object from the last remaining (nonempty) pile. While the positions in the two games in Examples 1 and 2 in the previous section could be described with a single nonnegative integer representing the size or monetary value of the single pile, the position in a Nim game requires a vector of nonnegative integers $(p_1, p_2, \ldots, p_m)$, the kth number p_k representing the current size of the kth pile.

Example 1: Game of Nim

Consider the Nim game with four piles of sticks: one stick in the first pile, two sticks in the second pile, three in the third, and four in the fourth. See Figure 10.4. We represent the initial position of this game with the vector (1, 2, 3, 4).

Let the first and second players be named A and B, respectively. A sample play of the game might go as follows. First A removes all four sticks from the fourth pile. The new position is (1, 2, 3, 0). Next B removes one stick from the third pile to produce position (1, 2, 2, 0). Now A removes the one stick in the first pile to produce position (0, 2, 2, 0). A has been playing a winning strategy, that is, moving into kernel positions, and is now about to win. If B removes all of the second or third pile, A will remove the other pile; or if B removes just one stick, A will remove one stick from the other pile and A will win on the next round. The reader is encouraged to play this Nim game with a friend. ■

Figure 10.4

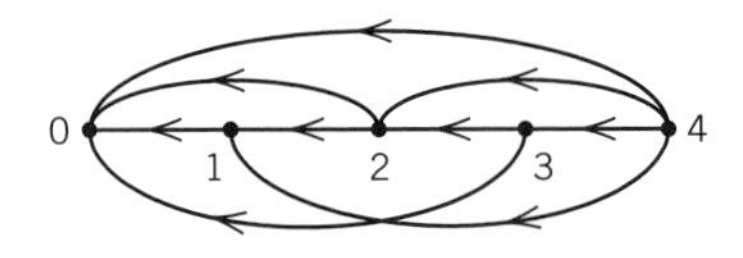

Figure 10.5

The game in Example 1 has $2 \times 3 \times 4 \times 5 = 120$ different positions (the ith pile has $i + 1$ possible sizes, $0, 1, \ldots, i$). Thus, it would be very cumbersome to draw the graph of this game and compute its kernel with a Grundy function, as in the previous section. In general, a Nim game with m piles and n_i objects in the ith pile will have $(n_1 + 1)(n_2 + 1) \cdots (n_m + 1)$ different positions. The only way that we could ever play winning Nim without a computer is if we could determine the Grundy number of a position (vertex) directly from the position vector $(p_1, p_2, \ldots, p_m)$. Fortunately, such direct computation is possible.

First we need to examine the Grundy function for a single-pile Nim game. Consider the Nim game with one pile of four sticks. See Figure 10.5. The graph of this game has edges $(i,\vec{} j)$, for all $0 \leq j \leq i \leq 4$. The Grundy number of vertex 0, the winning vertex, is 0; the Grundy number of vertex 1 is 1; and so on. In any one-pile Nim game, vertex i will have a Grundy number of i, since vertex i has edges to all lower-numbered vertices. Of course, the strategy of any one-pile Nim game is trivial: the first player removes the whole pile and wins. The nice form of this single-pile Grundy function will simplify the computation of Grundy functions for multi-pile Nim games.

Although the graph of a multi-pile Nim game is too complex to draw, we can still describe it symbolically. We have a vertex for each position $(p_1, p_2, \ldots, p_m)$, where $0 \leq p_i \leq n_i$ (n_i is the initial size of the ith pile). Since the only permissible moves are removing some amount from one pile, the associated graph has edges from vertex $(p_1, p_2, \ldots, p_m)$ to vertex $(q_1, q_2, \ldots, q_m)$ for each pair of vertices such that for one $j, q_j, < p_j$, and for all $i \neq j, q_i = p_i$. This graph is in some sense a "composition" of the graphs for each pile, for if we fix all p_i except one, say p_3, then the subgraph of vertices $(p_1, p_2, 0, p_4, \ldots, p_m), (p_1, p_2, 1, p_4, \ldots, p_m), \ldots, (p_1, p_2, n_3, p_4, \ldots, p_m)$ is exactly the graph for pile three alone.

This type of composition of graphs can be formalized as follows. The **direct sum** $H = H_1 + H_2 + \cdots + H_m$ of graphs $H_1, H_2, \ldots, H_m$ with vertex sets X_1, $X_2, \ldots, X_m$, respectively, has vertex set $X = \{(x_1, x_2, \ldots, x_m) \mid x_i \in X_i, 1 \leq i \leq m\}$ and edges defined by the successor sets

$$\begin{aligned} s((x_1, x_2, \ldots, x_m)) = {} & \{(y, x_2, x_3, \ldots, x_m) \mid y \in s(x_1)\} \\ & \cup \{(x_1, y, x_3, \ldots, x_{m1}) \mid y \in x(x_2)\} \\ & \vdots \\ & \cup \{(x_1, x_2, \ldots, x_{m-1}, y) \mid y \in s(x_m)\} \end{aligned}$$

It follows from this definition that the graph G of an m-pile Nim game is the direct sum of the graphs G_i of the ith pile: $G = G_1 + G_2 + \cdots + G_m$.

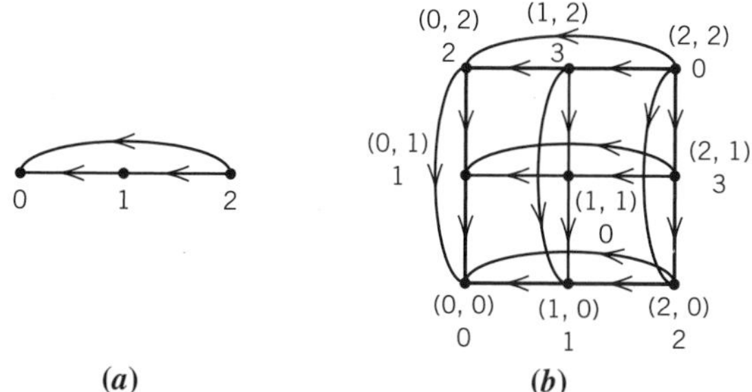

Figure 10.6 **(*a*)** **(*b*)**

Example 2: Graph of 2 × 2 Nim

Consider the simple Nim game with two piles of two objects each. Figure 10.6*a* shows the graph $G_1 = G_2$ of one pile alone. Figure 10.6*b* shows the graph $G = G_1 + G_2$ of the two-pile game. Find the Grundy function and kernel of this graph.

The Grundy numbers for the vertices are shown in Figure 10.6*b*. The vertices with Grundy number 0 are in the kernel. Since $g((2, 2)) = 0$, the second player has the winning strategy, according to Theorems 1 and 3 in the previous section. ■

Next we show how to compute the Grundy number of a vertex $(x_1, x_2, \ldots, x_m)$ in a direct sum of graphs from the Grundy numbers $g(x_i)$ of the vertices x_i in the component graphs G_i. The computation to be performed on these Grundy numbers is called a digital sum.

The **digital sum** c of nonnegative integers $c_1, c_2, \ldots, c_m$, written

$$c = c_1 \dot{+} c_2 \dot{+} c_3 \dot{+} \cdots \dot{+} c_m$$

is computed in the following manner. Let $c^{(k)}$ be the kth binary digit in a binary expansion of c, that is, $c = c^{(0)} + c^{(1)}2 + c^{(2)}2^2 + \cdots$, and similarly let $c_i^{(k)}$ be the kth digit in a binary expansion of c_i. Then $c^{(k)} \equiv c_1^{(k)} + c_2^{(k)} + \cdots + c_m^{(k)}$ (modulo 2). That is, the kth binary digit of c is 1 if the sum of the kth binary digits of the c_js is odd, and the kth binary digit of c is 0 if the sum of the kth binary digits of the c_js is even.

Example 3: Digital Sum

Compute the digital sum $2 \dot{+} 12 \dot{+} 15 \dot{+} 8$. We write the numbers 2, 12, 15, and 8 in binary form and determine the sum (modulo 2) of the digits in each column, as illustrated in Figure 10.7. Translating the value of the binary sum back into an integer, we see that this digital sum equals 9. ■

We now present a remarkable theorem. In essence, it says that digital sums are the only way to win at Nim.

Theorem

If the graphs $G_1, G_2, \ldots, G_n$ possess Grundy functions $g(x)$, then the direct sum $G = G_1 + G_2 + \cdots + G_m$ possesses a Grundy function $g(x)$, where $x = (x_1, x_2, \ldots, x_m)$,

	2^3	2^2	2^1	2^0
2 =	0	0	1	0
12 =	1	1	0	0
15 =	1	1	1	1
8 =	1	0	0	0
	1	0	0	1 = 9

Figure 10.7

defined by

$$g(x) = g((x_1, x_2, \ldots, x_m)) = g_1(x_1) \dotplus g_2(x_2) \dotplus \cdots \dotplus g_m(x_m)$$

Proof

We must show that $g(x)$ is the smallest nonnegative integer that is not equal to any number in the set $\{g(y) \mid y \in s(x)\}$. This definition of a Grundy function can be broken into two parts: (1) If $y \in s(x)$, then $g(x) \neq g(y)$: and (2) for each nonnegative integer $b < g(x)$, there exists some $y \in s(x)$ with $g(y) = b$.

Part (1): Show that if $y \in s(x)$, then $g(x) \neq g(y)$. Since $y = (y_1, y_2, \ldots, y_m) \in s((x_1, x_2, \ldots, x_m))$, then by the definition of direct sum, for some j, $y_j \in s(x_j)$, and for all $i \neq j$, $y_i = x_i$. If $c_k = g_k(x_k)$ and $d_k = g_k(y_k)$, then $c_j \neq d_j$ and $c_i = d_i$. So

$$g(x) = c_1 \dotplus c_2 \dotplus \cdots \dotplus c_j \dotplus \cdots \dotplus c_m = c' \dotplus c_j$$
$$g(y) = d_1 \dotplus d_2 \dotplus \cdots \dotplus d_j \dotplus \cdots \dotplus d_m = c' \dotplus d_j$$

where c' is the digital sum of all cs except c_j. It is not hard to show that $c' \dotplus c_j = c' \dotplus d_j$ if and only if $c_j = d_j$ (see Exercise 9). Since $c_j \neq d_i$, we conclude that $g(x) \neq g(y)$, as required.

Part (2): Show that for each nonnegative integer $b < g(x)$, there exists some $y \in s(x)$ with $g(y) = b$. A general proof of this part is fairly technical (see Berge [2], p. 25, for details). The practical side of this proof, discussed below, is finding a y with $g(y) = 0$ (a kernel vertex) and moving to it. A formal proof of part (2) is a generalization of the discussion below. ◆

Corollary

The vertex $(p_1, p_2, \ldots, p_m)$ in the graph of an m-pile Nim game has the Grundy number $p_1 \dotplus p_2 \dotplus \cdots \dotplus p_m$. Thus $(p_1, p_2, \ldots, p_m)$ is a kernel vertex if and only if $p_1 \dotplus p_2 \dotplus \cdots \dotplus p_m = 0$.

The reader should go back to the Nim game of two piles of two objects each in Example 2 and check that the Grundy numbers obtained for the vertices of the graph of that game are the digital sums of the pile sizes.

$$
\begin{array}{r@{\;=\;}ccc}
1 & 0 & 0 & 1 \\
2 & 0 & 1 & 0 \\
3 & 0 & 1 & 1 \\
4 & 1 & 0 & 0 \\
\hline
\multicolumn{1}{r}{} & 1 & 0 & 0 = 4
\end{array}
$$

Figure 10.8

Example 1 (continued)

The Grundy number of the starting vertex of the Nim game in Figure 10.4 is the digital sum $1 \dotplus 2 \dotplus 3 \dotplus 4 = 4$, as computed in Figure 10.8. The first player A wants to decrease the size of one of the piles so that the new digital sum is 0 (a kernel position). That is, A should alter the binary digits in one row of Figure 10.8 so that the sum (bottom) row is 0 0 0. For the sum row to become all 0s, every column now having an odd number of 1s should have one of its digits changed (either a 1 to a 0 or a 0 to a 1) to make the number of 1s in that column even.

Since the sum row has a 1 in just the 2^2 column, we can change the (single) 1 in that column to a 0. That is, pile 4's binary expansion should be changed from 1 0 0 to 0 0 0, and so player A should remove all sticks in the fourth pile. This new position (1, 2, 3, 0) has a Grundy number of 0. Note that to make the number of 1s even in the 2^2 column, we could not change any 0 to a 1, for the new binary expansion in the altered row would then be a larger number—an impossible move in Nim. ■

We now generalize the method of finding a vertex with Grundy number 0 given in the preceding example. Form a digital sum table as in Figure 10.8 for the current game position. *Pick a row e having a 1 in the leftmost column that has an odd number of 1s, that is, row e should have a 1 in a column with a 1 in the sum row. In row e, change the digit in every column having a 1 in the sum row.* After this change of digits, every column will have an even number of 1s, and so the sum row will be all 0s. Thus, this new position will be a kernel vertex. Note that since the leftmost digit that is changed in row e is a 1 (this is how row e was chosen), changing digits in row e will yield a smaller number, call it h. The first player should thus decrease the size of the eth pile to a size of h objects.

Example 4: Another Game of Nim

Consider the Nim game of four piles with 2, 3, 4, and 6 sticks shown in Figure 10.9*a*. The digital sum table for the initial position is shown in Figure 10.9*b*. The 2^1 column is the leftmost column with a 1 in the sum row, and so we must change a row with a 1 in the 2^1 column. We can use the first, second, or fourth row. Suppose that we choose the first row. Then since the 2^1 and 2^0 columns in the sum row have 1s, we change the digits in these columns in the first row. The new first row is 0 0 1. Thus, the first

(a)		2^2	2^1	2^0
| |	2 =	0	1	0
| | |	3 =	0	1	1
| | | |	4 =	1	0	0
| | | | | |	6 =	1	1	0
		0	1	1 = 3

Figure 10.9 **(*a*)** **(*b*)**

player should reduce the size of the first pile to 1. The reader should continue the play of this Nim game with a friend (or against oneself) to practice this kernel-finding rule. ∎

10.2 EXERCISES

1. Find the Grundy number of the initial position and make the first move in a winning strategy for the following Nim games:

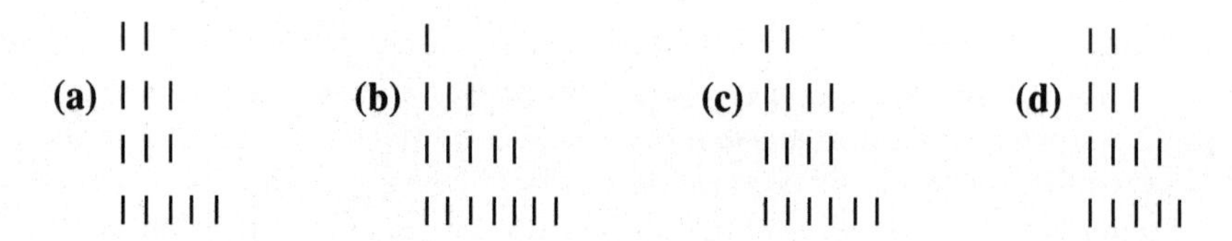

2. Suppose that no more than two sticks can be removed at a time from any pile. Repeat the games in Exercise 1 with this additional condition.

3. Suppose that no more than i sticks can be removed at a time from the ith pile (piles are numbered from top to bottom). Repeat Exercise 1 with this additional condition.

4. Suppose that only 1 or 2 or 5 sticks can be removed from each pile. Repeat the games in Exercise 1 with this constraint.

5. Suppose that only 1 or 4 sticks can be removed from each pile. Repeat the games in Exercise 1 with this constraint.

6. For the Nim game in Exercise 1(c), find moves that yield positions with Grundy number equal to:

(a) 1 **(b)** 2 **(c)** 3

7. Suppose three copies of the game in Example 2 of Section 10.1 are played simultaneously. Players stop adding money to a pile when the value of the pile is a square or exceeds 40. The player who adds the last amount to the last pile wins the money in all three piles. Note that there is a table of Grundy numbers for this game near the end of the previous section.

(a) If initially there are 2¢ in two piles and 1¢ in the third pile, what is the Grundy number of this position and what is a correct winning move?

(b) Find all kernel positions in which the sum of the money in the first two piles is less than 10¢ and the third pile is empty.

(c) Using the table of Grundy numbers for the one-pile game near the end of Section 10.1, write a computer program to compute the next winning move in this three-pile game.

8. Draw the direct sums involving the graphs shown below.

(a) $G_1 + G_1$ **(b)** $G_1 + G_2 + G_3$

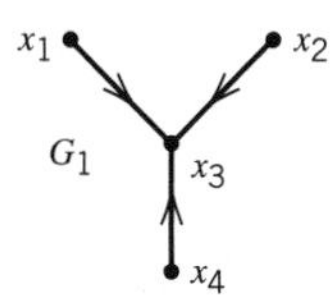

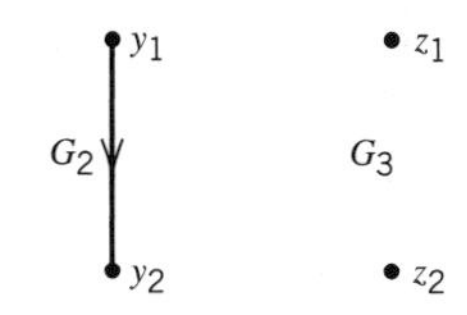

9. Prove the assertion in the proof of the theorem that $c' \dot{+} c_j = c' \dot{+} d_j$ if and only if $c_j = d_j$.

10. Generalize the argument for finding kernel vertices in Nim (preceding Example 4) to obtain part (2) of the proof of the theorem.

11. Consider the variation of Nim in which the last player to move *loses*. Show that the winning strategy is to play the regular last-player-wins Nim strategy but when only one pile has more than one stick now decrease that pile's size to 1 instead of 0 (or to 0 instead of 1).

12. Write a computer program to play winning Nim.

13. Consider the following variation on Nim. Place $C(n + 1, 2)$ identical balls in a triangle, like bowling pins. Two players take turns removing any number of balls in a set that all lie on a straight line. The player to remove the last ball wins.

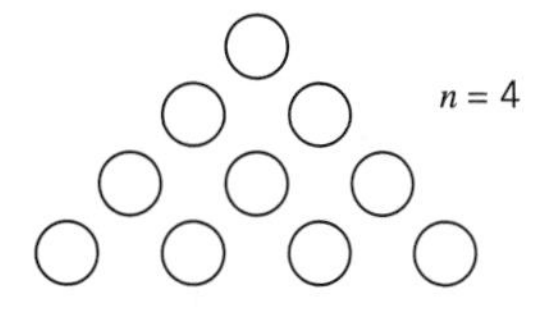

(a) Find a winning first move for this game when $n = 3$.

(b) Repeat for $n = 4$.

10.3 SUMMARY AND REFERENCES

The first published analysis of Nim by C. Bouton [3] appeared in 1902. The Grundy function for progressively finite games was presented by P. Grundy [5] in 1939. It was independently discovered a few years earlier by the German mathematician Sprague.

This chapter has only scratched the surface of the theory of finitely progressive games. Interested readers should turn to *On Winning Ways* by Berlekamp, Conway, and Guy [1]. The reader should also see the classic book about mathematics and games, *Mathematical Recreations and Essays* by Coxeter and Ball [4].

1. E. Berlekamp, J. Conway, and R. Guy, *On Winning Ways,* 2 volumes, Academic Press, New York, 1982.

2. C. Berge, *The Theory of Graphs,* Methuen-Wiley, New York, 1962.

3. C. L. Bouton, "Nim, a game with a complete mathematical theory," *Ann. Math.* 3 (1902), 35–39.

4. H. Coxeter and W. Ball, *Mathematical Recreations and Essays,* University of Toronto Press, 1972.

5. P. M. Grundy, "Mathematics and games," *Eureka,* January, 1939.

APPENDIX

A.1 SET THEORY

A set is a collection of distinct objects. In contrast to a sequence of objects, a set is unordered. Usually we refer to the objects in a set as the **elements,** or **members,** of the set. These elements may themselves be sets, as in the set of all 5-card hands; each hand is a set of 5 cards. A **family** is a collection in which multiple appearances of objects is allowed; for example, $\{a, a, a, b, b, c\}$.

A set S is a **subset** of set T if every element of S is also an element of T. In the case of a 5-card hand, we are working with a 5-card subset of the set of all 52 cards. Any set is trivially a subset of itself. A **proper** subset is a nonempty subset with fewer elements than the whole set. Two sets with no common members are called **disjoint** sets.

This text uses capital letters to denote sets and lowercase letters to denote elements (unless the elements are themselves sets). The number of elements in a set S is denoted by $N(S)$ or $|S|$. The symbol $\in$ represents set membership, for example, $x \in S$ means that x is an element of S; and $x \notin S$ means that x is not a member of S. The symbol $\subseteq$ represents subset containment, for example, $T \subseteq S$ means that T is a subset of S.

There are three ways to define a set with formal mathematical notation:

1. By listing its elements, as in $S = \{x_1, x_2, x_3, x_4\}$ or $T = \{\{1, -1\}, \{2, -2\}, \{3, -3\}, \{4, -4\}, \{5, -5\}, \ldots\}$—implicitly, T is the set of all pairs of a positive integer and its negative.
2. By a defining property, as in $P = \{p \mid p \text{ is a person taking this course}\}$ or $R = \{r \mid \text{there exist integers } s \text{ and } t \text{ with } t \neq 0 \text{ and } r = s/t\}$—$R$ is the set of all rational numbers;
3. As the result of some operation(s) on other sets (see below).

There are two special sets we often use: the **empty,** or **null,** set, written $\emptyset$; and the **universal** set of all objects currently under consideration, written $\mathcal{U}$.

It is important to bear in mind that in many real-world problems, sets cannot be precisely defined or enumerated. For example, the set of all ways in which a large computer program can fail is ill defined. A census of the population of the United States involves substantial error for several important subcategories of the populace, yet the official United States population was given in 1990 as 248,709,873. Beware of any calculations based on imprecise sets!

The three basic operations on sets that we will use are

1. The **intersection** of S and T, $S \cap T = \{x \in \mathcal{U} \mid x \in S \text{ and } x \in T\}$.
2. The **union** of S and T, $S \cup T = \{x \in \mathcal{U} \mid x \in S \text{ or } x \in T\}$.
3. The **complement** of S, $\overline{S} = \{x \in \mathcal{U} \mid x \notin S\}$.

A fourth operation that is sometimes useful is

4. The **difference** of S of minus T, $S - T = \{x \in \mathcal{U} \mid x \in S \text{ and } x \notin T\}$.

Observe that $S - T$ can be expressed in terms of the preceding operations as $S - T = S \cap \overline{T}$.

Example 1: Selecting Calculators

A sample of eight different brands of calculators consists of four machines that have rechargeable batteries and four that do not. Also, four out of the eight have memory. We want to select four calculators to be taken apart for thorough analysis. Half of this set of four should be rechargeable and half should have memory. How many ways can such a set of four machines be chosen from the eight brands?

To answer this question, we must know how many machines are rechargeable and have memory, how many are rechargeable and have no memory, and so on. That is, if R is the set of rechargeable machines and M the set of machines with memory, then we need information about the sizes of the intersections $N(R \cap M)$, $N(R \cap \overline{M})$, $N(\overline{R} \cap M)$, and $N(\overline{R} \cap \overline{M})$. Assume that there are two machines in each of these four subsets. See Figure A1.1. Then one strategy to get the desired mix of four machines would be to pick one machine from each category in Figure A1.1 ($2 \times 2 \times 2 \times 2 = 16$ choices). There are two other strategies for getting the four machines from Figure A1.1 (see Exercise 4 at the end of this section). ■

The study of set expressions involving the foregoing set operations and associated laws is called **Boolean algebra.** Three of the most important laws of Boolean algebra are

BA1. $S = \overline{\overline{S}}$.
BA2. $\overline{S \cap T} = \overline{S} \cup \overline{T}$.
BA3. $\overline{S \cup T} = \overline{S} \cap \overline{T}$.

To visualize set expressions, we use a picture called a **Venn diagram.** Figure A1.2*a* shows a Venn diagram for the sets S and T. The whole rectangle represents

	Memory	No memory
Rechargeable	2	2
Not rechargeable	2	2

Figure A1.1

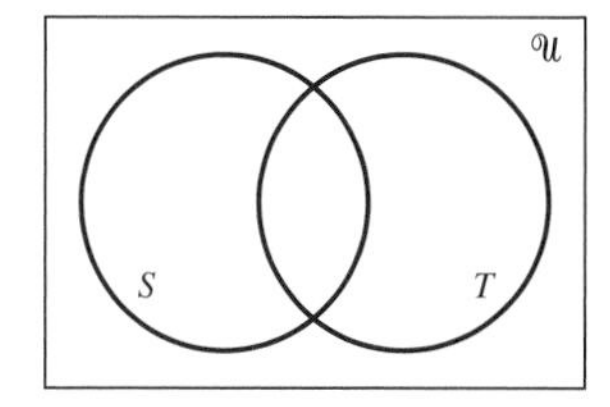

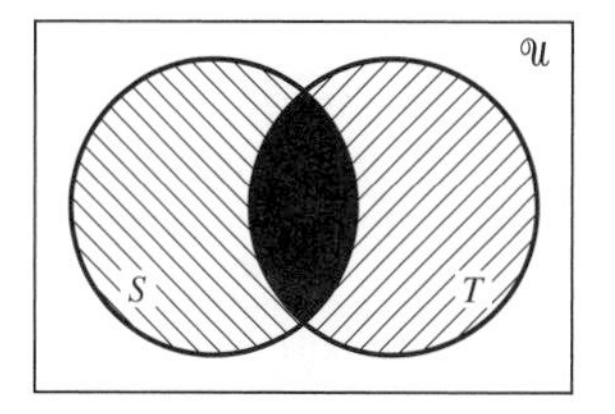

Figure A1.2

the universe $\mathcal{U}$ of all elements under consideration. The circles represent the sets S and T. In Figure A1.2*b*, the darkened region represents $S \cap T$ and the striped area $(S \cup T) - (S \cap T)$. Note that in particular problems, certain regions in a Venn diagram may be empty sets.

The next example illustrates how laws BA2 and BA3 can simplify a counting problem.

Example 2: Counting With Boolean Algebra

Consider the universe $\mathcal{U}$ of all 52^2 outcomes obtained by picking one card from a deck of 52 playing cards, replacing that card, and then again picking a second card (possibly the same card). Suppose that we want to compute the size of the set Q of outcomes with at least one spade or at least one heart. Let S be the set of outcomes with a void in spades, that is, in which no spade is chosen on either pick. Let H be the set of outcomes with a void in hearts. Write the set Q as an expression in terms of S and H, and calculate $N(Q)$.

The set Q equals $\bar{S} \cup \overline{H}$, the union of the set of outcomes with one or more spades (not a void in spades) and the set of outcomes with one or more hearts. The set $\bar{S} \cup \overline{H}$ is the shaded area in Figure A1.3. By BA3, $\bar{S} \cup \overline{H} = \overline{S \cap H}$, where $S \cap H$ is the set of outcomes with no spades and no hearts ($S \cap H$ is the unshaded region in Figure A1.3). That is, $S \cap H$ is the set of outcomes where each pick is one of the 26 diamonds or clubs. Thus $N(S \cap H) = 26^2$. By Figure A1.2, $N(\overline{S \cap H} = N(\mathcal{U}) - N(S \cap H) = 52^2 - 26^2 = 2704 - 676 = 2028$, and so $N(\bar{S} \cup \overline{H}) = N(\overline{S \cap H}) = 2028$. ■

Set theory, and more generally, nonnumerical mathematics, were studied little until the nineteenth century. G. Peacock's *Treatise on Algebra,* published in 1830, first suggested that the symbols for objects in algebra could represent nonnumeric entities. A. De Morgan discussed a similar generalization for algebraic operations

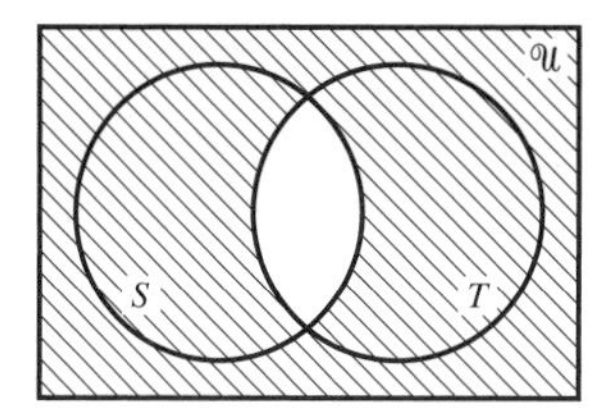

Figure A1.3

a few years later. G. Boole's *Investigation of the Laws of Thought* (1854) extended and formalized Peacock's and De Morgan's work to present a formal algebra of sets and logic. The great philosopher-mathematician Bertrand Russell has written, "Pure Mathematics was discovered by Boole." Subsequently it was found that numerical mathematics also needs to be defined in terms of set theory in order to have a proper mathematical foundation (see Halmos's *Naive Set Theory*). See Chapter 26 of Boyer's *History of Mathematics* for more details about the history of set theory.

EXERCISES

1. Let A be the set of all positive integers less than 30. Let B be the set of all positive integers that end in 7 or 2. Let C be the set of all multiples of 3. List the numbers in the following sets:

(a) $A \cap (B \cap C)$ **(b)** $A \cap (B \cup C)$ **(c)** $A \cap (\overline{B \cup C})$ **(d)** $A - (B \cap C)$

2. If $A = \{1, 2, 4, 7, 8\}$, $B = \{1, 5, 7, 9\}$, and $C = \{3, 7, 8, 9\}$ and $\mathcal{U} = \{1, 2, 3, 4, 5, 6, 7, 8, 9, 10\}$, then find set expressions (if possible) equal to:

(a) $\{7\}$ **(c)** $\{1\}$ **(e)** $\{4, 8, 10\}$
(b) $\{6, 10\}$ **(d)** $\{2, 7, 9\}$ **(f)** $\{3, 5, 6, 7, 9, 10\}$

3. Suppose that in Example 1 we were given only the additional information that two calculators have memory but are not rechargeable. Show that we can now deduce how many machines are in each of the other three boxes in Figure A1.1.

4. What are the other two strategies in Example 1 to pick a subset of four calculators, two of which are rechargeable and two of which have memory, given the numbers in Figure A1.1? How many choices are there for each of these ways?

5. Suppose we are given a group of 20 people, 13 of whom are women. Suppose 8 of the women are married. In each of the following cases, tell whether the additional piece of information is sufficient to determine the number of married men. If it is, give this number.

(a) There are 12 people who are either married or male (or both).

(b) There are 8 people who are unmarried.

(c) There are 15 people who are female or unmarried.

6. Label each region of the Venn diagram in Figure A1.2*a* with the set expression it represents (e.g., the region where both sets intersect would be labeled $A \cap B$).

7. Draw Venn diagrams and shade the area representing the following sets:

(a) $A - B$ **(b)** $\overline{A \cup B}$ **(c)** $(\overline{A \cup B}) \cap (\overline{A \cap B})$ **(d)** $A - (B - A)$

8. Write as simple a set expression as you can for the shaded areas in the following Venn diagrams.

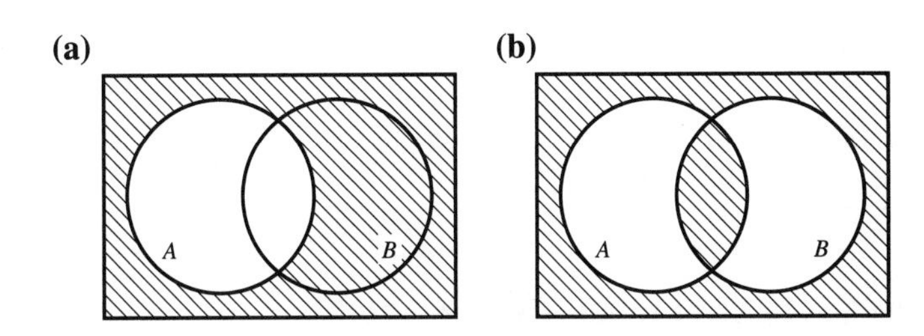

9. Create a Venn diagram for representing sets *A, B,* and *C*. The diagram should have regions for all possible intersections of *A, B,* and *C*. How many different regions are there? Label each region with the set expression it represents (e.g., the region where all three sets intersect would be labeled $A \cap B \cap C$).

10. Draw Venn diagrams for sets *A, B,* and *C* and shade the areas representing the sets in Exercise 1.

11. Verify with Venn diagrams the following laws of Boolean algebra:

(a) $(A \cup B) \cap A = A$

(b) $(A \cap B) \cup A = A$

(c) $A \cup \bar{A} = \mathcal{U}$

(d) $(A \cup B) \cap C = (A \cap C) \cup (B \cap C)$

(e) $(A \cap B) \cup C = (A \cup C) \cap (B \cup C)$

12. Use Venn diagrams to determine which pairs of the following set expressions are equivalent.

(a) $(A \cap B) \cup (\bar{A} \cap B)$

(b) $A - (B - (A - B))$

(c) $(\bar{A} - \bar{B}) \cap (\overline{\bar{A} \cup \bar{B}})$

(d) $(\bar{A} \cap (B - \bar{A}))$

13. Use Law BA2 to prove that $\overline{A \cap B \cap C} = \bar{A} \cup \bar{B} \cup \bar{C}$ [note that $A \cap B \cap C = (A \cap B) \cap C = A \cap (B \cap C)$].

14. Use Law BA3 to prove that $\overline{A \cup B \cup C} = \bar{A} \cap \bar{B} \cap \bar{C}$ [note that $A \cup B \cup C = (A \cup B) \cup C = A \cup (B \cup C)$].

15. Use Laws BA1, BA2, and BA3 plus the identities $A \cap A = A$, $A \cup A = A$, $A \cap B = B \cap A$, and $A \cup B = B \cup A$ to simplify the following set expressions:

(a) $(\overline{A \cup B}) \cap (A \cap B)$

(b) $A - (\bar{B} \cup \bar{A})$

(c) $(\overline{A \cup C}) \cap (\overline{A \cup B}) \cap (\overline{B \cup C})$

(d) $(A \cup B) \cap [(C \cup B) - (\bar{A} \cap \bar{B})]$

16. Suppose two dice are rolled. How many outcomes have a 1 or a 2 showing on at least one of the dice?

17. Suppose that a card is drawn from a deck of 52 cards (and replaced before the next draw) three times. Let *S* be the set of outcomes where the first card is a spade, *H* be the set where the second card is a heart, and *C* be the set where the third card is a club. For each of the following sets *E*, write *E* as a set expression in terms of *S*, *H*, *C*, and determine *N(E)*.

(a) Let *E* be the set of outcomes contained in at least one of the sets *S*, *H*, *C*.

(b) Let E be the set of outcomes contained in at least one but not all three of the sets S, H, C.

(c) Let E be the set of outcomes contained in exactly two of the sets S, H, C.

A.2 MATHEMATICAL INDUCTION

The most useful, and simplest, proof technique in combinatorial mathematics and computer science is **mathematical induction.** Let p_n denote a statement involving n objects. Then an induction proof that p_n is true for all $n \geq 0$ requires two steps:

1. Initial step: Verify that p_0 is true.
2. Induction step: Show that if $p_0, p_1, p_2, \ldots, p_{n-1}$ are true, then p_n must be true.

Sometimes p_n will be true only for $n \geq k$. Then the initial step is to verify that p_k is true, and in the induction step we assume only that $p_k, p_{k+1}, \ldots, p_{n-1}$ are true.

Example 1: Summation Formula

Let s_n denote the sum of the integers 1 through n, that is, $s_n = 1 + 2 + 3 + \cdots + n$. Show that $s_n = \frac{1}{2}n(n+1)$.

We use mathematical induction to verify this formula. Since s_0 is not defined, the initial step is to verify the formula for s_1. The formula says $s_1 = \frac{1}{2}1(1+1) = 1$—obviously correct. For the induction step, we assume that the formula is true for $s_1, s_2, \ldots, s_{n-1}$. In this problem (as in most induction problems), we need only to assume that the formula is true for s_{n-1}.

$$s_{n-1} = 1 + 2 + 3 + \cdots + (n-1) = \tfrac{1}{2}(n-1)[(n-1)+1] = \tfrac{1}{2}(n-1)n$$

We now use this expression for s_{n-1} to prove that the formula is true for s_n.

$$\begin{aligned} s_n &= [1 + 2 + \cdots + (n-1)] + n = s_{n-1} + n \\ &= \tfrac{1}{2}(n-1)n + n \\ &= \tfrac{1}{2}[(n-1)n + 2n] = \tfrac{1}{2}n(n+1) \end{aligned}$$

This completes the induction proof that $s_n = \frac{1}{2}n(n+1)$ for all positive n. ■

Example 2: Population Growth Model

Suppose that the population of a colony of ants doubles in each successive year. A colony is established with an initial population of 10 ants. How many ants will this colony have after n years?

Let a_n denote the number of ants in the colony after n years. We are given that $a_0 = 10$. Since the colony's population doubles annually, then $a_1 = 20$, $a_2 = 40$, and $a_3 = 80$. Looking at the first few values of a_n, we are led to conjecture that

$a_n = 2^n \times 10$. For the initial step, we check that $a_0 = 10 = 2^0 \times 10$, as required. For the induction step, we assume that $a_{n-1} = 2^{n-1} \times 10$. Now we use the annual doubling property of the colony:

$$a_n = 2a_{n-1} = 2(2^{n-1} \times 10) = 2^n \times 10 \quad \blacksquare$$

Example 3: Prime Factorization

A prime number is an integer $p > 1$ that is divisible by no other positive integer besides 1 and itself. Show that any integer $n > 1$ can be written as a product of prime numbers.

We prove this fact by mathematical induction. Since the assertion concerns integers $n > 1$, the initial step is to verify that 2 can be written as a product of prime numbers. But 2 is itself a prime. Thus 2 is trivially the product of primes (i.e., actually of a single prime), itself. Next assume that the numbers $2, 3, \ldots, n-1$ can be written as a product of primes, and use this assumption to prove that n can also be written as a product of primes. If n is itself a prime, then as in the case of 2, there is nothing more to do. Suppose n is not a prime, and so there is an integer m that divides n and for some integer k, $n = km$. Since k and m must be less than n, they can each be written as a product of primes. Multiplying these two prime products for k and m together yields the desired representation of n as a product of primes. ■

Although the Greeks used certain iterative arguments in geometric calculations and the principle of reductio ad absurdum, mathematical induction was first used explicitly by Maurolycus around 1550. Pascal used an induction argument in 1654 to verify the additive property of binomial coefficients in the array now known as Pascal's triangle [this property is identity (3) in Section 5.5]. The actual term *induction* was coined by De Morgan 200 years later. See Bussey, "Origin of mathematical induction," *American Math. Monthly,* 1917, for a fuller discussion of the history of mathematical induction.

EXERCISES

1. Prove by induction that $1 + 3 + \cdots + (2n+1) = (n+1)^2$.

2. Prove by induction that $1^2 + 2^2 + \cdots + n^2 = \dfrac{n(n+1)(2n+1)}{6}$.

3. Prove by induction that $-1^2 + 2^2 - 3^2 + \cdots + (-1)^n n^2 = (-1)^n \dfrac{n(n+1)}{2}$.

4. Prove by induction that $(1 \times 2) + (2 \times 3) + \cdots + n(n+1) = \dfrac{n(n+1)(n+2)}{3}$.

5. Prove by induction that $1^3 + 2^3 + \cdots + n^3 = \dfrac{n^2(n+1)^2}{4}$.

6. Prove by induction that $(1+2+\cdots+n)^2 = 1^3 + 2^3 + \cdots + n^3$ (assume the result in Example 1).

7. Prove by induction that $\frac{1}{1\times 2} + \frac{1}{2\times 3} + \cdots + \frac{1}{n\times(n+1)} = \frac{n}{n+1}$.

8. Prove by induction that $\frac{1}{1\times 4} + \frac{1}{4\times 7} + \frac{1}{7\times 10} + \cdots + \frac{1}{(3n-2)\times(3n+1)} = \frac{n}{3n+1}$.

9. Prove by induction that $(1\times 1!) + (2\times 2!) + \cdots + (n\times n!) = (n+1)! - 1$.

10. Prove by induction that for $a \neq 1$, $\frac{1-a^{n+1}}{1-a} = 1 + a + a^2 + \cdots + a^n$.

11. Prove by induction that $1^2 + 3^2 + 5^2 + \cdots + (2n-1)^2 = \frac{n(2n-1)(2n+1)}{3}$.

12. If the number a_n of calf births at Dr. Smith's farm after n years obeys the law $a_n = 3a_{n-1} - 2a_{n-2}$ and in the first two years $a_1 = 3$ and $a_2 = 7$, then prove by induction that $a_n = 2^{n+1} - 1$.

13. Prove that any positive integer has a *unique* factorization into primes. You may assume the result of Example 3; you need to use the fact that if a prime p divides a product of positive integers, then p divides one of the integers.

14. Write a computer program to find and print the prime factorizations of the first 50 integers.

15. Prove by induction that for any integer $m > 0$, $m \times n = n \times m$. By $r \times s$, we mean the sum of r copies of s.

16. Prove by induction that for integers $n \geq 5$, $2^n > n^2$.

17. Prove that the number of different subsets (including the null set and full set) of a set of n objects is 2^n.

18. Prove by induction that $\overline{A_1 \cap A_2 \cap \ldots \cap A_n} = \bar{A}_1 \cup \bar{A}_2 \cup \quad \cup \bar{A}_n$ (see Exercise 13 in Appendix A1).

19. Prove by induction that if n distinct dice are rolled, the number of outcomes where the sum of the faces is an even integer equals the number of outcomes with an odd sum.

20. Prove by induction that the number of n-digit binary sequences with an even number of 1s equals the number of n-digit binary sequences with an odd number of 1s.

21. Prove by induction that the sum of the cubes of three successive positive integers is divisible by 9.

22. The cat population in a dormitory has the property that the number of cats in one year is equal to the sum of the number of cats in the two previous years. If in the first year there was one cat and if in the second year there were two cats, then prove by induction that the number of cats in the nth year is equal to

$$\frac{1}{\sqrt{5}}\left[\left(\frac{1}{2}+\frac{1}{2}\sqrt{5}\right)^{n+1} - \left(\frac{1}{2}-\frac{1}{2}\sqrt{5}\right)^{n+1}\right].$$

23. Why cannot one prove by induction that the number of binary sequences of all finite lengths is finite?

24. What is wrong with the following induction proof that all elements $x_1, \ldots, x_n$ in a set of n elements are equal?

(a) Initial step ($n = 1$): The set has one element x_1 equal to itself.

(b) Induction step: Assume $x_1 = x_2 = \cdots = x_{n-1}$. Since also $x_{n-1} = x_n$ by induction assumption (when x_{n-1}, x_n are considered alone as a two-element set), then $x_1 = x_2 = \cdots = x_{n-1} = x_n$.

25. What is wrong with the following induction proof that for a $\neq 0$, $a^n = 1$?

(a) Initial step ($n = 0$): $a^0 = 1$—always true.

(b) Induction step: Assume $a^{n-1} = 1$ and now $a^n = \dfrac{a^{n-1}a^{n-1}}{a^{n-2}} = 1$.

A.3 A LITTLE PROBABILITY

Historically, counting problems have been closely associated with probability. Indeed, any problem of the form "how many hobbits are there who . . . " has the closely related form "what fraction of all hobbits . . . ," which in turn can be posed probabilistically as "what is the probability that a randomly chosen hobbit . . . ?" The famous Pascal's triangle for binomial coefficients was developed by Pascal around 1650 in the process of analyzing some gambling probabilities. The probability of getting at least 7 heads on 10 flips of a fair coin, the probability of being dealt a five-card poker hand (from a well-shuffled deck) with a pair or better, and the probability of finding a faulty calculator in a sample of 20 machines if 5 percent of the machines from which the sample was drawn are faulty—all these probabilities are essentially counting problems.

Two hundred years ago, the French mathematician Laplace first defined probability as follows:

$$\text{Probability} = \frac{\text{number of favorable cases}}{\text{total number of cases}}$$

This definition corresponds to the "person in the street's" intuitive idea of what probability is. In this text, we treat probability problems only where Laplace's definition of probability applies. Implicit in this definition is the assumption that each case is equally likely. If the total number of cases is infinite, for example, all real numbers between 0 and 1, then we would not be able to use Laplace's definition. A more subtle difficulty is that in some probability problems, "cases" have to be carefully defined or else they may not all be equally likely, as, for example, the possible numbers of heads observed when a coin is flipped 10 times. To clarify this difficulty, we need to introduce a little of the basic terminology of probability theory.

An **experiment** is a clearly defined procedure that produces one of a given set of outcomes. These outcomes are called **elementary events** and the set of all elementary events is called the **sample space** of the experiment. We are interested only in experiments where the elementary events are equally likely. An event that is a subset of several elementary events is called a **compound event.**

For example, when a single die is rolled, then obtaining a specific number, such as 5, is an elementary event, whereas obtaining an even number is a compound event. If S is the sample space of the experiment and E is an event in S, then Laplace's definition of probability says that the probability of event E, prob(E), is

$$\text{prob}(E) = \frac{N(E)}{N(S)}$$

In most instances, the size of S is easily determined, and so the problem of determining prob(E) reduces to counting the number of outcomes (elementary events) in event E. Returning to the die roll, the sample space of outcomes is $S = \{1, 2, 3, 4, 5, 6\}$. If the event E is obtaining an even number, then $E = \{2, 4, 6\}$, and $\text{prob}(E) = \frac{3}{6} = \frac{1}{2}$.

Many experiments we discuss involve a repeated (or simultaneous) series of simple experiments. Each round of the simple experiment is called a **trial.** For example, the experiment of flipping a coin three times involves three successive trials of the simple experiment of flipping a coin. Rolling two dice and recording the sum of the two values rolled involves the two simultaneous simple experiments of rolling a single die. In any experiment involving multiple trials, the elementary events are the sequences of outcomes of the simple experiments. If the simple experiments are performed simultaneously, we number the simple experiments and list their outcomes in order of the experiments' numbers.

The sample space of elementary events for the experiment of flipping a coin three times is

$$\{HHH, HHT, HTH, HTT, THH, THT, TTH, TTT\} \tag{1}$$

The sample space for the experiment of rolling two dice is

$$S = \{(i, j) \mid 1 \leq i \leq 6, 1 \leq j \leq 6\}$$

Let us consider more closely the experiment of rolling two dice and recording the sum of the two values on each die (the sample space is the set S listed above).

In terms of our formal definition of events, the sum of the two values equaling k is not an elementary event but rather a compound event. The event, the sum of the two dice equals 7, is the subset of elementary events

$$S_1 = \{(1, 6), (2, 5), (3, 4), (4, 3), (5, 2), (6, 1)\}$$

Thus, $\text{prob(sum} = 7) = N(S_7)/N(S) = 6/36$. Similarly, $S_2 = \{(1, 1)\}$ and so prob $(\text{sum} = 2) = N(S_2)/N(S) = 1/36$. Observe that a sum of 7 is six times as likely as a sum of 2. If we had considered the possible sums of the two dice as the elementary events, we would have had a sample space $S^* = \{2, 3, 4, 5, 6, 7, 8, 9, 10, 11, 12\}$ and each sum would have had probability $1/N(S^*) = 1/11$—clearly a mistake. The same type of error would have been made if we were to regard the number of heads as the elementary events in the experiment of flipping three coins [see the correct sample space listed above in Eq. (1)].

Suppose we have a box containing 5 identical red balls and 20 identical black balls. Our experiment is to draw a ball. What is the sample space of elementary events? All we can record as an outcome is the color of the ball, leading to the sample space $S = \{\text{red, black}\}$. But then by Laplace's definition of probability, picking a red ball will have probability $\frac{1}{2}$. Although this experiment consists of just a single trial, we are clearly making the same sort of mistake that occurs when the sums of dice are treated as elementary events. The reader is encouraged to pause a moment and try to supply his or her own explanation for resolving this red–black ball paradox before reading the next paragraph.

The logical solution is to consider not what color ball we withdraw from the box but where in the box we direct our hand to grasp a ball. There are 25 different spatial positions where balls are located in the box. So, in this sense, there are 25 different events. We resolve this complication about "identical" objects with the following rule.

Identical Objects Rule

In probability problems, there are no collections of identical objects; all objects are distinguishable.

In complex experiments one should always take a moment to be sure that the elementary events are properly identified.

The original motivation for studying probability was the same as used in this book, games of chance—rolling dice, flipping coins, and card games. Cardano calculated the odds for different sums of two dice in the 1500s. Around 1600, Galileo calculated similar odds for three dice. A series of letters about gambling probabilities between Pascal and Fermat, written around 1650, constitute the real beginning of probability theory. Jacques Bernoulli's *Ars Conjectandi* (1713) was the first systematic treatment of probability (and associated combinatorial methods). One hundred years later, Laplace published his epic treatise *Théorie Analytique des Probabilitiés*, a work containing both the definition of probability in a finite sample space used in this book and also advanced-calculus-based derivations of modern probability theory. See F. N. David, *Games, Gods, and Gaming: A History of Probability* (Dover, 1998) for an excellent history of probability theory.

EXERCISES

1. An integer between 5 and 12 inclusive is chosen at random. What is the probability that it is even?

2. In the experiment of flipping a coin three times, let E_k be the compound event that the number of heads equals k. Determine prob(E_k), for $k = 0, 1, 2, 3$.

3. In the experiment of rolling two distinct dice, find the probability of the following events:

(a) Both dice show the same value.

(b) The sum of the dice is even.

(c) The sum of the dice is the square of one of the die's value.

4. A die is rolled three times. What is the probability of having a 5 appear at least two times?

5. Find the probability that in a randomly chosen arrangement of the letters h, a, t, the following occurs:

(a) The letters are in alphabetical order.

(b) The letter h occurs somewhere after the letter t in the arrangement.

6. Find the probability that in a randomly chosen (unordered) subset of two numbers from the set 1, 2, 3, 4, 5 the following occurs:

(a) The subset is $\{1, 2\}$.

(b) 1 is not in the subset.

7. An urn contains six red balls and three black balls. If a ball is chosen, then returned, and a second ball chosen, what is the probability that one of the following is true?

(a) Both balls are black.

(b) One ball is black and one is red.

8. An urn contains two black balls and three red balls. If two different balls are successively removed, what is the probability that both balls are of the same color?

9. Two boys and two girls are lined up randomly in a row. What is the probability that the girls and boys alternate?

10. Five distinct dice are rolled. What is the probability of getting at least one 6?

11. If a school has 100 students of whom 50 take French, 40 take Spanish, and 20 take French and Spanish, then what is the probability that a randomly chosen student takes no language?

12. What is the probability that an integer between 1 and 50 inclusive is divisible by 3 or 4?

13. There are three urns each with two balls: One urn has two black balls, the second has two red balls, and the third a black and a red ball. If an urn is randomly chosen and a randomly chosen ball in that urn is red, what is the probability that the other ball in this urn is also red? (*Hint:* Let the sample space be the set of all experimental outcomes where a red ball is chosen first.)

14. What should the sample space be for the following problem: An urn has 10 black balls and 5 red balls; 4 balls are removed *but not seen* (our eyes are shut) and then a fifth is removed and observed; now what is the probability that the fifth ball is red?

15. What should the elementary events be in each of the following experiments so that they are equally likely?

(a) A fair coin is tossed until two heads occur or until the coin is flipped 10 times.

(b) A positive integer is chosen at random.

(c) Two positive integers are chosen at random and their difference is recorded.

(d) A ball is randomly chosen from an urn with two red and two black balls and this ball is replaced and a ball chosen again, if necessary, until a red ball is obtained.

A.4 THE PIGEONHOLE PRINCIPLE

The pigeonhole principle is one of the most simple-minded ideas imaginable, and yet its generalizations involve some of the most profound and difficult results in all of combinatorial theory. This topic is included in these appendices because it does not fit naturally into any chapter of this book. More mathematical texts on combinatorics devote a whole chapter to Ramsey theory—for example, see Brualdi's *Introductory Combinatorics*, 3rd ed. (Prentice-Hall, 1999). For a thorough treatment of Ramsey theory, see Graham, Rothschild, and Spencer, *Ramsey Theory* (John Wiley, 1990).

Pigeonhole Principle

If there are more pigeons than pigeonholes, then some pigeonhole must contain two or more pigeons. More generally, if there are more than k times as many pigeons as pigeonholes, then some pigeonhole must contain at least $k+1$ pigeons.

This principle is also called the Dirichlet drawer principle. An application of it is the observation that two people in New York City must have the same number of hairs on their heads. New York City has over 7,300,000 people, and the average scalp contains 100,000 hairs. Indeed, the pigeonhole principle allows us to assert that it is theoretically possible to fill a subway car with about 73 New Yorkers, all of whom have the same number of hairs on their head.

The following three problems suggest some of the diverse generalizations of the pigeonhole principle.

1. How large a set of distinct numbers between 1 and 200 is needed to assure that two numbers in the set have a common divisor?
2. How large a set of distinct numbers between 1 and n is needed to assure that the set contains a subset of five equally spaced numbers a_1, a_2, a_3, a_4, a_5 that is, $a_2 - a_1 = a_3 - a_2 = a_4 - a_3 = a_5 - a_4$?
3. Given a positive integer k, how large a group of people is needed to assure that either there exists a subset of k people in the group who all know each other or there exists a subset of k people none of whom know each other?

In Problem 3, it might be possible that for large values of k, there are groups of arbitrarily large size that do not meet the k mutual friends or k mutual strangers property. Such is not the case. One of the theorems of Ramsey theory is that there exists a finite number r_k such that any group with at least r_k people must satisfy this property. The theorem does not say what number r_k is. Actually, r_k has been determined for only a few small values of k.

Most generalizations of the pigeonhole principle require special, research-level skills rather than the combinatorial problem-solving logic and techniques that this text seeks to develop. However, we include the basic pigeonhole principle in this text because it can occasionally be very helpful. The following examples illustrate an obvious and two not so obvious uses of the principle.

Example 1: Pooling Responses

There are 20 small towns in a region of west Texas. We want to get three people from one of these towns to help us with a survey of their town. If we go to any particular town and advertise for helpers, we know from past experience that the chances of getting three respondents are poor. Instead, we advertise in a regional newspaper that reaches all 20 towns. How many responses to our ad do we need to assure that the set of respondents will contain three people from the same town?

By the pigeonhole principle, we need more than $2 \times 20 = 40$ responses. ■

Example 2: Connecting Computers with Printers

We have 15 minicomputers and 10 printers. Every five minutes, some subset of the computers requests printers. How many different connections between various computers and printers are necessary to guarantee that if 10 (or fewer) computers want a printer, there will always be connections to permit each of these computers to use a different printer?

Using a variant of the pigeonhole principle, we see that there must be at least 60 connections. Otherwise one of the 10 printers, call it printer A, would be connected to 5 (or fewer) computers and if none of the 5 computers connected to A were in the subset of 10 computers seeking printers, then printer A could not be used by any of these 10 computers. It happens that exactly 60 connections, if properly made, will solve the connection problem (see Exercise 13). ■

Example 3: Subsets Summing to 4

Show that any collection of 8 positive integers whose sum is 20 has a subset summing to 4.

We show that the collection must have one of the following four subsets with a sum of 4: (a) four 1s; (b) two 2s; (c) two 1s and a 2; (d) a 1 and a 3. That one of these possibilities must hold is a pigeonhole–type argument.

If S contained no 1 or 2, then its sum would be at least $3 \times 8 = 24$. So by the pigeonhole principle, S must contain a 1 or a 2.

Suppose S contains a 2. We are finished if there is a second 2 or if there are also two 1s. Possibly there is one 1, but all other integers are at least 3. If there is no 1, the other 7 integers, each at least 3, must sum to $20 - 2 = 18$—impossible by a pigeonhole principle–type argument. If there is one 1, the other 6 integers, each at least 3, must sum to $20 - 2 - 1 = 17$—again impossible.

Suppose S contains at least one 1 but no 2. If there is a 3 in S, case (d) applies. If there are four 1s in S, case (a) applies. The alternative is no 2s and no 3s in S and at most three 1s. Then there are at least five other integers in S of size at least 4 whose sum must be less than 20—impossible. ■

EXERCISES

1. Given a group of n women and their husbands, how many people must be chosen from this group of $2n$ people to guarantee the set contains a married couple?

2. Show that at a party of 20 people, there are 2 people who have the same number of friends.

3. In a round-robin tournament, show that there must be 2 players with the same number of wins if no player loses all matches.

4. Given 10 French books, 20 Spanish books, 8 German books, 15 Russian books, and 25 Italian books, how many books must be chosen to guarantee there are 12 books of the same language?

5. If there are 48 different pairs of people who know each other at a party of 20 people, then show that some person has 4 or fewer acquaintances.

6. A professor tells three jokes in her ethics course each year. How large a set of jokes does the professor need in order never to repeat the exact same triple of jokes over a period of 12 years?

7. Show that given any set of seven distinct integers, there must exist two integers in this set whose sum or difference is a multiple of 10.

8. Show that if $n + 1$ distinct numbers are chosen from $1, 2, \ldots, 2n$, then two of the numbers must always be consecutive integers.

9. Suppose the numbers 1 through 10 are randomly positioned around a circle. Show that the sum of some set of three consecutive numbers must be at least 17.

10. A computer is used for 99 hours over a period of 12 days, an integral number of hours each day. Show that on some pair of 2 consecutive days, the computer was used for at least 17 hours.

11. Show that any subset of eight distinct integers between 1 and 14 contains a pair of integers k, l such that k divides l.

12. Show that in any set of n integers, $n \geq 3$, there always exists a pair of integers whose difference is divisible by $n - 1$.

13. Show that any subset of $n + 1$ distinct integers between 2 and $2n$ ($n \geq 2$) always contains a pair of integers with no common divisor.

14. Show that any set of 16 positive integers (not all distinct) summing to 30 has a subset summing to n, for $n = 1, 2, \ldots, 29$.

15. In Example 2, find the required set of 60 computer–printer connections.

16. There used to be 6 computers and 10 printers in a large computing center. Each computer was connected to some subset of printers. Now the 10 old printers are being replaced by 6 more reliable printers, but temporarily the computers will still be allowed to believe that there are 10 printers. Four dummy printers will pass on computer requests to the 6 other real printers. How many connections between the 4 dummy printers and the 6 real printers are needed to handle any set of 6 printing requests?

17. Show that for any set S of 10 distinct numbers between 1 and 60, there always exist two disjoint subsets of S (not necessarily using all the numbers in S) both of whose numbers have the same sum.

18. Two circular disks each have 10 0s and 10 1s spaced equally around their edges in different orders. Show that the disks can always be superimposed on top of each other so that at least 10 positions have the same digit.

19. A student will study basketweaving for at least an hour a day for seven weeks, but never more than 11 hours in any one week. Show that there is some period of successive days during which the student studies a total of exactly 20 hours.

20. If G is an n-vertex graph in which each vertex has degree $\geq (n-1)/2$, show that G is connected, i.e., there exists a path joining every pair of vertices in G.

21. Show that any sequence of $n^2 + 1$ distinct numbers contains an increasing subsequence of $n + 1$ numbers or a decreasing subsequence of $n + 1$ numbers.

22. Show that at any party with at least six people, there exists either a set of three mutual friends or a set of three mutual strangers.

GLOSSARY OF COUNTING AND GRAPH THEORY TERMS

Arrangement: An arrangement is a sequence or ordered list of objects. An r-arrangement is an arrangement with r objects. An arrangement may or may not allow repetition of objects. There are $P(n, r)$ r-arrangements without repetition of r objects chosen from n objects. There are n^r r-arrangements with repetition of r objects chosen from n types of objects.

Binomial coefficient: A binomial coefficient is a coefficient in the polynomial expansion of a binomial expression such as $(a+x)^n$. The coefficient of x^r in $(1+x)^n$ is written $C(n, r)$ or $\binom{n}{r}$. This coefficient equals the number of distinct r-subsets of an n-set.

Bipartite graph: A bipartite graph $G = (X, Y, E)$ is a graph whose vertices are partitioned into two vertex sets, X and Y, and every edge in G joins a vertex in X with a vertex in Y.

Chromatic number: The chromatic number of a graph is the smallest number of colors that can be used in a coloring of a graph (see *Coloring a graph*).

Circuit: A circuit is a sequence of vertices $(x_1, x_2, x_3, \ldots, x_n)$ where $x_1 = x_n$, and x_i is adjacent to x_{i+1}. A vertex may not appear more than once in a circuit (except for the same vertex as the starting and ending vertex).

Coloring a graph: A coloring of a graph G assigns a color to each vertex so that adjacent vertices have different colors.

Combination: A combination is a subset of objects or, equivalently, an unordered collection of objects. Objects may or may not be repeated in a combination. There are $C(n, r)$ different combinations without repetition of r objects chosen from n objects. There are $C(n+r-1, r)$ different combinations with repetition of r objects chosen from n types of objects.

Complete graph K_n and complete bipartite graph $K_{m,n}$: K_n is a graph on n vertices with an edge joining every pair of vertices. K_3 is a triangle. K_2 is an edge. $K_{m,n}$ is a bipartite graph with m and n vertices in its two vertex sets and all possible edges between vertices in the two sets.

Complementary graph or complement: The complementary graph $\overline{G} = (V, \overline{E})$ of a graph $G = (V, E)$ has the same vertex set V as G does. A pair of vertices are joined by an edge in $\overline{G}$ if and only if they are not joined by an edge in G.

Component: An unconnected graph G consists of a collection of components or "connected pieces." A connected graph consists of a single component. Formally, a component of G consists of some particular vertex x and all vertices reachable from x by a path in G.

Connected graph: A graph is connected if there is a path joining any given pair of vertices. A directed graph is connected if it is connected when treated as an undirected graph (with all edge directions ignored).

Cycle: A cycle is a sequence of consecutively linked edges whose starting vertex is the ending vertex and in which no edge can appear more than once. Unlike a circuit, a vertex can be visited any number of times in a cycle.

Derangement: A derangement of a given arrangement of distinct objects is a re-arrangement such that no object is in the same position it had in the original arrangement.

Directed graph, directed edge: A graph is a directed graph if each edge $(a, \vec{b})$ is directed, going from a to b. A directed graph may contain two oppositely directed edges joining two vertices, such as $(a, \vec{b})$ and $(b, \vec{a})$.

Distribution: A distribution is an assignment of a given set of objects, which may be identical or distinct, to a set of distinct destinations. Unless explicitly prohibited, more than one object may go to the same destination.

Edge cover: An edge cover is a set S of vertices such that every edge in any graph is incident to one vertex in S.

Euler cycle, Euler trail: An Euler cycle (trail) is a cycle (trail) that contains all the edges in a graph. Further, it must visit each vertex at least once.

Generating function: A generation function $g(x)$ for a_n, the number of ways to do some procedure with n objects, is a polynomial or power series with the expansion $g(x) = a_0 + a_1x + a_2x^2 + \cdots + a_nx^n + \cdots$. Such a function is also called an *ordinary generating function* in contrast to an exponential generating function $g(x)$, which has the form $g(x) = a_0 + a_1\frac{x}{1!} + a_2\frac{x^2}{2!} + \cdots + a_n\frac{x^n}{n!} + \cdots$.

Graph: A graph $G = (V, E)$ consists of a finite set V of vertices and a finite set E of edges. Each edge $e = (a, b)$ joins two different vertices a, b $(a \neq b)$. Also, two edges cannot join the same pair of vertices. Unless G is a directed graph (see *Directed graph*), (a, b) and (b, a) are the same edge (order does not matter).

Hamilton circuit, Hamilton path: A Hamilton circuit (path) is a circuit (path) that contains every vertex of a graph.

Independent set: A set of vertices in a graph is independent if no pair of them are adjacent.

Interval graph: G is an graph interval G if there exists a collection of intervals on the line for which there is a one-to-one correspondence between the intervals and vertices of G so that two vertices are adjacent if and only if the corresponding intervals overlap.

Isomorphism, isomorphic graphs: Two graphs G_1 and G_2 are isomorphic if there exists a matching, called an isomorphism, of vertices in G_1 with the vertices in

G_2 so that two vertices in G_1 are adjacent if and only if the corresponding two vertices in G_2 are adjacent. Informally, two graphs are isomorphic if they are "the same graph" except that their vertices have different names.

Map coloring: A map of countries is properly colored by assigning a color to each country so that countries with a common border get different colors.

Matching: A matching in a bipartite graph $G = (X, Y, E)$ is a subset of edges that pair off, in a one-to-one fashion, some vertices in X with some vertices in Y.

Multigraph: A multigraph is a generalized graph in which (1) multiple edges are allowed—two or more edges can join the same two vertices; and (2) loops are allowed—edges of the form (a, a).

Network: A network is a graph, usually a directed graph, with a positive integer $k(e)$ assigned to each edge e of the graph.

Network flow: A flow is a function on the edges of a network that satisfies certain constraints listed at the start of Section 4.3.

Partition: A partition of a collection of identical objects divides the objects into a collection of groups of various sizes. One can also speak of a partition of an integer n as a collection of positive integers that sum to n.

Path: A path is a sequence of vertices $(x_1, x_2, x_3, \ldots, x_n)$ such that x_i is adjacent to x_{i+1}. A vertex may not appear more than once in a path, except possibly $x_1 = x_n$.

Permutation: A permutation of a set or sequence of objects is an arrangement of the set or sequence, normally with no repetition allowed. An r-permutation is an r-arrangement of r objects chosen from the set or sequence.

Planar graph: A graph is planar if there exists a way to draw it on a sheet of paper so that no edges cross. A **plane graph** is a planar graph drawn so that no edges cross.

Recurrence relation: A recurrence relation is an equation such as $a_n = 2a_{n-1} + 3a_{n-2}$, in which a_n, the number of ways to do some procedure with k objects, is expressed in terms of other a_ks, where $k < n$.

Selection: A selection is an unordered collection of objects. Objects may or may not be repeated in a selection. There are $C(n, k)$ different selections without repetition of k objects chosen from n objects. There are $C(n + k - 1, k)$ different selections with repetition of k objects chosen from n types of objects.

Subgraph: A subgraph is a graph that is contained in another graph. If $G' = (V', E')$ is a subgraph of $G = (V, E)$, then $V' \subseteq V$ and $E' \subseteq E$.

Trail: A trail is a sequence of consecutively linked edges in which no edge can appear more than once. Unlike a path, a vertex can be visited any number of times in a trail.

Tree: A tree is a graph with a special vertex called a *root* and for each vertex x, other than the root, there is a unique path from the root to x. An undirected tree is characterized as a connected graph with no circuits. For tree-related terms, see the following Subglossary of Tree Terminology.

Subglossary of Tree Terminology

Ancestors: Ancestors of vertex x are the set of vertices on the path from the root to x.

Backtracking search: See Depth-first search.

Balanced tree: A tree with all leaves at level h and $h - 1$, where h is the height of the tree.

Binary tree: A tree in which each internal vertex has exactly two children.

Breath-first search: From the root, find all vertices z adjacent to the root, then all vertices adjacent to one of the zs, and so on.

Child: Children of vertex x are vertices y with an edge $(x,\vec{}y)$ from x to y.

Depth-first search: Also known as *backtracking search,* builds a path from the root as far as possible; one backtracks from the current vertex when the path cannot be extended to a previously unvisited vertex and backs up the current path until a vertex is found at which a side path may be constructed to a new vertex.

Descendant: Descendants of vertex x are the set of vertices z whose path from the root to z contains x.

Height: The largest level number in a tree.

Inorder traversal: In a binary tree, starting from the root, this traversal recursively lists the vertices of the left subtree of a particular vertex x, then x, and then the vertices in the right subtree of x.

Internal vertex: A vertex with children; internal vertex and parent are equivalent terms.

Leaf: A vertex with no children; a leaf's only incident edge comes in from its parent.

Level or level number: The length of the path from the root to a given vertex; e.g., level 2 of a tree consists of all vertices whose path from the root has length 2.

***m*-ary tree:** A tree in which all internal vertices have m children.

Parent: The parent of vertex x is the unique vertex z with an edge $(z,\vec{}x)$ from z to x.

Postorder traversal: Lists vertices in the order they are last encountered in a depth-first search of a tree.

Preorder traversal: Lists vertices in the order they are first encountered in a depth-first search of a tree.

Root: The special vertex in a tree with a unique path to any other specified vertex; in a directed or rooted tree, the root is the unique vertex at level 0 that has no parent.

Rooted tree: A directed tree with all edges directed away from the root.

Sibling: Siblings of vertex x are those vertices with the same parent as x.

Spanning tree: A tree that is a subgraph of a connected graph and that contains all vertices of the graph.

Subtree: A connected subgraph of a tree.

BIBLIOGRAPHY

GRAPH THEORY AND ENUMERATION COMBINED

I. Anderson, *A First Course in Combinatorial Mathematics,* 2nd ed., Oxford University Press, New York, 1989.
V. K. Balakrishnan, *Theory and Problems of Combinatorics,* Schaum's Outline Series, McGraw-Hill, New York, 1995.
K. Bogart, *Introductory Combinatorics,* 2nd ed., Harcourt Brace Jovanovich, San Diego, 1990.
V. Bryant, *Aspects of Combinatorics: A Wide-Ranging Introduction,* Cambridge University Press, Cambridge, 1993.
R. Brualdi, *Introductory Combinatorics,* 3rd ed., North-Holland, Amsterdam, 1998.
P. Cameron, *Combinatorics: Topics, Techniques, Algorithms,* Cambridge University Press, Cambridge, 1994.
M. Erickson, *Introduction to Combinatorics,* John Wiley & Sons, New York, 1996.
R. Graham, D. Knuth, and O. Patashnik, *Concrete Mathematics: A Foundation for Computer Science,* 2nd ed., Addison-Wesley, Reading, MA, 1994.
F. Roberts, *Applied Combinatorics,* Prentice-Hall, Upper Saddle River, NJ, 1984.
D. Stanton, R. Stanton, and D. White, *Constructive Combinatorics,* Springer-Verlag, New York, 1986.
J. VanLint and R. Wilson, *A Course in Combinatorics,* Cambridge University Press, Cambridge, 1992.

GRAPH THEORY

C. Berge, *Graphs,* 2nd ed., Elsevier Science, New York, 1985.
N. Biggs, E. Lloyd, and R. Wilson, *Graph Theory 1736–1936,* Cambridge University Press, Cambridge, 1999.
M. Capobianco and J. Molluzzo, *Examples and Counterexamples in Graph Theory,* North Holland, New York, 1978.
G. Chartrand and L. Lesniak, *Graphs and Digraphs,* 3rd ed., Chapman and Hall, New York, 1996.
J. Gross and J. Yellen, *Graph Theory and Its Applications,* CRC Press, New York, 1998.

F. Harary, *Graph Theory,* Perseus Press, New York, 1995.
O. Ore, *Graphs and Their Uses,* Mathematics Association of America, Washington, DC, 1990.
R. Trudeau, *Introduction to Graph Theory,* Dover New York, 1994.
D. West, *Introduction to Graph Theory,* 2nd ed., Prentice-Hall, Upper Saddle River, NJ, 2001.
R. Wilson, *Introduction to Graph Theory,* 4th ed., Addison-Wesley, Reading, MA, 1997.
R. Wilson and J. Watkins, *Graphs: An Introductory Approach,* John Wiley & Sons, New York, 1990.

ENUMERATION

M. Hall, *Combinatorial Theory,* 2nd ed., Blaisdell, Boston, 1986.
G. Polya, R. Trajan and D. Woods, *Notes on Introductory Combinatorics,* Birkhauser, New York, 1983.
J. Riordan, *An Introduction to Combinatorial Analysis,* John Wiley & Sons, New York, 1958.
H. Ryser, *Combinatorial Mathematics,* Mathematics Association of America, Washington, DC, 1963.
R. Stanley, *Enumerative Combinatorics,* Wadsworth, Belmont, CA, 1986.
W. Whitworth, *Choice and Chance,* Hafner Press, New York, 1901.

SOLUTIONS TO ODD-NUMBERED PROBLEMS

CHAPTER ONE SOLUTIONS

Section 1.1

1. **(a)**

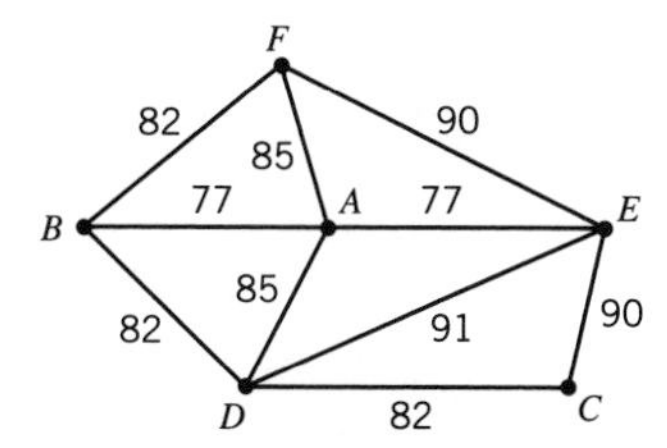

(b) $2(C, D), (C, E)$,

(c) yes, several routes.

3. **(a)** A 5-circuit.

5. **(a)** many possibilities,

(b) min = 4, several possibilities, e.g.,

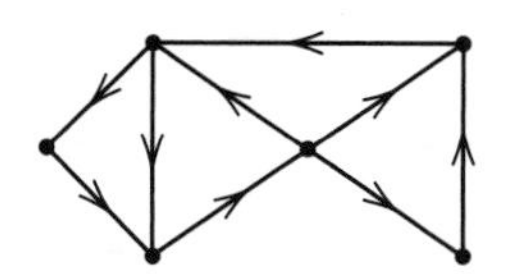

7. 7. 12.

9. **(a)** $\{A\text{-}a, B\text{-}b, C\text{-}d, D\text{-}c\}$.

(b) A and C each can fill only job b.

11. **(a)** $\{a\text{-}b, c\text{-}d\}$ or $\{a\text{-}c, b\text{-}d\}$,

(b) no possible (odd number of vertices),

(c) a, c, e collectively must be matched with just b and e.

15. **(a)** vertex = variety of chipmunk, if A splits into B and C, then make edges $(\vec{A, B})$ and $(\vec{A, C})$,

(b) 7 splits.

17. **(a)** 4 other pairs. **(b)** $\{b, j\}, \{c, h\}$ and 6 other pairs.

19. C, E, c, d (answer for both parts of the question).

21. minimal block surveillance—6, minimal corner surveillance—3.

23. **(a)** 5: squares (2, 4), (3, 4), (4, 4), (5, 4), (8, 4), **(b)** 10.

25. (a) (i) $\{b, c\}$, (ii) $\{A, a, B, b, D, e\}$, (iii) $\{a, f\}$ (other possibilities).
(b) (i) $\{a, d\}$, (ii) $\{C, c, d, E\}$, (iii) $\{b, c, d, e\}$.

27. only Figures 1.1a and 1.3.

29. (a)

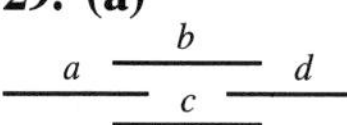

(b)

b e
a d
c f

(c)

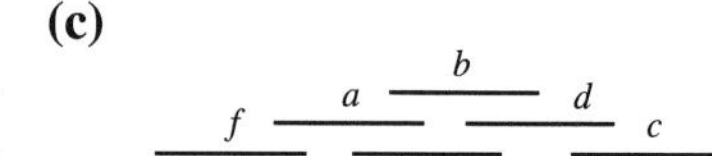

31.

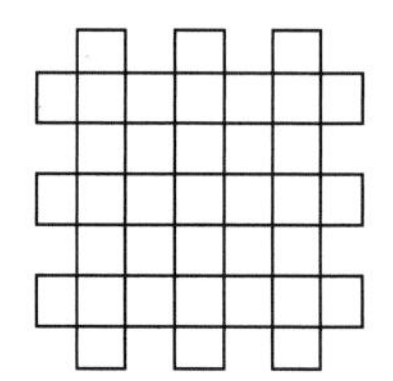

33. (b) 4-circuit a-b-c-d-a:

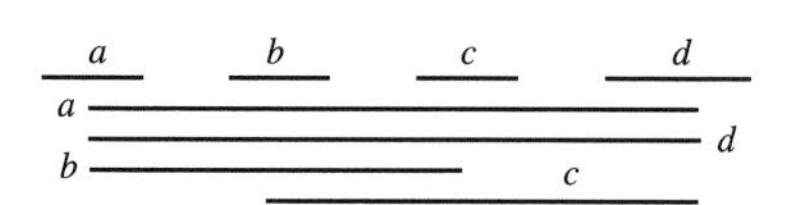

37. (a)

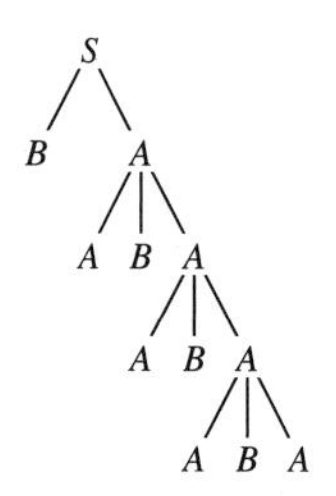

(b) not

Section 1.2

1.

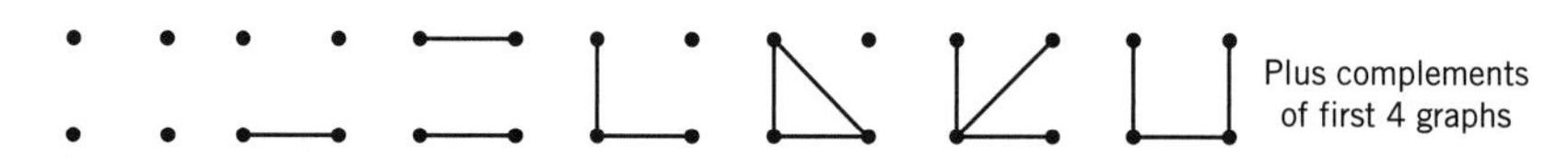

5. (a) yes, a-1, b-5, c-4, d-3, e-2, f-6,
(b) yes, a-6, b-1, c-3, d-5, e-2, f-4,
(c) yes, a-3, b-4, c-7, d-1, e-6, f-8, g-2, h-5, i-9,
(d) no, right graph has one edge,
(e) no, subgraphs of vertices of degree 3 do not match,
(f) yes, a-7, b-3, c-5, d-4, e-1, f-2, g-6,
(g) no, degree-2 vertex is part of triangle only in right graph,
(h) no, degree-2 vertices are adjacent only in left graph,
(i) no, degree-3 vertices are adjacent only in left graph,
(j) yes, a-1, b-2, c-3, etc.

7. Graphs 1-6, 13-18, 31-36, 37-42 mutually isomorphic, 7-12, 19-24, 25-30 isomorphic.

9. no, building on the isomorphism in Example 2, $b \to c$ but $5 \to 2$.

13. (a) two, **(b)** four, **(c)** one,

15. four in each.

Section 1.3

1. **(a)** 12, **(b)** 9, **(c)** 8 or 10 or 20 or 40.
3. 12.
5. if v vertices, then $e = \frac{1}{2}\, vp$ edges (where $\frac{1}{2}v$ is an integer since p is odd).
7. sum of in-degrees (or out-degrees) = number of edges, since sum counts each edge once.
11. **(a)** no, **(b)** no, **(c)** yes.

Section 1.4

1. **(a)**

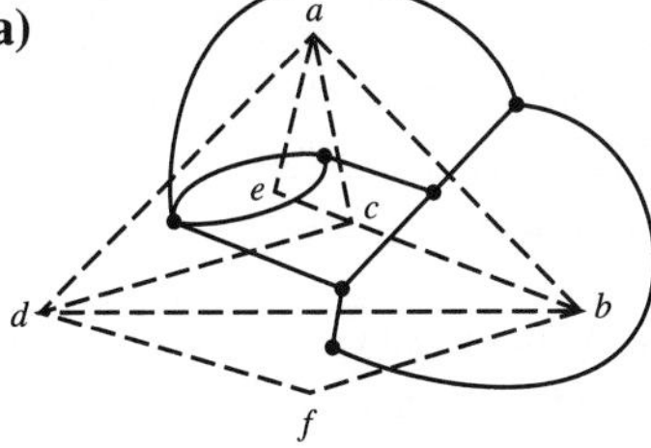

(b)

3. **(a)** yes, **(b)** no, delete a and (b, c), **(c)** yes **(d)** no, delete (a, e): $\{a, b, c, d, f\}$
(e) no, delete (b, c), (e, f): $\{a, e, f\}$, $\{b, c, d\}$,
(f) no, delete (b, j), (e, g): $\{a, d, h\}$, $\{c, f, i\}$,
(g) no, many possible $K_{3,3}$,
(h) no, delete (a, b), (b, c), (d, e), (f, g): $\{a, d, e\}$, $\{c, f, g\}$,
(i) no, delete (d, e): $\{a, c, e\}$, $\{b, f, g\}$,
(j) delete (f, g): $\{a, c, e\}$, $\{b, d, h\}$,
5. **(a)** $n \leq 4$, **(b)** r or $s \leq 2$.
7. **(a)** possible, $\mathbf{r} = 8$,
(b) possible, $\mathbf{e} = 12$,
(c) not possible, $\mathbf{v} = 6$,
(d) not possible, $\mathbf{r} = 10$,
(e) possible, $\mathbf{v} = 7$,
(f) possible, $\mathbf{e} = 12$, $\mathbf{r} = 8$,
(g) not possible (parity violation),
(h) possible, $\mathbf{e} = 10$, $\mathbf{v} = 7$,
(i) possible, $\mathbf{e} = 20$, $\mathbf{r} = 10$,
(j) not possible, $\mathbf{v} = 10$.
9. **(a)** degree of vertices in K_5 is $4 \Rightarrow$ degree of vertices in $L(K_5)$ is $2 \times (4 - 1) = 6$; $L(K_5)$ has v $= 10$ and $\mathbf{e} = \frac{1}{2}\sum\deg = 3\mathbf{v} = 30 \Rightarrow \mathbf{e} \not\leq 3\mathbf{v} - 6$;

(b)

11. **(a)** circuit length = sum of number of boundary edges of R_1 and $R_2 - 2$;

(b) circuit length = sum of number of boundary edges of each enclosed region minus 2 × (number of edges interior to circuit).

13. **(a)** $K_{3,3}$ and K_5 are critical nonplanar.

15. **(a)** $\mathbf{r} = \mathbf{e} - \mathbf{v} + \mathbf{c} + 1$,

(b) Using (a), corollary becomes $\mathbf{e} \leq 3\mathbf{v} - 3\mathbf{c} - 3$ ($\leq 3\mathbf{v} - 6$).

19. If false, deg ≥ 5 and so $5\mathbf{v} \leq$ sum of degrees $= 2\mathbf{e} \leq 2(3\mathbf{v} - 6) = 6\mathbf{v} - 12$, that is, $5\mathbf{v} \leq 6\mathbf{v} - 12$ or $12 \leq \mathbf{v}$—impossible.

25. $\mathbf{v} = p + 2l$, $\mathbf{e} = \frac{1}{2}\sum \deg = 2p + 3l$, answer $= \mathbf{r} - 1 = \mathbf{e} - \mathbf{v} + 1 = \mathbf{p} + l + 1$.

Supplement II

1. vertex = committee, edge = committee overlap (person). Graph is K_7 with $\mathbf{e} = \frac{1}{2}\sum \deg = \frac{1}{2}(6\mathbf{v}) = 21$.

3. $n = 12$.

7. yes, each component of G is a circuit.

9. **(a)** yes, trace out any sequence of edges and eventually a vertex z will be repeated, between first and second visit to z a circuit is formed,

(b) no, e.g.,

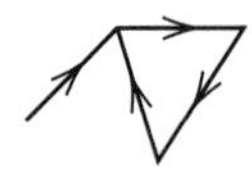

11. Vertices in different components of G are directly adjacent in $\overline{G}$, vertices in same component are joined in $\overline{G}$ by a path of length 2 via any vertex in other component of G.

15. *if:* suppose not strongly connected with no path from a to b—let V_1 consist of a and all vertices that can be reached by a directed path from a, V_2 is other vertices, *only if*: obvious.

17. A bridge edge cannot lie on a circuit.

19. **(a)** yes, see Exer. 7 in Section 1.2,

(b) no, odd number of vertices of odd degree,

(c) no; $\mathbf{e} = \frac{1}{2}\sum \deg = 13 \Rightarrow \mathbf{e} \nleq 3\mathbf{v} - 6$.

21. A took C's coat and B's hat, B took D's coat and C's hat.

23. **(a)** Repeatedly remove side circuits until trail from x to y has no repeated vertices.

(b)

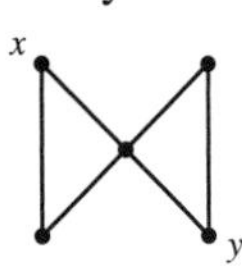

(c) similar argument to part (a).

(d) If odd number of edges of cycle are partitioned into circuits [part (d)], then some circuit must have odd number of edges.

25. **(a)** (h, a), (h, g).

27. Two-component, n-vertex graph with the most edges is a K_{n-1} plus an isolated vertex; it has $\frac{1}{2}(n-1)(n-2)$ edges.

31. **(a)** 5-circuit or 3-edge path.

(b) By hint, if G has n vertices, number of edges in $G = \frac{1}{2}$(number of edges in K_n) $= \frac{1}{2}[\frac{1}{2}n(n-1)] = \frac{1}{4}n(n-1)$. Since n and $n-1$ are not each divisible by 2, then either n or $n-1$ must be divisible by 4, i.e., $n = 4k$ or $n = 4k+1$.

33. *if:* obvious, *only if:* let x_n be vertex with 0 out-degree (if no such vertex, there is a directed circuit—Exer. 9(a), move x_n from graph and let x_{n-1} be vertex with 0 out-degree in remaining graph, continue indexing in this fashion.

35. (a)

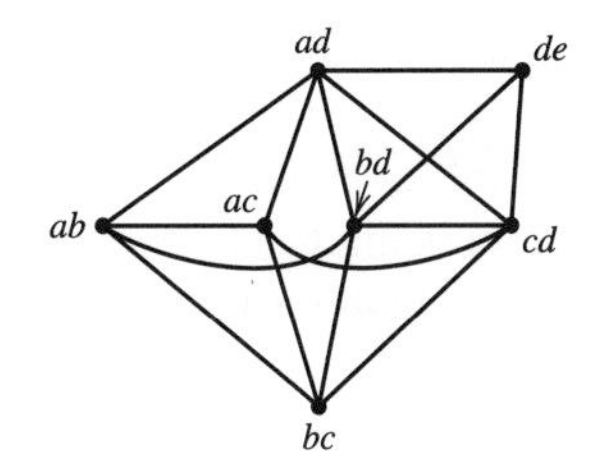

(b) each edge of K_n is incident to $n-2$ other edges at each end vertex, $2(n-2)$ incidences in all,

(c) can be no vertices of degree 0 or 1 and so all degrees ≥ 2; since $\mathbf{e} = \frac{1}{2}\sum \deg$, then to have $\mathbf{e} = \mathbf{v}$ (so that G and $L(G)$ have same number of vertices), all $\deg = 2 \Rightarrow G$ is any circuit.

37. (a) $b \leftrightarrow c$, a and d fixed,

(b) none,

(c) $a \leftrightarrow c$, $e \leftrightarrow f$, b and d fixed.

39. Consider a vertex y that beat x, [i.e., $(\vec{y, x})$]; if y beat every competitor that beat x, then y would have a greater score that x—not possible—and so for some w, $(\vec{x, w})$ and $(\vec{w, y})$.

41. Sketch of proof: show that each vertex in $(C_1 \cup C_2) - (C_1 \cap C_2)$ has even degree and then repeatedly trace a path (without repeating any edge) until a vertex x is revisited and remove the circuit formed between the first and second visit to x.

CHAPTER TWO SOLUTIONS

Section 2.1

1. (a) many possibilities,

(b) many possibilities with b and f as end vertices.

3. One example is a graph consisting of a circuit of 7 edges.

5. (a) No, once bridge crossed there is no way back to starting vertex,

(b) many possibilities, e.g., a 10-edge path.

7. An isolated vertex added to a connected graph with even degrees now has a Euler cycle but is not connected.

9. Build a graph with a vertex for each racer and an edge for each race; a Euler trail corresponds to a sequence of races in which each racer is in two consecutive races; this graph has the desired Euler trail because only A and F have odd degree.

11. A set of deadheading edges must have one edge at each odd-degree vertex; joining these edges at odd-degree vertices together (without changing the parity of the degree of other vertices) requires a set of paths.

13. A directed graph has a Euler trail if and only if at all but two vertices indegree = outdegree and at those two, indegree and outdegree differ by one. *Proof:* add an extra edge so that indegree = outdegree at two unbalanced vertices and resulting graph has Euler cycle; remove added edge yielding desired Euler trail.

15. no such Euler cycle containing all vertices.

17. **(a)** many possibilities,
(b) if at every stage graph of remaining edges is connected, then just before using last edge E and being forced to stop at starting point, E is only edge remaining and once taken there are no remaining edges,
(c) applies to Euler trails.

19. **(a)**

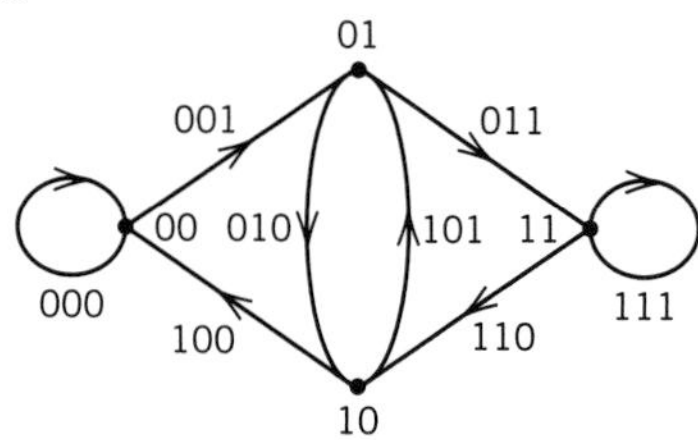

(b) concatenating the first digits of each node on an Euler cycle produces desired sequence,
(c) many possibilities, e.g., cycle 00-00-01-11-11-10-01 (-00) produces 00011101.

Section 2.2

1. many possibilities, e.g.,
(a)

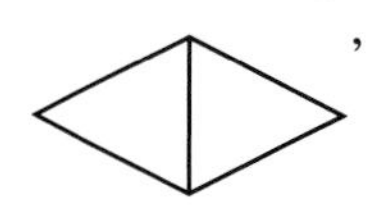

, **(b)**

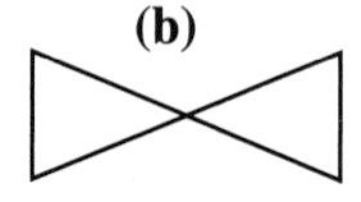

3. *a-g-c-b-f-e-i-k-h-d-j-a.*

5. path: *m-b-c-d-e-q-a-l-k-j-o-p-n-f-g-h-i*, no circuit: by symmetry at p, delete any edge, choose *p-m*, at m and c rule 1 forces subcircuit *m-b-c-d-m*.

7. **(a)** rule 1 at f, h, j forces subcircuit *e-f-g-h-i-j-e*,
(b) by symmetry at p, use *m-p-n* and delete *p-o* forcing *k-o-i*, at m and n, if both use edge going up (to g) then subcircuit *p-m-g-n-p*, if both m, n use edge down then subcircuit *p-m-k-o-i-n-p*, so by symmetry use *m-g* and *n-i*, deleted *i-h*, *i-c*, *i-j* forcing *g-h-b-c-d-j-k*, plus *k-o-i-n-p-m-g*, yielding a subcircuit.

9. **(a)** Hamilton circuit alternates between red and blue vertices and so must have equal numbers of each,
(b) follows from part (a),
(c) (i), (ii) have odd number of vertices, (iii) has 9 vertices in one part and 7 vertices in other part of bipartition.

11. (a)

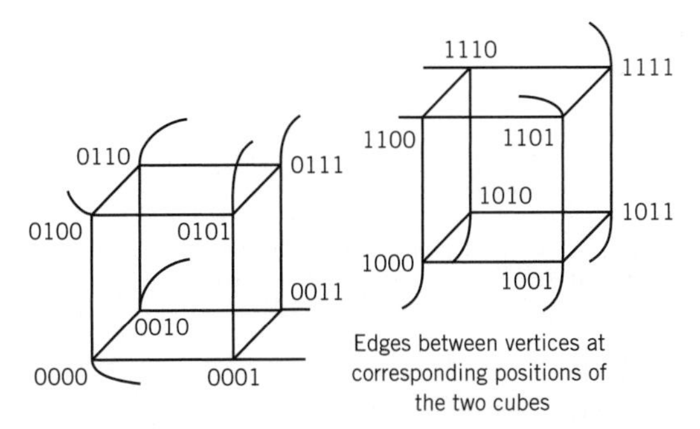

(b) many possibilities: 1-0000, 2-0001, 3-0011, 4-0010, 5-0110, 6-0111, 7-0101, 8-0100, 9-1100, 10-1101, 11-1111, 12-1110, 13-1010, 14-1011, 15-1001, 16-1000

13.

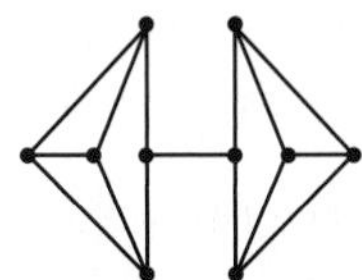

15. Many solutions but very tedious; the following heuristic works: at each stage, look at all possible squares (not yet visited) a knight's move from current square and move to a square with the minimum number of possible squares for the next move.

17. (a) *rook:* starting at upper left corner, go down first column, square by square, at bottom of first column move right to bottom of second column, move up second column, square by square, continue going down and up successive columns finishing at top square of right column from which one moves back to top square of left column, *king:* similar to rook, except in columns 2, 3, $\ldots, n-1$, avoid top square, when top square of right column reached, move left along top squares of each column to return to top left square,

(b) *rook:* not possible (variation on reasoning in Exer. 14), *king:* (1, 1)-(1, 2)-(2, 2)-(1, 3) ... (2, n), now go left and right covering rows 3 through n avoiding left square in each row except row n, finishing at (1, n) and now move up first column to starting square.

19. (a) $n!$,

(b) form the Hamilton circuits by placing vertices in a circle: first circuit formed by joining consecutive vertices, second circuit formed by joining vertices 2 positions apart (one intervening vertex), third circuit formed by joining vertices 3 positions apart, and so on (no subcircuits because n prime);

(c) form complete graph with professors as vertices, answer: 8 days, using part (b) to form 8 Hamilton circuits out of the edges of K_{17}.

21. if x, y nonadjacent, direct all edges inward at both x and y, now x and y can only be on a Hamilton path if they are the first vertex on the path—cannot be two first vertices.

23. many possibilities, e.g.,

Section 2.3

1. **(a)** 3, has K_3,
(b) 4, an attempt to 3-color outer 6 vertices gives b, d, f different colors so that c requires a fourth color,
(c) 2,
(d) 4, a, b, e, g form a K_4,
(e) 5, a, b, c, d, f form a K_5,
(f) 4, a, c, g, i form a K_4,
(g) 2,
(h) 2,
(i) 4, an attempt to 3-color vertices clockwise around the circle starting at a forces j to have same color as a,
(j) 3, outer pentagon (odd-length circuit) cannot be 2-colored,
(k) 4, an attempt to 3 color the sequence of vertices a, e, b, f, c, g, d forces d to have same color as a,
(l) 4, a 3-coloring forces a and h to have different colors, similalry for i and p so that a, h, i, p act like a K_4.

3. (b), (i), (j).

5. **(a)** $\{a, b, c\}$, $\{d, e, f\}$, $\{g, h\}$,
(b) $\{a, b, e\}$, $\{c, d\}$, $\{f, g\}$,
(c) $\{a, b, c\}$, $\{d, e, f\}$, $\{g, h, i\}$, $\{j, k\}$.

7. **(a)** $k(k-1)^3$
(b) smallest nonnegative integer for which $P_k(G)$ is positive is chromatic number of G.

9. No, Nevada (or Kentucky or West Virginia) and its neighboring states have duals that form odd-length wheels.

11. 3.

13. vertices = classes, edges = two classes with a common student, colors = class times.

15. vertices = banquets, edges = require a common room, colors = days.

17. **(a)** graph of contestants and matches is K_n, colors = days, edge coloring of K_n partitions edges (matches) into collections of matches that can be played on a common day,
(b) $n = 4$, requires 3 days (colors), each day pair player a with one of the other players and then pair off the remaining two players, $n = 5$, requires 5 days (10 matches to be played and at most 2 matches—4 players can play one day), draw K_5 in standard form as pentagon with all chords, give different color to each edge of outer pentagon, give same color to each chord as color of outer edge parallel to it (this method extends to K_n for any odd n).

Section 2.4

1. Proceed as in Theorem 5 using induction and the fact that there is a vertex x of degree 5, but since 6 colors are available, x is immediately colored with a color different from those used by its 5 neighbors.

3. Either x, y have the same color—which is forced when x, y, are coalesced to form G^c_{xy}—or they have different colors—which is forced when edge (x, y) is added to form G^+_{xy}.

5. **(a)** 3-color one triangle and then extend by successively coloring a vertex that lies on a triangle with two previously colored vertices,
(b) process in part (a) yields unique coloring.

7. **(a)** If not connected, then k colors are needed to color one of G's components and removing a vertex from another component will not reduce $\chi(G)$,
(b) if $\deg(x) \leq k-2$, then if $G-x$ could be $k-1$ colored, also G could be $k-1$ colored by giving x one of the $k-1$ colors not used by one of x's $k-2$ (or less) neighbors,
(c) if x disconnects so that $G-x$ has components G' and G'' then one of $G' \cup x$ or $G'' \cup x$ requires k colors and removing a vertex from the other component will not reduce $\chi(G)$.

9. **(a)** for $n=1$, $\chi(G)+\chi(\overline{G})=2$; assume for $n-1$ and consider an n-vertex graph G; by induction we can color $G-x$ and $\overline{G}-x$, for any given vertex x, with a total of n colors for both graphs (possibly fewer); x has a total of $n-1$ edges in G and $\overline{G}$ and so is not adjacent to one of the n color classes in one of G or $\overline{G}$ and can be added to that color class, although in the other graph x may require an additional color—for a total of $n+1$, as required,
(b) $\chi(\overline{G})) \geq$ size of largest complete subgraph in $\overline{G} = q$, size of largest independent set in G, and by Exer. 6(a), $\chi(G)q \geq n$,
(c) square both sides of inequality, new inequality follows immediately from (b) and the fact that $a^2+b^2 \geq 2ab$.

11. $(k^2-6k+8)=0$ when $k=4$ and so $P_k(G)=0$ for $k=4$, but the Four Color Theorem says that any planar graph can be 4-colored (i.e., has a positive number of 4-colorings).

13. *if:* label with numbers that are the length of the longest path starting at the vertex, adjacent vertices must have different-length longest paths since edge $(\vec{x, y})$ implies that x's longest path will be at least one greater that y's longest path length, *only if:* let colors be numbers 0, 1, ..., $k-1$ and direct edges from larger to smaller numbers.

15. Successively color intervals (i.e., vertices) from left to right according to the order of their left endpoints, giving each interval a color unused by previously colored interval; most colors needed will be maximum number of intervals overlapping at one point = size of largest complete subgraph.

Supplement

1.

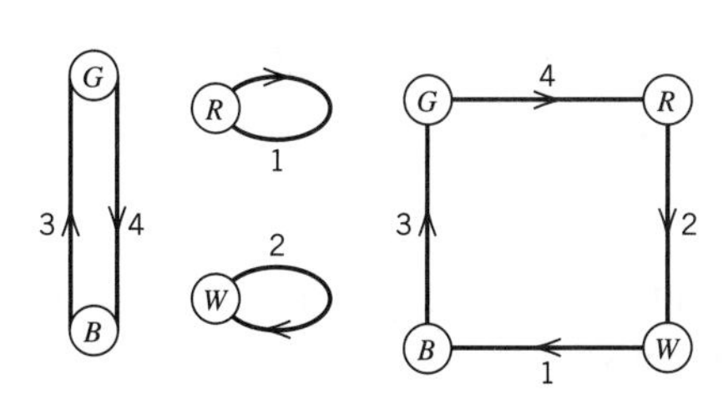

3. One labeled factor must be 1: W-B, 2: G-R, 3: B-G, 4: R-W, second labeled factor can be 1 & 2: G-B, 3: R-R, 4: W-W or 1: B-G, 2: W-B, 3: R-W, 4: G-R (or interchange 1 and 2).

5. **(a)** yes, a Hamilton circuit is a factor,
(b) having a Euler circuit has no relation to having a factor.

CHAPTER THREE SOLUTIONS

Section 3.1

1. **(a)** **(b)** **(c)**

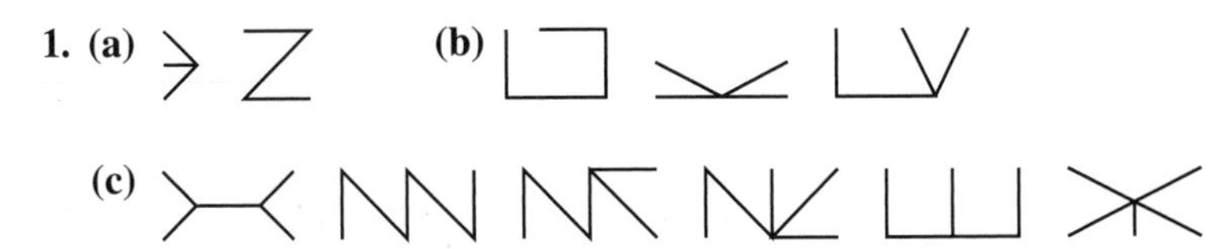

3. no odd circuits ⇒ bipartite (Theorem 2 of Sect. 1.3) ⇒ 2-colorable.

5. **(a)** see answer to part (b),
(b) since G is connected, a subset of edges can be chosen to form a tree containing all vertices of G, *but* this subgraph has one fewer edges that vertices (by Theorem 1) and if G contained other edges, besides those in this tree, it would have as many vertices as edges,
(c) if G has a circuit, removal of an edge an circuit does not disconnect $G \Rightarrow G$ has no circuits; now use part (a).

7. Start at any vertex and trace a trail (no repeated edges); since no vertex is ever visited twice (if so, circuit would result) and graph is finite, trail must end at a vertex x of degree 1; now start trail-building again at x to get a second vertex of degree 1.

9. height $h \Rightarrow$ each path to a leaf passes through at most h internal vertices with a choice of m children to go to at each internal vertex, for a total of at most m^h paths to a leaf.

11. $i/n =$ [by Corollary part (a)] $i/(mi+1) \approx i/mi = 1/m$.

13. largest $n-1$, smallest 2.

15. A tree is bipartite, one vertex class in the bipartition must have at least $n/2$ vertices.

17. Unbalancing a tree:

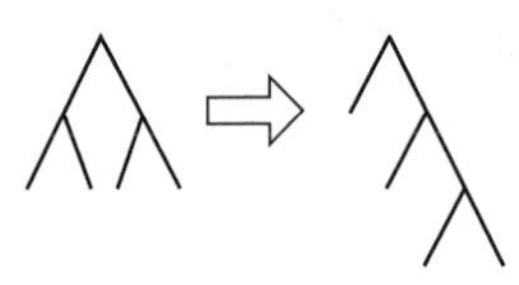

makes sum of level numbers larger and so smallest sum occurs when binary tree is balanced, in which case each level number is $\lfloor \log_2 l \rfloor$ or $\lfloor \log_2 l \rfloor + 1$, and sum of level numbers is at least $l \lfloor \log_2 l \rfloor$.

19. k choices for root, color parents before children and then each vertex when colored has $k-1$ choices (any color except color of parent).

21. **(a)** internal vertices are +s

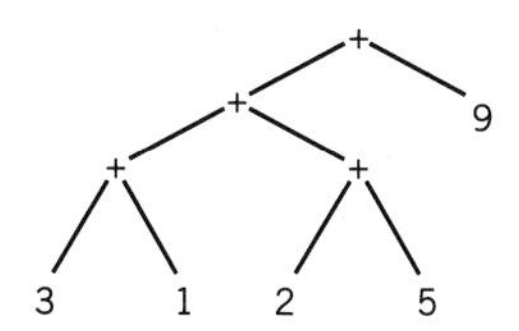

(b) $\lceil \log_2 100 \rceil = 7$.

23. leaves = letters, each leaf = n-digit binary sequence $\Rightarrow 2^n$ letters.

25. **(a)** 24, **(b)** 16, **(c)** 12,

(d) 9 tournaments (original and 8 losers' tournaments).

27. **(a)** $\left\lfloor \log_2 \frac{n+1}{2} \right\rfloor = \lfloor \log_2(n+1) \rfloor - 1$, **(b)**

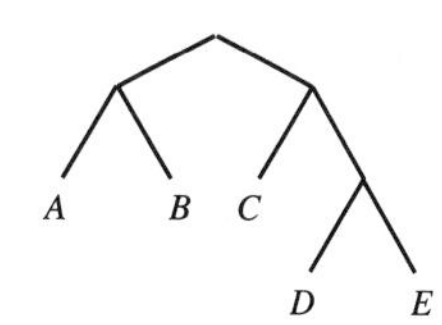

29. **(a)**

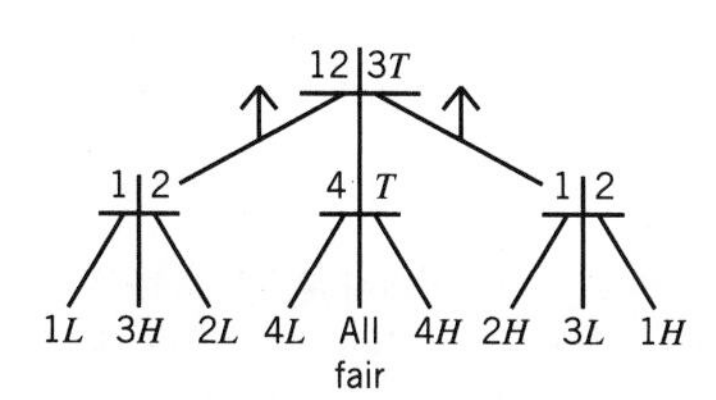

(b) first weighing is either one coin on either side or two coins on either side, in either case some outcome has more than 3 possibilities that must be distinguished in the one additional weighing (impossible, for example with one on each side, there are 5 possibilities if the scales balance).

Section 3.2

1. **(a)** any 8-vertex path,

(b) many possibilities, e.g., *a-b-d-c-h-k-i-g-e-f-j*,

(c) many possibilities, e.g., *a-b-c-d-e-f-g-h*.

3. all trees on 5 vertices [see solution to Exer. 1(b) in Section 3.1],

(b) all trees on 4 vertices [see solution to Exer. 1(a) in Section 3.1],

(c)

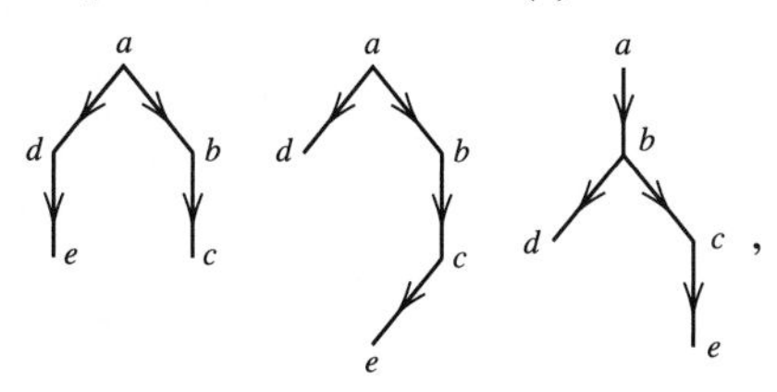

places of d and c can be interchanged in last tree.

5. 4 components, x_{17}, x_{19}, x_{23} isolated vertices, depth-first spanning tree for other component has path x_2-x_4-x_6-x_3-x_9-x_{12}-x_8-x_{10}-x_5-x_{15}-x_{20}-x_{14}-x_{16}-x_{18}-x_{22}-x_{24}-x_{26}-x_{13}, plus edges from x_{20} to x_{25}, from x_{14} to x_7 to x_{21}, and from x_{22} to x_{11}.

7. If the connected graph is not a tree, it has a circuit, which contains an edge e not on the one spanning tree; a second spanning tree must exist that contains e.

9. (a) If not all vertices reached, some reached vertex would have an edge to an unreached vertex, but a depth-first search would use such edge,

(b) immediate.

11. If C has no edge of a spanning tree T, removal of C could not disconnect graph (spanning tree's edges connected graph).

13.

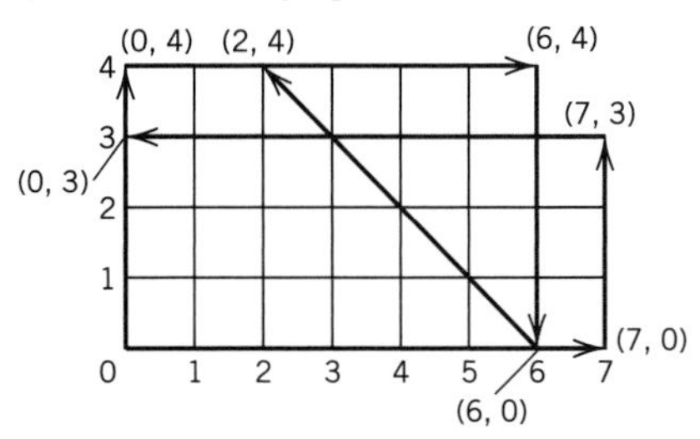

15. (0, 0)-(0, 4)-(4, 0)-(4, 4)—now 2 quarts in 10-quart pitcher.

17. right-hand-wall rule is same as depth-first search, which takes leftmost branch at every intersection (corner).

19. The ferryman takes goat across and returns alone; next he takes the dog across and brings the goat back; next he takes the tin cans across and returns alone, and finally he takes the goat across.

21. (a) Start with $M_1M_2M_3C_1C_2C_3$ on near shore; C_1C_2 across and C_1 return yields $M_1M_2M_3C_1C_3/C_2$; C_1C_3 across and C_1 return yields $M_1M_2M_3C_1/C_2C_3$; M_2M_3 across and M_2C_2 return yields $M_1M_2C_1C_2/M_3C_3$; M_1M_2 across and C_3 return yields $C_1C_2C_3/M_1M_2M_3$; C_2C_3 across, C_3 return, and C_1C_3 across completes crossing

(b) Let C_1 be the one cannibal that can row. Begin crossings as in part (a), then when $M_1M_2C_1C_2/M_3C_3$ on near shore, perform round trip of M_1C_1 across and M_3C_3 return to obtain $M_2M_3C_2C_3/M_1C_1$. Now continue as in part (a) with indices 1 and 3 interchanged.

23. A and B cross (2 min.) and A returns (1 min.); next C and D cross (10 min.) and B returns (2 min.); finally A and B cross again (2 min.).

25.

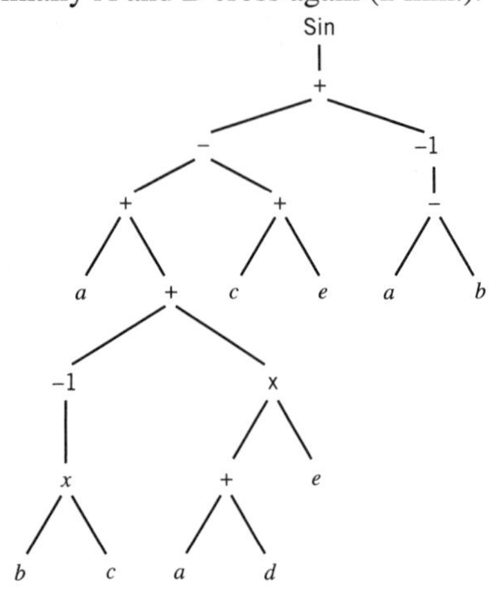

Section 3.3

1. Cost is 11: 1-3-2-4-1.
5. Cost is 14: T_1-T_5-T_4-T_3-T_2-T_1.
7. **(a)** 1-4-3-2, **(b)** 1-4-5-3-2-6-1, **(c)** 2-3-5-1-4-2.
9. 2 1 2
2 2 2
2 1 2

Section 3.4

1. Outcomes are ordered lists.
3. A binary comparison tree has $n!$ outcomes or leaves (as noted at the beginning of this section); so average number of comparisons = average leaf level = (by Exer. 13 in Section 3.1) $\log_2 n! = n \log_2 n$.
5. $1, 2, \ldots, n$.
7. If the initial heap is balanced [as decribed in Exer. 6 (a)], then largest level number is $\log_2 n$ and the number of comparisons to adjust the heap each time the root is removed will equal the largest level number, i.e., $\log_2 n$; for n iterations of removing the root and readjusting the heap, there will be $n\log_2 n$ comparison; constructing initial heap is similar.
11. **(b)** $O(n\log_2 n)$.

CHAPTER FOUR SOLUTIONS

Section 4.1

1. length 14: c-d-h-k-j-m.
3. **(a)** 31: L-c-d-f-g-k-W, **(b)** 32: L-b-h-j-m-W,
(c) 13: L-a-c-d-f-g-k-W, **(d)** L-c-d-f-g-k-W, 2 paths.
5. **(a)** 5: L-b-h-j-m-W,
(b) 6: L-c-d-f-g-k-W,
(c) 6: L-c-e-g-j-m-W.
7. If algorithm has found shortest paths to all vertices $\leq m$ unit from a, then a vertex x will be distance $m + 1$ from a if and only if the length $k(y, x)$ from a vertex y (y is closer to a than x) plus the distance $d(y)$ of y from a equals $m + 1$—this is exactly the test that the algorithm performs.
9. Graph with negative-length circuit has shortest path of unboundedly negative length (go around and around the negative circuit); algorithm relabels vertices and never terminates.
11. Define a tree by letting the first label of a vertex be its parent.

Section 4.2

1. **(a)** 59: path L-a-c-d-h-f-g-i-k-W plus edges (b, d), (d, e), (g, j), (k, l), (k, m),
(b) 60: replace (k, W) by (m, W),

(c) L, j (other possibilities),
(d) modify part (b) by deleting (a, c), (g, i), and adding (a, b), (g, k).

3. (a, b) 39: path N-b-c-d-e-g-j-m-R plus edges (d, h), (e, f), (f, i), (j, k).

5.

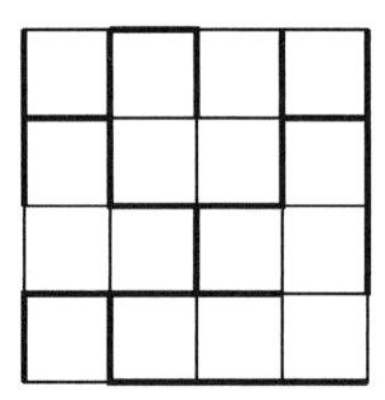

7. Modification: let initial edge be prescribed edge.

11. **(a)** If all edges of shortest length do not form circuit and they are not all in T', then add an omitted one and remove a longer edge in the resulting tree (as in proof of Prim's algorithm) to obtain a shorter minimal spanning tree—impossible.
(b) same reasoning as in part (**a**),
(c) *if:* part (b) is property of minimal spanning tree used to prove validity of Prim's algorithm; *only if:* verified in part (b).

Section 4.3

1. max flow $= 13$, $P = \{a, b, c\}$.

3. max flow $= 50$, $\overline{P} = \{f, z\}$.

5. **(a)** max flow $= 13$, $P = \{a, b, c\}$,
(c)

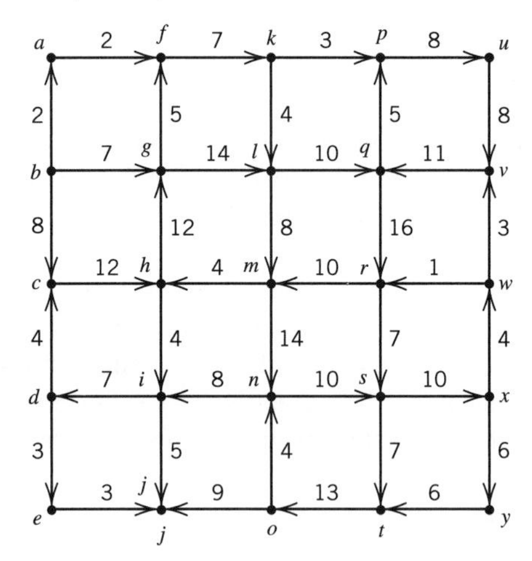

7. Set capacities of edges out of a and into z equal to 100 (equivalent to unlimited flow), max flow $= 150$, $P = \{a, b, c, d\}$.

9. 5, build paths by choosing leftmost unused edge leaving each vertex.

11. **(a)** 3 (3 edges leaving L),
(b) 15, quintuple flow in part (a).

13. See Exer. 12 for modeling the vertex capacity constraint; max flow $= 40$, $\overline{P} = \{f_i, f_o, g_o, z\}$.

15. max flow $= 2100$, route 400 on a_0-d_2-z_3-z_4, 300 on a_0-c_2-z_4, 400 on a_0-b_1-z_3-z_4, 200 on a_0-b_1-c_2-d_3-z_4, 400 on a_0, a_1-d_3-z_4, 400 on a_0-a_1-b_2-z_4.

17. max flow $= 36$.

19. **(a)** algorithm tries to reduce flow in incoming edge before adding flow to outgoing edge,
(b) yes.

21. Define tree with a vertex's parent being its first label.

23. **(a)** put flow of 1 in the two edges between c and e; other edges no flow,
(b) take flow in (a); and add a 2-unit flow path a-d-f-z.

27. If each flow path crosses a cut once, then the capacity of the cut $=$ sum of values of such flow paths $=$ value of flow.

29. **(a)** This is max flow–min cut theorem for model in Example 6,
(b) make edge capacities 1 and build vertex constraints of 1 (see Exer. 12), result is max flow–min cut theorem for this network.

31. One needs to build an initial feasible flow (requires at least one edge entering and one edge leaving each vertex, except a and z), then use same algorithm to reduce, instead of increase flow along a flow path.

33. **(a)** In step 2a of augmenting flow algorithm, incoming flow cannot be reduced below its lower bound value; otherwise use same algorithm.

Section 4.4

1. **(a)** several possibilities, see (b),
(b) using pairings A-G, Lo-J, first labels are (except for a, all second labels are 1): Bo-a^+, F-Bo^+, G-Bo^+, C-F^-, A-G^-, J-C^+, Bi-A^+, Lo-J^-, D-Bi^-, La-Lo^+, z-La^+, new matching A-G, D-Bi, Lo-La, Bo-F, C-J.

3. Give edges from a and into z the appropriate supplies and demands. Middle edges still ∞, one solution: A-Bi (3 dates), D-Bi (2), Bo-F (1), Lo-F (2), C-F (1), A-G (1), Bo-G (2), C-J (3), Lo-J (2), D-La (3).

5. no, schools with demand of 7 Ph.D.s can hire at most 6 (one from each university).

9. ***(a, b)*** Make complete bipartite graph, each vertex of degree n; by Example 3 there is a pairing for the first night; remove these edges; now again by Example 3, there is a pairing for the second night; continue in this fashion.

11. Start with a standard set-of-distinct-representatives matching network; replace the source a with 3 sources (one for each university), each with capacity $m/3$ and unit-capacity edges to each university's graduates.

13. Necessary and sufficient condition: for any set S of vertices, $|S| \leq |s(S)|$, where $s(S)$ is the set of successors of vertices in S (vertices with an edge coming in from a vertex in S); this condition guarantees A complete matching in hinted bipartite graph, which corresponds to a set of edges in original graph with one edge out of and one edge into each vertex.

15. For each left-side vertex, x, form a set of x's (right-side) neighbors.

CHAPTER FIVE SOLUTIONS

Section 5.1

1. **(a)** 12, 56, **(b)** 4, 15.
3. $8 \times 10 \times 6 \times 4$.
5. $26^4, 26 \times 25 \times 24 \times 23$.
7. **(a)** $8+7+5$,
(b) $8 \times 7 \times 5$,
(c) $3 \times \{8 \times 7 \times (7+5) + 7 \times 6 \times (8+5) + 5 \times 4 \times (8+7)\}$.
9. **(a)** $4 \times 47 = 188$,
(b) $(1 \times 48) + (12 \times 47)$.
11. 12.
13. **(a)** 9×10^4, **(b)** $5 \times 9 \times 10^3$,
(c) $9^4 + (4 \times 8 \times 9^3)$, **(d)** see Supplement at end of Chapter 5.
15. $52 \times 48/52 \times 51$.
17. **(a)** $10 \times 9 \times 8 \times 8, 10 \times 9 \times 8^{n-2}$,
(b) $9 \times 8 \times 1/9 \times 8 \times 8$,
(c) $9 \times 8 \times 7 \times 1/9 \times 9 \times 8 \times 8$.
19. $(6 \times 5 \times 4) - (5 \times 4 \times 3)$.
21. $(6 \times 5 \times 4) - (4 \times 3 \times 2)$ or $(2 \times 3 \times 4 \times 3) + (3 \times 2 \times 4)$.
23. **(a)** $9 \times 9 \times 8 \times 7$,
(b) $7 \times 8 \times 8 \times 8$,
(c) $9 \times 9 \times 8 \times 7 - 7 \times 7 \times 6 \times 5$.
25. $10 \times 9 \times (2^5 - 2)/2$
27. $4^5 - 3 \times 4^2$
29. $15 \times 10(14+9)$.
31. 4×10^3.
33. See Supplement at end of Chapter 5.
35. $2 \times 25/50 \times 49$.
37. $[(3 \times 3 \times 2) + (3 \times 3!)]/6^3$—pick the possible smallest value (3 choices), count ways of picking middle value and then arranging them (3×2 possibilities when middle value same as largest or smallest).
39. $9 \times 7 \times 5 \times 3$.
41. $16 \times (8 \times 7/2), n \times m \times (m+n-2)/2$.
43. $\{[28 \times (64-22)$ (Queen on edge of board)$] + [20 \times (64-24)$ (Queen one away from edge of board)$] + [12 \times (64-26)] + [4 \times (64-28)]\}/2$,
45. $2^{10} - 1$—each friend is or is not the subset, minus case of no one invited.
47. $2 \times 3 \times 6$.

Section 5.2

1. 52!
3. $P(15, 9)$.

5. $P(8, 5)$.
7. $C(10, 5)$, $C(8, 3)$.
9. $\{C(10, 8) + C(10, 9) + C(10, 10)\}/2^{10}$
11. **(a)** $C(10, 2) \times C(8, 2) \times C(6, 2) \times C(4, 2)$, **(b)** $5! \times 5!$.
13. $5 \times C(4, 2) \times 24^2$.
15. $C(n, 9)2^{n-9}$.
17. **(a)** $C(4, 2) \times \{C(6, 4) + C(6, 5) + C(6, 6)\} + [C(4, 3) \times C(6, 6)]$,
(b) $C(9, 3) + C(9, 4) + C(9, 5)$,
(c) $C(10, 5) - C(7, 2)$,
(d) $\{[C(4, 2) \times C(6, 2)] - (3 \times 5)\} + \{[C(4, 1) \times C(6, 3)] - C(5, 2)\} + C(6, 4)$.
19. $3! \times 6! \times 8! \times 5!$.
21. $C(21, 5) \times 10!; 2 \times 5!^2/10!$.
23. **(a)** $2 \times 5!/6! = 1/3$, **(b)** $C(6, 2) \times 4!/6!$
25. **(a)** $C(n, 3)$,
(b) $C(n - m, 3) + [m \times C(n - m, 2)]$
27. 12.
29. **(a)** 1-vote 1/6, 2-vote 1/2, **(b)** 1-vote 1/6, 2-vote 1/3,
(c) 1-vote 468/7!, 2-vote 1056/7!, **(d)** 1-vote 396/7!, 2-vote 864/7!.
31. **(a)** $26!/2$, **(b)** $26!/2^2$,
(c) $C(26, 5) \times 21!$.
33. $C(5, 3) \times P(6, 3) \times P(21, 3) + C(5, 4) \times P(6, 4) \times P(21, 2) + 21 \times 6!$.
35. $4 \times 3! \times C(7, 2) \times C(5, 2) \times 3!$
37. $C(30, 5)^{12}/C(360, 60)$.
39. **(a)** $\{[(3 \times 4) + (2 \times 5)] \times 8!\}/10!$,
(b) $\{(1 \times 9 \times 8!) + (8 \times 8 \times 8!)\}/10!$.
41. **(a)** $\{[C(4, 1)^4 \times C(36, 1)] + [4 \times C(4, 2) \times C(4, 1)^3]\}/C(52, 5)$,
(b) $\{C(26, 5) + C(13, 1)^2 C(26, 3) + C(13, 2)^2 \times C(26, 1)\}/C(52, 5)$.
43. See Supplement at end of Chapter 5.
45. $\{C(30, 2) \times C(28, 2) \times C(26, 2) \times C(24, 2)\}/4!$.
47. $4 \times 3 \times 8!/(2!2!)$.
49. **(a)** See Supplement at end of Chapter 5.
51. $C(64, 8) \times C(56, 8)$.
53. See Supplement at end of Chapter 5.
55. $C(6, 3) \times C(4, 2) + 2 \times [C(6, 2) \times C(4, 2) + C(6, 3) \times C(4, 1)] + [6 \times C(4, 2) + C(6, 2) \times C(4, 1) + C(6, 3)]$
57. See Supplement at end of Chapter 5.
59. **(a)** $C(10, 2) \times C(k - 10, 18)/C(k, 20)$,
(b) $k = 100$.
61. **(a)** $4!$,
(b) $4 \times 3 \times \{4 + 6\}$.
63. See Supplement at end of Chapter 5.
65. **(a)** $C(45, 3) + C(45, 2) \times C(45, 1)$,
(b) $[3 \times C(30, 3)$ (three integers with same value mod 3)$] + C(30, 1)^3$ (each integer with different value mod 3),

(c) as in (b), break into cases based on the value mod 4: $C(22,3) + [C(23,2) \times C(22,1)] + [2 \times C(23,2) \times C(45,1)] + [C(23,1) \times C(22,2)] + 22^2 \times 23$.

67. **(a)** $[(2 \times 2^3) + (3 \times 2^2)]/2^8$,
(b) $\{[2 \times 2^3) + (3 \times 2^2)] + [(2^2 + (2 \times 2)] + [(2 \times 2) + 1] + 2 + 1\}/2^8$.

69. $C(8,3)$, $C(8,3) - (8 \times 5)$.

71. See Supplement of Selected Solutions at end of Chapter 5.

73. **(a)** $C(10,6)$,
(b) $C(10,6) + [5 \times C(10,5)] + [4 \times C(10,4)] + C(10,3)$.

75. See Supplement of Selected Solutions at end of Chapter 5.

Section 5.3

1. $6!/1!2!3!$.

3. **(a)** 3^6,
(b) $(6!/2!^3)/3^6$

5. $C(10+4-1,10)$.

7. $C(9+3-1,9)$, subtract cases where one party has a majority (8 or more) $C(9+3-1,9) - [3 \times C(3+3-1,3)]$.

9. **(a)** $C(10+3-1,10)$,
(b) $C(15+3-1,15) - C(4+3-1,4)$.

11. $6!/3!2!1! + 6!/2!2!1!1!$.

13. $C(10,6) \times \{[C(6,2) \times 8!/2!2!1!^4] + [6 \times 8!/3!1!^5]\}$

15. Consider the cases of: (i) 4 of one letter and 1 of another letter, (ii) 3 of one kind and 2 of another letter, (iii) 3 of one letter and 1 of two others, (iv) 2 of two letters and 1 of another, and (v) 2 of one letter and 1 of three others—$(2 \times 3 \times 5!/4!1!) + (2 \times 2 \times 5!/3!2!) + [2 \times C(3,2) \times 5!/3!1!1!] + [C(3,2) \times 2 \times 5!/2!2!1!] + (3 \times 5!/2!1!^3)$.

17. $3 \times 10!/2!4!$.

19. $10!/2!2! - 9!/2!2!$.

21. **(a)** $10!/4!3!2!$,
(b) $2 \times 10!/4!3!2! - 9!/4!2!2!$.

23. See Supplement of Selected Solutions at end of Chapter 5.

25. Counts all $(1,2,3)$-sequences of length 10 two ways: left side looks at all cases of k_1 1s, k_2 2s, and k_2 3s, while right side counts all (unrestricted) 10-digit sequences of 1s, 2s, 3s.

27. See Supplement of Selected Solutions at end of Chapter 5.

29. See Supplement of Selected Solutions at end of Chapter 5.

31. Must have b or d after each c (except possibly last c); sum is over number of bs following cs (first sum when last c followed by b or d, second sum when last c is at end of sequence: $\sum C(3,k)21!/7!3!(8-k)![6-(3-k)]! + \sum C(2,k)20!/7!2!(8-k)![6-(2-k)]!$

Section 5.4

1. **(a)** $C(40+4-1,40)$,

(b) 1,
(c) $C(36+4-1, 36)$.

3. **(a)** $C(13, 4) \times C(13, 3)^3/C(52, 13)$,
(b) $[13!/5!5!2!1! \times 39!/8!8!11!12!]/52!/13!^4$,
(c) $4 \times C(48, 9)/C(52, 13)$,
(d) $4! \times (13!/4!3!^3)^4/52!/13!^4$.

5. $C(10+4-1, 10) \times C(4+4-1, 4) \times C(6+4-1, 6)$.

6. $3 \times C(24, 16) \times 2^8$.

7. $C[(5-2)+4-1, (5-2)] \times 5!$.

9. $C[(14-5)+(7-1), (14-5)]/C(20, 14)$.

11. $C(15+3-1, 15)$.

13. **(a)** $[52!/13!^4]/4!$,
(b) $[52!/8!^3 7!^4]/3! \times 4!$.

15. **(a)** $4^3 \times C(9+4-1, 9)$,
(b) $P(4, 3) \times C(9+4-1, 9)$,
(c) distribute teddies and fill out each child with lollipops: 4^3.

17. $4 \times C(4+4-1, 4)$.

19. $2 \times C(7+4-1, 7)$.

21. $C[(7-3)+5-1, (7-3)] \times 7!/4!2!1!$.

23. See Supplement of Selected Solutions at end of Chapter 5.

25. $[10!/2!^5]/5^{10}$.

27. **(a)** distributions of 8 distinct items into 3 boxes,
(b) distributions of 9 distinct items into 3 boxes with five items in the first box, etc.

29. **(a)** distributions of 6 identical objects into 31 boxes, $\sum_{i=1}^{31} x_i = 6$,
(b) distributions of 5 identical objects into 3 boxes with at most 4 objects in first box, etc., $\sum_{i=1}^{3} x_i = 5, x_1 \leq 5, x_2 \leq 4, x_3 \leq 2$.

31. $C(30+3-1, 30), 3 \times C[(30-16)+3-1, (30-16)]$.

33. **(a)** $C(7+4-1, 7)$,
(b) $C(7+5-1, 7)+1$
(c) $[C(13+4-1, 13)] - [4 \times C(3+4-1)]$.

35. $\sum_{k=0}^{6} C(k+2-1, k) \times C[(12-2k)+2-1, (12-2k)]$.

37. $\sum_{k=0}^{7} C(7+3-1, 7) \times C[(20-k)+4-1, (20-k)]$.

39. **(a)** $5 \times 6 \times 8$, **(b)** $(5 \times 6 \times 8) - 2$.

41. $4! \times C(4+5-1, 4)$.

43. $C(7+4-1, 7) \times 9!/6!$.

45. $C(3+4-1, 3) \times 5!/3!$

47. **(a)** $5! \times C(9, 5)$, **(b)** $21! \times 5! \times C(22, 5)$.

49. $\sum_{k=0}^{13} C(13, k) \times C(39, 13-k) \times C(26+k, k) \times C(26, 13)$.

51. $\sum 15!/a!b!c!$ summing over all a, b, c 3-tuples where $a+b+c=15$ with no letter greater than 7.

53. $(3! \times 25!/16!8!1!) + (3 \times 25!/14!7!4! \times 2^4) + \{3 \times [25!/12!6!7! \times (2^7-2) - 25!/6!6!1!]\} + 25!/10!5!^3$.

55. See Supplement of Selected Solutions at end of Chapter 5.

57. See Supplement of Selected Solutions at end of Chapter 5.

59. $C[(n-2m)+(2m+2)-1,(n-2m)]$ {simplifies to $C(n+1,2m+1)$.
61. $\sum_{i=0}^{5} C(k+m-1,k)\times C[(r-k)+(n-m)-1,(r-k)]$.

Section 5.5

8. $n=8$.
11. **(b)** $[C(n+1,2)]+[6\times C(n+1,3)]+[6\times C(n+1,4)]$.
13. **(a)** $2^n+(n\times 2^{n-1})$, **(b)** 1.5×2^n.
21. $\frac{1}{2}\times C(2n+2,n+1)-C(2n,n)$.
23. 0.

Section 5.6

1. 1234, 1243, 1324, 1342, etc.,
3. 1234, 1235, 1236, 1245, 1246, 1256, 1345, 1346, 1356, 1456, 2345, 2346, 2356, 2456, 3456.
5. **(a)** 123456789(10), (10)987654321,
(b) *abcd, wxyz.*

CHAPTER SIX SOLUTIONS

Section 6.1

1. **(a)** 8 products—$xxxx$, $x^2 1xx$, x^2x1x, x^2xx1, $1x^2xx$, $xx^2 1x$, xx^2x1, $x^2x^2 11$,
(b) 15 products—$11x^4$, $1xx^3$, $1x^2x^2$, $1x^3x$, $1x^4 1$, $x^2 1x^2$, $x^3 1x$, $x^4 11$, $xx^3 1$, $x^2x^2 1$, x^3x1, x^2xx, xx^2x, xxx^2, $x1x^3$,
(c) 10 products—$x^4 111$, $1x^4 11$, $x^2x^2 11$, $x^2 1x^2 1$, $x^2 1xx$, $x^2 11x^2$, $1x^2x^2 1$, $1x^2xx$, $1x^2 1x^2$, $11x^2x^2$,
(d) 15 products—$x^4 11$, x^3x1, $x^3 1x$, $x^2x^2 1$, x^2xx, $x^2 1x^2$, $xx^3 1$, xx^2x, xxx^2, $x1x^3$, $1x^4 1$, $1x^3x$, $1x^2x^2$, $1xx^3$, $11x^4$.
3. **(a)** $(1+x+x^2+x^3)(1+x+x^2+x^3+x^4)^2$,
(b) $(x+x^2+x^3+x^4+x^5)(x+x^2+x^3)(x+x^2+\cdots+x^7+x^8)$,
(c) $(1+x+x^2+\cdots)^4$,
(d) $(x+x^3+x^5+\cdots)^2(1+x+x^2+\cdots)^5$.
5. $(1+x+x^2+\cdots)^2(1+x)^2$, coef. of x^5.
7. $(x^3+x^4+x^5+\cdots)^4$, coefficient of x^{18}.
9. $(1+x+x^2+\cdots)^n$.
11. $(1+x^2+x^4+\cdots)(x+x^3+x^5+\cdots)(1+x+x^2+\cdots)^{n-2}$.
13. $(x+x^2+x^3+x^4+x^5+x^6)^n$.
15. $(x^{-3}+x^{-2}+x^{-1}+1+x+x^2+x^3)^4$.
17. $(1+x^5+x^{10}+\cdots)^8$.

19. $(1+x+x^2+\cdots)(x+x^2+x^3+\cdots)^4(1+x+x^2+\cdots)$, coefficient of x^{15}, generally x^{n-5}.

21. Cannot have a variable number of factors.

23. $(1+x+x^2+\cdots)(1+x^5+x^{10}+\cdots)(1+x^{10}+x^{20}+\cdots)$.

25. $(1+x+x^2+\cdots)^5(1+y+y^2+y^3)^5$.

27. $(xy+xz+yz)^8$.

29. $C(p,n), n!$.

Section 6.2

1. $C(8+n-1, 8)$.

3. $C(m,9)+C(m,7)+C(m,5)$.

5. $C(10+4-1,10)-4\times C(4+4-1,4)$.

7. $C(14+7-1,14)-7\times C(9+7-1,9)+C(7,2)\times C(4+7-1,4)$.

9. $C(15+4-1,15)-4\times C(8+4-1,8)+C(4,2)\times C(1+4-1,1)$.

11. **(a)** $C(10+10-1,10)$,
(b) $C(10+4-1,10)-3\times C(11+4-1,11)$,
(c) 0,
(d) $C(11+2-1,11)+3\times C(12+2-1,12)$,
(e) $b^{12}\times C(12,m)$.

13. 0.

15. **(a)** 0,
(b) 1,
(c) $C(12+8-1,12)$,
(d) $4^{12}\times C(12+5-1,12)$,
(e) $C(4+4-1,4)$.

17. **(a)** $(x^2+x^3+x^4+\ldots)^3$, $C((10-6)+3-1,(10-6))$,
(b) $(1+x+x^2)(1+x+x^2+\ldots)^2$, $C(10+3-1,10)-C(7+3-1,7)$,
(c) $(1+x^2+x^4+\ldots)(1+x+x^2+\ldots)^2$, $\sum C((10-2k)+2-1,(10-k))$.

19. $C(12+5-1,12)-5\times C(7+5-1,7)+C(5,2)\times C(2+5-1,2)$.

21. **(a)** $C(15+n-1,15)+C(10+n-1,10)$,
(b) $C(n,15)+C(n,10)$.

23. $C(8+7-1,8)-7\times C(3+7-1,3)$.

25. $C(6+3-1,6)-C(3+3-1,3)-C(2+3-1,2)$.

27. $C(14+10-1,14)-4\times C(10+10-1,10)-6\times C(7+10-1,7)+C(4,2)\times C(6+10-1,6)+4\times 6\times C(3+10-1,3)-4\times C(2+10-1,2)+C(6,2)$.

29. $[C(10+4-1,10)-4\times C(4+4-1,4)]\times[C(15+4-1,15)-4\times C(9+4-1,9)+C(4,2)\times C(3+4-1,3)]$.

33. **(a)** $C(m+n,m+r)$,
(b) 0, if r odd, $C(n,r/2)$, r even,
(c) 3^n.

35. **(b)** 2.

39. **(b)** $n/2$, pn, 10, m/p.

41. $P_x(t) = \left(\frac{1}{2}\right)^m \left(\frac{1-(t/2)^{s+1}}{1-(t/2)}\right)^m$.

Section 6.3

1. **(a)** 5 partitions—4, $3+1$, $2+1+1$, $2+2$, $1+1+1+1$,
(b) 11 partitions—6, $5+1$, $4+2$, $4+1+1$, $3+3$, $3+2+1$, $3+1+1+1$, $2+2+2$, $2+2+1+1$, $2+1+1+1+1$, $1+1+1+1+1+1$.

3. $(1+x+x^2+x^3)(1+x^2+x^4+x^6)(1+x^3+x^6+x^9)\ldots$.

5. $(1-x)^{-1}(1-x^5)^{-1}(1-x^{10})^{-1}(1-x^{25})^{-1}$.

7. **(b)** $(1+x)(1+x^2)(1+x^3)\ldots = \left(\frac{1-x^2}{1-x}\right)\left(\frac{1-x^4}{1-x^2}\right)\left(\frac{1-x^6}{1-x^3}\right)\left(\frac{1-x^8}{1-x^4}\right)\cdots$
$= \text{(after canceling)}\ \frac{1}{(1-x)(1-x^3)(1-x^5)\ldots}$.

19. **(a)** Multiply the partition generating function (given just before Example 1) times $\frac{1}{1-x}$;
(b) $(x^3+x^6)/(1-x^3)(1-x^4)(1-x^6)$,
(c) same as **(b)**.

21. **(a)** Let the number of dots forming the first row and first column ($= 2k-1$ if first row and column have length k) in a self-conjugate Ferrers diagram be the length of the first row in the distinct, odd-parts diagram; delete the first row and column of the self-conjugate diagram and use the number of dots in the reduced self-conjugate diagram to define the second row of the distinct, odd-parts diagram, etc.

Section 6.4

1. $\left(1+x+\frac{x^2}{2!}+\frac{x^3}{3!}+\frac{x^4}{4!}+\frac{x^5}{5!}\right)^5$.

3. $(1+x)^5 e^{21x}$.

5. $\left(1+x+\frac{x^2}{2!}+\frac{x^3}{3!}+\frac{x^4}{4!}\right)^{13}$, coefficient of $x^{13}/13!$.

7. **(a)** $1/2(3^r+1)$,
(b) $1/4[3^r+2+(-1)^r]$,
(c) $3^r - 2\times 2^r + 1$.

9. **(a)** $22^{10}+4\times P(10,1)\times 22^9+6\times P(10,2)\times 22^8+4\times P(10,3)\times 22^7+P(10,4)\times 22^6$,
(b) $26^{10}-4\times 25^{10}+6\times 24^{10}-4\times 23^{10}+22^{10}$.

11. $\frac{1}{2}4^r$.

13. **(a)** $e^x - 2\times e^{(n-1)x/n} + e^{(n-2)x/n}$,
(b) $[n^r - 2(n-1)^r + (n-2)^r]/n^r$.

15. **(a)** $(e^x-1)/x$,
(b) $(1-x)^{-1}$.

21. $e^{\mu(x-1)}$.

Section 6.5

1. **(a)** $x/(1-x)^2$,
(b) $13/(1-x)$,
(c) $3x(1+x)/(1-x)^3$,
(d) $[3x/(1-x)^2]+[7/(1-x)]$,
(e) $4!x^4/(1-x)^5$.
3. $x(3-x)/(1-x)^3$.
5. **(a)** $(4x^2-3x+1)/(1-x)^3$,
(b) $\log_e(1-x)$.

CHAPTER SEVEN SOLUTIONS

Section 7.1

1. $a_n = 5a_{n-1}, a_0 = 1$ or $a_1 = 5$.
3. $a_n = a_{n-1} + 2a_{n-2}$.
5. $a_n = 3a_{n-1} + 2a_{n-5} + a_{n-10} + a_{n-25}$.
7. $a_n = a_{n-1} + a_{n-2}$.
9. $a_n = a_{n-1} + a_{n-2}$.
11. $a_n = a_{n-1} + a_{n-2}$.
13. **(a)** $a_n = a_{n-1} + n, n > k$, initial condition: $a_k = k + 1$,
(b) 43.
15. $a_n = 1.06(a_{n-1} + 50)$.
17. **(a)** $a_n = a_{n-1} + a_{n-3} + a_{n-5}$,
(b) $a_n = a_{n-1} + a_{n-3} + a_{n-5} - a_{n-6}$.
19. $a_n = a_{n-1} + 3^{n-1}$.
21. $a_n = 2a_{n-1} + 4^{n-1}$.
23. $a_n = 3a_{n-1} - a_{n-3}$.
25. **(a)** $a_n = 2a_{n-1}, a_n = ma_{n-1}$,
(b) $a_n = a_{n-1} + 2^n, a_n = a_{n-1} + m^n$.
27. **(a)** $P(k,4) + P(k,3)$,
(b) $P(k,4) + P(k,3) + P(k,2)$,
(c) $P(k,5) + 3P(k,4) + 2P(k,3)$.
29. $a_{n,k} = a_{n,k-1} + a_{n-1,k-1} + a_{n-2,k-1} + a_{n-3,k-1}$.
31. $a_{n,k} = a_{n-1,k-1} + a_{n-k,k}$.
33. $a_n = b_{n-1} + c_{n-1}, b_n = c_n = 2^{n-1} - b_{n-1} - c_{n-1}$.
35. $a_n =$ such sequences starting with a 0, $b_n =$ such sequences starting with a 1, $c_n = n$-digit binary sequences with no consecutive 0s, $a_n = b_{n-1} + c_{n-3}, b_n = a_{n-1} + b_{n-1}, c_n = c_{n-1} + c_{n-2}$.
37. $p_n =$ ways to hand out a penny, nickel, or dime on successive days with a penny on the first day, n_n and d_n are defined similary, $p_n = n_{n-1} + d_{n-1}, n_n = p_{n-5} + d_{n-5}, d_n = p_{n-10} + n_{n-10}$.
39. $a_n = a_{n-2}\,a_2 + a_{n-4}\,a_4 + a_{n-6}\,a_6 + \cdots + a_2a_{n-2}$.
41. $a_n = a_{n-1}\,a_1 + a_{n-2}\,a_2 + \cdots + a_1a_{n-1}$.

43. **(a)** $a_n = 2a_{n-1}$
(b) $a_n = 2^{n-1}$,
(c) the first k integers in the sequence will form a set of consecutive integers, the $(k+1)$-st integer can be the next larger or the next smaller number to extend this consecutive set.

45. $a_n = a_{n-1}$, n not a multiple of 5, $a_n = a_{n-1} + 2$, n a multiple of 5 but not 10, $a_n = a_{n-1} + 3$, n a multiple of 10.

47. $a_n = a_{n-3}$, n even, $a_n = a_{n-3} + \left\lfloor \frac{n+1}{4} \right\rfloor$.

Section 7.2

1. **(a)** $An - 5$, **(b)** $An^{1/2} + 2n$, **(c)** $A + 4n - \lfloor \log_2(n) \rfloor$,
(d) $An - 2$, **(e)** $a_n = An^4 - 5n/7$, **(f)** $a_n = An^2 - 3n$.

3. **(a)** $a_n = a_{n/10} + 1$, $a_n = \log_{10} n$,
(b) $a_n = 10a_{n/10} + 1$, $a_n = \frac{1}{9}n - \frac{1}{9}$.

5. $a_n = 2a_{n/2} + n - 1$, $a_n = n \log_2 n - n + 1$.

9. **(a)** Pick largest from first half and largest from second half and compare,
(b) $a_n = 2a_{n/2} + 1$,
(c) $a_n = n - 1$.

Section 7.3

1. $(1.08)^n \times 1000$.

3. **(a)** $a_n = \frac{2}{5}4^n + \frac{3}{5}(-1)^n$, **(b)** $a_n = 1$,
(c) $a_n = 2$, **(d)** $a_n = \frac{1}{2}n^2 - \frac{1}{2}n + 1$.

5. $a_n = 2a_{n-1} + a_{n-2}$, $a_n = \frac{2+\sqrt{2}}{4}(1+\sqrt{2})^n + \frac{2-\sqrt{2}}{4}(1-\sqrt{2})^n$.

7. $p_n - p_{n-1} = 2(p_{n-1} - p_{n-2})$, $p_n = (3 \times 2^n) - 2$.

11. $c_1 = 9$, $c_2 = -18$.

Section 7.4

1. **(a)** $a_n = 3C(n, 2) + 1$,
(b) $a_n = 2C(n+1, 3) + 3$,
(c) $a_n = 6C(n+2, 3) - 3C(n+1, 2) + 10$.

3. $a_n = a_{n-1} + 2C(n, 2) + n$, $a_n = 2C(n+1, 3) + C(n+1, 2)$.

5. $a_n = a_{n-1} + \sum_{k=1}^{n-3}\{k \times (n-2-k) + 1\}$, $a_n = C(n, 4) + C(n, 2) - n + 1$.

7. $a_n = 1250(1.08)^n - 1250$.

9. **(a)** $-3^n + 1$, **(b)** $\frac{5}{3}2^n + \frac{1}{3}(-1)^n$,
(c) $(3 \times 2^n) - n - 2$, **(d)** $(15 \times 2^n) - 2n^2 - 8n - 12$.

11. $3 \times 2^n - 3n - 2$.

13. $a_n = 2a_{n-1} - a_{n-2} + (10 \times 2^k)$, $a_n = 960n - 20 + (40 \times 2^n)$.

17. $A \times 2^n + B \times 2^n + \frac{3}{2}n + \frac{25}{4}$.
19. **(a)** $a_n = \sqrt{2^{n+1} - 1}$,
(b) $a_n = n!$, n even, $a_n = 0$, n odd.

Section 7.5

1. **(a)** $g(x) - 1 = xg(x) + \dfrac{2x}{1-x}$,
(b) $g(x) - x - 1 = 3xg(x) - 3 - 2x^2g(x) + \dfrac{2x^2}{1-x}$,
(c) $g(x) - 1 = xg(x) + \dfrac{2x^2}{(1-x)^3}$,
(d) $g(x) - 1 = 2xg(x) + \dfrac{2x}{1-2x}$.
(e) $g(x) - 1 - x = \dfrac{1}{1-2x}[g(x) - 1]$.
7. $a_n = a_2a_{n-1} + a_3a_{n-2} + \cdots + a_{n-1}a_2, n \geq 3, a_1 = 1, a_n = \dfrac{1}{n-1}C(2n-4, n-2)$.
9. $a_n = a_1a_{n-1} + a_2a_{n-2} + \cdots + a_{n-1}a_1, n \geq 2, a_1 = 1, a_n = \dfrac{1}{n}C(2n-2, n-1)$.
11. $a_{n,k} = pa_{n-1,k-1} + qa_{n-1,k}$, $F_n(x) = (q + px)^n$.
13. similar to recurrence relations in Example 5 except with 4^{n-1} replacing 3^{n-1} and $a_1 = 0$ instead of $a_1 = 1$ (still $b_1 = c_1 = 1$), $a_n = \frac{3}{15}(4^n - 1)$, n even, $= \frac{2}{15}(4^n - 4)$, n odd.
15. **(a)** $a_n = \sum C(n-1, k-1)a_{n-k}$,
(b) $g(x) = e^{e^x - 1}$.

CHAPTER EIGHT SOLUTIONS

Section 8.1

1. $2 \times 5 \times 26^4 - 5^2 \times 26^3$.
3. $3^n - 3 \times 2^{n-1}$.
5. $\{C(52, 5) - C(13, 5) \times C(4, 1)^5\}/C(52, 5)$.
7. $4 \times (9^3 + 9 \times 8 \times C(3, 2)$.
9. $600 - 200 - 200 - 100$.
11. **(a)** $200 - 70 - 100 + 30$.
(b) $200 - 100 - 60$.
13. **(a)** $200 - 3 \times 80 + 3 \times 30 - 15$,
(b) $80 - 2 \times 30 + 15$.
15. $30 - 15 - 10 - 6 + 5 + 3 + 2 - 1$.
17. $3^{20} - 3 \times 2^{20} + 3 \times 1$.
19. $6!/2!^3 - 3 \times 5!/2!^2 + 3 \times 4!/2! - 3!$.

21. $C(37, 10) - [C(27, 10) + C(25, 10) + C(22, 10)] + [C(15, 10) + C(12, 10) + 1]$.
23. $C(52, 6) - 3 \times C(48, 6) + 3 \times C(44, 6) - C(40, 6)$.
25. $3 \times C(6, 2) \times 4! - \{2 \times C(6, 3) \times 3! + 6!/2!^2\} + C(6, 4) \times 2!$.
27. $(15!/3!^5 - 3 \times 5! \times 10!/2!^5 + 3 \times 5!^3 - 5!^3)/(15!/3!^5)$.
29. 20.
31. $N(Y) = N(Y - K) + N(Y \cap K) = 50 + 20$.
33. $N(\text{Salad}) = 70$, N (sandwiches) $= 45$.

Section 8.2

1. $10^n - 3 \times 9^n + 3 \times 8^n - 7^n$.
3. $13 \times C(48, 6) - C(13, 2) \times C(44, 2)$.
5. **(a)** $\{C(52, 13) - C(4, 1) \times C(39, 13) + C(4, 2) \times C(26, 13) - C(4, 3) \times C(13, 13)\}/C(52, 13)$,
(b) $\{C(4, 1) \times C(39, 13) - C(4, 2) \times C(26, 13) + C(4, 3) \times C(13, 13)\}/C(52, 13)$,
(c) $\{C(52, 13) - C(4, 1) \times C(48, 13) + C(4, 2) \times C(44, 13) - C(4, 3) \times C(40, 13) + C(36, 13)\}/C(52, 13)$.
7. $9!/3!^3 - 3 \times 7!/3!^2 + 3 \times 5!/3! - 3!$.
9. $26! - \{3 \times 23! + 24!\} + \{2 \times 20! + 2 \times 21!\} - 18!$.
11. $C(25 + 6 - 1, 25) - 3 \times C(18 + 6 - 1, 18) + 3 \times C(11 + 6 - 1, 11) - C(4 + 6 - 1, 4)$.
13. $C(12 + 4 - 1, 12) - C(4, 1) \times C(10 + 3 - 1, 10) + C(4, 2) \times 9 - C(4, 3) \times 1$.
15. 27.
17. $\approx 10!^2/e$.
19. $10!/2!^5 \times \{10!/2!^5 - C(5, 1) \times 8!/2!^4 + C(5, 2) \times 6!/2!^3 - C(5, 3) \times 4!/2!^2 + C(5, 4) - C(5, 5)\}$.
21. $15!/3!^5 - 5 \times 5 \times 12!/3!^4 + C(5, 2) \times 5 \times 4 \times 9!/3!^3 - C(5, 3) \times 5 \times 4 \times 3 \times 6!/3!^2 + 5 \times 5! - 5!$.
23. **(a)** $n^5 - C(5, 1) \times n^4 + C(5, 2) \times n^3 - C(5, 3) \times n^2 + \{C(5, 4) - C(5, 5)\} \times n$;
(b) $n^5 - C(5, 1) \times n^4 + C(5, 2) \times n^3 - \{(C(5, 3) - 1) \times n^2 + n^3\} + \{(C(5, 4) - 2) \times n + 2 \times n^2\} - n$.
(c) $n^5 - C(7, 1) \times n^4 + C(7, 2) \times n^3 - \{C(7, 3) - 3) \times n^2 + 3 \times n^3\} + \{(C(7, 4) - 14) \times n + 14 \times n^2\} - \{(C(7, 5) - 2) \times n + 2 \times n^2\} + \{C(7, 6) - C(7, 7)\} \times n$.
27. if $21 = C(2 + 6 - 1, 2)$, then $\sum(-1)^k \times C(6, k) \times (21 - k)^n$.
29. $5!^n \times \{(2n - 1)! - \sum C(n, k) \times (2n - 1 - k)!\}$.
31. $P(C(7, 3), 7) - 7 \times P(C(6, 3), 7) + C(7, 2) \times P(C(5, 3), 7)$.
33. $C(P(6, 3), 8) - 6 \times C(P(5, 3), 8) + C(6, 2) \times C(P(4, 3), 8)$.
35. $\left\{\sum(-1)^k \times C(n, k) \times (n - k)^r\right\}/n!$.
37. $\sum_{k=3}^{n-1}(-1)^{k+1} \times C(k, 3) \times C(n, k) \times (n - k)^r$, $\sum_{k=3}^{n-1}(-1)^{k+1} \times C(k - 1, 2) \times C(n, k) \times (n - k)^r$.
39. $[C(5, 2) \times 9!/2!^3] - [2 \times C(5, 3) \times 8!/2!^2] + [3 \times C(5, 4) \times 7!/2!] - [4 \times 6!]$.
47. $\sum_{k=0}^{n} k \times \left\{\sum_{j=k}^{n}(-1)^{j-k} \times C(j, k) \times n!/j!\right\}$.

Section 8.3

1. 5×5 board with darkened squares on main diagonal.
3. $5! - 8 \times 4! + 20 \times 3! - 16 \times 2! + 4 \times 1!$.
5. $7! - (9 \times 6!) + (30 \times 5!) - (46 \times 4!) + (32 \times 3!) - (8 \times 2!)$.
7. $5! - (7 \times 4!) + (16 \times 3!) - (13 \times 2!) + (2 \times 1!)$.
9. 3.
11. **(a)** 4×5 board with darkened squares in 4 positions just to right of main diagonal,
(b) $(x+1)^4$,
(c) $\sum_{j=k}^{5}(-1)^{k+j} \times C(j,k) \times C(n-1,j) \times (n-j)!$.
13. 2×2 array of darkened squares and "L" (a column of 3 squares beside a single square both have $1 + 4x + 2x^2$).

CHAPTER NINE SOLUTIONS

Section 9.1

1. **(a)** not symmetric,
(b) yes,
(c) not transitive,
(d) not transitive,
(e) not transitive.
3. **(a)** 6 symmetries (as in Example 3),
(b) 4 symmetries,
(c) 1 symmetry.
5. **(a)** all C_j left fixed,
(b) $\begin{pmatrix} C_1C_2C_3C_4C_5C_6C_7C_8C_9C_{10}C_{11}C_{12}C_{13}C_{14}C_{15}C_{16} \\ C_1C_3C_4C_5C_2C_7C_8C_9C_6C_{11}C_{10}C_{13}C_{14}C_{15}C_{12}C_{16} \end{pmatrix}$
(c) $\begin{pmatrix} C_1C_2C_3C_4C_5C_6C_7C_8C_9C_{10}C_{11}C_{12}C_{13}C_{14}C_{15}C_{16} \\ C_1C_3C_2C_5C_4C_6C_9C_8C_7C_{11}C_{10}C_{15}C_{14}C_{13}C_{12}C_{16} \end{pmatrix}$
(d) $\begin{pmatrix} C_1C_2C_3C_4C_5C_6C_7C_8C_9C_{10}C_{11}C_{12}C_{13}C_{14}C_{15}C_{16} \\ C_1C_2C_5C_4C_3C_9C_8C_7C_6C_{10}C_{11}C_{14}C_{13}C_{12}C_{15}C_{16} \end{pmatrix}$
11. **(a)** a, b, c are rotations or 0°, 120°, and 240°, respectively d, e, f are flips around vertical axis, axis 30° clockwise of vertical, and axis 30° counterclockwise of vertical; row is first symmetry, column second symmetry:

	a	b	c	d	e	f
a	a	b	c	d	e	f
b	b	c	a	f	d	e
c	c	a	b	e	f	d
d	d	e	f	a	b	c
e	e	f	d	c	a	b
f	f	d	e	b	c	a

(b) straightforward.
(c) let $a = \binom{1\ 2\ 3\ 4}{1\ 2\ 3\ 4}$, $b = \binom{1\ 2\ 3\ 4}{2\ 1\ 4\ 3}$, $c = \binom{1\ 2\ 3\ 4}{3\ 4\ 1\ 2}$, $d = \binom{1\ 2\ 3\ 4}{4\ 3\ 2\ 1}$,

	a	b	c	d
a	a	b	c	d
b	b	a	d	c
c	c	d	a	b
d	d	c	b	a

13. only left structure (right structure has 6 isomers).
21. **(b)** $\{\pi_1, \pi_3, \pi_5, \pi_6\}$ or $\{\pi_1, \pi_3, \pi_7, \pi_8\}$,
(c) in addition to subgroups in (b) and G, G', G'', other subgroups are $\{\pi_1\pi_i\}$, for $i = 3, 5, 6, 8$.

Section 9.2

1. **(a)** 11, **(b)** 24.
3. $\frac{1}{3}[3^{15} + (2 \times 3^5)]$.
5. 51.
7. $\frac{1}{2}(5^n + 5^{n/2})$ n even, and $\frac{1}{2}[5^n + (3 \times 5^{(n-1)/2})]$ n even.
9. **(a)** in cycle form: $\pi_1 = (1)(2)(3)$, $\pi_2 = (12)(3)$, $\pi_3 = (13)(2)$, $\pi_4 = (1)(23)$, $\pi_5 = (123)$, $\pi_6 = (132)$,
(b) $\psi(\pi_1) = C(12 + 3 - 1, 12) = 91$, $\psi(\pi_2) = \psi(\pi_3) = \psi(\pi_4) = 7$, $\psi(\pi_5) = \psi(\pi_6) = 1$, answer: $\frac{1}{6}(91 + 7 + 7 + 7 + 1 + 1) = 19$.
11. $\psi(\pi_1) = 18$, $\psi(\pi_3) = 6$, $\psi(\pi_7) = \psi(\pi_8) = 12$, and other $\psi(\pi_i)' = 0$, answer $\frac{1}{8}(18 + 6 + 12 + 12) = 6$;
13. **(b)** $\{\pi_1, \pi_3, \pi_7, \pi_8\}$, **(c)** $\{\pi_1, \pi_6\}$.

Section 9.3

1. 55.
3. **(a)** 130,
(b) 92,
(c) cyclic color sequence on hexagon of R-W-B-R-W-W and R-B-W-R-W-W.
5. **(a)** $\frac{1}{6}(m^4 + 2m^2 + 3m^3)$,
(b) $\frac{1}{8}(m^9 + 2m^3 + m^5 + 4m^6)$,
(c) $\frac{1}{2}(m^5 + m^3)$,
(d) $\frac{1}{4}(m^8 + 3m^4)$,
(e) $\frac{1}{12}(m^7 + 2m^2 + 2m^3 + 4m^4 + 3m^5)$.
7. **(a)** $\frac{1}{6}(m^6 + 2m^2 + 3m^4)$,
(b) $\frac{1}{8}(m^{12} + 2m^3 + 3m^6 + 2m^7)$,
(c) $\frac{1}{2}(m^6 + m^4)$.
9. $\frac{1}{4}[7^4 + (3 \times 7^2)] = 637$.

11. **(a)** $\frac{1}{2}(2^8+2^4)=136$,
(b) $\frac{1}{4}(2^8+2^4+0+2^4)=72$.
13. $\psi(\pi_i)=$ number of cycles of length 1.
15. **(a)** $\frac{1}{p}[m^p+(p-1)\times m]$,
(b) $\frac{1}{2p}\{m^p+[(p-1)\times m](p\times m^{(p+1)/2})\}$.

Section 9.4

1. $b^5+b^4w+2b^3w^2+2b^2w^2+bw^4+w^5$.
3. $b^4+w^4+r^4+b^3w+b^3r+bw^3+w^3r+br^3+wr^3+2b^2w^2+2b^2r^2+2w^2r^2+2b^2wr+2bw^2r+2bwr^2$.
5. **(a)** $\frac{1}{6}\{(b+w)^4+2(b^3+w^3)(b+w)+3(b^2+w^2)(b+w)^2\}$,
(b) $\frac{1}{4}\{(b+w)^6+2(b^2+w^2)^3+(b^2+w^2)^2(b+w)^2\}$,
(c) $\frac{1}{8}\{(b+w)^{12}+2(b^4+w^4)^3+3(b^2+w^2)^6+2(b^2+w^2)^5(b+w)^2\}$.
7. **(a)** $\frac{1}{6}\{(b+w)^6+2(b^3+w^3)^2+3(b^2+w^2)^2(b+w)^2\}$,
(b) $\frac{1}{4}\{(b+w)^7+2(b^2+w^2)^3(b+w)+(b^2+w^2)^2(b+w)^3\}$.
9. **(a)** $\frac{1}{12}\{(b+w)^4+8(b^3+w^3)(b+w)+3(b^2+w^2)^2\}$,
(b) $\frac{1}{24}\{(b+w)^6+6(b^4+\omega^4)(b+w)^2+3(b^2+w^2)^2(b+w)^2+6(b^2+w^2)^3+8(b^3+w^3)^2\}$.
11. $\frac{1}{24}\{(b+w)^4+6(b^4+b^4)+8(b^3+w^3)(b+w)+3(b^2+w^2)^2+6(b^2+w^2)(b+w)^2\}$.
13. **(a)** if not a cyclic rotation of all corners, the length of the cycle would have to divide p—impossible,
(b) $C(p,k)/p$.
15. **(a)** 36, **(b)** 216.

CHAPTER TEN SOLUTIONS

Section 10.1

1. **(a)** $\{a,c\}$ or $\{b,d\}$,
(b) f,
(c) no kernel, consider directed 5-circuit b,a,d,g,h,b,a—if a is K (kernel), then d not in K, then g in K, then h not in K, then b in K—impossible since a in K; similar sort of argument (also involving c, f, e) if a not in K.
3. $\{3,4,9,11,12,16,17,21,25,26,27,31,32,36,\text{over } 40\}$.
5. A goes to 2, B must go to 4 or else A will win.
7. Move to multiples of $5k+1$ (initial position is win for second player).
9. By symmetry assume $g(a)=0$, then $g(e)=1$, then $g(d)=0$, then $g(c)=1$, then $g(b)=0$, but now two kernel vertices are adjacent.
11. S is a kernel if and only if all vertices not in S have an edge to a vertex in S while no vertex in S has an edge to a vertex in S, that is, if and only if $W(S)=S$.

13. Follows from parts (a) and (b) of Exercise 12 since $g(x) = k$ means there is a path of length k starting at x while $l(x)$ is length of longest path starting at x; longest path is at least length k (maybe there is another longer path starting at x).

15. Suppose x and y are adjacent because there is an edge from x to y; then x must have a larger level number than y and a different Grundy value from its successor y.

17. If there were an infinite number of vertices, then one of the finite number of starting vertices, call it x_1, must have an infinite number of vertices reachable from it, and one of the finite number of successors of x_1, call it x_2, must have an infinite number of vertices reachable from it, and one of the finite number of successors of x_2, call it x_3, must have an infinite number of vertices reachable from it, and so on, without end.

19. Let $a = 0000$, $b = 000$, $c = 0\_00$, $d = 00\_0$, $e = 00$, $f = 0\_0$, $g = 0$, $h = \_$(win); $s(a) = \{b, c, d, e, f\}, s(b) = \{e, f, g\}, s(c) = s(d) = \{e, f, g\}, s(e) = \{g, h\}, s(f) = \{g\}, s(g) = h, s(h) = \emptyset$; $g(f) = g(h) = 0, g(a) = g(g) = 1, g(e) = 2, g(b) = g(c) = g(d) = 3$.

Section 10.2

1. (a) 7, remove 3 from 4th pile,
(b) 0,
(c) 4, remove 4 from 2nd, 3rd, or 4th pile,
(d) 0.

3. (a) 3, remove 3 from 3rd pile or 2 from 4th pile,
(b) 2, remove 2 from 3rd or 4th pile,
(c) 0,
(d) 0.

5. (a) 0,
(b) 0,
(c) 1, remove 1 from 4th pile,
(d) 3, remove 1 from 3rd pile.

7. (a) 3, add nickel to 3rd pile,
(b) (0, 0), (0, 4), (0, 6), (0, 9), (1, 1), (2, 2), (2, 5), (2, 8), (3, 3), (3, 7), (4, 4), (4, 6), (5, 5).

9. If $c_j = d_j$, then trivially $c' \dot{+} c_j = c' \dot{+} d_j$; if $c' \dot{+} c_j = c' \dot{+} d_j$, then c_j and d_j must have 1s in the same positions in their binary representations, that is, they must be equal.

11. Immediately the proposed strategy works.

13. (a) Remove 3 balls along one of the 3 lines formed by ball on one of the 3 sides of the arrangement.

APPENDICES SOLUTIONS

Section A.1

1. (a) 12, 27

(b) 2, 3, 6, 7, 9, 12, 15, 17, 18, 21, 22, 24, 27,
(c) 1, 4, 5, 8, 10, 11, 13, 14, 16, 19, 20, 23, 25, 26, 28, 29,
(d) all $1 \le k \le 29$ except 12, 27.

3. We are given $N(\overline{R} \cap M) = 2$ as well as that $N(M) = N(R) = N(\overline{M}) = N(\overline{R}) = 4$; then $N(R \cap M) = N(M) - N(\overline{R} \cap M) = 4 - 2 = 2$, $N(\overline{R} \cap \overline{M}) = N(\overline{R}) - N(\overline{R} \cap M) = 4 - 2 = 2$, and clearly $N(R \cap \overline{M}) = 2$.

5. **(a)** impossible,
(b) yes, $20 - 8 - 8 = 4$,
(c) $20 - 15 = 5$.

7. **(a)**

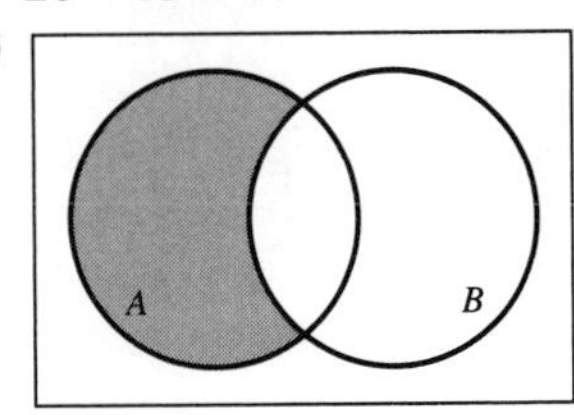

(b)

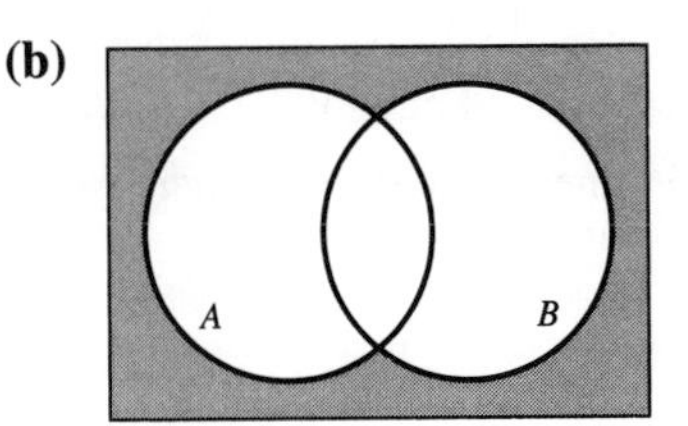

(c) $(\overline{A \cup B}) \cap (\overline{A \cap B}) = (\overline{A \cap B})$, see Figure A1.3,
(d) $A - (B - A) = A$; here all expressions involve $\cap$, so that $A\overline{B}C = A \cap \overline{B} \cap C$.

9.

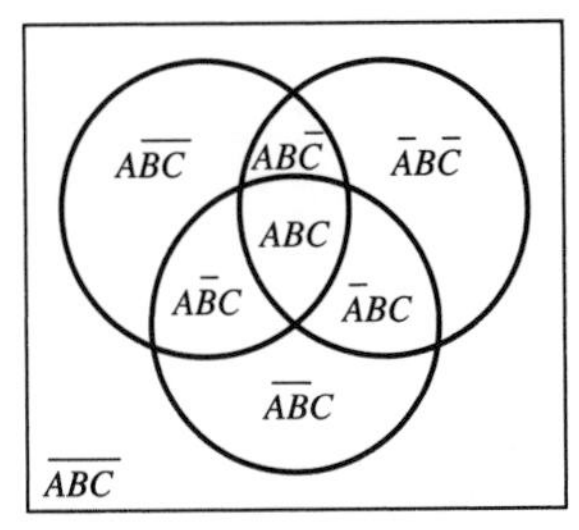

17. **(a)** $E = S \cup H \cup C$, $52^3 - 39^3$,
(b) $E = (S \cup H \cup C) \cap (\overline{S \cap H \cap C})$, $52^3 - 39^3 - 13^3$,
(c) $E = (S \cap H \cap \overline{C}) \cup (S \cap \overline{H} \cap C) \cup (\overline{S} \cap H \cap C)$, $3 \times 13^2 \times 39$.

Section A.2

23. One can prove by induction only a property that is a function of n; for example, one can prove that there are a finite number of binary sequences of length $\le n$.

25. The initial step only assumes that $n \ge 1$, not $n \ge 2$, but for $n = 1$, a^{n-2} is undefined.

Section A.3

1. 1/2.

3. **(a)** 1/6,
(b) $18/36 = 1/2$,
(c) $3/36 = 1/12$.

5. **(a)** 1/6, **(b)** 1/2.

7. **(a)** $1/3 \times 1/3 = 1/9$, **(b)** $2 \times 1/3 \times 2/3 = 4/9$.

9. $(2 \times 2 \times 2)/4! = 1/3$.
11. $1 - [(50 + 40 - 20)/100] = 3/10$.
13. $2/3$.
15. **(a)** all sequences with k tails, $0 \leq k \leq 8$, and one head followed by a head,
(b) all positive integers,
(c) all ordered pairs of positive integers,
(d) all sequences of k black balls, $k \geq 0$, followed by a red ball.

Section A.4

1. $n + 1$.
15. Printer i is connected to computers $i, i+1, i+2, i+3, i+4, i+5$.

INDEX